Global Neoproterozoic Petroleum Systems: The Emerging Potential in North Africa

Geological Society books refereeing procedures

The Society makes every effort to ensure that the scientific and production quality of its books matches that of its journals. Since 1997, all book proposals have been refereed by specialist reviewers as well as by the Society's Books Editorial Committee. If the referees identify weaknesses in the proposal, these must be addressed before the proposal is accepted.

Once the book is accepted, the Society Book Editors ensure that the volume editors follow strict guidelines on refereeing and quality control. We insist that individual papers can only be accepted after satisfactory review by two independent referees. The questions on the review forms are similar to those for *Journal of the Geological Society*. The referees' forms and comments must be available to the Society's Book Editors on request.

Although many of the books result from meetings, the editors are expected to commission papers that were not presented at the meeting to ensure that the book provides a balanced coverage of the subject. Being accepted for presentation at the meeting does not guarantee inclusion in the book.

More information about submitting a proposal and producing a book for the Society can be found on its web site: www.geolsoc.org.uk.

It is recommended that reference to all or part of this book should be made in one of the following ways:

Craig, J., Thurow, J., Thusu, B., Whitham, A. & Abutarruma, Y. (eds) 2009. *Global Neoproterozoic Petroleum Systems: The Emerging Potential in North Africa.* Geological Society, London, Special Publications, **326**.

Bhat, G. M., Ram, G., Koul, S. 2009. Potential for oil and gas in the Proterozoic carbonates (Sirban Limestone) of Jammu, northern India. *In*: Craig, J., Thurow, J., Thusu, B., Whitham, A. & Abutarruma, Y. (eds) *Global Neoproterozoic Petroleum Systems: The Emerging Potential in North Africa.* Geological Society, London, Special Publications, **326**, 245–254.

GEOLOGICAL SOCIETY SPECIAL PUBLICATION NO. 326

Global Neoproterozoic Petroleum Systems: The Emerging Potential in North Africa

EDITED BY

J. CRAIG
Eni Exploration and Production Division, Italy

J. THUROW
University College London, UK

B. THUSU
University College London, UK

A. WHITHAM
CASP, Cambridge, UK

and

Y. ABUTARRUMA
Earth Science Society of Libya, Libya

2009
Published by
The Geological Society
London

THE GEOLOGICAL SOCIETY

The Geological Society of London (GSL) was founded in 1807. It is the oldest national geological society in the world and the largest in Europe. It was incorporated under Royal Charter in 1825 and is Registered Charity 210161.

The Society is the UK national learned and professional society for geology with a worldwide Fellowship (FGS) of over 9000. The Society has the power to confer Chartered status on suitably qualified Fellows, and about 2000 of the Fellowship carry the title (CGeol). Chartered Geologists may also obtain the equivalent European title, European Geologist (EurGeol). One fifth of the Society's fellowship resides outside the UK. To find out more about the Society, log on to www.geolsoc.org.uk.

The Geological Society Publishing House (Bath, UK) produces the Society's international journals and books, and acts as European distributor for selected publications of the American Association of Petroleum Geologists (AAPG), the Indonesian Petroleum Association (IPA), the Geological Society of America (GSA), the Society for Sedimentary Geology (SEPM) and the Geologists' Association (GA). Joint marketing agreements ensure that GSL Fellows may purchase these societies' publications at a discount. The Society's online bookshop (accessible from www.geolsoc.org.uk) offers secure book purchasing with your credit or debit card.

To find out about joining the Society and benefiting from substantial discounts on publications of GSL and other societies worldwide, consult www.geolsoc.org.uk, or contact the Fellowship Department at: The Geological Society, Burlington House, Piccadilly, London W1J 0BG: Tel. +44 (0)20 7434 9944; Fax +44 (0)20 7439 8975; E-mail: enquiries@geolsoc.org.uk.

For information about the Society's meetings, consult *Events* on www.geolsoc.org.uk. To find out more about the Society's Corporate Affiliates Scheme, write to enquiries@geolsoc.org.uk.

Published by The Geological Society from:
The Geological Society Publishing House, Unit 7, Brassmill Enterprise Centre, Brassmill Lane, Bath BA1 3JN, UK

(*Orders*: Tel. +44 (0)1225 445046, Fax +44 (0)1225 442836)
Online bookshop: www.geolsoc.org.uk/bookshop

British Library Cataloguing in Publication Data

A catalogue record for this book is available from the British Library.
ISBN 978-1-86239-287-8

Typeset by Techset Composition Ltd, Salisbury, UK
Printed by MPG Books Ltd, Bodmin, UK

Distributors

North America
For trade and institutional orders:
The Geological Society, c/o AIDC, 82 Winter Sport Lane, Williston, VT 05495, USA
Orders: Tel. +1 800-972-9892
Fax +1 802-864-7626
E-mail: gsl.orders@aidcvt.com

For individual and corporate orders:
AAPG Bookstore, PO Box 979, Tulsa, OK 74101-0979, USA
Orders: Tel. +1 918-584-2555
Fax +1 918-560-2652
E-mail: bookstore@aapg.org
Website: http://bookstore.aapg.org

India
Affiliated East-West Press Private Ltd, Marketing Division, G-1/16 Ansari Road, Darya Ganj, New Delhi 110 002, India
Orders: Tel. +91 11 2327-9113/2326-4180
Fax +91 11 2326-0538
E-mail: affiliat@vsnl.com

Contents

Preface

The papers published here are a selection of those presented at a conference held at the Geological Society of London in November 2006. The conference was organised by convenors Jonathan Craig (Eni), Juergen Thurow and Bindra Thusu (MPRG, UCL), Andy Whitham (CASP), and Yousef Abutarruma (Earth Science Society of Libya), with support from the Moroccan Society of Petroleum Geologists and the Geological Society of Oman.

This volume is introduced by the editors' review of existing and potential petroleum plays around the world and their relationship to the controversial 'Snowball Earth' hypothesis. The papers in the first part of the book set the scene by defining the current state of knowledge of the Neoproterozoic – stratigraphy, palaeogeography, geochronology and a global review of Infracambrian petroleum systems. The succeeding papers represent research focusing on the Neoproterozoic in North Africa, India, Spain, Namibia and South America.

Many people helped to make the meeting in London a success and have assisted in the completion of this volume. In particular we thank Faraj Said, NOC Libya, R. Abourawi, Libyan Oil Holdings, Dr B. Belgaseum, Libyan Petroleum Institute, A. Asbali, Arabian Gulf Oil Company. Eni, Shell, Chevron, Hydro, Petro-Canada, BP, Baker Hughes and Lynx gave financial support for the conference and the production of this volume.

In the completion of this volume the editors are particularly indebted to Angharad Hills at the Geological Society who has steered and encouraged us through the publication process. Sebastian Lüning, Dan Le Heron, Bashir Ahmad, and Heather Cheshire gave invaluable assistance during the editing process.

Finally, we thank the authors and reviewers – Philip Allen, Ian Bartholemew, Thilo Bechstädt, Anthony Cohen, Alan Collins, Andrea Cozzi, Kath Gray, Omar S. Hammuda, Andrew Knoll, Ed Landing, Daniel Le Heron, Graham Logan, Arne Thorshøj Nielsen, Ganti Rao, Greg Ravizza, Graham Shields, Alan Smith, Guy Spence and Charlie Underwood – for adhering to the deadlines imposed and making this volume what it is. We hope that their efforts will encourage and inspire further innovative research into the exciting potential of Neoproterozoic rocks.

Variant placename spellings

Placename	*Variants*	
Cyrenaica	Cyrenica	
Al Kufrah	Al Kufra	Kufra
Jabal Arkenu	Jebel Arkenu	Jbel Arkenu
Sirte	Sirt	
Mourizidie	Mourizide	
Hassaouna	Hasswina	Hasswanah

Global Neoproterozoic petroleum systems: the emerging potential in North Africa

JONATHAN CRAIG[1]*, JUERGEN THUROW[2], BINDRA THUSU[2], ANDY WHITHAM[3] & YOUSEF ABUTARRUMA[4]

[1]*Eni Exploration and Production Division, Via Emilia 1, 20097 San Donato Milanese, Milan, Italy*

[2]*MPRG (Maghreb Petroleum Research Group), University College London, Gower Street, London WC1E 6BT, UK*

[3]*CASP (Cambridge Arctic Shelf Programme), Huntingdon Road, Cambridge CB3 0DH, UK*

[4]*Earth Science Society of Libya, Tripoli, Libya*

**Corresponding author (e-mail: jonathan.craig@eni.it)*

Abstract: The Neoproterozoic Eon is relatively poorly known from a petroleum perspective, despite the existence of producing, proven and potential plays in many parts of the world. In tectonic, climatic and petroleum systems terms, the Neoproterozoic to Early Cambrian period can be divided into three distinct phases: a Tonian to Early Cryogenian phase, prior to about 750 Ma, dominated by the formation, stabilization and initial break-up of the supercontinent of Rodinia; a mid Cryogenian to Early Ediacaran phase (*c.* 750–600 Ma) including the major global-scale 'Sturtian' and 'Marinoan' glaciations and a mid Ediacaran to Early Cambrian (*c.* post 600 Ma) phase corresponding with the formation and stabilization of the Gondwana Supercontinent. There is increasing evidence that deposition of many mid to late Neoproterozoic (to Early Palaeozoic) organic-rich units was triggered by strong post-glacial sea level rise on a global scale, following the 'Snowball Earth' type glaciations, coupled with basin development and rifting on a more local scale.

Fieldwork in North Africa including the Taoudenni Basin in Mauritania, Algeria and Mali; the Anti-Atlas region of Morocco and the Cyrenaica, Kufra and Murzuk basins in Libya has added to the understanding of reservoir, source and seal relationships and confirmed the widespread presence of Precambrian stromatolitic carbonate units of potential reservoir facies. Current research on the chronostratigraphy, distribution and quality of source rocks, controls on reservoir quality and distribution of seals in the Precambrian–Early Cambrian hydrocarbon plays throughout South America, North Africa, the Middle East and the Indian Subcontinent is documented in this Special Publication.

One might, quite reasonably, ask why, when there are already more than enough challenges in exploring for conventional hydrocarbons in the Phanerozoic succession, we should want to turn our attention to the much more complex and challenging Precambrian succession. Of course, the reality is that, much as exploration has moved progressively into deeper water and more hostile environments in recent years, it has also begun to address deeper, older and, in many ways, more difficult reservoirs. In short, much of the 'easy exploration' around the world has been done and we are gradually being forced to focus on more difficult exploration targets that we have ignored in the past because there were easier things to do!

In the specific context of northern Africa, several recent publications have described in detail the work undertaken over the past two decades to unravel the complexities of the Lower Palaeozoic sequences in the region and, in particular, to understand the Upper Ordovician glacigenic hydrocarbon reservoirs and the overlying Lower Silurian hydrocarbon source rock (Sutcliffe *et al.* 2005; Lüning *et al.* 2000*a*; Le Heron *et al.* 2004; Le Heron & Craig 2008; Craig *et al.* 2008). This work ultimately led to the discovery and successful development of the giant El Feel ('Elephant') Field in the Murzuq Basin in Libya, and now forms the foundation for the continuing highly successful exploration of the prolific Late Ordovician–Early Silurian hydrocarbon plays in North Africa and the Middle East. During the course of this work, it became increasingly apparent that below the Palaeozoic there is a thick sedimentary succession in many parts of North Africa about which we know very little, but which frequently contains tantalizing evidence of

From: Craig, J., Thurow, J., Thusu, B., Whitham, A. & Abutarruma, Y. (eds) *Global Neoproterozoic Petroleum Systems: The Emerging Potential in North Africa.* Geological Society, London, Special Publications, **326**, 1–25.
DOI: 10.1144/SP326.1 0305-8719/09/$15.00

active petroleum systems and which has clear analogies with some major proven and producing petroleum systems elsewhere in the world.

The goal of the Global Infracambrian Petroleum Systems Conference held at the Geological Society of London in November 2006, which was the inspiration for this publication, was to review current knowledge about Neoproterozoic–Early Cambrian petroleum systems worldwide and to demonstrate that the Late Precambrian (Neoproterozoic) succession in North Africa is worthy of more attention than we have given it in the past.

The core subject of this Geological Society Special Publication – the period of Earth's history we call the Neoproterozoic Era – began 1000 Ma ago, lasted for some 458 Ma and ended at the start of the Cambrian 542 Ma ago. In many ways the publication of this volume represents the opening of a new chapter in petroleum exploration in North Africa and the Middle East. This new chapter is focused on the Neoproterozoic–Early Cambrian sequences underlying the prolific Palaeozoic petroleum systems that have themselves, in the last two decades, passed from frontier exploration concepts to one of the main targets of hydrocarbon exploration across the region. With time, and with an appropriate level of focus and active research, the Neoproterozoic–Early Cambrian successions in North Africa and the Middle East could prove to be a new challenging frontier for hydrocarbon exploration across this vast region.

Global climate and petroleum source rock distribution

A common theme that runs through this Special Publication is the role of global climate and glaciation in the occurrence and distribution of petroleum source rocks in the Neoproterozoic successions.

A plot of global climate through time for the last billion years and extending some 100 Ma into the future (Fig. 1) shows that the Earth has experienced alternating periods of greenhouse and icehouse climate (Coppold & Powell 2000). There appears to be cyclicity in this global climate record, with the greenhouse periods lasting some 250 Ma and the icehouse periods lasting around 100 Ma. These cycles can themselves be grouped into three longer Supercycles of 300–350 Ma each. It is, of course, well recognized that these long-period cycles in global climate are linked to plate tectonic processes, and to cycles in the formation and subsequent 'break-up' of supercontinents through time. In an ideal greenhouse world, the continental configuration is such that equatorial currents can encircle the globe, and there is exchange between tropical and polar waters. This configuration leads to a climate too warm for polar ice caps to develop. Conversely, in an ideal icehouse world, the continents are generally grouped at equatorial latitudes (and, perhaps, also at the poles). In this configuration, any currents encircling the globe tend to be polar rather than equatorial. This limits heat exchange between tropical and polar regions, and, so, promotes the formation of polar ice caps (e.g. Fensome & Williams 2001).

Comparison of the global climate record with the main periods of global glaciation (Crowell 1999) and the concentration of carbon dioxide in the atmosphere (Royer *et al.* 2004) during the Phanerozoic (Fig. 1) shows that the Permo-Carboniferous glacial era and the current glacial interval correspond with periods of low carbon dioxide concentration (low greenhouse gas). Anomalously, the Late Ordovician glaciation occurs in the middle of a period of apparent greenhouse climate and at a time of high CO_2 levels, possibly some 14 times the level of today, although there is a substantial degree of uncertainty in this value (± 5 or greater). The graph of atmospheric CO_2 concentration (Fig. 1) has not been extended back to the Precambrian because it exhibits large and comparatively rapid variations in this time period (Hoffman *et al.* 1998; Halverson *et al.* 2005).

It is interesting from a petroleum perspective to consider the relationships between global climate, sea level and distribution of source rocks through time. Figure 2 shows the temporal distribution of the main effective petroleum source rocks of the world in terms of the percentage of world hydrocarbon reserves generated from them, together with a generalized plot of eustatic sea level. In broad terms, the eustatic sea-level curve exhibits the same cyclicity as the global climate record, with periods of high sea level corresponding with periods of greenhouse climate (and low ice volumes). The deposition of many of the world's major petroleum source rocks appears intimately linked to periods of marine transgression and at least some of these transgressions are, predominantly, glacially driven. There is a growing body of evidence to suggest that even the smaller and more frequent cyclical, or at least episodic, eustatic sea-level oscillations throughout geological time are caused by fluctuations in ice volume (e.g. Weissert & Erba 2004; Simmons *et al.* 2007; Bornemann *et al.* 2008; Stephenson *et al.* 2008).

Interestingly, when the distribution of *Effective Petroleum Source Rocks of the World* shown in Figure 2 was originally published by Klemme & Ulmishek (1991), they estimated that only 0.2% of world hydrocarbon reserves were derived from Neoproterozoic source rocks.

An estimation of reserves per source rock for North Africa (updated from Macgregor 1996) shows a Mesozoic–Cenozoic petroleum system,

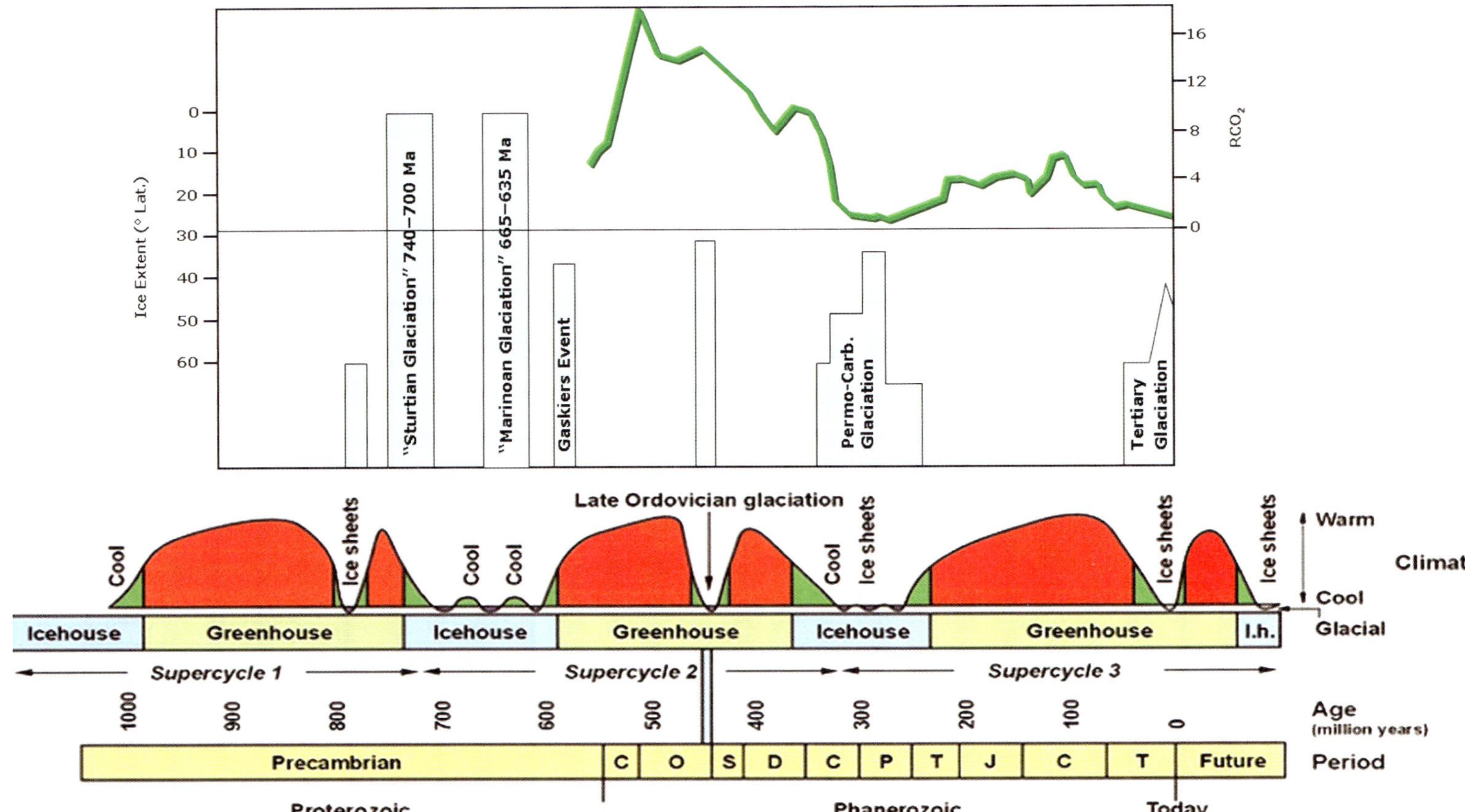

Fig. 1. Global climate, glaciations and atmospheric carbon dioxide levels through time from 1000 Ma to 100 Ma in the future. Carbon dioxide levels are shown as a ratio compared to present-day levels. The maximum extent of ice cover during the main periods of glaciation, as inferred from the preservation of glacigenic sediments and climate modelling, is shown in degrees of latitude from the poles. Ice extent data in past after Crowell (1999); global climate change based on geological data as summarized by Coppold & Powell (2000).

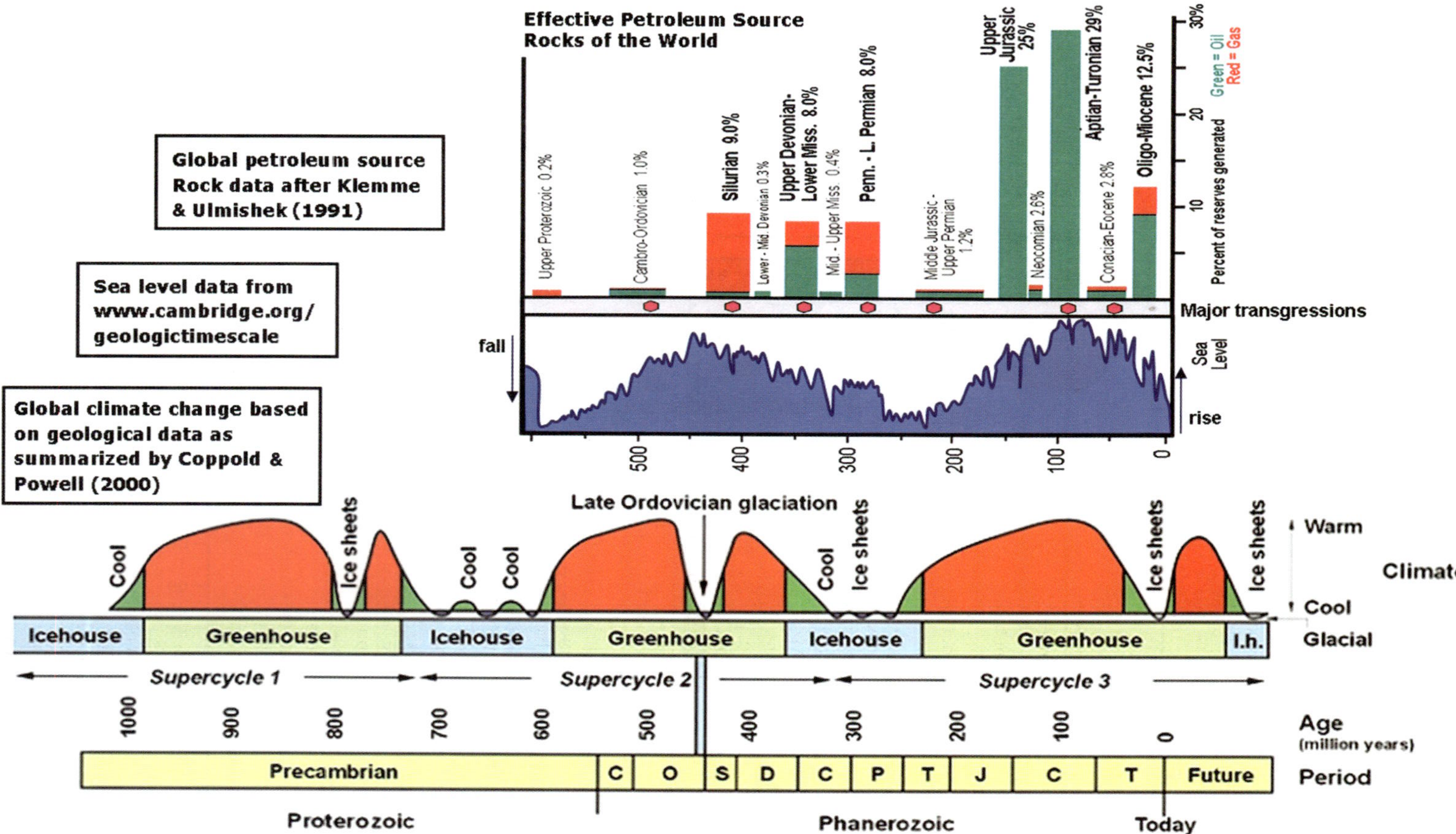

Fig. 2. Global climate, sea level and the distribution of the major effective petroleum source rocks of the world through time from 1000 Ma to 100 Ma in the future.

with total reserves of some 57 Bboe (billion barrels of oil equivalent) and a Palaeozoic petroleum system with total reserves of around 50 Bboe (see **Lottaroli *et al.*** 2009). The Palaeozoic petroleum system is dominated by the prolific post-glacial source rock at the base of the Silurian succession, which immediately overlies the Late Ordovician glacigenic reservoir system. It would seem logical to test whether this glacial reservoir–post-glacial source rock relationship is also valid for the major Neoproterozoic glaciations as we explore older petroleum systems in North Africa and, indeed, elsewhere in the world.

Neoproterozoic stratigraphy, tectonic events and global correlation

In the geological timescale published by Harland *et al.* (1990), the term Neoproterozoic was not used in the formal timescale, but rather the subdivisions of the Proterozoic, proposed by the Precambrian Subcommission of the ICS 1988 (fig. 2.2, p. 17), were quoted (Smith pers. comm.). These proposals defined the Neoproterozoic as extending from the base of the Cambrian at 542 Ma down to an arbitrary base at 1000 Ma with subdivision into Vendian and Late Riphean. In the more recent Gradstein *et al.* (2004) timescale, the Neoproterozoic Era covers the same time interval, but is subdivided into three periods, named from oldest to youngest, Tonian, Cryogenian and Ediacaran (Fig. 3). The term Tonian is derived from *Tonos* meaning 'stretch', Cryogenian comes from *Cryos* for 'ice' and *genesis* for 'birth', this being the period of global-scale glaciations, and the Ediacaran is named after the Ediacara Hills in South Australia, the type locality for the Ediacara biota. A thorough review of Neoproterozoic timescales, stratigraphy, current nomenclature, and the challenges of regional and local correlation of Neoproterozoic successions is given by **Smith** (2009).

In the past, the whole of the stratigraphic section between the base of the Cambrian and the igneous or metamorphic basement was commonly assigned, rather loosely, to the 'Infracambrian'. This was for the good practical reason that, until very recently, there was very little biostratigraphic analysis on which to base robust age dating, let alone to make regional or local stratigraphic correlations. With careful and rigorous sampling, it is sometimes possible to recover distinctive assemblages of acritarchs (organic-walled microfossils, probably related to algae) from these Neoproterozoic rocks (see, e.g. **Bhat *et al.*** 2009; **Lottaroli *et al.*** 2009). This period of geological time corresponds with the 'dawn of life' on Earth and it is only during the Late Neoproterozoic that animal size and complexity increased to the point that the diversity of soft-bodied fossils allows the definition of a distinct biostratigraphic period. This, the Ediacaran Period, is characterized by the wonderful Ediacaran biota that has been recorded from several key localities around the world, with Ediacara in South Australia, Charnwood Forest in Leicestershire, England, and Mistaken Point on the Avalon Peninsula, Newfoundland being, perhaps, the most famous (e.g. Seldon & Nudds 2004; Nudds & Seldon 2008). The Ediacaran creatures were soft-bodied and frequently grew to large size. Some can be classified as jellyfish and sea-pens, and, although many do not seem to be directly related to modern plants or animals, they are generally considered to represent the 'precursors' at the explosion of life that occurred in the Cambrian (e.g. Vidal & Moczydlowska-Vidal 1997). The Ediacaran fauna can be subdivided into three broad, regional groups: one characteristic of Baltica, Siberia, northern Laurentia and Australia; the second diagnostic of Namibia, South America and southern Laurentia; and the third restricted to the Avalonia terrane, including both Newfoundland and Charnwood Forest (Waggoner 1999, 2003; Malone *et al.* 2008). The fossils offer a tantalizing glimpse of life in the Neoproterozoic oceans, and provide hope for robust biostratigraphic correlation and palaeogeographic reconstruction for the latest Neoproterozoic. However, they remain rare and somewhat enigmatic and they have, as a result, achieved almost iconic status, even appearing on recent sets of Australian (Vickers-Rich & Trusler 2006) and Namibian postage stamps.

Given the practical difficulties in definition and correlation of the Neoproterozoic successions outlined above, the term 'Infracambrian' has been retained in this publication, where appropriate (e.g. **Benshati *et al.*** 2009; **Hlebszevitsch *et al.*** 2009; **Le Heron *et al.*** 2009; **Lüning *et al.*** 2009), to represent sequences of undefined, but most probably Late Precambrian–earliest Cambrian age, which occur below the lowest definitively dated Cambrian successions and above igneous or metamorphic basement. The terms Tonian, Cryogenian and Ediacaran are preferred, but are only applied where dating is sufficiently robust to allow them to be used with some confidence (e.g. **Bechstädt *et al.*** 2009; **Lottaroli *et al.*** 2009).

The absence of a robust biostratigraphic framework for most of the Late Precambrian makes global correlations of these sequences very difficult. Historically, such correlations have come to rely on isotope-based schemes, involving a variously weighted combination of litho-, bio- and chemo- and sequence stratigraphy, underpinned, where possible, by relevant radiometric ages (Gradstein *et al.* 2004; Ogg *et al.* 2008). The most reliable

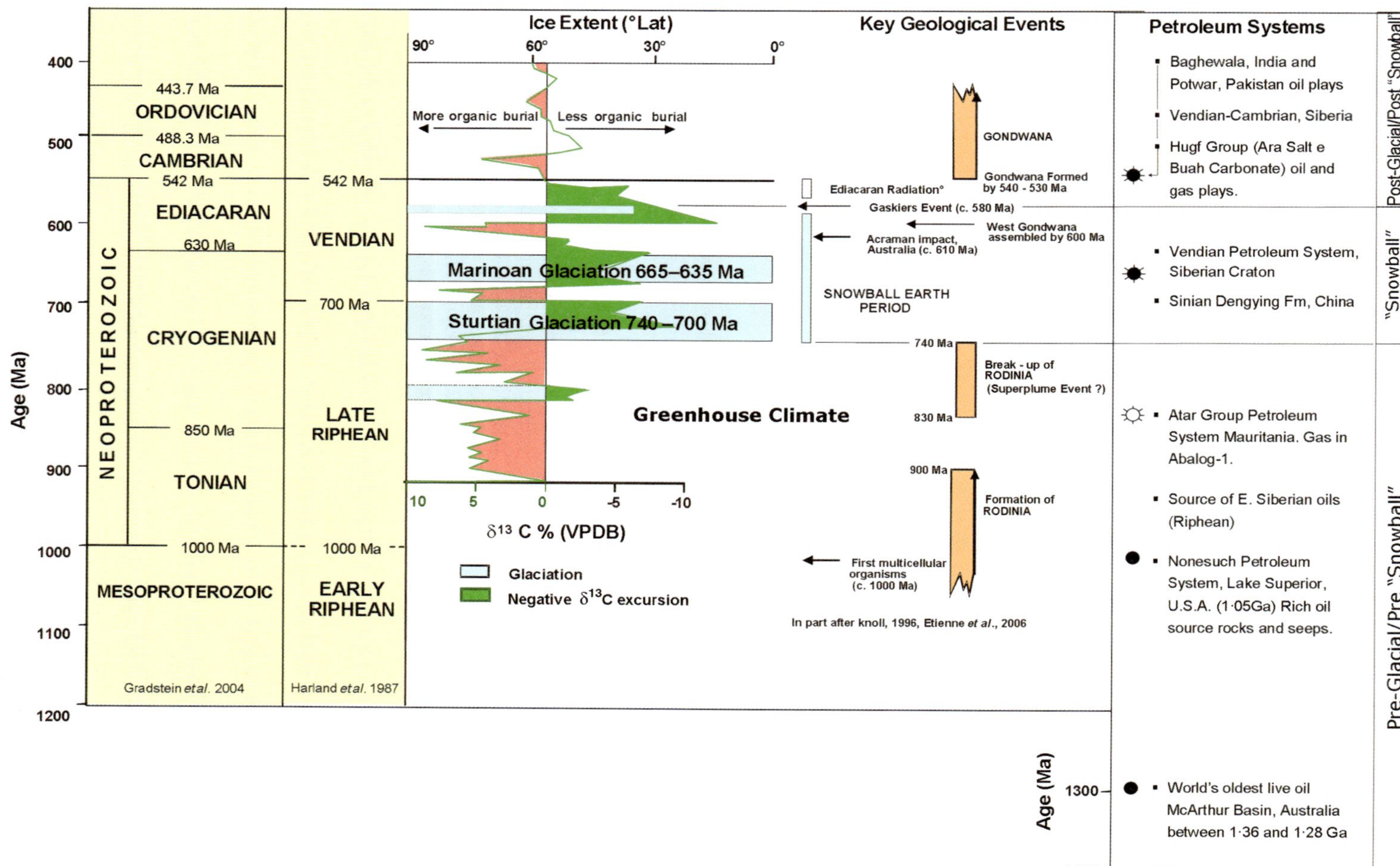

Fig. 3. Neoproterozoic timescale, age and extent of the main glaciations, key geological events and main Neoproterozoic petroleum systems of the world.

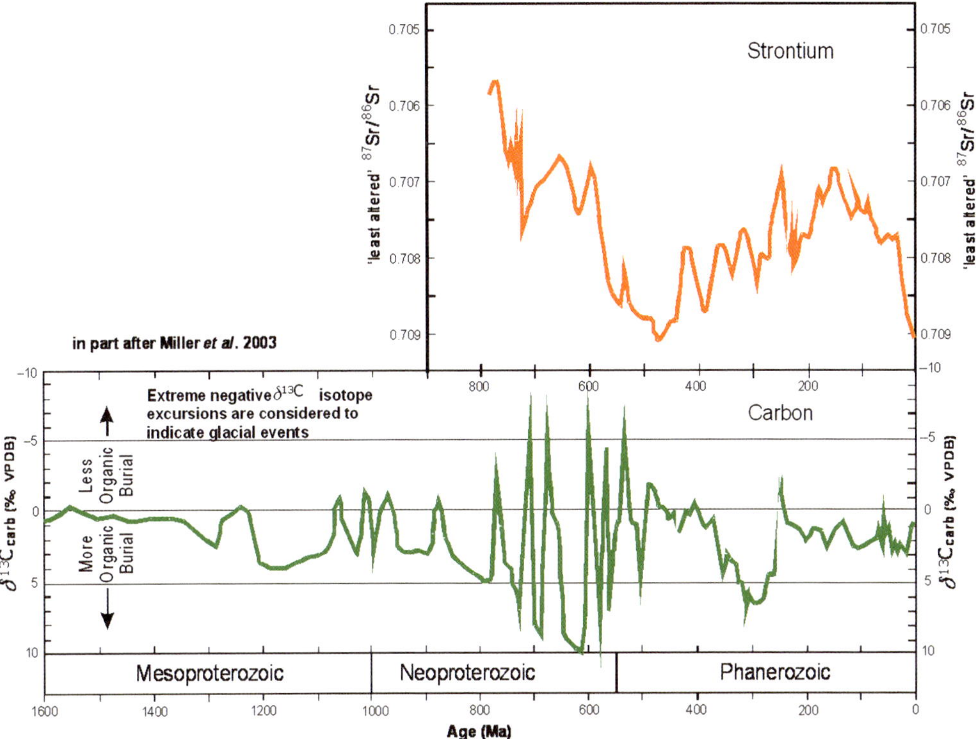

Fig. 4. Secular variation in carbon and strontium isotopic composition in shallow-marine carbonates from 1000 Ma to the present day (in part after Miller *et al.* 2003).

radiometric dates obtained for the Precambrian are from U–Pb (uranium–lead) dating of individual zircons, although there can be a problem with Pb loss causing the ages to be underestimated. This is particularly true of sensitive high-resolution ion microprobe (SHRIMP) analyses, which often give concordant dates, but where the effect of Pb loss is difficult to determine. Analyses of chemically abraded zircons by isotope dilution mass spectrometry (Bowring *et al.* 2007) are generally considered to have yielded the most reliable dates so far, including the 582 Ma date for the Gaskiers Glaciation in Newfoundland, and the 635 Ma date for the end of the Ghaub Glaciation in Namibia and the Nantuo Glaciation in south China (Allen pers. comm.). The recent development of an additional, apparently robust, Re–Os (rhenium–osmium) depositional-age geochronometer for organic-rich sedimentary rocks (e.g. black shales) holds considerable potential for improving the chronostratigraphic calibration of Precambrian successions. A comprehensive review of the method and its application to dating Neoproterozoic black shales from central Australia and from south China is given by **Kendall *et al.*** (2009).

The two most important and commonly used isotopic ratios for correlation and dating purposes are $^{87}Sr/^{86}Sr$ and $\delta^{13}C$ (Fig. 4) (Miller *et al.* 2003). Strontium isotopic ratios are used because they are believed to reflect a truly global signal and because the so-called 'least-altered' ratios exhibit a significant and fairly steady increase through the Neoproterozic. $\delta^{13}C$ isotopic ratios are used because they exhibit large variations during the mid-Neoproterozoic. These are considered to reflect rapid, glacially driven changes in redox cycling of carbon, with the large $\delta^{13}C$ negative excursions to a zero organic productivity, reflecting periods of 'photosynthetic shut-down', basin anoxia and stratification, although methanogenesis and organogenesis must be at least locally important where the negative $\delta^{13}C$ excursions exceed the zero organic productivity, mantle-derived CO_2 value of about $-5‰$ (e.g. the 'Shuram' excursion, which reaches $-12‰$, and the similar 'Wonoka' and 'Reynella' excursions).

It is these large-scale negative $\delta^{13}C$ excursions, in particular, that are used as the key stratigraphic correlation tool in the Neoproterozoic and which have been used to define two global-scale

Neoproterozoic glaciations (the so-called 'Snowball Earth' or 'Slushball Earth' periods); the older 'Sturtian Glaciation' occupying the period from about 740 to 700 Ma and the younger 'Marinoan Glaciation' occupying the period from about 665 to 635 Ma (Fig. 3) (Etienne *et al.* 2006). There is at least one older glaciation, at around 800 Ma, and a younger one, the Gaskiers Event, at approximately 582 Ma (thought to have been short lived and, perhaps, lasting less than 1 Ma), but these are generally considered to be the products of regional, and potentially diachronous, glaciation rather than more synchronous global 'Snowball' or 'Slushball' ice ages. This summary gives an impression of rather definite and distinct glacial events within the Neoproterozoic, but, in reality, there is little consensus about the number, duration or, indeed, the severity of the glaciations (e.g. Kennedy *et al.* 1998; Etienne *et al.* 2007; Allen & Etienne 2008), with perhaps the exception of the 'Marinoan' in this case, there does appear to be reasonable consistency in the dating with the Ghaub Glaciation and the Nantuo Glaciation, both ending at about 635 Ma (Allen pers. comm.) as well as good evidence from Australia that grounded ice did reach equatorial latitudes. With rapidly improving geochronology (e.g. **Kendall *et al.*** 2009), the simple 'two-epoch model' for Neoproterozoic glaciations is becoming increasingly untenable and it may be more appropriate to consider a long Cryogenian period of broadly icehouse conditions extending from about 725 to *c.* 580 Ma, with alternating glacial and interglacial phases (Fig. 1). Palaeogeographic reconstructions for the Cryogenian period presented by **Scotese** (2009) suggest that even the major Neoproterozoic glacial phases may not have been truly global in extent, as there is little evidence of other (preserved) glacigenic rocks in a wide belt around the palaeo-equator. This remains a highly controversial and much debated subject, and there are alternative palaeogeographic reconstructions that favour more global-scale glaciations during the Cryogenian period (e.g. Collins & Pisarevsky 2005). Certainly, it appears that some of the Cryogenian glaciations were unusually severe and extensive. On this basis, it is possible to divide the Neoproterozoic broadly into three phases: a Tonian–pre-Cryogenian pre-glacial phase (prior to *c.* 750 Ma), a Cryogenian glacial phase (from *c.* 750 to *c.* 600 Ma) and a post-Cryogenian–Ediacaran post-glacial phase (from *c.* 600 Ma to the base of the Cambrian at 542 Ma). In fact, we can also consider the various Neoproterozoic petroleum systems on the basis of this threefold division (Fig. 3).

Interestingly, and almost certainly not coincidently, the same threefold division is reflected in the global tectonic events during the Neoproterozoic. The 'Pre-glacial' period corresponds with the formation, stabilization and initial break-up of the supercontinent of 'Rodinia', while the 'Post-glacial' period corresponds with the amalgamation and stabilization of 'Gondwana', leaving the intervening 'Glacial' phase as a period of active extensional tectonics in Laurentia, Namibia, South Australia, south China, northern India and Baltica, when the major cratonic fragments were dispersing and reorganizing between the two supercontinent configurations (Fig. 3).

In summary the key characteristics of the Neoproterozoic period of Earth history are as follows.

- The Neoproterozoic Eon (1000–542 Ma) was a period of massive atmospheric, climatic and tectonic change.
- It was dominated by the Cryogenian 'Snowball Earth' glaciations, which probably consisted of a series of distinct glacial–interglacial cycles between approximately 750 and about 600 Ma.
- Evidence for glaciation is found in mid to Late Neoproterozoic successions in many parts of the world.
- Deposition of Neoproterozoic ('Infracambrian') strata occurred during the interval between the break-up of the Tonian supercontinent of Rodinia and the Palaeozoic supercontinent of Gondwana.
- The evolution of life is marked by the emergence of the first recognizable animal life around 600 Ma, before the latest Neoproterozoic–Early Cambrian 'metazoan explosion'. There was a diversification after the last of the main Cryogenian glaciations (Moczydlowska 2008), but most evolutionary steps occurred after 575 Ma with the appearance of complex spiny acritarch assemblages (in contrast to the older, simple, non-spiny spheromorph-dominated assemblages) and the evolution of the distinctive Ediacaran fauna (see **Butterfield** 2009).

Global Precambrian and 'Infracambrian' petroleum systems

The main Precambrian and 'Infracambrian' (Neoproterozoic–Early Cambrian) petroleum systems in the world (Fig. 5) can be classified as either 'producing or proven' (those that either do, or could soon, produce commercial volumes of hydrocarbons) or 'potential' (where all the elements of a Neoproterozoic play are known to exist, but where there is, as yet, no commercial production). While the map shown in Figure 5 may not be comprehensive, it does at least illustrate that Precambrian and 'Infracambrian' petroleum systems are relatively abundant and widespread. The oldest live oil recovered to date is sourced from Mesoproterozoic rocks

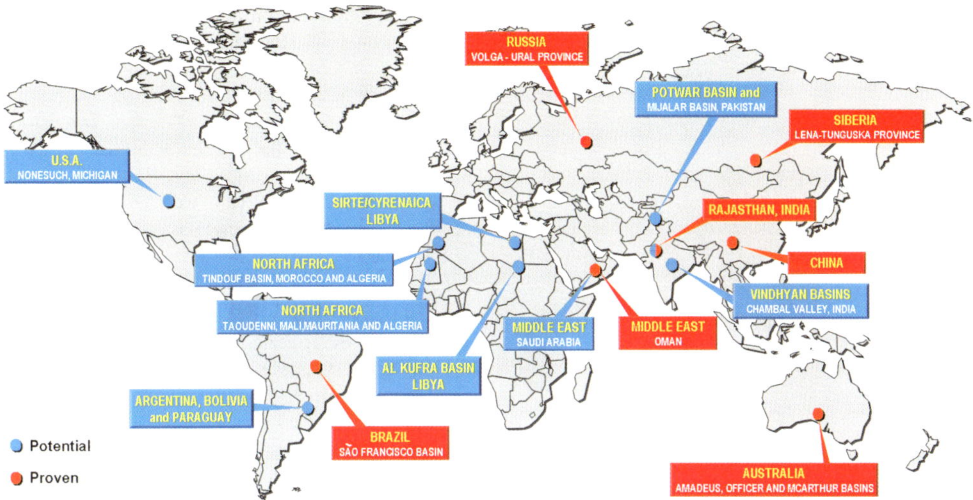

Fig. 5. Proven/producing and 'potential' Precambrian petroleum systems of the world.

within the Velkerri Formation (Roper Group) of the McArthur Basin of northern Australia (Jackson *et al.* 1986; Crick *et al.* 1988) dated at 1361 $\pm$ 21 and 1417 $\pm$ 29 Ma (Re–Os dates), with at least the initial phase of oil generation and migration having taken place before 1280 Ma (see **Kendall *et al.*** 2009), followed closely by the Nonesuch Oil of Michigan. However, the geologically oldest commercial production is probably from the somewhat younger mid to Late Neoproterozoic (Cryogenian–Ediacaran) petroleum systems of the Lena–Tunguska province in East Siberia and in southern China, and from the latest Neoproterozoic–Early Cambrian Huqf Supergroup in Oman. **Ghori *et al.*** (2009) give a comprehensive review of both proven and potential global Neoproterozoic petroleum systems.

Correlation of the main Neoproterozoic lithostratigraphic units along a section extending from North Africa to the Middle East (Fig. 6) illustrates some interesting and highly significant relationships. These include the broad threefold division of the 'Infracambrian' succession with a Tonian–Cryogenian sequence consisting largely of carbonate and shale, preserved in Mauritania, a dominantly clastic, Cryogenian sequence preserved patchily in a series of individual graben and half-graben across much of North Africa, followed by a rather uniform and laterally extensive, mixed facies, Ediacaran sequence. Given the difficulty of correlation within the Neoproterozoic, such regional-scale correlations are inevitably subject to a significant level of uncertainty and are being refined continuously as new lithostratigraphic, biostratigraphic and chemostratigraphic data become available. However, it is clear that this broad threefold division is characteristic of Neoproterozoic successions in many parts of the world.

Neoproterozoic and Lower Palaeozoic geology of the Peri-Gondwanan Margin

The 'Peri-Gondwanan Margin' occupies the broad region of the Gondwana supercontinent from present-day northern South America, through North Africa, the Middle East and the Indian Subcontinent to northern Australia. The Gondwana supercontinent formed through the collisional amalgamation of the African, South American, Indian, Australian and Antarctic terranes during the late Precambrian (see, e.g. **Hlebszevitsch *et al.*** 2009; **Scotese** 2009; **Smith** 2009) and consisted of the old stable cratonic blocks (including the West African and Chad cratons) separated by Pan-African mobile belts, which in North Africa have a dominant north–south structural grain (Fig. 7). The assembly of both western and eastern Gondwana continued until the Cambrian, and occurred in two main stages: at approximately 640–600 Ma (e.g. Amazonia colliding with the Congo/Saõ Francisco continent, and the amalgamation of northern Africa); and at about 570–510 Ma with the collision of Kalahari with the South American continents, and of the Congo/Saõ Francisco continent, India and Australia with nascent Gondwana (e.g. Jacobs & Thomas 2002; Collins & Pisarevsky 2005; Li *et al.* 2006; Pisarevsky *et al.* 2008). The collisional amalgamation of Gondwana and the associated delamination of the underlying mantle resulted in

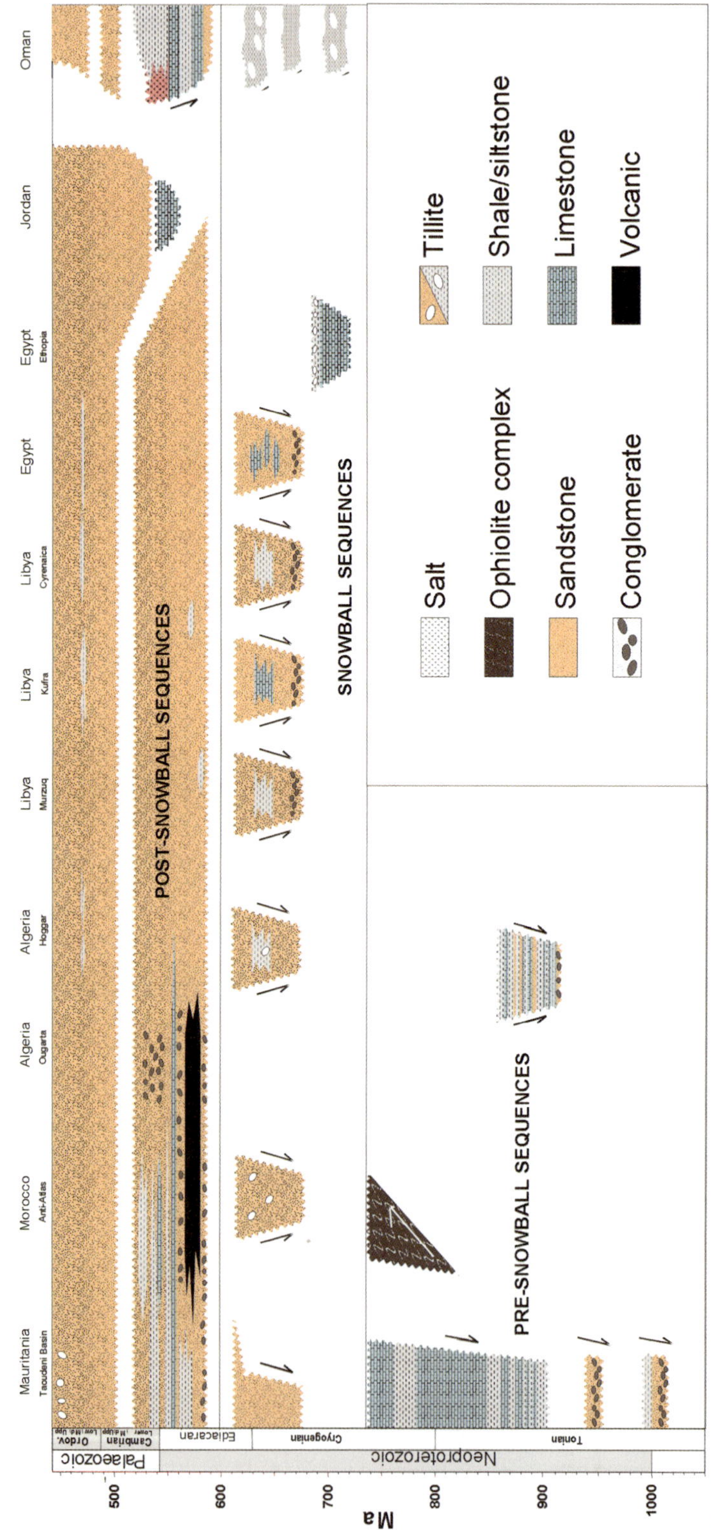

Fig. 6. Summary of the lithostratigraphic and chronostratigraphic correlation of selected Neoproterozoic successions from North Africa to the Middle East.

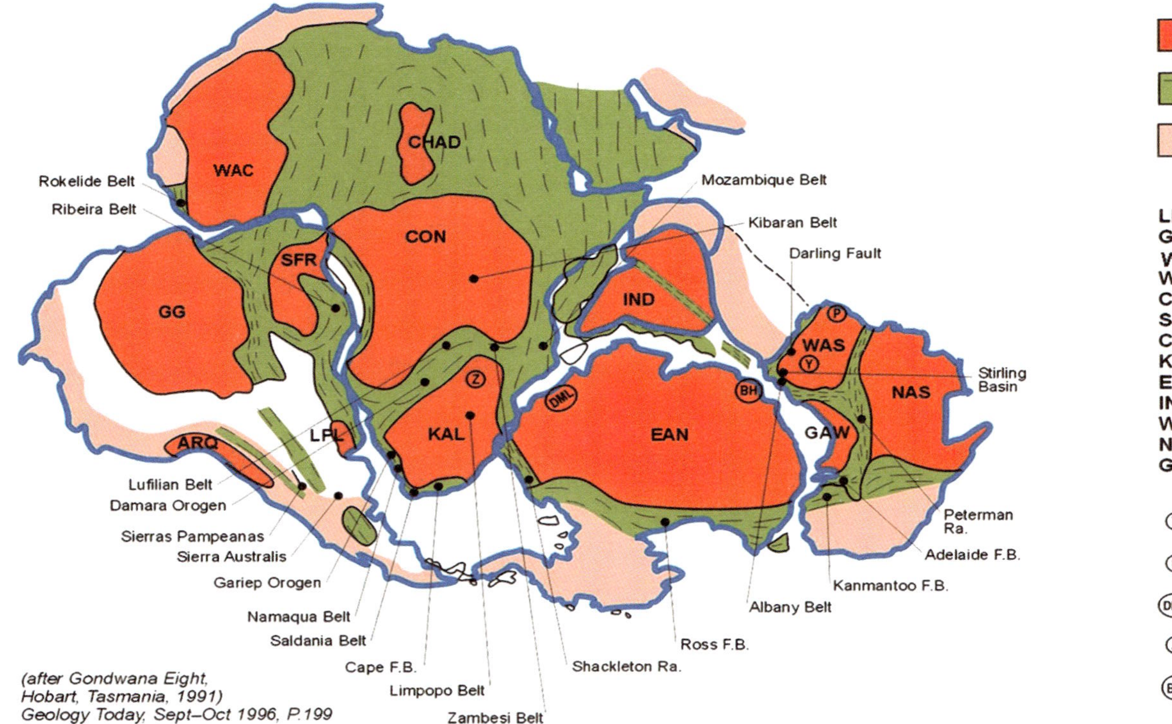

Fig. 7. Palaeogeographic reconstruction of the Gondwana Supercontinent at the end of the Neoproterozoic Era.

massive uplift, unroofing and peneplanation of the supercontinent, and the deposition of vast quantities of clastic sediment across North Africa, the Middle East, the Indian subcontinent and Australia, much of it was derived from erosion of the Pan-African mountain belts to the south. The northern margin of the Gondwana supercontinent was periodically flooded by eustatic transgressions and formed a broad, shallow-marine continental shelf throughout the latest Neoproterozoic and much of the Early Palaeozoic. During the Early Palaeozoic, reactivation of mainly north–south Pan-African structures across North Africa and the Middle East triggered the development of broad, intra-cratonic sag basins that remained active depocentres throughout the Palaeozoic. However, the stress conditions responsible for the development of these basins remains poorly understood. In the proximal areas most of the Early Palaeozoic succession consists of belts of shallow-marine sandstone, which migrated laterally with changing sea level and passed offshore into marine shales (Craig *et al.* 2008).

The Palaeozoic intra-cratonic basins and their associated distinctive Palaeozoic petroleum systems occupy a belt 500–1000 km wide along the entire northern margin of the Gondwana Supercontinent. The core area of the Lower Palaeozoic petroleum systems lies in the Palaeozoic basins of North Africa and the Middle East. However, some elements of these plays extend further east through India and Australia and, potentially, also west into South America, including Brazil and Argentina.

The peneplanation surface at the base of the Cambrian succession is easily seen both at outcrop throughout North Africa – for example, in the Algerian Tassili, where flat-lying Cambrian sediments rest unconformably on Neoproterozoic synorogenic sediments – and, perhaps even more dramatically in seismic data from areas such as the Al Kufrah Basin in SE Libya, where a remnant Neoproterozoic? Basin, containing 'Infracambrian?' strata with a thickness of more than 1500 m, appears to be preserved beneath the subhorizontal Palaeozoic succession (Craig *et al.* 2008, fig. 10; Klitsch *et al.* 2008; **Benshati *et al.*** 2009, Fig. 11).

Nearly all the Early Palaeozoic sag basins along the Peri-Gondwanan Margin are underlain by Neoproterozoic basins, which contain either a proven or a potential Neoproterozoic petroleum play. It is possible that these rather enigmatic 'sag' basins formed initially as a result of thermal subsidence following Neoproterozoic rifting, although the fact that in many there appears to be a long period of either relative stability or uplift and peneplanation during the latest Neoproterozoic and earliest Cambrian suggests that any such relationship is not simple. In Oman, for example, rifting ceased at about 640 Ma and was followed by a phase of extensive, mostly shallow-water deposition (Nafun Group) until about 540 Ma when major platform-basin variations indicate the formation of the salt basins of the Ara Group through tectonic reactivation of a pre-existing north–south structural grain (Allen 2007). This chronology implies a period of at least 100 Ma between the Neoproterozoic stretching and the development of the Palaeozoic basins – too long for a conventional thermal subsidence mechanism. One possibility is that the Palaeozoic basins developed as a result of stretching of thick continental lithosphere at a very low strain rate over a long period of time, driven by Early Palaeozoic plate reorganization and reactivating the underlying Neoproterozoic structure. This might prolong the basin subsidence sufficiently to account for the difference in timing (Allen pers. comm.).

The stratigraphy, sedimentology and structural relationships of the 'Infracambrian' rocks encountered at outcrop and in the subsurface, in the Murzuq, Al Kufrah and Sirte basins, and on the Cyrenaica Platform in Libya are described in some detail by **Aziz & Ghnia** (2009), **Benshati *et al.*** (2009) and **Le Heron *et al.*** (2009), while the hydrocarbon prospectivity of the Neoproterozoic–Early Cambrian ('Infracambrian') successions in these basins and in other parts of northern and western Africa is discussed by **Lottaroli *et al.*** (2009) and **Lüning *et al.*** (2009). In addition, the tectonic and stratigraphic evolution of the terminal Neoproterozoic–Middle Cambrian (intra-Vendian/Ediacaran–intra-Tremadocian) succession of the Cadenas Ibéricas in NE Spain – a rifted fragment of the NE Africa Gondwana margin – is described by **Gámez Vintaned *et al.*** (2009).

The Late Ordovician–Early Silurian petroleum system in North Africa – an analogue for Neoproterozoic reservoir–source rock relationships?

Unfortunately, we know relatively little about the Neoproterozoic successions in the basins developed along the Peri-Gondwanan Margin, partly because they are rarely penetrated in the subsurface and partly because, while there are good surface exposures in some areas, these are frequently in remote, difficult to access and, in some cases, potentially dangerous locations. In these circumstances, we have to rely on analogues in order to develop possible new hydrocarbon plays. These analogues tend to be either proven or producing Neoproterozoic petroleum systems elsewhere in the world or, in the case of understanding broader geological concepts – such as the relationships between glaciation, reservoir and source rock distribution – analogues based on other parts of the geological record. In this latter case, we are fortunate to have the Late Ordovician–Early Silurian petroleum system in North Africa, for which we have a thorough understanding of the reservoir, source and seal relationships as a result of nearly two decades of intensive study (e.g. Lüning *et al.* 2000*a*; Le Heron & Craig 2008 and references therein).

The Late Ordovician glaciation is probably not a direct analogue for the major globally extensive Late Neoproterozoic glaciations. For example, there are good reasons to believe that the atmospheric conditions during the Late Neoproterozoic glaciations were more variable and more extreme than those during the Late Ordovician glaciation. However, the latter was certainly extensive and extreme, and, as such, is probably the best analogue available.

The Late Ordovician ice sheet was centred over the remnant Pan-African Mountains in central Africa, and expanded outwards onto the surrounding continental shelves (Vaslet 1990; Le Heron & Dowdeswell 2009). At its maximum extent, it was of comparable size to the present-day Antarctic Ice Sheet, covering nearly 12×10^6 km^2 over 65° of palaeo-latitude and extending as far north as 30°S. The glacigenic sediments deposited by the Late Ordovician ice sheet crop out in South America, North and South Africa, the Arabian Peninsula and parts of SW Europe (Craig *et al.* 2008, fig. 16).

In North Africa, the Upper Ordovician glacigenic sequence contains one of the most important and widespread reservoir horizons: the Mamuniyat Formation in Libya and the equivalent Unit IV in Algeria (Davidson *et al.* 2000; Echikh & Sola 2000; Hirst *et al.* 2002; Le Heron *et al.* 2004; Le Heron & Craig 2008). In simple terms, there are two distinct facies belts: a dominantly sandy belt, and hence reservoir, in the south, and a dominantly shaley belt, and hence seal, in the north (Fig. 8).

The glacigenic sandstones of the Mamuniyat Formation and its equivalents are typically overlain by black and grey Silurian shales belonging to the Tanezzuft Formation. These shales record flooding of the North Gondwana continental shelf as a result of glacio-eustatic sea-level rise linked to the collapse of the Late Ordovician ice sheet. Locally, the basal part of the Tanezzuft Formation consists of a highly organic-rich unit – generally less than 25 m thick – of black, graptolite shale, which forms one of the major hydrocarbon source rocks of the region (Lüning *et al.* 2000*a*, *b*). Although the formation is widely distributed across North Africa, the basal organic-rich 'hot' shale source facies is quite patchy. It is more widespread and continuous in the 'outboard' areas of central Algeria (south of the Atlas Front), but much more restricted and discontinuous in the more proximal 'inboard' areas, such as the northern Murzuq Basin, where the topography of the underlying Late Ordovician glacial landscape was probably more pronounced (Lüning *et al.* 2000*a*; Craig *et al.* 2008).

Ultimately, we can integrate all of the sedimentological, palaeogeographical and biostratigraphic information into detailed chronostratigraphic charts for the entire Late Ordovician–Early Silurian glacial–post-glacial system. Typically, there are two clear, regionally extensive cycles of glacial advance and retreat, although up to four separate cycles are preserved in some areas. The phases of glacial advance are indicated by the subglacial erosion surfaces, while the deposition of 'ice-contact fans' mark the phases of glacial retreat. The advance–retreat cycles are followed by a period of reworking associated with isostatic rebound (the result of unloading), intimately coupled with marine transgression, reworking and the deposition of the post-glacial, organic-rich, 'hot' shales in the remnant topographic lows and onlapping the adjacent palaeo-highs.

We know that the entire Late Ordovician glaciation in North Africa occurred within the time span of a single graptolite biozone (the *extraordinarius* zone); a period of about 500 000 years, and through spectral analysis of the cyclicity recorded in the compositional variations of age-equivalent Late Ordovician evaporites in the Canning Basin of Western Australia, we suspect that the individual cycles of glacial advance represent the 100 000-year eccentricity cycles of the Milankovitch series (Sutcliffe *et al.* 2000*b*; Kaljo *et al.* 2003). If this is correct, it appears that full glacial conditions during the Late Ordovician may have lasted for as little as 200 000 years (two cycles) or, perhaps, 400 000 years (four cycles), but certainly a very short period of time, given the thickness, complexity and extensive nature of the associated sediments.

Several key characteristics of 'glacigenic reservoir–source systems' can be inferred from detailed examination of the Late Ordovician–Early Silurian succession in North Africa. Such systems are likely to include the following:

- Spatially complex, heterogeneous reservoir systems controlled by the distribution of highly erosive ice-streams and associated ice-grounding lines.
- Complex, but organized, distribution of glacial landforms, including subglacial tunnel valleys, lateral and terminal moraines, streamlined bedforms, and subglacial and intra-sediment striated surfaces.

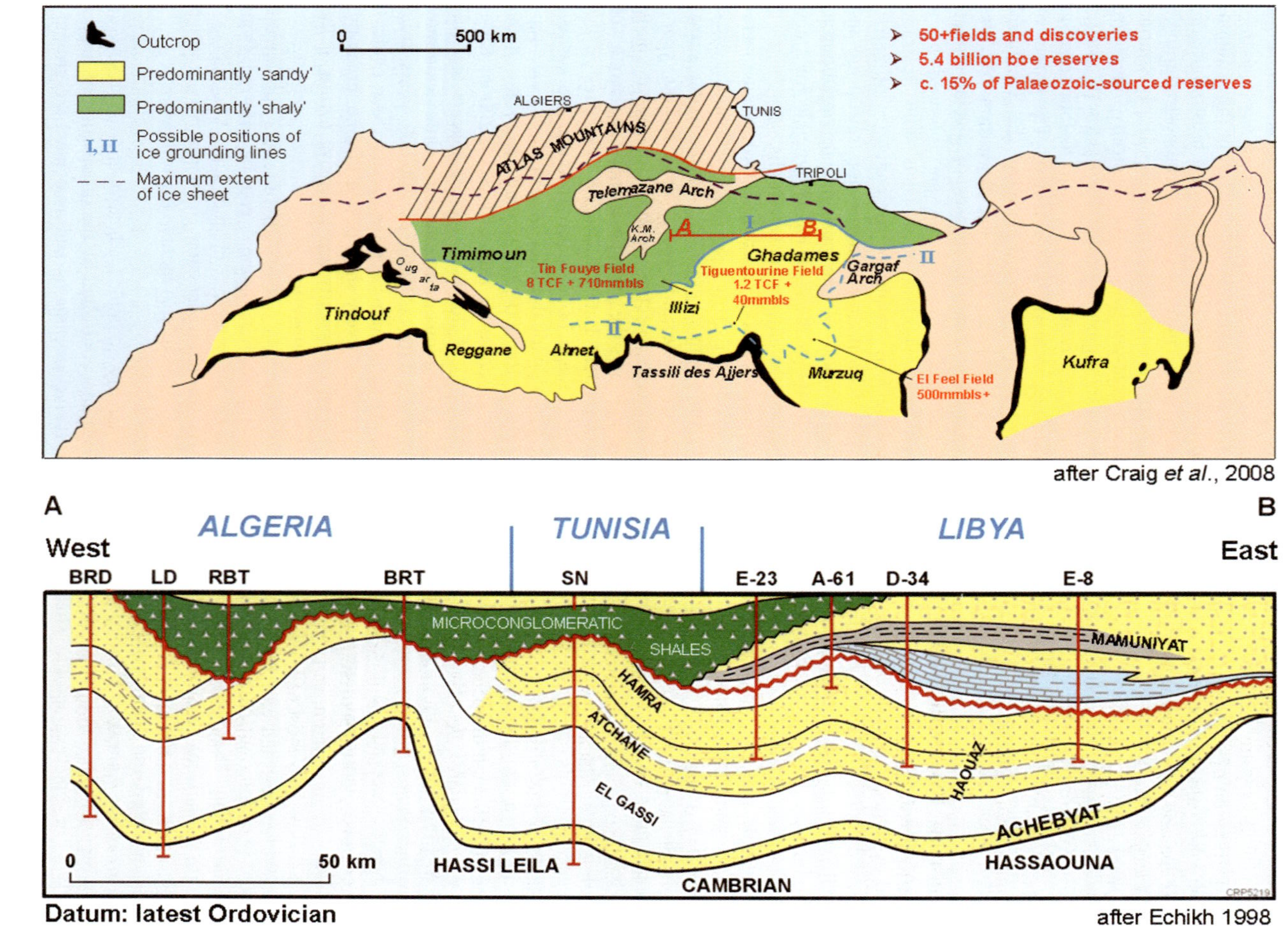

Fig. 8. Distribution and dominant lithology of the Upper Ordovician (Hirnatian) glacigenic sediments in North Africa (after Craig *et al.* 2008).

- Multiple phases of glacial advance and retreat, with associated sediment packages, separated by prominent erosion surfaces.
- Deposition of post-glacial, transgressive sequences, strongly controlled by remnant glacial topography (locally accentuated or ameliorated by post-glacial isostatic rebound), with locally patchy distribution of organic-rich source rocks in topographic palaeo-lows and the progressive onlap of palaeo-highs during continued post-glacial transgression.

It seems likely that similar characteristics should be a feature of depositional systems associated with each of the major global glaciations that have occurred throughout Earth history, including those in the Neoproterozoic Era.

'Infracambrian' (Neoproterozoic–Early Cambrian) petroleum systems of the Peri-Gondwanan Margin

Tonian–Early Cryogenian: Taoudenni Basin, Mauritania, Mali, Algeria (c. 1000–750 Ma)

The Tonian–Cryogenian phase of the Neoproterozoic (the 'Pre-glacial' phase) comprises petroleum systems developed between about 1000 and 750 Ma. Perhaps the best example of these systems in North Africa lies in the rather poorly known, remote and underexplored Taoudenni Basin, which extends across Mauritania, Mali and southern Algeria.

The Taoudenni Basin is developed over one of the old pre-Pan-African cratonic blocks: the West African Craton. Although the palaeomagnetic constraints on the position of the West African Craton at this time are rather poor, most modern palaeogeographic reconstructions place it as a separate continental fragment, located close to the South Pole (e.g. Collins & Pisarevsky 2005; **Scotese** 2009).

The West African Craton and the overlying Taoudenni Basin occupy most of West Africa, and are flanked to the west and east by north–south-trending Pan-African 'mobile belts' formed during the accretion of the Gondwana Supercontinent (Li *et al.* 2006). The Neoproterozoic succession is well exposed in an 1100 km-long outcrop belt along the northern margin of the Taoudenni Basin through Mauritania, the NW corner of Mali and into southern Algeria (e.g. Moussine-Pouchkine & Bertrand-Sarfati 1997). The outcrops at the SW end of this belt, in the Atar Region, are relatively well known, and there are comprehensive and accurate regional geological maps and sections available in the public domain (Deynoux *et al.* 2006). The Neoproterozoic–Early Cambrian succession is well defined and reasonably well dated radiometrically (Deynoux *et al.* 2006). It consists of a uniformly east-dipping succession of 'Pre-glacial' sediments, ranging from approximately 1000 Ma to somewhat younger than 775 Ma, unconformably overlain by a subhorizontal succession of younger Neoproterozoic sediments with, at the base, a glacigenic sequence including diamictites. These are considered to be younger than 630 Ma, which, if correct, suggests they are 'Marinoan' or younger. Below the main regional unconformity the succession mainly consists of interbedded carbonates and shales, while above it there are glacial diamictites overlain by marine and fluvial sandstones and shales. The succession below the unconformity includes a superbly exposed sequence of stromalolitic carbonates, for which a comprehensive depositional model can be constructed using analogues exposed in the younger Nafun Group in Oman (e.g. Cozzi & Al-Siyabi 2004).

Given the lithostratigraphy of the succession, the geometric relationships observed in the field, the sparse seismic data available in the basin and the associated radiometric age constraints, it is possible to relate the sequence to a threefold division in the evolution of the basin. This includes an early 'Pre-glacial', pre-Pan-African phase, characterized by a relatively flat cratonic platform, followed by a period of Pan-African extension corresponding to the 'Glacial' period and ending with a period of 'Post-glacial' Pan-African foreland basin development. Again, there appears to be an interesting correlation between glaciation and continental extension. The interesting part of the succession, from the perspective of this publication, is the lower part–the Hank Group and its lateral equivalent, the Atar Group – because these contain hydrocarbons.

Cross-sections through the northern margin of the basin presented by **Lüning *et al.*** (2009) and **Rahmani *et al.*** (2009) show the same threefold structural evolution, with active growth faulting during the 'Glacial' equivalent extensional phase, followed by Pan-African foreland basin development, and then severe inversion, uplift, erosion and peneplanation during the final stages of the Pan-African Orogeny, before the deposition of the overlying Palaeozoic succession.

The subsurface portion of the Taoudenni Basin is currently very poorly known, although a new phase of exploration is underway. At present, there are only six wells in the entire basin. One of the most interesting, from the Neoproterozoic perspective, is Abolag-1, drilled in 1973. This penetrated more than 600 m of succession assigned to the 'Infracambrian', but undated at the time, and from which gas was recovered on test at a rate of 13 600 m^3/day. This 'Infracambrian' succession is subdivided into an upper clastic sequence and a lower carbonate sequence, both of which appear to be gas-bearing.

When the Abolag well was drilled, the 'Infracambrian' succession was considered to be unfossiliferous but, with careful resampling and rigorous preparation work, a diverse and distinctive assemblage of acritarchs has recently been recovered from this interval. By comparison with equivalent assemblages from Siberia, Australia and southern Poland, this assemblage is considered to be Tonian–Early Cryogenian in age (see **Lottaroli *et al.*** 2009). The new acritarch assemblage from Abolag includes distinctive cylindrical forms, algal, filamentous cyanobacterial sheaths and 'amorphous organic matter' (AOM). The AOM could potentially be the type of organic material from which the gas tested in the well was generated. In fact, there is very good evidence for the existence of oil-prone black shales with TOC (total organic carbon) content of 10–20% in the Neoproterozoic Atar Group, both at outcrop (where they are known as the 'burning shales') and in the subsurface where they have been encountered in shallow boreholes. The rather broad biostratigraphic age assignment has also been supported by new carbon isotope analysis (Thurow pers. comm.), which shows the existence of the important isotopic excursions characteristic of the Neoproterozoic glacial events and also the presence of several important stratigraphic breaks within the Neoproterozoic succession.

Overall, there is little doubt that the lower, predominantly stromatolitic, carbonate succession in the Abolag well is Tonian–Early Cryogenian in age (*c.* 1000–750 Ma), and that the gas-bearing section is the lateral equivalent of the Atar Group at outcrop. The succession in the well has a low TOC content, and appears to be of high thermal maturity, although this is somewhat inconsistent with the golden brown colour of the kerogen and the well-preserved state of the acritarchs, and may be a local thermal effect associated with the abundant dolerite intrusions. Certainly, at outcrop, the equivalent black shales of the Atar Group are known to be of low thermal maturity. The carbonates in the well are tight, but heavily fractured and contain bitumen, indicative of an original oil charge. Further details of this interesting, but challenging, Early Neoproterozoic hydrocarbon play in the Taoudenni Basin are given by **Lottaroli *et al.*** (2009) and **Rahmani *et al.*** (2009). **Hlebszevitsch *et al.*** (2009) describe a broadly time-equivalent Early Neoproterozoic (Riphean) petroleum system developed on the western margin of the Saõ Francisco Craton, in the Saõ Francisco Basin of Brazil, while **Bhat *et al.*** (2009) describe the potential for oil and gas in similar Proterozoic stromatolitic carbonates in the Himalayan foothills of NW India.

Mid-Cryogenian–Mid-Ediacaran petroleum systems (750–600 Ma)

The 'Glacial' or 'Snowball' phase petroleum systems occur during the period from the Mid-Cryogenian to the Early–Mid-Ediacaran (*c.* 750–600 Ma). This period encompasses the two Neoproterozoic, allegedly global, glaciations (the 'Sturtian' and 'Marinoan'), which are the subject of the somewhat controversial and still hotly debated 'Snowball Earth' hypothesis. Hoffman *et al.* (1998) and Kirschvink (1992) propose that, owing to a combination of very unusual continental configurations and atmospheric conditions at this time, the Earth oscillated rapidly between almost total ice cover with mean surface temperatures of $-50\,^{\circ}\mathrm{C}$ and 'super-greenhouse' conditions with temperatures of perhaps $+50\,^{\circ}\mathrm{C}$. The key observations presented in support of this hypothesis include: widespread distribution of Late Neoproterozoic glacial deposits on virtually every continent; palaeomagnetic evidence that the glacial ice line reached sea level close to the equator for long periods; stratigraphic evidence that glacial events began and ended abruptly; the reappearance of banded iron formations after an absence of 1.2 billion ($\times 10^9$) years; worldwide occurrence of cap carbonates, with unusual features, in sharp successive contact with underlying Late Neoproterozoic glacial deposits; and the existence of very large positive and negative $\delta^{13}C$ anomalies, respectively, before and after each glacial event.

Some of these observations have been hotly debated, although there remains broad agreement that the Neoproterozoic glacial deposits are widespread and that at least some of the glaciations reached sea level at low latitude (the Elatina Glaciation in Australia being, perhaps, the best example). Other observations, such as the abrupt end to the glacial events, have been challenged and, although there is frequently a sharp contact between the glacial diamictites and the overlying cap carbonate horizons, it now seems more likely that the carbonate deposition was delayed, diachronous and relatively slow, tracking the rising post-glacial sea level. Similarly, the occurrence of Neoproterozoic banded iron formations is considered to be rare (and, where they do occur, can often be explained by local oceanographic effects) and the original interpretation of some of the $\delta^{13}C$ anomalies has also been challenged (Allen 2006; Fairchild & Kennedy 2007; Allen & Etienne 2008 and references therein).

Irrespective of whether the 'Snowball Earth' hypothesis ultimately proves to be correct, it has at least had the benefit of stimulating much new research and scientific debate about this period of geological history. From a hydrocarbon exploration perspective, it has also raised the possibility that there are sufficient similarities between the Neoproterozoic and the Late Ordovician glacial systems to allow the latter to be used to develop generic petroleum systems models for the former and so predict, for example, the likely distribution of Neoproterozoic source rocks.

The Ghaub glacigenic sequence of the Fransfontain Ridge in Namibia is one of the most intensely studied 'Snowball Earth' successions in the world (Hoffman *et al.* 1998; Hoffman & Schrag 2000, 2002). A 60 km-long section through this succession, illustrated by Hoffman (2006), extends from what was, at that time, a shallow-water carbonate platform in the west, offshore and down a distally tapered submarine foreslope wedge to the east. Interestingly, the Fransfontain Ridge section exhibits glacial features of similar scale and character to those observed in the Late Ordovician of North Africa (Fig. 9). These include:

- A strong, continuous glacial erosion surface at the base, with relief of more than 50 m on the platform.
- Two discrete cycles of glacial advance and retreat, recorded as older and younger diamictite (tillite)–cap carbonate cycles.
- In the older sequence, a 20 km-wide and 500 m-deep trough, filled with a complex of submarine channel and levee deposits, partially controlled by active growth faulting and, in the younger sequence, a second 18 km-wide and 100 m-deep trough. These resemble classic stacked, subglacial tunnel valleys carved beneath a long-lived, fast-flowing palaeo-ice stream.
- A double-crested build-up of massive diamictite, which Hoffman (2006) interprets as a 'medial moraine deposited near the mouth of the relatively narrow palaeo-ice-stream that eroded the trough'.
- Finally, a lowstand wedge ... or possibly an ice-contact fan.

Overall, the Neoproterozoic (Late Cryogenian) glacigenic sequence at Fransfontain Ridge seems to have many of the characteristics of a classic wet-based glacial system, with multiple phases of glacial advance and retreat, and strong subglacial incision

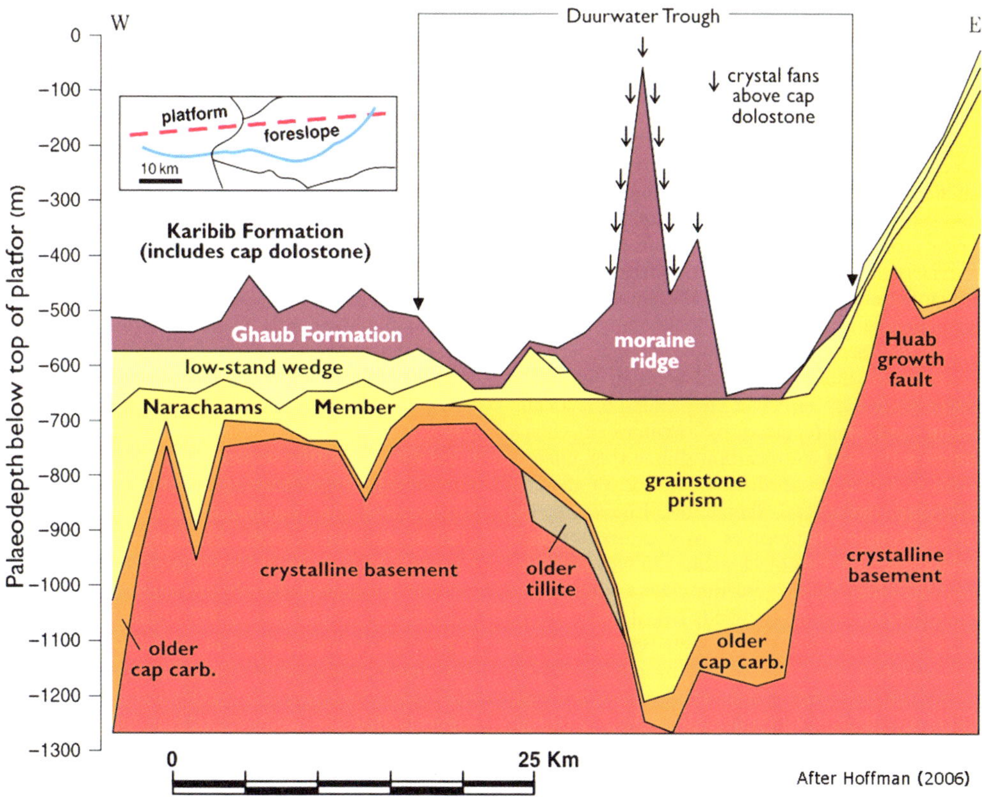

Fig. 9. A Neoproterozoic (Cryogenian) ice stream in Namibia? Cross-section through the Abenab and Lower Tsumeb subgroups (Otavi Group), Fransfontein Ridge, Namibia (after Hoffman 2006).

associated with fast-flowing ice streams defining the glacial maxima, filled by a complex variety of glacigenic sediments deposited during recession and collapse of the ice sheets. This suggests strong similarities between the Neoproterozoic and the Late Ordovician glacial systems, and, possibly therefore, also between the nature and distribution of Neoproterozoic post-glacial source rocks and those deposited during the Early Silurian post-glacial transgression. In Namibia, the post-glacial Neoproterozoic succession occurs primarily within the so-called 'cap-carbonate sequence'. Such cap-carbonate sequences around the world share several characteristics. Typically, they were deposited during post-glacial sea-level rise; are transgressive and typically extend far beyond the preceding glacial deposits, disconformably blanketing the pre-glacial rocks; comprise deep-water to shelfal to supra-tidal facies, including microbial bioherms and biostromes (stromatolites); grade across a marine flooding surface into deeper water limestone or shale; commonly pass upwards into organic-rich black shales (e.g. the Sheepbed Formation in Canada); contain antiformal structures that have been attributed to wave action (e.g. the Keilberg cap carbonate; Hoffman & Allen 2007), tepee formation or soft sediment deformation; are associated with barite concentrations; contain gas (methane?) escape features such as pipes, deformation features and cementation (e.g. the Reynella cap carbonate in Australia: Kennedy *et al.* 2001), and 'tubes' that have been attributed to the vertical growth of columnar stromatolites, but could also be due to methane gas escape (e.g. Noonday dolomite: Kennedy 1996); span several magnetic reversals (particularly, in the case of the Elatina cap carbonate: Raub *et al.* 2007). Finally, the associated alkalinity is inferred to have been supplied by intense carbonate and silicate weathering.

However, there are several areas in the world where such cap carbonate sequences are locally absent and are replaced by organic-rich, black shales with good hydrocarbon source rock characteristics, either interbedded with, and/or directly overlying, the glacial diamictites. Elsewhere, the cap carbonates themselves pass either upwards and/or laterally into black shales. In one example, from the older Neoproterozoic glacial sequence in the Saõ Francisco Basin in SE Brazil, post-glacial black shales within the Vazante Group have a TOC content that is, locally, in excess of 3% (Olcott *et al.* 2005, 2006; **Hlebszevitsch *et al.*** 2009). In addition to their hydrocarbon potential, these organic-rich post-glacial shales represent attractive targets for Re–Os geochronology because they provide a minimum age constraint for the end of the associated glaciation (see **Kendall *et al.*** 2009). The relationship of post-glacial sediments with the underlying 'cap carbonates' and the glacial diamictite units within the Late Cryogenian successions in northern Namibia, together with the influence of rift-related uplift on their associated petroleum prospectivity, are described in detail by **Bechstädt *et al.*** (2009).

These relationships allow us to construct a conceptual model for Neoproterozoic post-glacial source rock deposition in which, during the glacial maximum, a deep trough is carved by a palaeo-ice stream, possibly controlled, or assisted, by active extensional faulting along the flanks (see Le Heron *et al.* 2009*b*, fig. 10). Cap carbonates are deposited in a variety of shallow- to deeper-water environments on the flanks of this valley during post-glacial or post-interglacial flooding events. In the centre of the trough, the cap carbonates pass both laterally and, eventually, vertically into organic-rich black shales, while towards the base of the glacigenic sequence they are intimately associated with the glacial diamictites deposited during the final retreat of the ice sheet.

Late Ediacaran–Early Cambrian petroleum systems (600–c. 500 Ma)

There are several proven and potential petroleum systems of Late Ediacaran–Early Cambrian ('Post-glacial') age around the Peri-Gondwana Margin, most notably in Oman, India and Pakistan (Fig. 5). These occur in rocks that range from approximately 600 to 500 Ma, but they are primarily associated with successions that span the Neoproterozoic–Cambrian boundary at 542 Ma. During the later stages of the collisional amalgamation of Gondwana, east–west compression resulted in the disruption of the East Gondwana portion of the new supercontinent by a series of crustal-scale sinistral transcurrent faults and the development of a series of associated basins (Husseini & Husseini 1990; Allen 2007). These basins, which are largely filled with evaporitic sequences of latest Neoproterozoic–earliest Cambrian age, contain the main Late Ediacaran–Early Cambrian petroleum systems (Talbot & Alavi 1996; Sharland *et al.* 2001; Kusky & Matsah 2003; Grosjean *et al.* 2009).

Late Ediacaran–Cambrian magmatism in the Himalaya (Cawood *et al.* 2007), Iran (Ramezani & Tucker 2003; Hassenzadeh *et al.* 2008), SE Turkey (Ustaömer *et al.* 2008), west Turkey (Compston *et al.* 2002; Strachan *et al.* 2007) and into Avalonia suggests that the northern margin of Gondwana was very active until at least the mid-Cambrian (Collins pers. comm.), and that subduction took place, at least locally, along parts of the margin. Deposition on the Peri-Gondwana Margin was dominated by repeated transgressions and regressions of the Palaeo-Tethys Ocean during the Late Neoproterozoic–Early Cambrian (Fig. 10). A wide continental shelf extended all along the Arabian–African margin, with a belt of ?transcurrent faults extending through present-day Arabia and the western part of the Indian Subcontinent. A series of basins, including the Rub Al'Khali, Hormuz, South Oman, Miajalar and South Punjab/Naguar–Ganganagar basins, form a distinctive elongate 'Salt Basin Domain' on the shelf, extending across present-day Oman and Saudi Arabia, Iran, southern and central Pakistan, and western and northern India. The area of the South Oman Basin and associated Ghaba Salt Basin in northern Oman is particularly instructive from a hydrocarbon perspective because it contains a highly prolific and well-understood Late Neoproterozoic–Early Cambrian petroleum system. This system includes proven hydrocarbon accumulations in two contrasting plays: intrasalt carbonates (referred to as 'stringers', but are the disrupted remnants of carbonate ramps and platforms) and silicilytes (organic-rich microcrystalline quartz rocks with a sheet-like pore network) in the South Oman Basin itself; and karstified carbonates (the Buah carbonates) on the so-called 'Huqf Highs' in North Oman.

Neoproterozoic–Early Cambrian Huqf Supergroup rocks form the main petroleum system (source, reservoir and seal) in Oman. More than 90% of Oman's current oil production is derived from Neoproterozoic–Early Cambrian source rocks.

The geological setting of central and southern Oman is well known (e.g. Droste 1997). It consists of a series of separate 'salt basins' (see **Ghori *et al.*** 2009, Fig. 3) with localized outcrops of Neoproterozoic rocks to the east (Huqf, Mirbat) and the north (Jabal Akhdar). The main basins are filled with thick sequences of Neoproterozoic–Early Cambrian sediments, consisting of a lower fault-controlled syntectonic 'Abu Mahara Group', containing the glacial sequence of the region, an overlying, more uniform, Nafun Group deposited in a post-rift, thermal subsidence phase, followed by a second rift-related system containing the main, hydrocarbon-bearing Ara Group carbonate–evaporite cycles (Allen 2007).

Immediately to the east, in what is now the border area between Pakistan and India, there is a less well-known continuation of the Middle East basin system, with a similar configuration of rhombohedral (?pull-apart) basins flanked by regional highs containing outcrops of Neoproterozoic rocks. A regional NE–SW-oriented cross-section through the border region between Sindh Province in eastern Pakistan and Rajasthan State in western India shows a well-developed regional high in the vicinity of Jaisalmer, and, to the north and south, two, apparently extensional, basins containing Neoproterozoic–Early Cambrian sediments. The northern basin, the South Punjab (or Naguar–Ganganagar) Basin contains the giant Neoproterozoic-sourced Baghewala heavy oil field.

Comparison of the age-equivalent Neoproterozoic–Early Cambrian sequences in Oman and Pakistan/West India indicates:

- A similar age for the pre-sedimentary basement, with the Malani Volcanic suite of India (750 Ma) being coeval with much of the crystalline and volcanic basement in the Huqf and Mirbat areas of Oman (820–720 Ma: Allen 2007).
- A much reduced sediment thickness in India (1 km) compared with Oman (4 km), suggesting a more cratonic setting for India.
- An apparent absence of glacigenic sediments in the Indian basins, of equivalent age to the Abu Mahara Group in Oman, except in the Lesser Himalaya where the Krol/Blaini succession is relatively thick and contains probable diamictites.
- A possible correlation between the carbonate-dominated Bilara Group in India, which records two major negative $\delta^{13}C$ shifts, and the Nafun and Ara groups in Oman.
- Lateral facies changes in the Nagaur–Ganganagar Basin in Rajasthan from Bilara carbonates on the basin margins to Hanseran Group evaporites in the basin centre, similar to the facies changes observed within the Ara Group in the South Oman Salt Basin.
- A possible correlation between the six or seven refreshing–desiccation, carbonate–evaporite cycles in the 600 m-thick Salt Range Formation of Pakistan, and the age-equivalent Hanseran Group in India (Kumar & Chandra 2005), with the A0–A6 cycles of the Ara Group in Oman.

The Baghewala Field in Rajasthan is estimated to contain around 628 million barrels (in place) of non-biodegraded, viscous, heavy oil in four separate reservoirs: two Neoproterozoic, one latest Neoproterozoic–Early Cambrian and one Late Cambrian. The presence of such a large field suggests a world-class Neoproterozoic source rock in this region. The oil from the Baghewala Field has a very distinctive geochemical signature, in common with other Neoproterozoic source rocks globally (Peters *et al.* 1995). In fact, there is some evidence that two different oil source systems are active in these basins: (1) 'oil shales' that produce low sulphur, light oil (42–50° API), which with adequate maturation can migrate relatively long distances; and (2) 'laminated organic-rich dolomites' that produce heavy, high sulphur oil during early maturation and which can only migrate a short

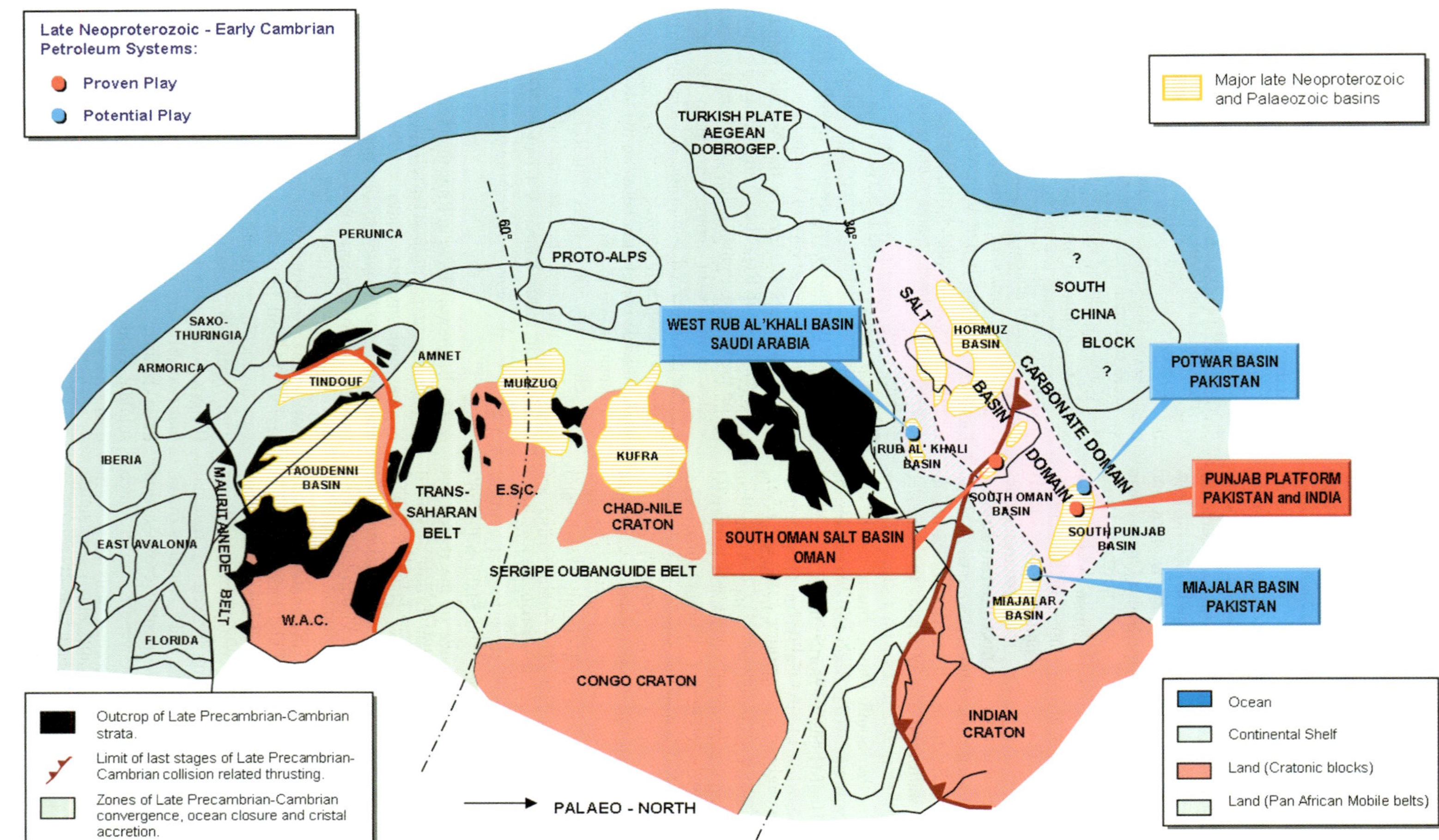

Fig. 10. Generalized Late Neoproterozoic (Ediacaran)–Early Cambrian palaeogeography of the 'Peri-Gondwanan Margin' (*c.* 610–520 Ma).

distance from the source. Both sources are recognized in Oman and, locally, in Pakistan.

Summary and conclusions

In summary:

- For the last billion years global climate has been dominated by a cyclical series of Greenhouse (250 Ma) and Icehouse (100 Ma) phases.
- Hydrocarbon source rock deposition is intimately linked to climate and, in some cases, specifically to periods of post-glacial marine transgression.
- The Neoproterozoic–Early Cambrian period can be broadly divided into three distinct phases related to global tectonics and climate:

  (1) Tonian–Early Cryogenian: *c.* 1000–750 Ma;
  (2) Mid-Cryogenian–Mid-Ediacaran: *c.* 750–600 Ma;
  (3) Late Ediacaran–Early Cambrian: *c.* 600–500 Ma.

- 'Pre-glacial' Neoproterozoic petroleum systems on the Peri-Gondwana Margin are largely restricted to old cratonic blocks. They consist predominantly of stromatolitic carbonate reservoirs, charged from interbedded and laterally equivalent black shales containing organic matter of algal origin.
- 'Glacial' Neoproterozoic petroleum systems are controlled by the deposition of organic-rich shale source rocks deposited during periods of post-glacial transgression. The Late Ordovician–Early Silurian Glacial–Post-glacial petroleum system provides a good analogue for reservoir, seal and source distribution in these Neoproterozoic 'Glacial' systems.
- 'Post-glacial' Neoproterozoic–Early Cambrian petroleum systems on the Peri-Gondwana Margin in the Middle East and the Indian Subcontinent are mainly associated with fault-bounded basins in East Gondwana, which are filled with mixed carbonate, evaporate and shale successions of latest Neoproterozoic and earliest Cambrian age. Oman is the best known of these latter systems, but very similar systems occur, or are likely to occur, in other basins in Arabia, the western Indian Subcontinent and, possibly, also in some parts of North Africa.

Conclusions

Neoproterozoic–Early Cambrian petroleum systems are widely developed globally and our knowledge of them is improving rapidly. Recent fieldwork in the Taoudenni Basin in Mauritania, the Anti-Atlas region of Morocco, the Al Kufrah Basin in Libya, the Naguar–Ganganagar Basin of Rajasthan, the Son Valley of central India and the Himalayan foothills of NW India by members of the Maghreb Petroleum Research Group has added substantially to our understanding of Neoproterozoic–Early Cambrian reservoir, source and seal relationships. This has confirmed the widespread presence of stromatolitic carbonate units of potential reservoir facies and of black shales with potential source rock characteristics in many Neoproterozoic successions across North Africa, the Middle East and the Indian Subcontinent. Work is now underway to establish a robust biostratigraphic and chronostratigraphic framework for the Neoproterozoic and Early Palaeozoic succession along the entire Peri-Gondwana Margin; to characterize the distribution, quality, kinetic parameters, biomarker characteristics and maturation history of the key source rock horizons, the controls on reservoir quality, the distribution and integrity of regional seals, and to quantify risk and uncertainty in these highly underexplored Neoproterozoic hydrocarbon plays. There is already a widespread and growing perception that these plays will form an important target for future exploration, not only on the Peri-Gondwana Margin, but also worldwide; a perception strongly reinforced by the contributions to this Special Publication.

The authors thank P. Allen, A. Collins and A. Smith for stimulating and insightful discussion. Their comprehensive reviews of this paper added significantly to its quality and content.

References

ALLEN, P. A. 2006. Snowball Earth on trial. *Eos, Transactions of the American Geophysical Union*, **87**, 495.

ALLEN, P. A. 2007. The Huqf Supergroup of Oman: basin development and context for Neoproterozoic glaciation. *Earth Science Reviews*, **84**, 139–185.

ALLEN, P. A. & ETIENNE, J. L. 2008. Sedimentary challenge to Snowball Earth. *Nature Geoscience*, **1**, 1–10.

AZIZ, A. & GHNIA, S. 2009. Distribution of Infracambrian rocks and the hydrocarbon potential within the Murzuq and Al Kufrah basins, NW Africa. *In*: CRAIG, J., THUROW, J., THUSU, B., WHITHAM, A. & ABUTARRUMA, Y. (eds) *Global Neoproterozoic Petroleum Systems: The Emerging Potential in North Africa*. Geological Society, London, Special Publications, **326**, 211–219.

BECHSTÄDT, T., JÄGER, H., SPENCE, G. & WERNER, G. 2009. Late Cryogenian (Neoproterozoic) glacial and post-glacial successions at the southern margin of the Congo Craton, northern Namibia: facies, palaeogeography and hydrocarbon perspective. *In*: CRAIG, J., THUROW, J., THUSU, B., WHITHAM, A.

& ABUTARRUMA, Y. (eds) *Global Neoproterozoic Petroleum Systems: The Emerging Potential in North Africa*. Geological Society, London, Special Publications, **326**, 255–287.

BENSHATI, H., KHOJA, A. & SOLA, M. 2009. Infracambrian sediments in Libyan sedimentary basins. *In*: CRAIG, J., THUROW, J., THUSU, B., WHITHAM, A. & ABUTARRUMA, Y. (eds) *Global Neoproterozoic Petroleum Systems: The Emerging Potential in North Africa*. Geological Society, London, Special Publications, **326**, 181–191.

BHAT, G. M., RAM, G. & KOUL, S. 2009. Potential for oil and gas in the Proterozoic carbonates (Sirban Limestone) of Jammu, northern India. *In*: CRAIG, J., THUROW, J., THUSU, B., WHITHAM, A. & ABUTARRUMA, Y. (eds) *Global Neoproterozoic Petroleum Systems: The Emerging Potential in North Africa*. Geological Society, London, Special Publications, **326**, 245–254.

BORNEMANN, A., NORRIS, R. D. *ET AL*. 2008. Isotopic evidence for glaciation during the Cretaceous Supergreenhouse. *Science*, **319**, 189–192.

BOWRING, S. A., GROTZINGER, J. P., CONDON, D. J., RAMEZANI, J., NEWALL, M. J. & ALLEN, P. A. 2007. Geochronologic constraints on the chronostratigraphic framework of the Neoproterozoic Huqf Supergroup, Sultanate of Oman. *American Journal of Science*, **307**, 1097–1145.

BUTTERFIELD, N. J. 2009. Macroevolutionary turnover through the Ediacaran transition: ecological and biogeochemical implications. *In*: CRAIG, J., THUROW, J., THUSU, B., WHITHAM, A. & ABUTARRUMA, Y. (eds) *Global Neoproterozoic Petroleum Systems: The Emerging Potential in North Africa*. Geological Society, London, Special Publications, **326**, 55–66.

CAWOOD, P. A., JOHNSON, M. R. W. & NEMCHIN, A. A. 2007. Early Palaeozoic orogenesis along the Indian margin of Gondwana: Tectonic response to Gondwana assembly. *Earth and Planetary Science Letters*, **255**, 70–84.

COLLINS, A. S. & PISAREVSKY, S. A. 2005. Amalgamating eastern Gondwana. The evolution of the Circum-Indian Orogens. *Earth Science Reviews*, **71**, 229–270.

COMPSTON, W., WRIGHT, A. E. & TOGHILL, P. 2002. Dating the Late Precambrian volcanicity of England and Wales. *Journal of the Geological Society, London*, **159**, 323–339.

COPPOLD, M. & POWELL, W. 2000. *A Geoscience Guide to the Burgess Shale: Geology and Paleontology in Yoho National Park*. Burgess Shale Geoscience Foundation, British Columbia, Canada.

COZZI, A. & AL-SIYABI, H. A. 2004. Sedimentology and play potential of the Late Neoproterozoic Buah Carbonates of Oman. *GeoArabia*, **9**(4), 11–36.

CRAIG, J., RIZZI, C. *ET AL*. 2008. Structural styles and prospectivity in the Precambrian and Palaeozoic Hydrocarbon systems of North Africa. *In*: *Sedimentary Basins of Libya, Third Symposium, Geology of East Libya*, Volume 4. Earth Science Society of Libya, Tripoli, 51–122.

CRICK, I. H., BOREHAM, C. J., COOK, A. C. & POWELL, T. G. 1988. Petroleum geology and geochemistry of Middle Proterozoic McArthur Basin, northern Australia II: Assessment of source rock potential. *AAPG Bulletin*, **72**, 1495–1514.

CROWELL, J. C. 1999. *Pre-Mesozoic Ice Ages: Their Bearing on Understanding the Climate System*. Geological Society of America Memoir, Boulder, Colorado, **192**, 1–106.

DAVIDSON, L., BESWETHERICK, S. *ET AL*. 2000. The structure, stratigraphy and petroleum geology of the Murzuq Basin, southwest Libya. *In*: SOLA, M. A. & WORSLEY, D. (eds) *Geological Exploration of the Murzuq Basin*. Elsevier, Amsterdam, 295–320.

DEYNOUX, M., AFFATON, P., TROMPETTE, R. & VILLENEUVE, M. 2006. Pan-African tectonic evolution and glacial events registered in Neoprotozoic to Cambrian cratonic and foreland basins of West Africa. *Journal of African Earth Science*, **46**, 397–426.

DROSTE, H. H. J. 1997. Stratigraphy of the Lower Paleozoic Haima Supergroup of Oman. *GeoArabia*, **2**, 419–472.

ECHIKH, K. & SOLA, M. 2000. Geology and hydrocarbon occurrences in the Murzuq Basin, S.W. Libya. *In*: SOLA, M. A. & WORSLEY, D. (eds) *Geological Exploration in the Murzuq Basin*. Elsevier, Amsterdam, 175–222.

ETIENNE, J. L., ALLEN, P., LE GUERROUE, E. & RIEU, R. 2006. *Snowball Earth 2006*. ETH, Zurich.

ETIENNE, J. L., ALLEN, P. A., RIEU, R. & LE GUERROUÉ, E. 2007. Neoproterozoic glaciated basins: a critical review of the Snowball Earth hypothesis by comparison with Phanerozoic glaciations. *In*: HAMBREY, M. J., CHRISTOFFERSEN, P., GLASSER, N. F. & HUBBARD, B. (eds) *Glacial Sedimentary Processes and Products*. International Association of Sedimentologists, Special Publications, **39**, 343–399.

FAIRCHILD, I. J. & KENNEDY, M. J. 2007. Neoproterozoic glaciation in the Earth system. *Journal of the Geological Society, London*, **164**, 895–921.

FENSOME, A. & WILLIAMS, G. L. (eds). 2001. *The Last Billion Years: A Geological History of the Maritime Provinces of Canada*. Atlantic Geoscience Society/ Nimbus Publishing, Halifax, Nova Scotia.

GÁMEZ VINTANED, J. A., SCHMITZ, U. & LIÑÁN, E. 2009. Upper Vendian–lowest Ordovician sequences of the western Gondwana margin, NE Spain. *In*: CRAIG, J., THUROW, J., THUSU, B., WHITHAM, A. & ABUTARRUMA, Y. (eds) *Global Neoproterozoic Petroleum Systems: The Emerging Potential in North Africa*. Geological Society, London, Special Publications, **326**, 231–244.

GHORI, K. A. R., CRAIG, J., THUSU, B., LÜNING, S. & GEIGER, M. 2009. Global Infracambrian petroleum systems: a review. *In*: CRAIG, J., THUROW, J., THUSU, B., WHITHAM, A. & ABUTARRUMA, Y. (eds) *Global Neoproterozoic Petroleum Systems: The Emerging Potential in North Africa*. Geological Society, London, Special Publications, **326**, 109–136.

GRADSTEIN, F. M., OGG, J. & SMITH, A. G. 2004. *A Geologic Time Scale 2004*. Cambridge University Press, Cambridge.

GROSJEAN, E., LOVE, G. D., STALVIES, C., FIKE, D. A. & SUMMONS, R. E. 2009. Origin of petroleum in the Neoproterozoic–Cambrian South Oman Salt Basin. *Organic Geochemistry*, **40**, 87–110.

HALVERSON, G. P., HOFFMAN, P. F., SCHRAG, D. P., MALOOF, A. C. & RICE, A. H. 2005. Toward a Neoproterozoic composite carbon-isotope record. *Geological Society of America, Bulletin*, **117**, 1181–1207.

HARLAND, W. B., ARMSTRONG, R. L., COX, A. V., CRAIG, L. E., SMITH, A. G. & SMITH, D. G. 1990. *A Geologic Time Scale 1989*. Cambridge University Press, Cambridge.

HASSENZADEH, J., STOCKLI, D. F. *ET AL*. 2008. U–Pb zircon geochronology of late Neoproterozoic-Early Cambrian granitoids in Iran: implications for paleogeography, magmatism, and exhumation history of Iranian basement. *Tectonophysics*, **451**, 71–96.

HIRST, J. P. P., BENBAKIR, A., PAYNE, D. F. & WESTLAKE, I. R. 2002. Tunnel valleys and density flow processes in the upper Ordovician glacial succession, Illizi Basin, Algeria: influence on reservoir quality. *Journal of Petroleum Geology*, **25**, 297–324.

HLEBSZEVITSCH, J. C., GEBHARD, I., CRUZ, C. E. & CONSOLI, V. 2009. The 'Infracambrian System' in the southwestern margin of Gondwana, southern South America. *In*: CRAIG, J., THUROW, J., THUSU, B., WHITHAM, A. & ABUTARRUMA, Y. (eds) *Global Neoproterozoic Petroleum Systems: The Emerging Potential in North Africa*. Geological Society, London, Special Publications, **326**, 289–302.

HOFFMAN, P. F. 2006. A Cryogenian ice stream in Namibia, ice-sheet dynamics on Snowball Earth and the limitations of the glacial sedimentary record. (Abstract.) *In*: *Snowball Earth 2006, Monte Verità, Ticino, Switzerland, 16–21 July 2006*. ETH Zurich, Ascona.

HOFFMAN, P. F. & ALLEN, P. A. 2007. Snowball Earth on trial; discusson and reply. *Eos, Transactions of the American Geophysics Union*, **88**, 110.

HOFFMAN, P. F. & SCHRAG, D. P. 2000. Snowball Earth. *Scientific American*, **285**, 50–57.

HOFFMAN, P. F. & SCHRAG, D. P. 2002. The snowball earth hypothesis: testing the limits of global change. *Terra Nova*, **14**, 129–155.

HOFFMAN, P. F., KAUFMAN, A. J., HALVERSON, G. P. & SHRAG, D. P. 1998. A Neoproterozoic snowball Earth. *Science*, **281**, 1342–1346.

HUSSEINI, M. I. & HUSSEINI, S. I. 1990. Origin of the Infracambrian Salt Basins of the Middle East. *In*: BROOK, J. (ed.) *Classic Petroleum Provinces*. Geological Society, London, Special Publications, **50**, 279–292.

JACKSON, M. J., POWELL, T. G., SUMMONS, R. E. & SWEET, I. P. 1986. Hydrocarbon shows and petroleum source rocks in sediments as old as 1.7×10^9 years. *Nature*, **322**, 727–729.

JACOBS, J. & THOMAS, R. J. 2002. The Mozambique Belt from an East Antarctic perspective. *Bulletin of the Royal Society of New Zealand*, **35**, 3–18.

KALJO, D., MARTMA, T., MANNIK, P. & VIIRA, V. 2003. Implications of Gondwana glaciations in the Baltic late Ordovician and Silurian and a carbon isotopic test of environmental cyclicity. *Bulletin de la Société Géologique de France*, **174**, 59–66.

KENDALL, B., CREASER, R. A. & SELBY, D. 2009. ^{187}Re–^{187}Os geochronology of Precambrian organic-rich sedimentary rocks. *In*: CRAIG, J., THUROW, J., THUSU, B., WHITHAM, A. & ABUTARRUMA, Y. (eds) *Global Neoproterozoic Petroleum Systems: The Emerging Potential in North Africa*. Geological Society, London, Special Publications, **326**, 85–107.

KENNEDY, M. J. 1996. Stratigraphy, sedimentology, and isotopic geochemistry of Australian Neoproterozoic postglacial cap dolostones: deglaciation, δ^{13}C excursions, and carbonate precipitation. *Journal of Sedimentary Research*, **66**, 1050–1064.

KENNEDY, M. J., CHRISTIE-BLICK, N. & SOHL, L. E. 2001. Are Proterozoic cap carbonates and isotopic excursions a record of gas hydrate destabilization following Earth's coldest intervals? *Geology*, **29**, 443–446.

KENNEDY, M. J., RUNNEGAR, B., PRAVE, A. R., HOFFMANN, K.-H. & ARTHUR, M. A. 1998. Two or four Neoproterozoic glaciations? *Geology*, **26**, 1059–1063.

KIRSCHVINK, J. L. 1992. Late Proterozoic low latitude glaciation: the Snowball Earth. *In*: SCHOPF, J. W. & KLEIN, C. (eds) *The Proterozoic Biosphere; A Multidisciplinary Study*. Cambridge University Press, Cambridge, 51–52.

KLEMME, H. D. & ULMISHEK, G. F. 1991. Effective petroleum source rocks of the world: stratigraphic distribution and controlling depositional factors. *AAPG Bulletin*, **75**, 1809–1851.

KLITSCH, M., BUSSERT, R. & PAZZI, C. 2008. Main results of geological fieldwork in Jabal Arknù, Jabal Babèin and Jabal Asbah Areas (S.E. Al Kufrah Basin) as a base for interpretation of new 2D seismic data in Concession 201, Al Kufrah Basin. (Abstract.) *In*: *Sedimentary Basins of Libya, Fourth Symposium: Geology of Southern Libya, Tripoli, Libya, 17–20 November 2008*. Earth Science Society of Libya, 61.

KUMAR, V. & CHANDRA, R. 2005. *Geology and Evolution of the Nagaur–Ganganagar Basin with Special Reference to Salt and Potash Mineralisation*. Geological Survey of India, Special Publications, **62**.

KUSKY, T. M. & MATSAH, M. I. 2003. Neoproterozoic dextral faulting on the Najd Fault System, Saudi Arabia, preceded sinistral faulting and escape tectonics related to closure on the Mozambique Ocean. *In*: YOSHIDA, M., WINDLEY, B. F. & DASGUPTA, S. (eds) *Proterozoic East Gondwana: Supercontinent Assembly and Breakup*. Geological Society, London, Special Publications, **206**, 327–361.

LE HERON, D. P. & CRAIG, J. 2008. First-order reconstructions of a Late Ordovician Saharan ice sheet. *Journal of the Geological Society, London*, **165**, 19–29.

LE HERON, D. P. & DOWDESWELL, J. A. 2009. Calculating ice volumes and ice flux to constrain the dimensions of a 440 Ma North African ice sheet. *Journal of the Geological Society of London*, **166**, 277–281.

LE HERON, D., SUTCLIFFE, O., BOURGIG, K., CRAIG, J., VISENTIN, C. & WHITTINGTON, R. 2004. Sedimentary architecture of Upper Ordovician tunnel valleys, Gargaf Arch, Libya: implications for the genesis of a hydrocarbon reservoir. *GeoArabia*, **9**(2), 137–159.

LE HERON, D. P., SUTCLIFFE, O. E., WHITTINGTON, R. J. & CRAIG, J. 2005. The origins of glacially related soft-sediment deformation structures in Upper Ordovician glaciogenic rocks: implication for ice sheet dynamics. *Palaeogeography, Palaeoclimatology, Palaeoecology*, **218**, 75–103.

Le Heron, D. P., Howard, J. P., Alhassi, A. M., Anderson, L., Morton, A. & Fanning, C. M. 2009*a*. Field-based investigations of an 'Infracambrian' clastic succession in SE Libya and its bearing on the evolution of the Al Kufrah Basin. *In*: Craig, J., Thurow, J. W., Thusu, B., Whitham, A. G. & Abutarruma, Y. (eds) *Global Neoproterozoic Petroleum Systems: The Emerging Potential in North Africa*. Geological Society, London, Special Publications, **326**, 193–210.

Le Heron, D. P., Craig, J. & Etienne, J. L. 2009*b*. Ancient glaciation and hydrocarbon accumulations in North Africa and the Middle East. *Earth-Science Reviews*, **93**, 47–76.

Li, X.-H., Li, Z.-X., Sinclair, J. A., Li, W.-X. & Carter, G. 2006. Revisiting the 'Yanbian Terrane': implications for Neoproterozoic tectonic evolution of the western Yangtze Block, South China. *Precambrian Research*, **151**, 14–30.

Lottaroli, F., Craig, J. & Thusu, B. 2009. Neoproterozoic–Early Cambrian (Infracambrian) hydrocarbon prospectivity of North Africa: a synthesis. *In*: Craig, J., Thurow, J., Thusu, B., Whitham, A. & Abutarruma, Y. (eds) *Global Neoproterozoic Petroleum Systems: The Emerging Potential in North Africa*. Geological Society, London, Special Publications, **326**, 137–156.

Lüning, S., Craig, J., Loydell, D. K., Storch, P. & Fitches, B. 2000*a*. Lower Silurian 'hot shales' in North Africa and Arabia: regional distribution and depositional model. *Earth Science Reviews*, **49**, 121–200.

Lüning, S., Kolonic, S., Geiger, M., Thusu, B., Bell, J. S. & Craig, J. 2009. Infracambrian hydrocarbon source rock potential and petroleum prospectivity of NW Africa. *In*: Craig, J., Thurow, J., Thusu, B., Whitham, A. & Abutarruma, Y. (eds) *Global Neoproterozoic Petroleum Systems: The Emerging Potential in North Africa*. Geological Society, London, Special Publications, **326**, 157–180.

Lüning, S., Loydell, D. K., Sutcliffe, O., Ait Salem, A., Zanella, E., Craig, J. & Harper, D. A. T. 2000*b*. Silurian–Lower Devonian black shales in Morocco, which are the organically richest horizons? *Journal of Petroleum Geology*, **23**, 293–311.

Macgregor, D. S. 1996. The hydrocarbon systems of North Africa. *Marine Petroleum Geology*, **13**, 329–340.

Malone, S. J., Meert, J. G. *et al.* 2008. Paleomagnetism and detrital zircon geochronology of the Upper Vindhyan sequence, Son Valley and Rajasthan, India: a ca. 1000 Ma age for the Purana Basins. *Precambrian Research*, **164**, 137–159.

Miller, N. R., Alene, M., Sacchi, R., Stern, R. J., Conti, A., Kröner, A. & Zuppi, G. 2003. Significance of the Tambien Group (Tigrai, N. Ethiopia) for Snowball Earth events in the Arabian–Nubian Shield. *Precambrian Research*, **121**, 263–283.

Moczydlowska, M. 2008. The Ediacaran microbiota and the survival of Snowball Earth conditions. *Precambrian Resarch*, **167**, 1–15.

Moussine-Pouchkine, A. & Bertrand-Sarfati, J. 1997. Tectonosedimentary subdivisions in the Neoproterozoic to Early Cambrian cover of the Taoudenni Basin (Algeria–Mauritania–Mali). *Journal of African Earth Sciences*, **24**, 425–443.

Nudds, J. R. & Selden, P. A. 2008. Fossils explained 56, Fossil – Lagerstäten. *Geology Today*, **24**, 153–158.

Ogg, J. G., Ogg, G. & Gradstein, F. M. 2008. *The Concise Geologic Time Scale*. Cambridge University Press, Cambridge.

Olcott, A. N., Sessions, A. L., Corsetti, F. A. & Kauffman, A. J. 2006. Photosynthesis during Neoproterozoic glaciation. (Abstract.) *In*: *Snowball Earth 2006, Monte Verità, Ticino, Switzerland, 16–21 July 2006*, ETH Zurich, Ascona.

Olcott, A., Sessions, A., Corsetti, F., Kauffman, A. & de Oliviera, T. 2005. Biomarker evidence for photosynthesis during Neoproterozoic glaciation. *Science*, **310**, 471–474.

Peters, K. E., Clark, M. E., Das Gupta, S., McCaffrey, U. & Lee, C. Y. 1995. Recognition of an Infracambrian source rock based on biomarkers in the Baghewala-1 oil, India. *AAPG Bulletin*, **79**, 1481–1494.

Pisarevsky, S. A., Murphy, J. B., Cawood, P. A. & Collins, A. S. 2008. Late Neoproterozoic and Early Cambrian palaeogeography: models and problems. *In*: Pankhurst, R. J., Trouw, R. A. J., de Brito Neves, B. B. & de Wit, M. J. (eds) *West Gondwana: Pre-Cenozoic Correlation Across the South Atlantic Region*. Geological Society, London, Special Publications, **294**, 9–31.

Rahmani, A., Goucem, A., Boukhallat, S. & Saadallah, N. 2009. Infracambrian petroleum play elements of the NE Taoudenni Basin (Algeria). *In*: Craig, J., Thurow, J., Thusu, B., Whitham, A. & Abutarruma, Y. (eds) *Global Neoproterozoic Petroleum Systems: The Emerging Potential in North Africa*. Geological Society, London, Special Publications, **326**, 221–229.

Ramezani, J. & Tucker, R. D. 2003. The Saghand region, central Iran: U–Pb geochronology, petrogenesis and implications for Gondwana tectonics. *American Journal of Science*, **303**, 622–665.

Raub, T. D., Evans, D. A. D. & Smirnov, A. V. 2007. Siliciclastic prelude to Elatina–Nuccaleena deglaciation: lithostratigraphy and rock magnetism through the base of the Ediacaran system. *In*: Vickers-Rich, P. & Komarower, P. (eds) *The Rise and Fall of Ediacaran Biota*. Geological Society, London, Special Publications, **286**, 53–76.

Royer, D. L., Berner, R. A., Montanez, I. P., Tabar, N. J. & Beerliong, D. J. 2004. CO_2 as a primary driver of Phanerozoic climate. *GSA Today*, **14**, 4–10.

Scotese, C. R. 2009. Late Proterozoic plate tectonics and palaeogeography: a tale of two supercontinents, Rodinia and Pannotia. *In*: Craig, J., Thurow, J., Thusu, B., Whitham, A. & Abutarruma, Y. (eds) *Global Neoproterozoic Petroleum Systems: The Emerging Potential in North Africa*. Geological Society, London, Special Publications, **326**, 67–83.

Selden, P. A. & Nudds, J. R. 2004. *Evolution of Fossil Ecosystems*. Wiley, London.

Sharland, P. R., Archer, R. *et al.* 2001. *Arabian Plate Sequence Stratigraphy*. GeoArabia, Special Publications, **2**, 371.

SIMMONS, M. D., SHARLAND, P. R., CASEY, D. M., DAVIES, R. B. & SUTCLIFFE, O. 2007. Arabian Plate sequence stratigraphy: potential implications for global chronostratigraphy. *GeoArabia*, **12**, 101–130.

SMITH, A. G. 2009. Neoproterozoic timescales and stratigraphy. *In*: CRAIG, J., THUROW, J., THUSU, B., WHITHAM, A. & ABUTARRUMA, Y. (eds) *Global Neoproterozoic Petroleum Systems: The Emerging Potential in North Africa*. Geological Society, London, Special Publications, **326**, 27–54.

STEPHENSON, M. H., MILLWARD, D., LENG, M. J. & VANE, C. H. 2008. Palaeological and possible evolutionary effects of early Namurian (Serpukhovian, Carboniferous) glacioeustatic cyclicity. *Journal of the Geological Society, London*, **165**, 993–1006.

STRACHAN, R. A., COLLINS, A. S., BUCHAN, C., NANCE, R. D., MURPHY, J. B. & D'LEMOS, R. D. 2007. Terrane analysis along a Neoproterozoic active margin of Gondwana: insights from U–Pb zircon geochronology. *Journal of the Geological Society, London*, **164**, 57–60.

SUTCLIFFE, O. E., DOWDESWELL, J. A., WHITTINGTON, R. J., THERON, J. N. & CRAIG, J. 2000. Calibrating the late Ordovician glaciation and mass extinction by the eccentricity cycles of the Earth's orbit. *Geology*, **23**, 967–970.

SUTCLIFFE, O. E., CRAIG, J. & WHITTINGTON, R. 2005. Late Ordovician glacial pavements revisited: a reappraisal of the origin of striated surfaces. *Terra Nova*, **17**(5), 486–487.

TALBOT, C. J. & ALAVI, M. 1996. The past of a future syntaxis across the Zagros. *In*: ALSOP, G. I., BLUNDELL, D. J. & DAVIDSON, I. (eds) *Salt Tectonics*. Geological Society, London, Special Publications, **100**, 89–109.

USTAÖMER, P. A., USTAÖMER, T., COLLINS, A. S. & ROBERTSON, A. H. F. 2008. Cadomian (Ediacaran–Cambrian) arc magmatism in the Bitlis Massif, SE Turkey: magmatism along the developing northern margin of Gondwana. *Tectonophysics*, **4**, 99–112.

VASLET, D. 1990. Upper Ordovician glacial deposits in Saudi Arabia. *Episodes*, **13**, 147–161.

VICKERS-RICH, P. & TRUSLER, P. 2006. Images of the Precambrian: where art and scientific theory converge. (Abstract.) *In*: *Snowball Earth 2006, Monte Verità, Ticino, Switzerland, 16–21 July 2006*, ETH Zurich, Ascona.

VIDAL, G. & MOCZYDLOWSKA-VIDAL, M. 1997. Biodiversity, speciation and extinction trends of Proterozoic and Cambrian phytoplankton. *Paleobiology*, **23**, 230–246.

WAGGONER, B. 1999. Biogeographic analyses of the Ediacara biota: a conflict with paleotectonic reconstructions. *Paleobiology*, **25**, 440–458.

WAGGONER, B. 2003. The Ediacaran biotas in space and time. *Integrative and Comparative Biology*, **43**, 104–113.

WEISSERT, H. & ERBA, E. 2004. Volcanism, CO_2 and palaeoclimate: a Late Jurassic–Early Cretaceous carbon and oxygen isotope record. *Journal of the Geological Society, London*, **161**, 695–702.

Neoproterozoic timescales and stratigraphy

ALAN G. SMITH

Department of Earth Sciences, University of Cambridge, Downing Street, Cambridge CB2 3EQ, UK
(e-mail: ags1@cam.ac.uk)

Abstract: The Infracambrian is a term for mostly Neoproterozoic successions in North Africa and areas to the east. Its base lies within the middle Neoproterozoic period, or Cryogenian, includes the youngest Neoproterozoic period, or Ediacaran, and continues into the early Cambrian to the level at which trilobites first appear. The Cryogenian lacks any biostratigraphic zonation; and no global biostratigraphic schemes exist for the Ediacaran. The formal classification of the Neoproterozoic is currently under review.

The Cryogenian–Ediacaran (CE) interval includes at least three prominent diamictite horizons that are clearly linked to penecontemporaneous glaciations. The oldest is 'Sturtian', next oldest is 'Marinoan' (probably the most extensive), both names are Australian in origin but are used internationally. The Gaskiers glaciation is the youngest and probably the least extensive. There are important unresolved problems of the precise number, age, extent and nomenclature of the Neoproterozic glaciations in Australia.

Several palaeomagnetic poles in the age range 600–550 Ma place glacial deposits of that age range in Australia in tropical latitudes. These data, together with older poles from Laurentia, gave rise to the notion of the Snowball Earth, in which the Earth froze over completely, but the profound refrigeration required appears to have had very little effect on biological evolution.

Biostratigraphic zonation with the precision attainable throughout the Phanerozoic does not appear possible for the CE interval. Thus, most correlations are based on about 40 U–Pb and Re–Os dates. These confirm the existence of at least three glacial sequences (sequence is used here as an informal term), but it is possible that the Sturtian and Marinoan were essentially one glacial unit. Deglaciation was accompanied by the unique 'cap carbonates'.

The glacial sequences all show a characteristic $\delta^{13}C$ pattern, but present knowledge is inadequate to use these patterns for detailed global correlation. The most reliable chemostratigraphic correlations are likely to be based on strontium isotope variations.

Black shale horizons commonly follow deglaciation. A few basins produce Neoproterozoic hydrocarbons; others are potential producers. As a whole, the Neoproterozoic represents both a scientific and an exploration frontier.

This paper started as a paper on the so-called Infracambrian, but it soon became clear that it needed to deal with the stratigraphic subdivisions and dating of the Neoproterozoic first, because the Infracambrian itself is not formally defined, although its extent is well understood in areas where the term is still in widespread use.

The Neoproterozoic itself is one of the most interesting geological eras. It includes three important geological events: (1) the break-up of Rodinia; (2) glacial deposits that have given rise to the notion of the 'Snowball Earth'; and (3) the biological transition from the pre-Ediacaran to the Phanerozoic, regarded by many palaeontologists as the most important biological change in the Earth's history. The literature on each of these topics is vast. Some recent references on particular topics are: Rodinia and East Gondwana and their evolution (Cawood 2005; Collins & Pisarevsky 2005); on the age and distribution of glacial deposits (Evans 2000); arguments for (Hoffman 2005) and against the Snowball Earth (Eyles & Januszczak 2004); on cap carbonates (Shields 2005; Corsetti & Lorentz 2006); on dating and dates (MacGabhann 2005; Halverson 2006; Kendall *et al.* 2006); on the Ediacaran (McCall 2006); on palaeobiology (Knoll *et al.* 2004; Grey 2005; Narbonne 2005; Butterfield 2007; Xiao & Kaufman 2007); and on geochemistry (Halverson *et al.* 2005, 2007). A recent general review is given in Etienne *et al.* (2006) and Fairchild & Kennedy (2007). Several of these papers deal with more than one topic, all of them have extensive reference lists to previous literature and many other papers could have been cited as an entrée into these themes. Current and recent work is summarized on the website www.IGCP512.com.

Formal stratigraphy

The present method for defining Phanerozoic stages, epochs and periods is to formally define a Global Stratotype Section and Point (GSSP) marking the base of each of these units (e.g. chapter 2 in Gradstein *et al.* 2004). The top of each unit is

From: CRAIG, J., THUROW, J., THUSU, B., WHITHAM, A. & ABUTARRUMA, Y. (eds) *Global Neoproterozoic Petroleum Systems: The Emerging Potential in North Africa.* Geological Society, London, Special Publications, **326**, 27–54.
DOI: 10.1144/SP326.2 0305-8719/09/$15.00

defined by the GSSP at the base of the next overlying unit. Each GSSP is marked by a 'golden spike' in a rock sequence somewhere (throughout this paper the term sequence is used informally rather than in its seismic stratigraphic sense). Thus, the problem of deciding whether something is Cambrian or not, is initially a problem of correlating to the GSSP. Of course, once there is a precise determination of the age of the boundary, age dates can be used as a proxy for correlation.

The base of the Cambrian, which is defined by a GSSP on the Burin Peninsula in SE Newfoundland (Brasier *et al.* 1994; Landing 1994), is dated at 542 ± 1 Ma (see the discussion in Shergold & Cooper 2004, pp. 159–163). There are good biostratigraphic arguments for an alternative placement of this GSSP at the base of the Tommotian stage of Siberia, which lies somewhere within the current early Cambrian (Khomentovskii & Karlova 2005; Butterfield 2007), but the sections proposed appear to contain hiatuses and are therefore unsuitable (G. Geyer pers. comm.).

The Neoproterozoic, which is the youngest Precambrian era, is divided into three periods, Ediacaran (youngest), Cryogenian and Tonian, each with the rank of a Phanerozoic period, such as the Cambrian (Fig. 1). Until recently, all Precambrian stratigraphic boundaries have been defined by arbitrary numerical ages and not by GSSPs. The only exception at the present time is the definition of the base of the Ediacaran, named after the town of Ediacara in the Flinders Range in south Australia, where abundant Ediacaran fossil assemblages were discovered (Sprigg 1947). This recently agreed GSSP is in the Nuccaleena Formation in the Adelaide Rift Complex (Knoll *et al.* 2006) (Fig. 2). The section at the GSSP has no igneous material that could be used for dating. Its age is estimated by correlation to Namibia and China, where igneous material suggests its age is <635 Ma and >632 Ma (Knoll *et al.* 2006), but this estimate depends on assuming lithostratigraphic and isotopic correlations are also of chronostratigraphic significance, discussed later in this chapter.

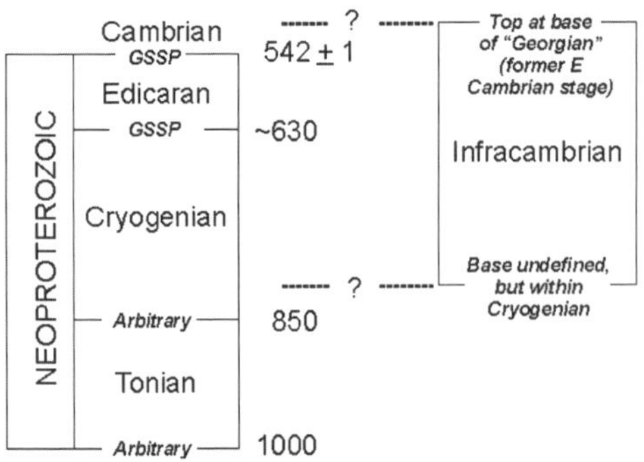

Fig. 1. Schematic representation of the formal stratigraphy of the Neoproterozoic–Early Cambrian interval and the approximate relationships of the Infracambrian to it. Ages in millions of years (Ma).

The Cryogenian is the middle period of the Neoproterozoic. Its name is derived from *cryos* (ice) and *genesis* (birth), because it includes at least two major glaciations, the younger known as the 'Marinoan', and the older as the 'Sturtian' (see later). There is no GSSP, but the age of its base is set at 850 Ma (Plumb 1991). In practice, some workers informally place the base of the Cryogenian at the beginning of the Sturtian glaciation, which may be as young as 735–715 Ma, more than 100 Ma younger than the current formal definition (e.g. Hoffman 2005; Halverson 2006).

The oldest Neoproterozoic period is the Tonian, from *tonas* or stretch, supposedly because it marks a time when older platforms increased in area (see Robb *et al.* 2004, table 9.1 p. 133), but the changes are not at all obvious globally. The age of its base is set at 1000 Ma (Plumb 1991). These formal numerical definitions for the base of the Cryogenian and for the base of the Tonian are currently under review.

Definition of the Infracambrian

The term 'Infracambrien' was first proposed by Menchikoff (1949, p. 309) in footnote 3 of his paper on the stratigraphy of the Western Sahara. He wrote: 'On pourrait donner le nom *d'Infra-Cambrien* aux formations situées à la base du Géorgien, peut-être même au-dessous de sa limite inférieure, mais nettement liées au Paléozoïque certain'. Which translates approximately to: 'one may give the name of *Infra-Cambrien* to formations situated at the base of the Georgian [essentially the base of the trilobite-bearing Cambrian] perhaps even beneath its lower limit, but conformably bedded under definite Palaeozoic [strata]'. This definition was made at a time when the beginning of the Cambrian was placed generally at the level at which the first macrofossils, i.e. trilobites, appeared.

It was named the 'Système Infracambrien' by Pruvost (1951), and included Precambrian sedimentary sequences that underlay known Cambrian rocks and unconformably overlay generally metamorphosed rocks (Fig. 3). Such a definition included at its top strata now regarded as early Cambrian in age. It also included rocks as old as the Mesoproterozoic Belt–Purcell Supergroup of western North America (Pruvost 1951, p. 56). The term has persisted in informal use in the Anti-Atlas basins of North Africa and in adjacent areas all the way to Pakistan (see below). Several other informal terms, such as Eo-Cambrian and sub-Cambrian, were coined for similar sequences elsewhere, but have largely fallen into disuse.

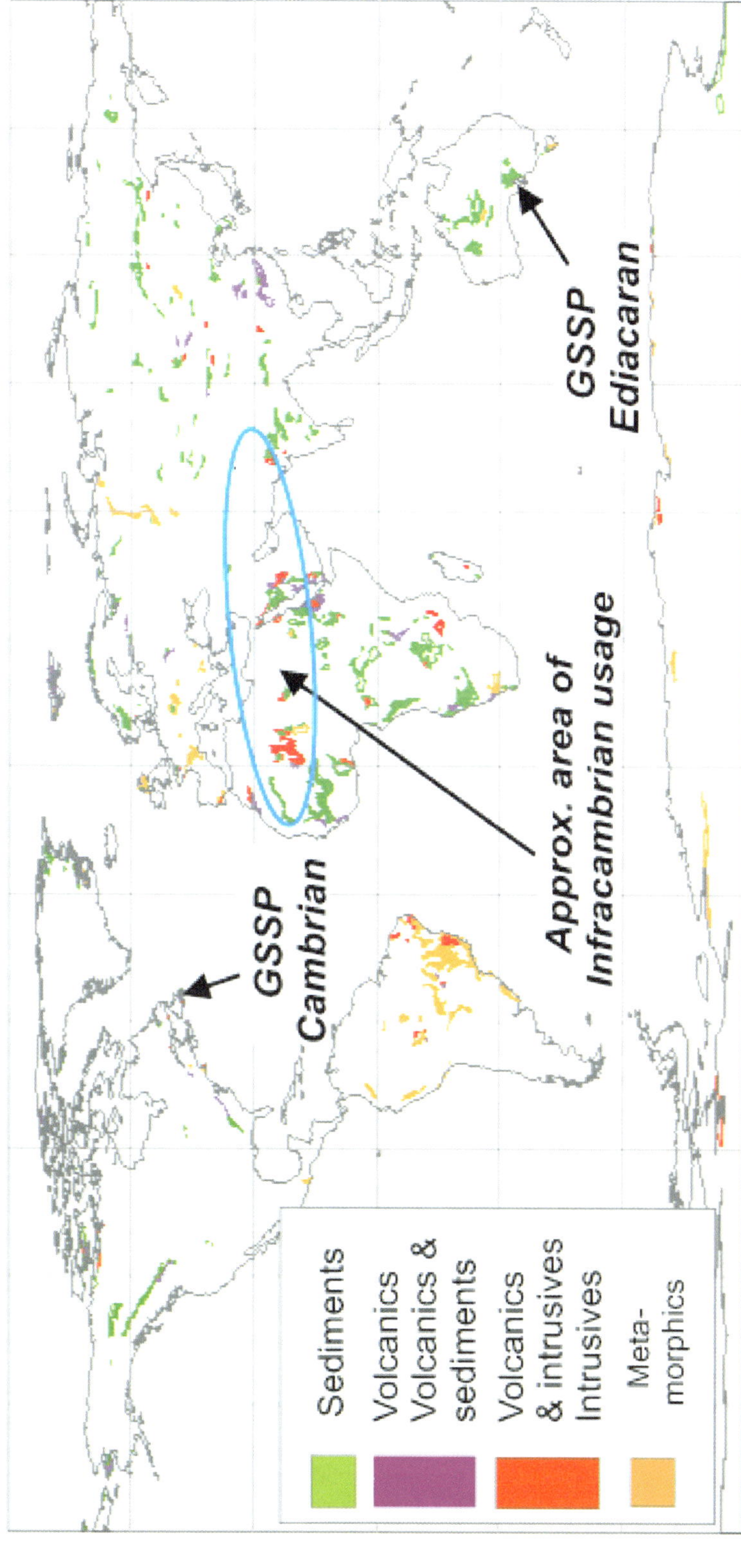

Fig. 2. Neoproterozoic outcrops, Ediacaran and Cambrian GSSPs, and approximate area of current Infracambrian usage. The outcrops are from a database provided by L. Chorlton of the National Research Council of Canada. The Neoproterozoic rocks are subdivided into six categories: metamorphics; sediments; sediments and volcanics (here grouped with volcanics); volcanics; and intrusives (here grouped with intrusives); volcanics; and intrusives. In addition, some areas are outcrops of mixed Neoproterozoic and Mesoproterozoic rocks. These have been given paler shades, but are not readily recognizable as such on the figure. Cylindrical equidistant projection.

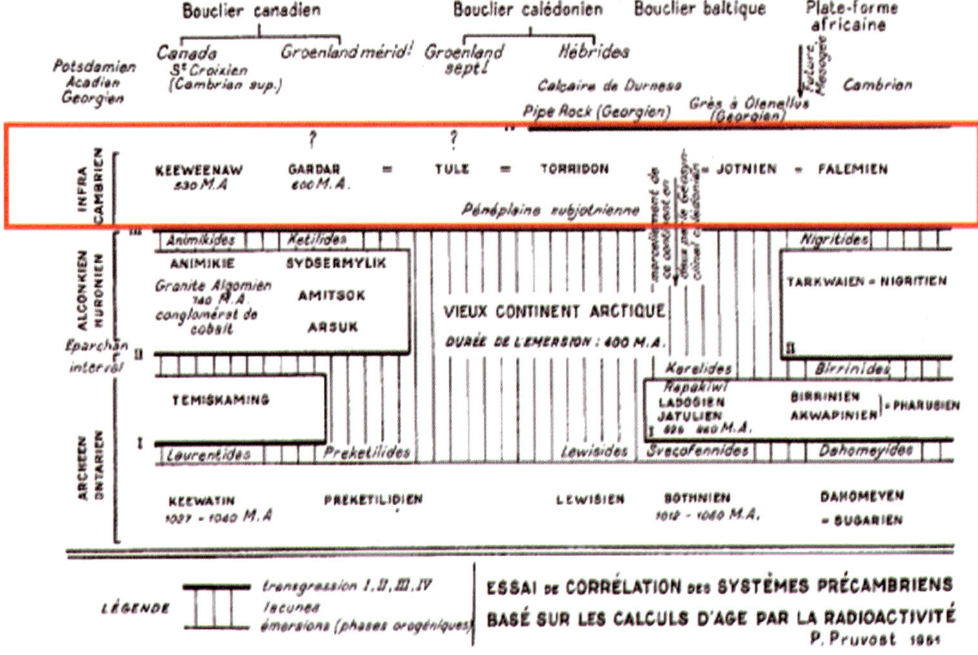

Fig. 3. The coloured box outlines the Infracambrian as envisaged by Pruvost (1951).

Infracambrian and formal stratigraphy

As currently used, the Infracambrian is confined to the later Neoproterozoic–earliest Cambrian interval. The Georgian stage of the early Cambrian, which Pruvost (1951) suggested marked the top of the Infracambrian, fell into disuse some time ago, but subsequent replacements have also been abandoned. The current situation for formally defining the top of the Infracambrian is therefore unsatisfactory, because there are no modern definitions of the stages that make up the early (and middle) Cambrian epochs (see the discussion in Shergold & Cooper 2004, pp. 147–153). The Cambrian Sub-Commission is actively working on these problems and will probably propose Cambrian GSSPs that meet modern criteria within the next 1–2 years, but currently it is not possible to link the top of the Infracambrian to any formal international stratigraphic scheme.

The Infracambrian base is similarly undefined, but presumably lies at or below the base of the Huqf Supergroup (e.g. Allen *et al.* 2004). As noted above, the base of the Cryogenian is arbitrarily defined by a numerical value of 850 Ma (Plumb 1991), placing the Huqf within that period. Thus, the Infracambrian lies wholly within the early Cambrian–Cryogenian interval.

The Infracambrian can also be regarded as embracing the Vendian (younger) and later Riphean (older) of Neoproterozoic nomenclature used in the former Soviet Union (Keller 1979; Ye *et al.* 1980). The Riphean–Vendian scheme is still widely used in Russia, but must be regarded as representing regional rather than international practice. The top of the Vendian is the base of the Cambrian as defined by Russian geologists, which differs from that of the international scale. Similarly, the base of the Vendian is close to, but not coincident with, the base of the Ediacaran. Neither boundary is marked by a GSSP.

The present usefulness of the term Infracambrian is due to the gross similarities of the Neoproterozoic sequences in North Africa, Iran, NE Arabia, Pakistan and India to one another, in turn reflecting their deposition in similar tectonic settings. All were laid down in elongate basins that were probably parallel to a northward-facing continental margin, although the details are unclear. However, Neoproterozoic rocks as a whole have a very wide distribution, as shown in Figure 2. In the future, the rapid increase in age dates and other correlation markers may make it possible to correlate all Infracambrian sections with the emerging standard international stratigraphic scale, and for clarity this informal term could then be abandoned.

Table 1. *Neoproterozoic diamictites (Gaskiers, Marinoan and Sturtian). Data from Evans (2000, tables 1 and 2) [E<number>]; Hoffman (2005, table 1) [H05]; and Hambrey & Harland (1981) [H&H, p. < >]. See also MacGabhann (2005) and Eyles & Januszczak (2004) [E&S04] for further comments. Banded iron formations shown as *(well developed) or †(not well developed). Negative latitudes are S; negative longitudes are W. Plotted in Figures 7 & 9–11*

Lat. (°)	Long. (°)	Name	Area	References
Gaskiers				
18.80	126.90	Egan	Kimberleys, NW Australia	E43; H05
40.00	90.00	Hankalchough	Tarim	E33; H05
70.00	28.50	Mortensnes	Varanger peninsula, Norway	E20; H05
61.20	11.50	Moelv	Oslo, Norway	E23; H05; H&H, p. 624
56.00	−5.80	Inishowen-Loch na Cille	Ireland & Scotland	E12; H05: H&H, p. 632; age from McCay *et al.* (2006)
42.30	−72.20	Squantum	New England, USA	E74; H05; H&H, p. 762
47.00	−53.50	Gaskiers Fm	Newfoundland, Canada	E75; H05; H&H, p. 760
Marinoan				
−26.00	131.00	Olympic	Central Australia	H05
−29.80	138.00	Elatina	Adelaide, SE Australia	E40; H05
−32.60	139.50	Elatina	Adelaide, SE Australia	E40; H05
−35.00	138.90	Elatina	Adelaide, SE Australia	E40; H05
−31.00	141.00	Elatina	Adelaide, SE Australia	E40; H05
−18.80	126.90	Landrigan	Kimberleys, NW Australia	E44; H05
29.80	78.50	Blaini	Himalayas, India	E37; H05
−24.00	15.00	Blasskranz	Witvlei-Naukluft, Namibia	E47; H05
−28.00	16.50	Numees	Gariep, Namibia*	E49; H05
−11.51	27.97	Petit conglomerate	Zambia	E53; H05
−3.07	12.14	Upper tilloid	W Congo	E62; H05
−18.50	15.50	Ghaub	N Namibia	E64 (see Figs 4 & 11); H05
−21.00	15.50	Ghaub	N Namibia	E64; H05
−19.50	16.50	Ghaub	N Namibia	E64; H05
20.53	−11.39	Jbeliat	Taoudenni, NW Africa	E65; H05
10.00	1.00	Kodjari	Volta, Africa	E69; H05
21.00	58.40	Unnamed	Mirbat, Arabia	E36; H05
22.70	58.60	Fiq	Oman	Sturtian according to E&J04
−18.00	−42.00	Jequital/Macaubas	Sao Francisco, S America†	E57; H05; Sturtian according to E&J04
−18.00	−55.00	Puga	Alto Paraguay†	E77; H05
40.00	90.00	Tereeken	Tarim	E33; H05
24.70	107.30	Nantuo	Guizhou*	E35; H05
70.00	28.50	Smalfjord	Varanger peninsula	E20; H05
54.80	−8.20	Stralinchy-Reelan	Ireland	E12; H&H p. 632; Age from McCay *et al.* (2006)
79.90	18.30	Wilsonbreen	Polarisbreen Group, E Svalbard (=Sveanor Fm)	E17; H05
73.00	−24.00	Storeelv	Tillite Group, E Greenland	E15; H05
36.20	−116.70	Wildrose diamictite, Kingston Peak Formation	California	H05, age given as Sturtian; see Halverson *et al.* (2005), p. 1190
36.40	−115.50	Wildrose diamictite, Kingston Peak Formation	Nevada	H05, age given as Sturtian; see Halverson *et al.* (2005), p. 1190
64.60	−129.60	Ice Brook Fm	NW Territories	E2; H05 refers to as Stelfox member
62.80	−127.20	Ice Brook Fm	NW Territories	E2; H05 refers to as Stelfox member
Sturtian				
−26.00	131.00	Areyonga*	Central Australia	E39; H05
−29.70	136.10	Pualco-Appila*	Adelaide	E40; H05
−33.00	139.50	Pualco-Appila*	Adelaide	E40; H05

(Continued)

Table 1. *Continued*

Lat. (°)	Long. (°)	Name	Area	References
−35.40	138.50	Pualco-Appila*	Adelaide	E40; H05
−30.80	138.7	Pualco-Appila*	Adelaide	E40; H05
−18.80	126.90	Walsh	Kimberleys	E42; H05
−23.70	17.00	Blaubekker	Witvlei-Naukluft	E46; H05
−28.00	16.50	Kaigas	Gariep	E48; H05
−11.51	27.97	Grand conglomerate	Zambia	E52; H05
−3.07	12.14	Lower tilloid	W Congo	E61; H05
−18.50	15.50	Chuos*	N Namibia	E63, see Figures 4 & 11; H05
−21.00	15.50	Chuos*	N Namibia	E63; H05
−22.00	14.70	Chuos*	N Namibia	E63; H05
21.00	58.40	L Mirbat	Mirbat, Arabia	E36; H05
22.70	58.60	Gubrah	Oman	H05
40.00	90.00	Bayisi(?)	Tarim	E33; H05
24.70	107.30	Chang'an	Guizhou	E35, see Fig. 9; H05
48.00	96.00	Tsagaan Oloom Fm	Mongolia	E32; H05
55.80	−6.10	Port Askaig	Scotland	E12; H05; H&H p. 632; age from McCay *et al.* (2006)
55.80	−6.10	Port Askaig	Scotland	E12; H05; H&H p. 632
57.50	−2.50	Port Askaig	Scotland	E12; H05; H&H p. 632
79.90	18.30	Petrovbreen	Polarisbreen Gp, E Svalbard	E16; H05
73.00	−24.00	Ulveso	Tillite Group, E Greenland	E15; H05
36.20	−116.70	Surprise diamictite, Kingston Peak Formation	California	Age given as Marinoan in H05. See Halverson *et al.* (2005), p. 1190
36.40	−115.50	Surprise diamictite, Kingston Peak Formation	Nevada	Age given as Marinoan in H05. See Halverson *et al.* (2005), p. 1190
43.10	−112.30	Scout Mountain member diamictite, Pocatello Formation	Idaho	E7; H05
40.60	−111.00	Mineral Fork Fm	Utah	E8
40.80	−114.10	Sheeprock Group	Utah	E8
51.40	−116.40	Toby cgl	Mt. Law, British Columbia	E8; H05
49.60	−116.70	Toby cgl	Columbia Point, British Columbia	E8; H05
48.70	−117.90	Toby cgl	Monk Creek area, Washington	E8 H05
36.60	−81.50	Franklin (=Konnarock?)	Blue Ridge, Appalachians	E11; H05;
56.10	−121.80	Mt Vreeland Fm	British Columbia	E5; H05; E&J04 make it Sturtian
58.70	−123.40	Mt Lloyd George diamictite	British Columbia	E5; H05
61.70	−126.30	Rapitan Group*	Thundercloud Range, NW Territories	E5; H05
62.70	−127.50	Rapitan Group*	Mountain River, NW Territories	E1; H05; H&H p. 375
65.40	−134.80	Rapitan Group*	Snake River area, Yukon	E1; H05; H&H p. 375
65.30	−141.90	Upper Tindir Group	Alaska	E3; H05; H&H p. 375
Neoproterozoic diamictites with poorly known ages				
−10.00	−41.00	Bebedouro/Rio Preto	S America	E58, equated with Jequital
−17.00	−47.00	Ibia/Cristilina*	S America	E59, equated with Jequital
−22.00	−48.00	Carandai	S America	E60, equated with Jequital
61.13	92.18	Chingasan*	Siberia	E28
59.08	113.51	Patom	Siberia	E29
57.44	91.53	Sayan rift	Siberia	E30
45.00	75.00	W Altaids*	Siberia	E31
64.50	15.20	Langmarkberg Fm	Sweden	E22
54.58	27.59	Unnamed	E Europe	E24, see Figure 7 for names of Baltica tillites

(*Continued*)

Table 1. *Continued*

Lat. (°)	Long. (°)	Name	Area	References
56.27	41.55	Yablonovka	E Europe	E24
59.82	58.04	Churochnaya	Central Urals	E25
58.90	58.79	Churochnaya	Central Urals	E26
53.70	57.26	Kurgashlya*	S Urals	E27
81.00	−74.00	Pearya	Arctic Canada	E19

The application of the standard tools used for Phanerozoic correlation and dating – such as biostratigraphy, stable isotope variations, palaeomagnetism, stratigraphic correlation and age dating – has produced a variety of global correlation schemes for the Cryogenian–Ediacaran (CE) interval. But, even with many new isotopic measurements, the interval is still 'incomplete and poorly calibrated radiometrically' (Halverson *et al.* 2005, p. 1204). Possible subdivisions of the CE interval are given later in this paper. It is just over 300 Ma in duration, or about as long as the Palaeozoic, and some 50 Ma longer than the Mesozoic and Cenozoic together (Gradstein *et al.* 2004). Thus, the CE is very much a scientific frontier and, simultaneously, a frontier for hydrocarbon exploration. As might be expected, the divergence of interest between academia and industry means that the primary concerns of one endeavour are rarely mentioned by the other. Of course, the two are intimately related, as is clear from the exploration work and scientific results that have followed, for example, from the support of Petroleum Development Oman (PDO) for basic research in Oman.

The main purpose of this paper is to outline the methods available for the litho- and chronostratigraphic correlation of the CE interval, and the general results obtained so far. Its purpose is not to discuss the detailed evidence for a particular model – papers citing a great range of evidence for different models appear in the literature with a high frequency – but it is necessary to outline the end members of the current models. These will be referred to as the Snowball Earth and the Phanerozoic models.

Neoproterozoic rocks

Diamictites and 'cap carbonates'

Neoproterozoic rocks make up a widely distributed global patchwork of outcrops (Fig. 2). As in the

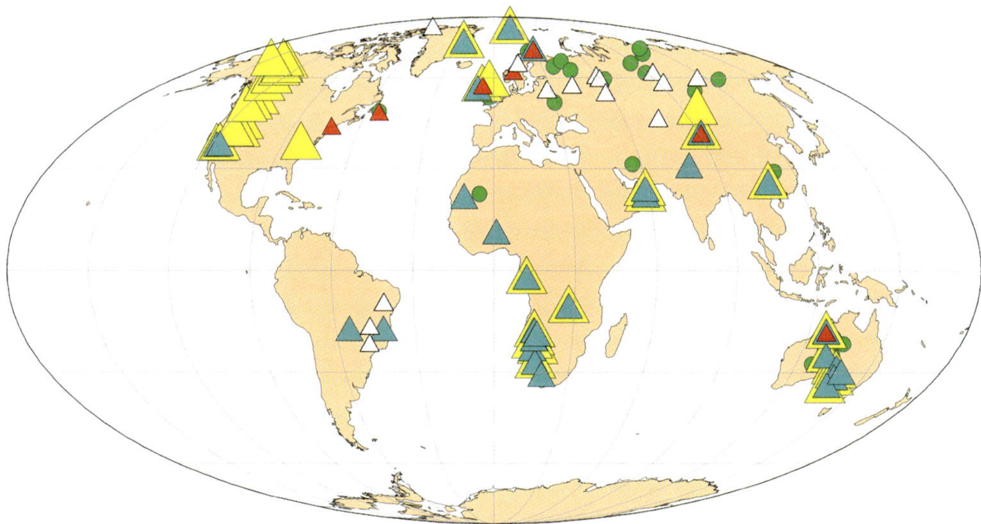

Fig. 4. Present-day locations of presumed Neoproterozoic tillites (Table 1) and Ediacaran fossil locations (Narbonne 2005) on a Mollweide projection with the geographic grid drawn at intervals of 30°. The oldest tillites (Sturtian) are shown as large yellow triangles; the Marinoan as smaller light blue triangles; the Gaskiers tillites as small red triangles. Neoproterozoic tillites of uncertain age are shown as white triangles. Ediacaran fossil localities are green circles.

Phanerozoic, they include clastic, carbonate and mixed sequences, but what appears to characterize the Cryogenian–Ediacaran interval is the presence of two widespread diamictite horizons (Kennedy *et al.* 1998) and at least one less extensive horizon (e.g. Halverson 2006) (see Table 1 and Fig. 4) and their associated 'cap carbonates'. A diamictite is a lithified sediment made up of coarse, poorly sorted, angular to well-rounded clasts supported typically by a clayey argillaceous matrix. Although there is no genetic connotation in the term, the Neoproterozoic diamictites are commonly interpreted as glacigenic, with the cap carbonates representing deglaciation (see later). Diamictites and cap carbonates are together the principal features used for late Neoproterozoic lithostratigraphic correlation.

Diamictites as rift sequences

A radically different interpretation of the diamictites is that of Eyles & Januszczak (2004, p. 13), who state that it is 'still a commonly held belief among many geologists that massive Neoproterozoic diamictites are necessarily glacial in origin', and go on to assert that 'most Neoproterozoic diamictites are deepwater debris flows indicative of tectonically active basins and carry no palaeoclimate interpretation' (p. 16). However, they recognize that 'direct evidence for a glacial source of sediment is provided by the occasional striated and outsized clasts in fine-grained laminated facies'. Although their review has been characterized as an 'inordinately sceptical synthesis' (Fanning & Link 2004), it is a timely reminder of the indirect link between diamictites and glaciation. More recently, Eyles *et al.* (2007, pp. 308–309) have argued that several examples of Australian 'tillites' are actually 'the product of rain-out and mass flow in offshore settings, rather than terrestrial subglacial deposits formed below grounded glaciers'.

Nevertheless, it will be assumed here that the common association of these Neoproterozoic diamictites with glacial sediments implies that all the major Neoproterozoic diamictites, even if they are of strictly non-glacial origin, were penecontemporanous with nearby glacial activity.

Extent of the diamictites

Although there were earlier suggestions about an extensive Neoproterozoic ice age (e.g. Eyles 2005), it is convenient to start with the evidence for 'a great infra-Cambrian glaciation' (Harland 1964) (Fig. 4 below). In Australia, the Sturtian and Marinoan diamictites are major stratigraphic units (Dunn *et al.* 1971), whose names are now applied globally (see Figs 5 & 6), although this usage has been questioned (Corsetti & Lorentz 2006). Each diamictite is itself a relatively small part of the much thicker Sturtian and Marinoan stratigraphic units preserved in a series of Neoproterozoic–Middle Cambrian rift basins and on their margins in the Adelaide area of South Australia (Preiss 2000). According to Preiss (2000), there is only one major Sturtian diamictite horizon, the Bolla Bollana Tillite. Equivalent horizons formed in adjacent troughs and on adjacent shelves are known

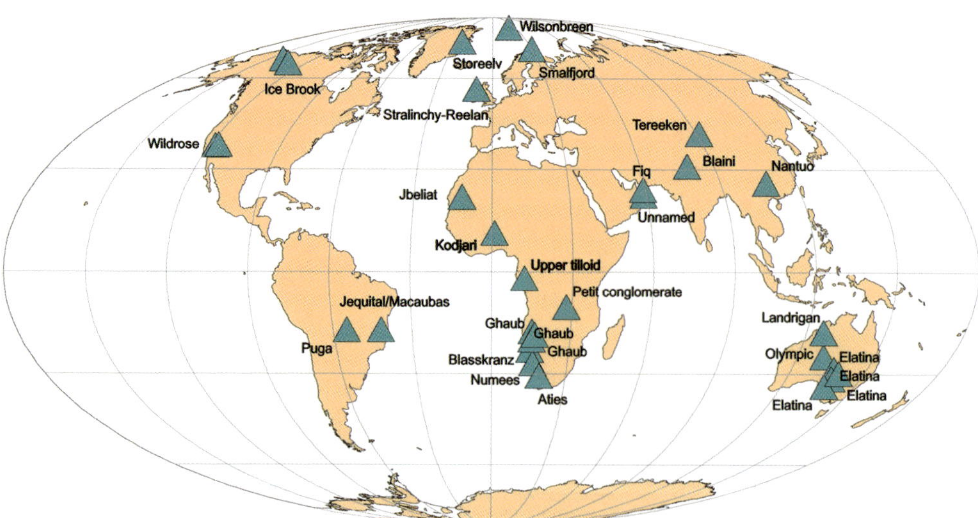

Fig. 5. Marinoan tillites are shown as blue triangles. Names are taken from Table 1.

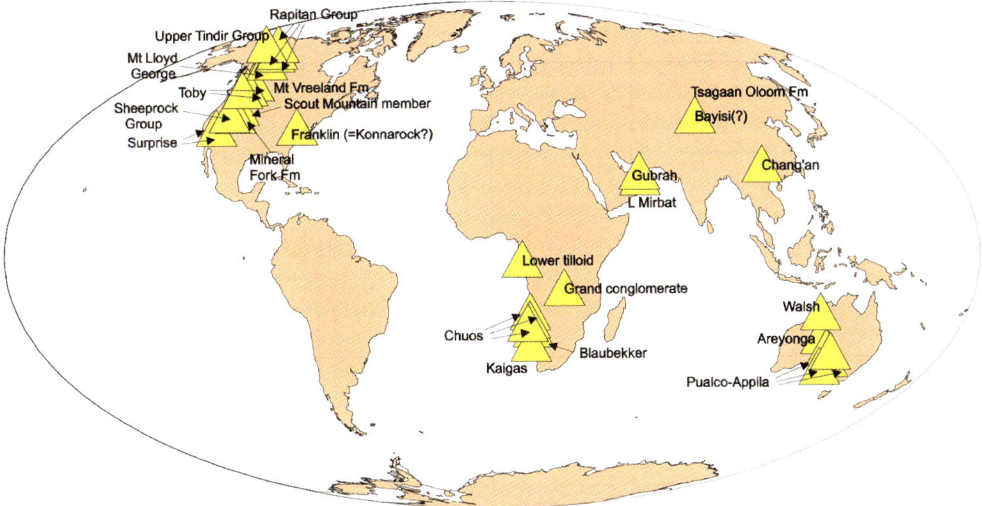

Fig. 6. Sturtian tillites are shown as yellow triangles. Names are taken from Table 1.

variously as the Pualco, Apilla and Sturt tillites, and are accompanied by other glacigenic units (Fig. 7).

The Marinoan is a name for a stage in South Australian stratigraphy that extends from immediately above the top of the Sturtian, well below the actual glacial units, to the Cambrian boundary (Preiss 2000, table 3) (Fig. 7). It includes much of the Cryogenian and all of the Ediacaran, and is subdivided into four groups/subgroups. The tillites within Marinoan 2 are known as the Mount Curtis and Pepuarta tillites, and in some areas are preceded and followed by dropstone horizons. Dunn *et al.* (1971) suggested the possible use of the Sturtian and Marinoan 2 glacigenic sequences as global Neoproterozoic time markers (Figs 5 & 6).

The Gaskiers glacial sequence in the Gaskiers Formation is named after the village of Gaskiers in the Avalon Peninsula of SE Newfoundland (Anderson & King 1981, p. 760). It is also known as the Ediacaran glaciation (Hoffman 2005, table 1). The Gaskiers is the youngest Neoproterozoic glacigenic sequence and appears to be much less extensive than either the Marinoan or Sturtian sequences (Halverson *et al.* 2005) (Fig. 8). The Egan Formation, a glacial unit in the Kimberley area, lies towards the top of the Marinoan Stage (Grey & Corkeron 1998) and may be of Gaskiers age (Jenkins pers. comm. 2007).

The current stratigraphic and diamictite nomenclature causes a problem because the timespan of the Gaskiers lies within the Marinoan 4 unit, whereas the Marinoan glaciation may be restricted to the Marinoan 2 unit. To add to the confusion, the glacial Elatina Formation, a key Neoproterozoic glacial unit, is not part of a basal Marinoan glaciation, as often stated, because it occurs in the upper part of Marinoan 2, well above the base of the Marinoan and, in particular, above the Marino Arkose Member of Marinoan 1, after which the whole stage is named (Eyles *et al.* 2007). However, there is still uncertainty about its age. Preiss (2000) places it within Marinoan 2, an assignment accepted by Hoffman (2005), but Eyles *et al.* (2007) (Fig. 7) argue for a much younger age, placing it within Marinoan 4 and essentially giving it a Gaskiers age (see later).

The Kaigas glacial sequence, lying at the base of the presumed Sturtian glacial sequence in Namibia (Hoffman 2005, table 1), is regarded by some as a distinct fourth diamictite unit, which may have equivalents in eastern Brazil and South China (e.g. MacGabhann 2005).

Some workers recognize five glaciations (see comments in Kennedy *et al.* 1998). All of the glacial sequences have synonyms, many of which are listed by Evans (2000) and MacGabhann (2005), but their multiplicity makes the literature confusing.

Palaeolatitudes, diamictites and Snowball Earth

What dominates the current approach to Neoproterozoic stratigraphy springs from some quite unexpected and unique palaeomagnetic results. Apparently reliable palaeomagnetic poles place some Neoproterozoic tillites in tropical latitudes. This result led to the notion that during these glacigenic periods the entire Earth was covered by ice and snow – or a 'Snowball' Earth (Kirschvink

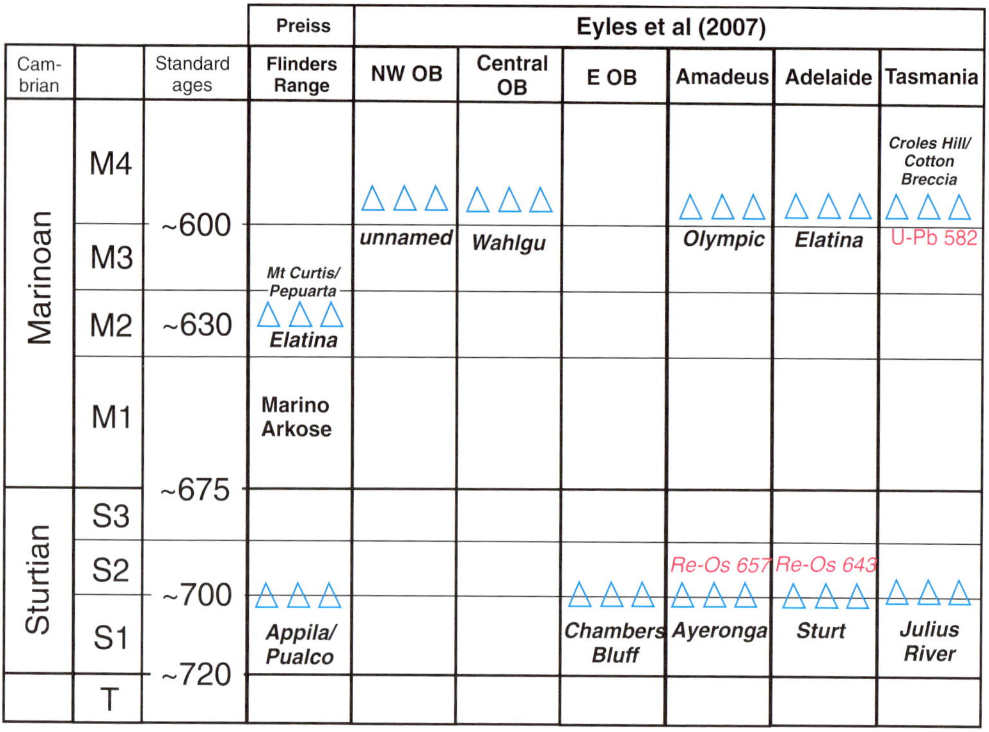

			Preiss	Eyles et al (2007)					
Cam-brian		Standard ages	Flinders Range	NW OB	Central OB	E OB	Amadeus	Adelaide	Tasmania
Marinoan	M4								*Croles Hill/ Cotton Breccia*
		~600							
	M3		*Mt Curtis/ Pepuarta*	*unnamed*	*Wahlgu*		*Olympic*	*Elatina*	U-Pb 582
	M2	~630	*Elatina*						
	M1		Marino Arkose						
		~675							
Sturtian	S3								
	S2						Re-Os 657	Re-Os 643	
		~700							
	S1		*Appila/ Pualco*			*Chambers Bluff*	*Ayeronga*	*Sturt*	*Julius River*
		~720							
	T								

Fig. 7. Australian stratigraphic correlations and dates. The correlations are an abstract of data in Preiss (2000, tables 2 and 3) in the first stratigraphic column and Eyles *et al.* (2007, fig. 2) in the remainder. In the first column are the Australian stratigraphic names, Sturtian and Marinoan. The second column shows how these units have been subdivided into groups and subgroups, e.g. S2, M4, etc. The 'standard ages' are those widely used in the literature but contrast with new dates, three of which are shown and are discussed in the text. The names in bold italics are those of some of the Neoproterozoic glacial deposits in Australia, their locations are shown in their correct relative stratigraphic position, but the names are displaced up or down. The location of the Marino Arkose, which gives its name to the Marinoan unit is shown in normal type. OB, Officer Basin, part of the Centralian Superbasin. T, Torrensian. The figure shows at least six important features: (1) stratigraphically, the Marinoan includes the entire Ediacaran, terminating at the base of the Cambrian; (2) new Re–Os dates that suggest the top of the Sturtian is much younger than its standard age; (3) the major disagreements about the age of the Elatina Formation (M2? or M4?); (4) evidence for the M4 Elatina age via correlations to Tasmania; (5) the general scarcity of 'Marinoan' glacial deposits in Australia; and (6) some current uncertainties in correlations and in stratigraphic nomenclature.

1992). A critical review of the palaeomagnetic evidence for the Snowball Earth is beyond the scope of this paper, but it is worth pointing out that, of the many poles cited by Hoffman (2005, p. 558), some for the Sturtian glaciation place part of ice-covered Laurentia on the palaeoequator (e.g. Park 1997). Similarly, three independent pole determinations from the glacially influenced Elatina Formation of SE Australia, part of the overlying Marinoan glacial period (or possibly younger), place it very close to the palaeoequator (Fig. 9). Additional Australian poles from the same interval give a mean pole that keeps the region in the equatorial region. Poles from a later Neoproterozoic glacial sequence in Oman, place that sequence in tropical latitudes too (Kilner *et al.* 2005).

The Snowball model has been developed in detail by Hoffman and his colleagues (Hoffman *et al.* 1998; Hoffman & Schrag 2002; Halverson *et al.* 2005; Hoffman 2005). As Hoffman (2005, p. 561) put it; 'In a nutshell, the Snowball Earth hypothesis ... postulates that the ocean froze over from pole to pole for long periods during the Sturtian and Marinoan glaciations', with global temperatures as low as $-50\,^{\circ}C$. The Snowball model predicts that Neoproterozoic continental diamictites were almost ubiquitous, subsequently reduced to a patchwork by erosion and burial under younger rocks.

The model has been criticized, most recently in Snowball Earth 2006 (Etienne *et al.* 2006, 2007). Nevertheless, the stratigraphic implications of the

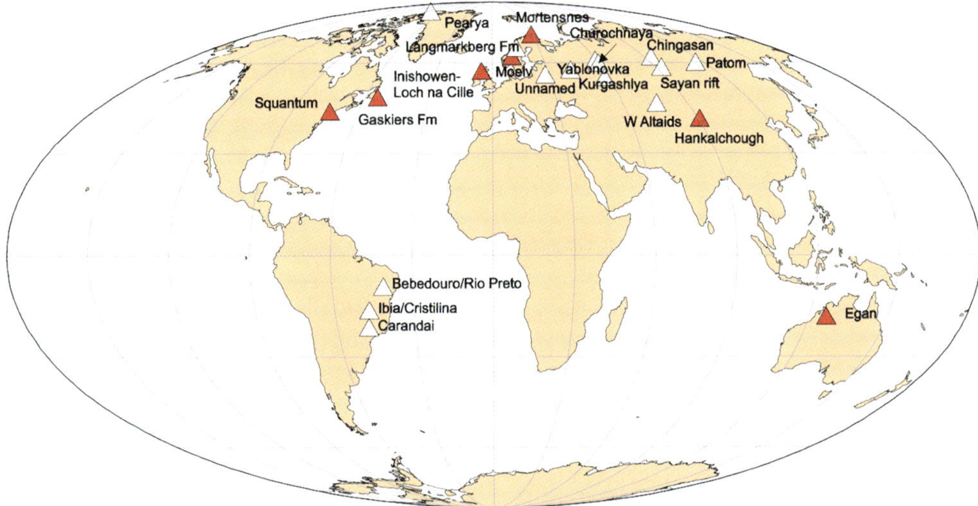

Fig. 8. Gaskiers tillites are shown as small red triangles. Neoproterozoic tillites of uncertain age are shown as white triangles. Names are taken from Table 1.

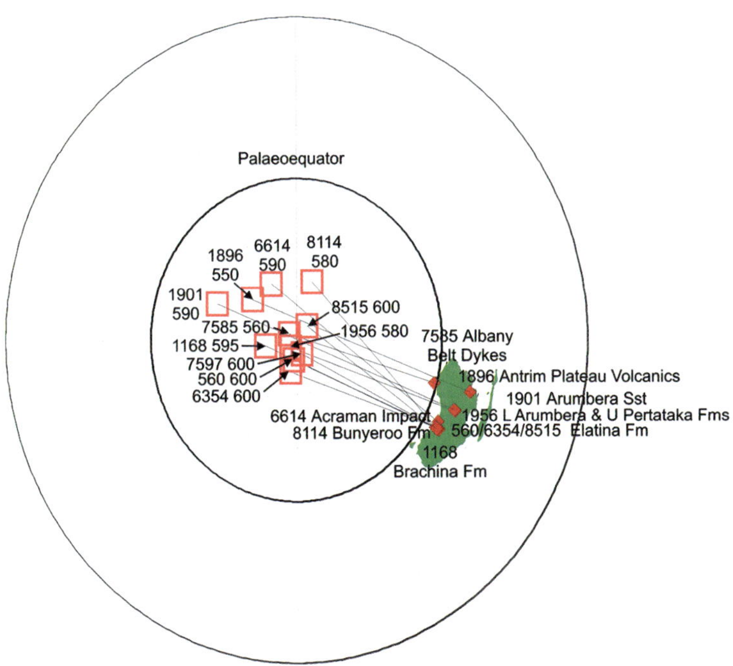

Fig. 9. Late Neoproterozoic palaeomagnetic poles (*c*. 600–550 Ma) from Australia plotted on an azimuthal equidistant projection. The palaeoequator is the bold circle in the middle of the figure. The poles are from the global palaeomagnetic database, shown as red open squares with their catalogue number and age indicated. All poles are considered to be of good quality and record primary magnetization. The corresponding sites are shown as red diamonds and join their respective pole positions. Australia has been repositioned so that the mean pole for *c*. 580 Ma is the projection pole for the map, placing southernmost Australia on the equator, one of the key pieces of evidence for the Snowball Earth model.

model require discussion. In particular, if the model is correct then Neoproterozoic diamictites should provide excellent chronostratigraphic and lithostratigraphic markers. In principle, Neoproterozoic stratigraphic correlation is simple: it involves determining how many Snowball glaciations there are, dating them and then identifying to which glaciation a given diamictite corresponds.

Phanerozoic glaciations

There are three major glacial episodes in the Phanerozoic: the Cenozoic, which may have existed since *c.* 42 Ma (Tripati *et al.* 2005) and in which we are still living; Wegener's classic Permo-Carboniferous glacial period, about 85 Ma in duration; and the Late Ordovician glaciation, whose main phase may have been as short as 1 Ma. The inferred latitudinal distribution of the maximum ice cover for all three glaciations is shown in Figure 10. Despite a marked difference in their durations, and in the corresponding continental configurations, all three Phanerozoic glacial episodes show a remarkably uniform inferred latitudinal ice distribution: no significant ice lies equatorward of 30°, and there is no evidence at all for any ice sheets in the equatorial zone itself (Smith & Pickering 2003). The reason why Neoproterozoic continental ice sheets were able to reach the equator, whereas those of all subsequent ice ages were not, is a problem that remains to be solved.

Neoproterozoic diamictite stratigraphy

The tropical location of some ice sheets suggests that, whether the Earth did freeze over entirely or whether there were simply continental ice sheets at the equator, the first-order Neoproterozoic stratigraphic framework is an alternation of ice-free and glacial intervals. There is a continuum between the two possibilities: Snowball Earth becomes a tropical Phanerozoic glacial model by a reduction in the size of frozen oceanic areas. In turn, this evolves into a normal Phanerozoic model by a progressive restriction of ice to higher latitudes.

Hoffman (2005, p. 564) recognizes that virtually all students of the Cryogenian glacial and glacial marine sediments have been impressed by their lithological similarity to Quaternary and Recent deposits. The production and deposition of large amounts of glaciclastic sediment in a marine environment necessitates the presence of water in facilitating subglacial sliding onland, the erosion and release of sediment, and its transport to offshore basins by water (Eyles *et al.* 2007).

At first sight, the existence of a vigorous hydrological cycle seems contrary to the Snowball model. However, Snowball Earth was essentially a time of global sediment mummification. During the transition to Snowball Earth there would presumably have been abundant wet-based glaciers and a full hydrological cycle; followed by a period of stasis, terminating with abundant wet-based glaciers and a full hydrological cycle as the ice sheets melted. Thus, the fact that Neoproterozoic glacial deposits have no characteristics requiring extreme cold (e.g. Etienne *et al.* 2007; Eyles *et al.* 2007) is not necessarily evidence against Snowball Earth. In fact, Hoffman (2005, p. 564) agrees that there is direct evidence for open water within the Cryogenian marine sequences, but he points out that all the lines of evidence cited share a common limitation, which is the degree to which they record the ice maximum. Although several important geochemical consequences may follow from the Snowball model (e.g. Hoffman & Schrag 2002), it is unclear how a study of diamictites and glacial sediments, as opposed to the associated cap carbonates (e.g. Shields 2005), can provide support for or against this model.

Biostratigraphy of the CE interval

Biostratigraphic zonation is the most important generally applicable method for subdividing and correlating the Phanerozoic. However, the pre-Ediacaran biosphere was made up almost exclusively of microscopic organisms with low diversity, no evidence of faunal provinciality and an almost constant morphology, i.e. evolution was at a very slow rate (Butterfield 2007). A coarse biostratigraphic stromatolite zonation is possible in the Neoproterozoic of Australia (Grey & Corkeron 1998), but there is no global biostratigraphic scheme for the Cryogenian.

Acritarchs may have the potential to provide global biostratigraphic zonation. Acritarchs are small, organic-walled microfossils with a central non-mineralized vesicle and are of unknown taxonomic affinity, classified solely by their morphology. Cryogenian acritarchs show lower taxonomic diversity and lower morphological disparity than both older and younger assemblages: there are fewer large Cryogenian acritarchs and fewer complex acanthomorphic (spiny) acritarchs (Huntley *et al.* 2006). These changes do not yet provide a usable biostratigraphic zonation.

However, coarse acritarch zoning has been demonstrated in the early Ediacaran. Early Ediacaran acritarchs show the first biological radiation of the fossil record, accompanied by the 'first (more-or-less) measurable extinction event in the fossil record' of the distinctive pre-Ediacaran acritarchs (Butterfield 2007). In their place are large,

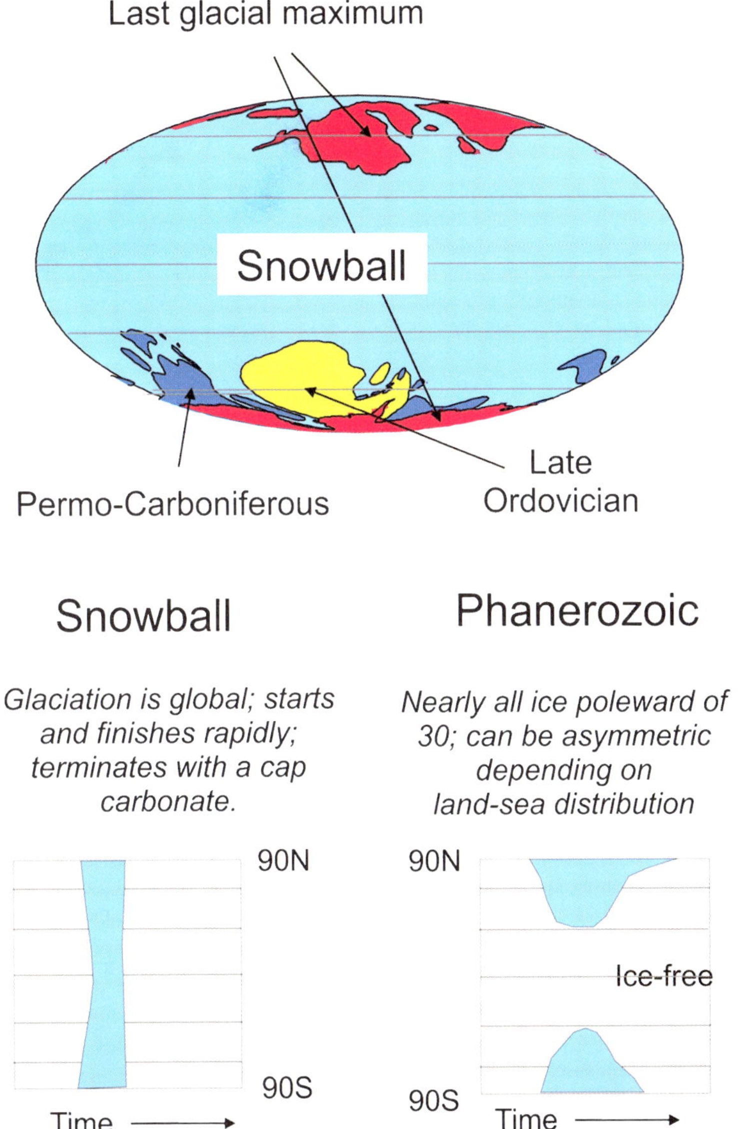

Fig. 10. Generalized distributions of the inferred maximum extent of ice for the major Phanerozoic glaciations (modified from Smith & Pickering 2003) and schematic latitudinal distributions in time for a 'Snowball' and Phanerozoic glaciation.

conspicuously ornamented acritarchs that evolved much more rapidly than any of the pre-Ediacaran biota. Those in the Centralian Superbasin of Australia may allow the first 50 Ma of the Ediacaran period to be divided into five biozones (Grey 2005), providing an Ediacaran biostratigraphic scheme where macrofossils are rare or absent altogether. The scheme is as yet untested outside Australia and has been criticized by Butterfield (2007) on the grounds that the microfossil succession in the basin may mostly reflect a fluctuating environment rather than evolution (Butterfield & Chandler 1992). It is unclear whether acritarch zonation is possible for the later Ediacaran.

The later Ediacaran Period boasts the first macrofossils, preserved as casts and moulds in contemporaneous sediments. Collectively, they make up the widespread Ediacara Biota (Narbonne 1998, 2005; Narbonne & Gehling 2003; Knoll *et al.* 2006) and are, with very few exceptions, unique to the

Ediacaran period (Figs 7 & 8). However, these famous fossils were present as diverse assemblages only from the end of the youngest (Gaskiers) Neoproterozoic glaciation to the base of the Cambrian. They do not provide precise biostratigraphic markers, and are unlikely to do so in the future (Knoll *et al.* 2006).

In short, Neoproterozoic biostratigraphic zonation currently allows only the coarsest subdivision of Neoproterozoic rocks. With regard to the Snowball model, Hoffman (2005, p. 562) notes that it 'presents the greatest challenge to phototrophs and their dependents – where liquid water exists there is no light, where light exists there is no liquid water' and speculates that crack systems in the ice would offer extensive refugia. However, from their analysis of deep cores from the Officer Basin and the Adelaide Rift Complex in Australia, Eyles *et al.* (2007) cannot recognize any effects of severe cold in the Australian Neoproterozoic fossil record. Similar conclusions have been reached elsewhere, such as in Death Valley, California (Corsetti *et al.* 2006). How this can happen on a Snowball Earth is unclear.

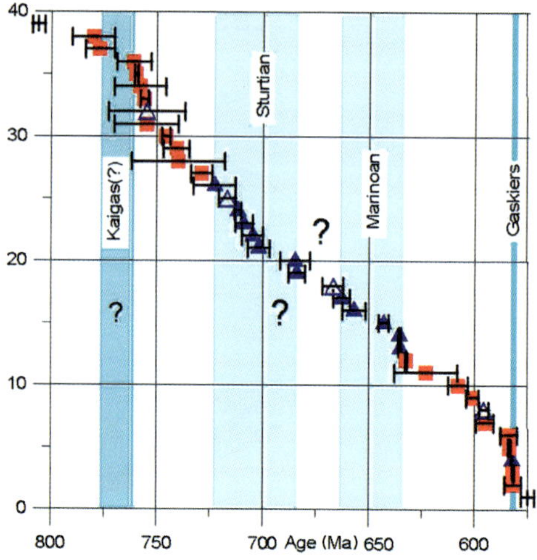

Fig. 11. Non-glacial dates shown as filled red squares and dates within glacial successions as filled blue triangles, together with error bars. Two intrusion dates are shown as crosses (at the beginning and end of the graph), but provide poor age constraints. Four dates from clasts and reworked tuffs lying within glacial successions are shown as open blue triangles. Some of these assignments may be open to a different interpretation. The queries indicate data gaps.

Age dates

Isotopic age dating is by far the most important method available for dating the Neoproterozoic. The available U–Pb zircon dates are taken mostly from MacGabhann's (2005) compilation with a few additions from the literature (Fig. 11). There are also three rhenium–osmium (Re–Os) dates on shales (Kendall *et al.* 2004, 2006).

Rhenium–osmium dating has only recently become available and has the potential to revolutionize timescale work in unfossiliferous black shales in general and the dating of the Neoproterozoic in particular. It depends on the decay of ^{187}Re to ^{187}Os. The principles involved are very similar to Rb–Sr dating (Creaser *et al.* 2002; Creaser 2003). When black shales are deposited they contain very little initial Re and Os, but the reducing environment concentrates the Re and Os that is in the sea water into the shales, although precisely how the isotopes are reset, equilibrated and transferred to the hydrocarbons is not fully known (Selby & Creaser 2005). Recently developed methods give dates from Neoproterozoic black shales in western Canada with errors as small as $\pm 0.8\%$, 4.7 Ma (Kendall *et al.* 2004), or even $\pm 0.4\%$, 2.4 Ma, in Australian examples (Kendall *et al.* 2006). The dates obtained are presumably the age of some phase of diagenesis, but appear to be the same, within the limits of error, to U–Pb dates.

There are 38 dates (Table 2) listed in chronological order, with the youngest first, shown graphically in Figure 11 and globally in Figure 12. In the field, the Neoproterozoic glacial deposits are widespread and are clearly divisible into discrete units on each continent in which they are found, i.e. on a given continent they do not form a continuum distributed through the CE interval. In Hoffman's (2005, table 1) compilation only 11 exist as isolated deposits; most occur in pairs, with 15 Sturtian–Marinoan examples and three Marinoan–Gaskiers pairs; triplets are rare but do occur in the same general region, e.g. in the Kimberley area of Australia and the Tarim block in Asia. The Dalradian of the British Isles also shows a triplet (McCay *et al.* 2006).

The oldest glaciation, the Kaigas, is undated, but may lie in the data gap between *c.* 777 and *c.* 761 Ma. MacGabhann (2005) puts it in the range of approximately 780–735 Ma. The longest glaciation appears to be the Sturtian, with an age range of *c.* 723–684 Ma, a duration of approximately 39 Ma, although the date of 723 Ma has a large asymmetric error (−10 to +16 Ma) not shown in Figure 11. There is a data gap within the Sturtian from *c.* 702 to *c.* 684 Ma. Halverson *et al.* (2005, fig. 15) give two possible ranges for the Sturtian, *c.* 742–735 Ma or *c.* 720–700 Ma, depending on differing interpretations of the dates and carbon isotope anomalies, whereas MacGabhann (2005) suggests *c.* 715–680 Ma. The succeeding glaciation, the Marinoan, has an age range of about *c.* 663–635 Ma. The Gaskiers, the

Table 2. *Dates, errors, approximate locations, and references of dated Cryogenian and Ediacaran rocks. Data mostly from MacGabhann (2005, table 1), referred to as BAM in the table. Plotted on Figures 11 & 12. Negative latitudes are S; negative longitudes are W*

Date	Error	Method	Lat. (°)	Long. (°)	Material and comments	References
575.00	3.00	U–Pb	−39.90	144.01	City of Melbourne Bay, Tasmania; sills intrude Elatina diamictite (Cottons Breccia) believed Marinoan	Calver *et al.* (2004)
582.00	4.00	U–Pb	−40.75	144.50	King Island, Tasmania; rhyodacite flow underlies Croles Hill Diamictite. Has a cap carbonate. If Croles Hill = Cottons, and if Marinoan, then Marinoan much younger than elsewhere. BAM suggests a miscorrelation and that these diamictites are Gaskiers. Other alternatives exist	Calver *et al.* (2004)
582.10	0.40	U–Pb	47.00	−53.50	Post-Gaskiers, Newfoundland; locality not given	BAM, pers. comm. from Bowring
582.40	0.40	U–Pb	47.00	−53.50	Within-Gaskiers, Newfoundland; locality not given	BAM, pers. comm. from Bowring
583.70	0.60	U–Pb	47.00	−53.50	Pre-Gaskiers, Newfoundland; locality not given	BAM, pers. comm. from Bowring
584.00	4.00	U–Pb	48.60	−3.20	Granville Fm, supposedly glacial lies on dated rocks. Probably correlative of Gaskiers. Not shown on Gaskiers tillite figure	Guerrot & Peucat (1990)
595.00	4.00	U–Pb	56.00	−5.60	Tayvallich keratophyre in Dalradian at the base of Argyll Group. Inishowen diamictite lies 'not far above' the dated rock; MacDuff Tillite overlies the Inishown	Halliday *et al.* (1989); Brasier & Shields (2000); Condon & Prave (2000); Condon *et al.* (2002)
595.50	2.00	U–Pb	42.10	−70.70	Welded tuff clast in Squantum 'tillite'. Tillite is younger	Thompson & Bowring (2000)
601.00	3.00	U–Pb	56.00	−5.60	Tayvallich tuffs. Port Askaig tillite lies below	Dempster *et al.* (2002)
607.80	4.70	Re–Os	52.87	−118.30	Thickness of overlying rock and chemostratigraphy suggests it is equivalent to older N American tillite (Sturtian). Black shale just above Horsethief Creek. Minimum age for 2nd Windermere (Marinoan) glaciation	Kendall *et al.* (2004)
623.00	15.00	U–Pb	−18.00	−55.00	Volcanics underlying Puga diamictite	Trompette *et al.* (1998)
635.23	0.57	U–Pb	27.50	110.00	Tuff within Nantuo cap carbonate. Right at the top of assumed Marinoan tillite. Key date.	Condon *et al.* (2005)
635.50	1.20	U–Pb	−21.99	15.73	Ash bed within Kachab Dropstone unit, Ghaub Fm, Namibia	Hoffmann *et al.* (2004)

(Continued)

Table 2. *Continued*

Date	Error	Method	Lat. (°)	Long. (°)	Material and comments	References
643.00	2.40	Re–Os	−32.00	138.40	Tindelpina post-glacial shale, Adelaide rift complex. Lies on top of Sturtian. Possible conflict with 663 Ma below if 663 Ma date shows that Marinoan glaciation had already begun	Kendall *et al.* (2006)
657.20	5.40	Re–Os	−24.30	130.00	Aralka Fm post-glacial shale, Amadeus Basin. Provides tight minimum age on Sturtian and Areyonga glaciations. Possible conflict with 663 Ma below if 663 Ma date shows that Marinoan glaciation had already begun	Kendall *et al.* (2006)
663.00	4.00	U–Pb	27.50	110.00	Tuff in basal Datango Fm, E Guizhou. Lies immediately below Nantuo diamictite and above Tiesiao diamictite, S China. Regarded as Marinoan	Zhou *et al.* (2004)
667.00	5.00	U–Pb	42.90	−112.40	Upper Scott Mountain Member, Pocatello Fm, southern Idaho. Reworked fallout tuff. 20 m above upper diamictite, immediately below 2nd cap-like carbonate. Sturtian had finished, Marinoan had not yet begun. Key date	Fanning & Link (2004)
684.00	4.00	U–Pb	45.20	−115.30	Flows in roof pendant in Windermere. Glaciation (Sturtian) ongoing. Key date	Lund *et al.* (2003)
685.00	7.00	U–Pb	45.20	−115.30	Flows in roof pendant in Windermere. Glaciation (Sturtian) ongoing. Key date	Lund *et al.* (2003)
702.00	5.00	U–Pb	38.90	−78.10	Robertson River igneous suite shown as within Mechum River Fm (diamictite). Pre-glacial, only one glacial event in Appalachians; assumed correlative with older Windermere (Rapitan = Sturtian). Error assumed	Tollo & Hutson (1996), p. 61, quoting a GSA abstract
705.00	5.00	U–Pb	38.90	−78.10	Robertson River igneous suite shown as within Mechum River Fm (diamictite) – see above	Tollo & Hutson (1996), p. 61, quoting a GSA abstract
709.00	5.00	U–Pb	42.2	−112.10	Scott Mountain Member, Pocatello Fm, Oxford Mountain, southern Idaho. Tuff immediately below upper diamictite. Sets max. age. Lies between two glacial episodes	Fanning & Link (2004)
711.80	1.60	U–Pb	23.30	57.30	Ash bed within Ghubrah Fm (older diamictite), Oman	Allen *et al.* (2002)
717.00	4.00	U–Pb	42.90	−112.40	Upper Scott Mountain Member, Pocatello Fm, southern Idaho. Clast in glacials, probably from Bannock volcanic member of Pocatello Fm. Probably dates rifting event. Right at the base of glacials.	Fanning & Link (2004)

723.00	10.00	U–Pb	23.30	57.30	Tuffaceous gwk within Ghubrah Fm, Oman. Any cap carbonate truncated by erosion. Overlain by 2nd glacial (Fiq) assumed Marinoan	Brasier & Shields (2000)
729.00	5.00		38.40	−78.60	Robertson River granitoid under Mechum River Fm. See above. Date is pre-glacial, only one glacial event in Appalachians; assumed correlative with older Windermere (Rapitan) glaciation. Error assumed at 5 Ma.	Tollo & Hutson (1996), p. 61
740.00	22.00	U–Pb	−18.00	−42.00	Jequitai diamictite on Sao Francisco Craton overlain by dated volcanics. Possible equivalent of Kaigas in Namibia	De Alvarenga *et al.* (2003)
741.00	6.00	U–Pb	−27.60	16.50	Rosh Pinah Fm, Gariep belt, Namibia Rhyolite between Kaigas (below) and Numees (above)	Frimmel *et al.* (1996)
746.00	2.00	U–Pb	−20.50	15.00	Naauwport volcanics. Underlies older Chuos diamictite; gives a maximum age of Chuos, equated with Sturtian	Hoffmann & Prave (1996)
755.00	15.00	U–Pb	41.00	90.00	Bayisi Fm, probably glaciogenic, underlain by these dated volcanics	Xu *et al.* (2005)
755.00	18.00	U–Pb	65.00	−129.60	Granite clast in Sayunei Fm = L Rapitan (Sturtian) diamictite. Key date for maximum age of Sturtian	Ross & Villeneuve (1997)
756.00		U–Pb	−20.00	15.00	Ash bed within Ombombo Subgroup, Namibia	Hoffmann & Prave (1996)
758.00	12.00	U–Pb	36.90	−80.30	Konnarock (=Mt Rogers) Fm is glacial; underlain by dated Mt Rogers volcanics.	Bailey & Peters (1998)
760.00	1.00	U–Pb	−20.10	15.00	Ash bed within Ombombo Subgroup, Namibia. Predates the syn-rift, glaciogenic Chuos Formation (=Sturtian glaciation) and defines the base of the Abenab Subgroup. See p. 1193 for further comment	Halverson *et al.* (2005)
761.00	8.00	U–Pb	25.00	109.50	Volcanics in Liantuo Fm in N Guangxi, S China, give a maximum age of diamictites.	Li *et al.* (2003)
777.00	7.00	U–Pb	−32.50	138.80	Boucat Volcanics, at the base of Burra Gp, Adelaide succession, S Australia. Burra Group immediately underlies Sturtian glacials. See also Evans p.381 & Calver *et al* 2004.	Preiss (2000)
780.00	10.00	U–Pb	−27.00	16.00	Basement predates Kaigas glaciation in Namibia	Allsopp *et al.* (1979)
806.00	3.00	U–Pb	57.50	−3.60	Port Askaig tillite lies above. Shear zone in base of Grampian Group in Dalradian	Noble *et al.* (1996)

youngest and shortest glaciation, was ongoing at 582 Ma, and no more than about 2 Ma in duration (Table 2).

There is also a significant data gap between a date of 663 Ma in the Tiesiao diamictite at or just below the base of the Marinoan and 684 Ma at the top of the Sturtian. Zhou *et al.* (2004) note that the available dates seem to partition the Neoproterozoic glacial interval into three windows with clustered glacial deposits – the Sturtian (or Changan-Tiesiao in South China, *c.* 750–660 Ma), Marinoan (or Nantuo, *c.* 660–600 Ma) and Gaskiers (*c.* 580 Ma) – implying a temporal continuity between the Sturtian and Marinoan glaciations.

One interpretation of Figure 11 is that, as in the Snowball model, the Neoproterozoic continents alternated between an essentially ice-free and an essentially ice-covered state. What is perhaps surprising is that the available dates show little evidence of glaciations coexisting with areas where there are no glaciations, although this could simply be due to an emphasis on sampling material for dating from the glacial intervals alone. However, given that at any one location the glacial intervals are distinct, and separated by significant intervals of non-glacigenic sediments, another interpretation of Figure 11 is that the dated glacial intervals are globally diachronous. These uncertainties are expressed by the conclusion of Halverson *et al.* (2005 p. 1201) that 'many gaps and ambiguities in the record remain, including ... the duration and timing of the Marinoan and Sturtian glaciations'. They also suggest (Halverson *et al.* 2005, p. 1200) that because the Gaskiers glaciation appears to lack a global cap carbonate (see later) it may not have been a snowball event.

Some dating problems

U–Pb dates from the Pocatello Formation of Idaho suggest that the Sturtian glaciation there may have lasted until about 670 Ma (Fanning & Link 2004), rather than *c.* 700 Ma, again implying continuity of the Marinoan with the Sturtian and an arbitrary boundary between the two at *c.* 670–660 Ma.

Correlations based on published dates differ from author to author. For example, MacGabhann (2005) correlates the Numees of Namibia and the Port Askaig of Scotland with the Sturtian, whereas Hoffman (2005) makes them Marinoan.

There are also puzzling dating problems for type Marinoan glacial rocks in south Australia and Tasmania. The Marinoan Elatina Formation of the Adelaide Rift Complex has been correlated with the Cottons Breccia of King Island, Tasmania. This, in turn, is correlated with the Croles Hill Diamictite, which is underlain by rhyodacite that has a U–Pb date of 582 $\pm$ 4 Ma, or the same age as the much younger Gaskiers glaciation in Newfoundland (Fig. 11). These relationships could be accounted for by postulating that one of the type Marinoan glacial sequences is actually Gaskiers in age, or that the Cottons Breccia and Croles Hill Diamictite are not stratigraphic equivalents.

Kendall *et al.* (2006) obtained an Re–Os date of 643.0 $\pm$ 2.4 Ma from black shales from just above the top of the 'Sturtian' glaciation in south Australia

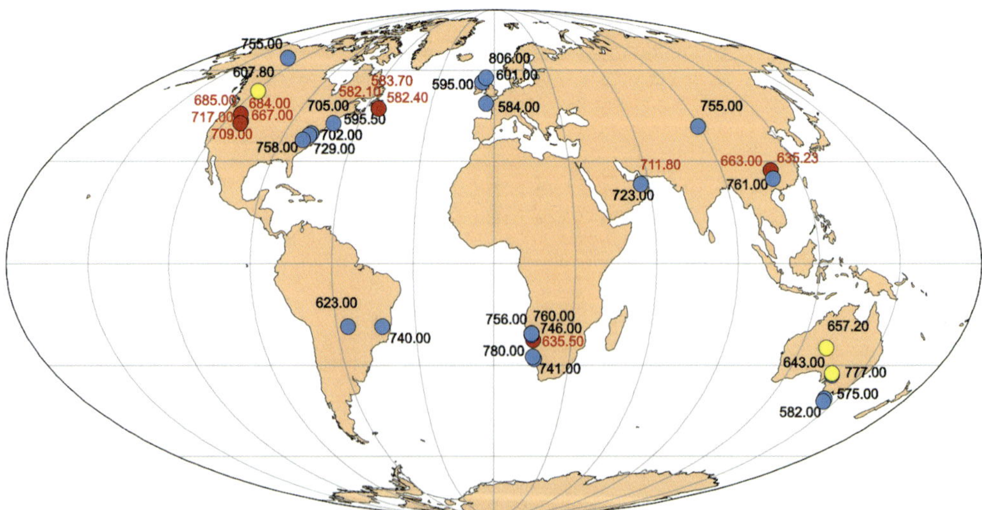

Fig. 12. U–Pb and Re–Os dates from Neoproterozoic basement, volcanics, tuffs and shales (see Table 2). U–Pb dates as blue circles; key U–Pb dates as red circles with red text. Re–Os dates as yellow circles.

and one of 657.2 ± 5.4 Ma from the presumed Sturtian correlative in central Australia (Areyonga). Kendall *et al.* (2006) suggest that the 'Sturtian' ice age was either markedly diachronous, and/or there was more than one 'Sturtian'-type glaciation. A logical alternative is that the dated rocks are actually 'Marinoan' in age.

In reality, the stratigraphy of the Neoproterozoic rift basins in SE Australia is complex (Preiss 2000; Jenkins pers. comm. 2007). According to Jenkins, there are at least six distinct glacial horizons there. Thus, to regard the glacial sequences as consisting of just two (Sturtian and Marinoan) is an oversimplification. As Corsetti & Lorentz (2006) point out, the Sturtian and Marinoan diamictites in Australia are not dated via U–Pb, and, as they show that cap carbonate style is misleading (see later), they suggest that the broad use of the terms Sturtian and Marinoan outside of Australia should be abandoned, but is unclear what should replace them.

There are diamictites in the North Atlantic region of East Greenland, NE Svalbard, north Norway and the northern British Isles (Halverson *et al.* 2004; MacGabhann 2005). Stratigraphic relationships and chemostratigraphy suggest that the diamictite pairs in Svalbard (Polarisbreen Group) and in Greenland (Tillite Group) are of Marinoan age, and that the younger of two of diamictites in Norway (Mortensnes Formation) is Gaskiers in age. However, none of these diamictites are isotopically dated. Sokolov & Fedonkin (1990) report isotopically undated Vendian diamictites from several areas in European Russia and the Urals, and also show the possible extent of the ice at the time.

The iridium concentration in the basal cap carbonates the eastern Congo has been used to estimate the duration of the Marinoan glacial period (Bodiselitsch *et al.* 2005), but the assumptions used have been criticized by Fairchild & Kennedy (2007).

In summary, the existence of two major Neoproterozoic glaciations – Sturtian and Marinoan – and two minor glaciations – Kaigas and Gaskiers – is in general agreement with the available age dates. However, some age dates are in conflict with this model and several important stratigraphic correlations are not yet supported by precise age dates.

Cap carbonates

Some Neoproterozoic diamictites immediately overlie thick platform carbonates, which contain ooids and thick displacive cement crusts and other features that are particularly suggestive of rapidly evaporating, warm conditions (Fairchild & Kennedy 2007). The abrupt transition to glacigenic deposition suggests a sudden change to a much colder climate.

Similarly rapid changes from a cold to a warm climate are suggested by the carbonates that generally cap Neoproterozoic glacial deposits. These cap carbonates may have high negative $\delta^{13}C$ values (see later and recent reviews by Shields 2005; Corsetti & Lorentz 2006; Fairchild & Kennedy 2007; Hoffman *et al.* 2007). Shields (2005) refers to the dolostone (= dolomite) units that lie on marine–terrestrial periglacial facies as cap dolostones. Any locally developed limestones that overlie cap dolostones are referred to as post-glacial limestones. Cap dolostones are remarkably uniform, pale, pinkish to buff-coloured, thin, generally less than 5 m, but locally up to 27 m thick, laterally extensive units of microcrystalline dolomite. Regionally extensive cap dolomite units cover vast areas but are sometimes discontinuous laterally, with preferential preservation on topographic highs and thicknesses dwindling to zero in both landward and seaward directions (James *et al.* 2001). Where several discrete glacial intervals occur in a single succession, each is commonly overlain by a cap carbonate that may constitute the only carbonate within an otherwise siliciclastic succession (Fairchild & Kennedy 2007). Shields (2005) notes that the term has been confusingly generalized in the literature 'to include sedimentary units that show no direct association with glacigenic facies'.

Most contacts between glacigenic deposits and cap carbonates are conformable, implying a genetic relation between them during deglaciation. Estimates of the duration of cap deposition range from as little as some tens of thousands of years to the elapsed time for the generation of several palaeomagnetic reversals observed within some cap carbonates, i.e. >1 Ma (e.g. Grotzinger & Knoll 1995; Kennedy 1996; Hoffman *et al.* 1998; Kennedy *et al.* 2001; Trindade *et al.* 2003; Shields 2005; Fairchild & Kennedy 2007). The longer timescale is also supported by stratigraphic studies that show complex interfingering relations of cap carbonate and siliciclastic intervals along a basin margin (Dyson & von der Borch 1994). Eyles & Januszczak (2004, pp. 23–27) suggest that those cap carbonates that appear to be separated by considerable thicknesses from the underlying diamictites, interpreted as of glacial origin, imply a time difference of up to several million years, although Shields (2005) would probably not regard these as cap carbonates. Because of their likely relationship to deglaciation, the differences in timescale estimates lead to different models for cap carbonate deposition (Shields 2005; Fairchild & Kennedy 2007; Hoffman *et al.* 2007). From a detailed study of the Keilberg cap dolostone in Namibia, Hoffman *et al.* (2007, p. 129) conclude that irrespective of the timescale involved, the evidence for its diachroneity and a base-level rise of over 0.5 km means that deposition of this dolostone took place at the same time as the melting of grounded ice sheets.

Cap carbonates assigned to the Marinoan and Sturtian glaciations have been distinguished from one another by a cladistic analysis based on differences in their sedimentary and isotopic features (Kennedy *et al.* 1998). These age assignments have not changed much in the light of later work, e.g. Hoffman (2005). Lithologically, Sturtian cap carbonates are generally dark, organic-rich, finely laminated carbonates with rhythmic laminae, and some contain roll-up structures. Marinoan cap carbonates are generally lighter in colour and include unusual features, such as sea-floor crystal fans (pseudomorphs of aragonite and/or barite), tubestones, sheetcrack cements and tepee-like structures (Corsetti & Lorentz 2006). These differences, together with carbon isotope behaviour (not discussed here), supposedly allow a Sturtian or a Marinoan age to be assigned to a given glacial–cap carbonate sequence where chronometric data are absent, as is commonly the case.

When combined with carbon isotope profiles (see later), cap carbonates would appear to be a potentially powerful tool for making global stratigraphic correlations of undated Neoproterozoic glacial sequences. However, Corsetti & Lorentz (2006) have examined in detail 10 cap carbonate successions that have some form of radiometric dating control. They show that cap carbonates with Sturtian characters occur in both the Sturtian and the Marinoan glacial intervals. Those with Marinoan characters occur in all three glacial intervals (Fig. 13). Thus, from a global viewpoint, Neoproterozoic cap carbonates showing a particular style of cap development cannot be used as a chronostratigraphic marker, even in cases where carbon isotope data are also available.

Gaskiers glacial deposits are not followed everywhere by an isotopically uniform cap carbonate. Thin Gaskiers cap carbonates are known from sporadic occurrences in the type area in Newfoundland (Myrow & Kaufman 1999; MacGabhann 2005), and the Hankalchough diamictite of North China (Xiao *et al.* 2004). The Gaskiers glaciation could have been diachronous and more typical of Phanerozoic-type ice ages.

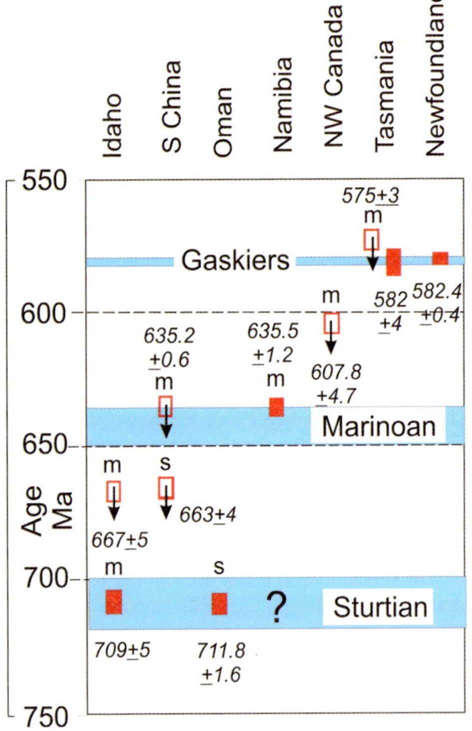

Fig. 13. Cap carbonate styles and dates, modified from Corsetti & Lorentz (2006 fig. 8). Filled red rectangles are the positions of dated samples (shown numerically with errors) within glacial sequences, whose height is the error. Open red rectangles are samples lying above glacial sequences or associated with cap carbonates. 's' and 'm' are the cap carbonate 'Sturtian' or 'Marinoan' styles said to distinguish the two, but both exist in Sturtian glacial sequences, and Marinoan-style cap carbonates exist in all three glacial sequences shown.

Banded iron formations

Banded iron formations, unknown anywhere else in rocks younger than approximately 1.7 Ga (Holland 2003), are found in the Sturtian glacials of western North America, Namibia and Australia, and in the Marinoan glacials of Namibia and Brazil (Hoffman 2005, table 1). They are regarded as one of the key, but not universal, features of Sturtian and Marinoan glacial sequences (Halverson *et al.* 2005). Although important locally as sedimentary deposits of hematite (Fe_2O_3) and occasional manganiferous deposits, they do not have a significant volume.

It is generally agreed that they indicate anoxic bottom waters, but the reasons for their reappearence after a time lapse of 1 Ga are not agreed. For some (e.g. Hoffman & Schrag 2002) they represent synchronous events reflecting a major geochemical change in the oceans and reinforce the idea of the Neoproterozoic glacial deposits as timelines. For others, these banded iron formations, together with the 'glacials', are both expressions of the rifting of continental crust during the break-up of Rodinia, the banded iron formations being associated with hydrothermal activity from incipient mid-ocean ridges, whose formation need not be synchronous (Eyles & Januszczak 2004). It is unclear why they do not reappear in Phanerozoic glacial deposits. Clearly, they make local rather than global stratigraphic markers.

Isotopic variations

In the absence of abundant radiometric data and of well-defined biostratigraphic zones in the Neoproterozoic, stratigraphic correlation in this time interval is largely dependent on using diamictites and cap carbonates, as discussed earlier. Large-scale glaciations of all ages are accompanied by isotopic changes, particularly in oxygen, strontium and carbon preserved in carbonates. Potentially, isotopic variations of strontium and carbon, in particular, have the power to refine this coarse subdivision. The $\delta^{18}O$ and $^{87}Sr/^{86}Sr$ values in carbonates are commonly altered by meteoric diagenesis and by dolomitization (Halverson *et al.* 2005). Shields (2005) and Jaffrés *et al.* (2007) discuss $\delta^{18}O$ variations and their possible implications.

Halverson *et al.* (2007) have presented a composite marine chemostratigraphic record of global Neoproterozoic $^{87}Sr/^{86}Sr$ variations that is also linked to high-resolution $\delta^{13}C$ data in carbonates (Halverson *et al.* 2005; Halverson 2006). Halverson *et al.* (2007) conclude that the record remains poorly resolved and time-calibrated, but that all changes in $^{87}Sr/^{86}Sr$ correspond to major biogeochemical or climatic events. The irregular fluctuations of the $^{87}Sr/^{86}Sr$ ratio familiar on a timescale of about 50 Ma in the Phanerozoic record can be used for high-resolution stratigraphy and stratigraphic correlation (e.g. McArthur *et al.* 2004), but Halverson *et al.* (2007) note that it will require much more work before the Neoproterozoic strontium isotope record has this level of precision.

Carbon isotopes

The present concensus is that $\delta^{13}C$ ratios in Neoproterozoic carbonates that are clearly diagenetically altered and dolomitized carbonates may still preserve their original $\delta^{13}C$ composition because the high concentration of C in carbonate rocks, relative to meteoric fluids, makes the $\delta^{13}C$ ratio much more resistant to diagenesis than $\delta^{18}O$ (Halverson *et al.* 2005). However, some cap carbonates (rather than Neoproterozoic carbonates in general) show large lateral and vertical variations both in $\delta^{18}O$ and $\delta^{13}C$, suggesting significant diagenetic modification of the isotopic ratios, possibly by dolomitization (Jiang *et al.* 2003).

Generally, $\delta^{13}C$ ratios vary quite markedly, before, during and after a glaciation. For the Neoproterozoic, the carbon isotope variations have the potential to act as global timelines (Knoll *et al.* 1986; Knoll & Walter 1992; Pyle *et al.* 2004). In general, the $\delta^{13}C$ ratio declines quite rapidly from values in excess of 5 to values more negative than −5, then makes an even more rapid recovery to small negative or positive values after the glaciation (Halverson 2006). For the ratios to act as timelines, the Neoproterozoic oceans need to have been part of a global ocean and the variations need to be globally contemporaneous. Comparison with the changes that took place in the late Ordovician–early Silurian 'icehouse' interval, from *c.* 445 to *c.* 425 Ma (e.g. Page *et al.* 2007), provide support for the use of Neoproterozoic $\delta^{13}C$ variations associated with glacial periods as global chronostratigraphic markers. Nevertheless, Fairchild & Kennedy (2007) follow Melezhik *et al.* (2001) in arguing that the diagenetic complexity of the carbon isotope results, and the difficulty of correlating Neoproterozoic rocks, means that Sr isotope stratigraphy probably provides the best potential for unique chemostratigraphic correlations.

Palaeomagnetism

Continental palaeomagnetic data are important for all global stratigraphic work for two reasons. In the Neoproterozoic they are fundamental because the low-latitude poles of some glacial sequences gave rise to the notion of the Snowball Earth. In general, pole positions allow continents to be oriented correctly and placed in their former latitudinal position, but available tectonic and palaeomagnetic data give rise to non-unique models for the break-up and evolution of Rodinia and the formation of Gondwana (e.g. Collins & Pisarevsky 2005).

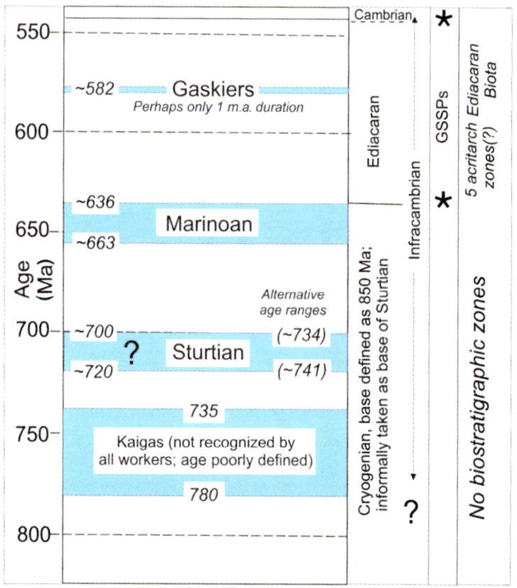

Fig. 14. Summary of Neoproterozoic stratigraphy.

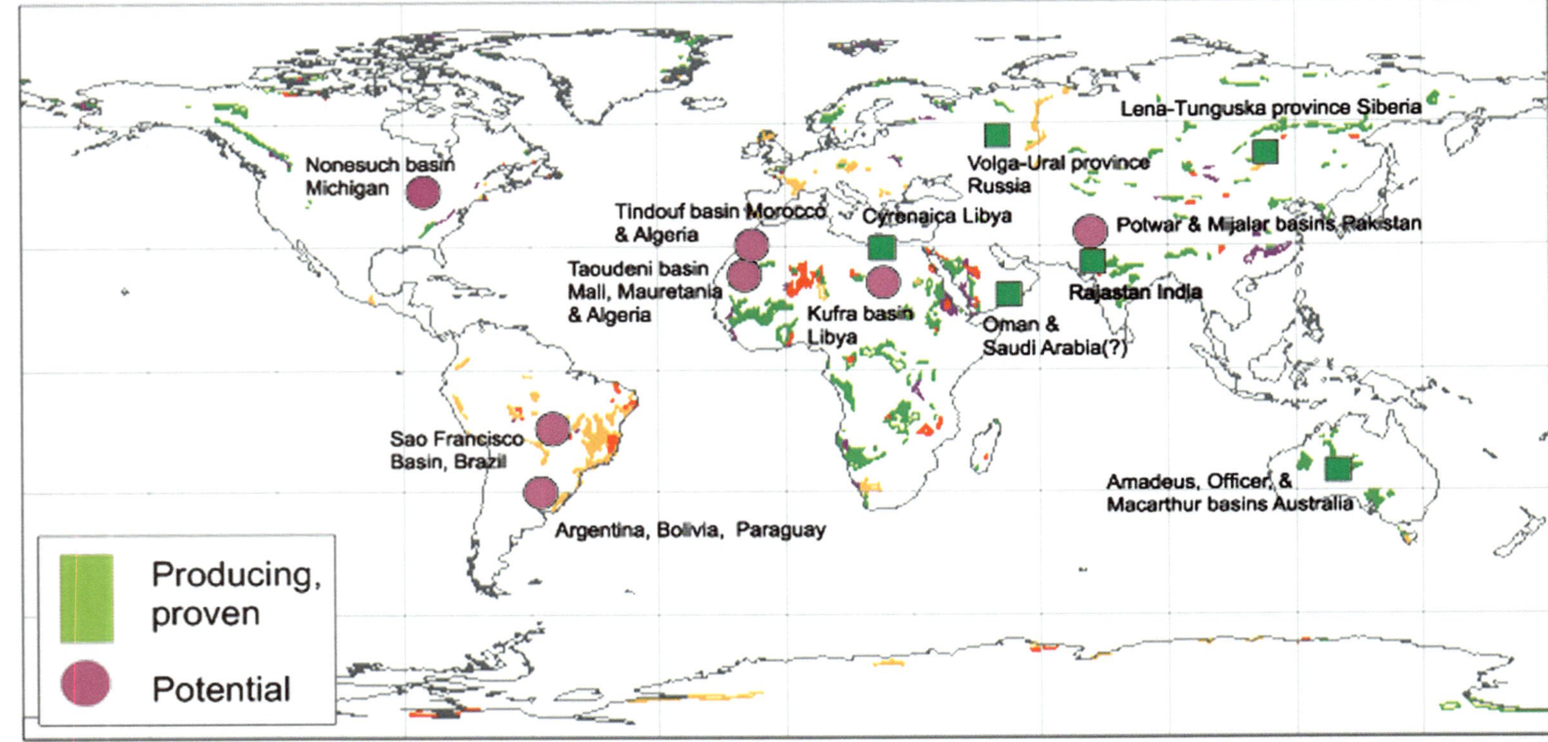

Fig. 15. Neoproterozoic petroleum systems and rock outcrops. The systems are modified from a figure in the conference booklet; the outcrops are those in Figure 2. Cylindrical equidistant projection.

The second reason for the importance of palaeomagnetic data is that polarity reversals provide a bar code that potentially allows high-precision stratigraphic correlation. In theory, polarity reversals will eventually allow precise Neoproterozoic correlations to be made. Some polarity reversals have been documented (e.g. Maloof *et al.* 2006), but the currently available poles are inadequate for palaeomagnetic correlations to be attempted.

A general summary of Neoproterozoic stratigraphy is given in Figure 14.

Hydrocarbon generation and potential hydrocarbon reservoirs

Sources and productivity

According to Butterfield (2007), a major impact of the early evolution of eumetazoans (i.e. all animals except sponges) in the early Ediacaran, together with further changes in the later Ediacaran, would been a rapidly increased standing biomass, which in turn would have had a profound impact on the biogeochemistry of the time. Eumetazoans also caused a major shift in the source of marine primary productivity – from predominantly cyanobacteria in the Proterozoic to predominantly (eukaryotic) algae in the Phanerozoic (Summons *et al.* 1999). All of these changes are presumably reflected in the hydrocarbons originally present in any Neoproterozoic source rocks.

Like Phanerozoic glaciations, the Neoproterozoic glaciations commonly ended with a highstand in sea level and the deposition of organic-rich black shales. These black shales together with other Neoproterozoic black shales are potential hydrocarbon source rocks.

One model for cap carbonate formation suggests that they may be a record of gas hydrate destabilization, largely from terrestrial permafrost (Kennedy *et al.* 2001; Jiang *et al.* 2003) rather than from marine sources. Details of the sedimentary structures in a 5 m-thick cap carbonate in south China are given in the caption to figure 1 of Jiang *et al.* (2003). The distinctive sedimentary structures there are collectively known only in methane seep environments, leading to the identification of cap carbonates as submarine sites of Neoproterozoic methane seeps. Jiang *et al.* (2003) estimate that, in order to form, the Marinoan cap carbonates must have taken place during 'the largest hydrate-dissociation event in Earth's history' (Jiang *et al.* 2003, p. 824). How any of this methane could have been trapped and preserved to the present day is unclear. Shields (2005) has criticized this model because it is difficult to understand how material from point sources, such as methane seeps, could distribute the ions required for cap carbonates homogeneously around the globe.

Reservoirs

The erosion at the base of ice sheets depends on the nature of the bedrock. If it is hard, eskers may develop, but, if the bedrock is soft, then so-called 'tunnel valleys' (e.g. Lonergan *et al.* 2006) may be cut by the ice. These can be tens of metres or more in depth, several kilometres long, with V-shaped rather than U-shaped profiles, and are potential reservoirs for hydrocarbon storage, whether derived from Neoproterozoic black shales or from other source rocks.

The tectonic setting of many Neoproterozoic sequences is attributed to rifts or passive margins (e.g. Eyles & Januszczak 2004), some of which contain producing horizons or potential producers (Fig. 15). Phanerozoic passive margins, and former Phanerozoic passive margins, together with rifts, contain many major present-day hydrocarbon reservoirs. Thus, significant Neoproterozoic oil and gas fields may remain to be discovered, despite the fact that, compared with their Phanerozoic equivalents, many more will have been destroyed by tectonism and erosion.

Conclusions

Until the stratigraphic definitions for the base of the Cryogenian and for the stages in the early Cambrian are agreed, it is not possible to suggest how the Infracambrian can be precisely defined in terms of an international stratigraphic scale. Thus, the Infracambrian can be retained as a useful stratigraphic name for the later Neoproterozoic rocks of northern Gondwana until such definition is possible.

CE palaeomagnetic data from Australia, Laurentia and Oman all suggest a low-latitude position for these continental areas during at least one glaciation. These results provide the basic evidence for the Snowball Earth model. Palaeomagnetism also shows polarity reversals in some cap carbonates, suggesting that the duration of cap carbonate formation may at times exceed one million years.

Although diamictites per se have little palaeoclimatic significance, the close association of most, or possibly all, CE diamictites with glacigenic sediments suggests a direct connection between the two. The diamictites allow the subdivision of the glacial intervals into Kaigas (minor), Sturtian, Marinoan and Gaskiers.

However, precise correlations from the type areas of the Sturtian and Marinoan in Australia to other continents are difficult to make, in part because the detailed stratigraphy in Australia is disputed. In particular, the age of the Elatina Formation is not well established: it could be of Marinoan or of Gaskiers age.

The distribution of age dates suggests that the Gaskiers is very short, but those that are currently regarded as distinct Marinoan and Sturtian glaciations may represent a glacial continuum during which ice was always present somewhere on the planet. The presence of significant non-glacigenic successions in the same continuum shows that ice was not present everywhere, and may indicate significant diachroneity of the ice distribution. Re–Os dating of black shales has the potential to revolutionize the dating of the CE interval.

The stratigraphic usefulness of cap carbonates may be limited, or at least requires further investigation. 'Sturtian'-style cap carbonates, defined by specific sedimentary features, are found in Marinoan cap carbonates, and 'Marinoan'-style cap carbonates are also found in Sturtian and Gaskiers cap carbonates.

Unlike chemostratigraphic correlation in parts of the Phanerozoic, chemostratigraphic correlations for the CE interval are commonly relatively coarse or ambiguous. Strontium isotopes appear the most likely to provide the first generally available chemostratigraphic correlation method, but eventually the most likely precise correlation tool will involve a combination of Sr and $\delta^{13}C$ variations with Re–Os dates.

Globally, the evolution of Neoproterozoic organisms was at a very slow rate. No effects of severe cold can be recognized in the Australian Neoproterozoic fossil record and elsewhere. How this could have happened on a Snowball Earth is unclear.

The author thanks the Organizing Committee of the conference for their invitation to write this paper. Many colleagues have made helpful suggestions and criticisms, among them are A. Whitham, A. Tripati, N. Stubblefield, A. Page and P. Gibbard. He also thanks the librarians of the Department of Earth Sciences for their help in tracking down several obscure references, and two referees for their pithy but helpful comments.

References

ALLEN, P. A., BOWRING, S. *ET AL.* 2002. Chronology of Neoproterozoic glaciations: new insights from Oman. *In*: KNOPER, M. & CAINCROSS, B. (eds) *16th International Sedimentological Congress Abstracts Volume, Rand Afrikaans University, Johanesburg, South Africa*, 7–8.

ALLEN, P. A., LEATHER, J. & BRASIER, M. D. 2004. The Neoproterozoic Fiq glaciation and its aftermath, Huqf Supergroup of Oman. *Basin Research*, **16**, 507–534.

ALLSOPP, H., KÖSTLIN, E. O., WELKE, H. J., BURGER, A. J., KRÖNER, A. & BLIGNAULT, H. J. 1979. Rb–Sr and U–Pb geochronology of Late Precambrian–early Palaeozoic igneous activity in the Richtersveld (South Africa) and southern South West Africa. *Transactions of the Geological Society of South Africa*, **82**, 185–204.

ANDERSON, M. M. & KING, A. F. 1981. Precambrian tillites of the Conception Group on the Avalon Peninsula, southeastern Newfoundland. *In*: HAMBREY, M. J. & HARLAND, W. B. (eds) *Earth's Pre-Pleistocene Glacial Record.* Cambridge University Press, Cambridge, 760–767.

BAILEY, C. M. & PETERS, S. E. 1998. Glacially influenced sedimentation in the late Neoproterozoic Mechum River formation, Blue Ridge province, Virginia. *Geology*, **26**, 623–626.

BINGEN, B., GRIFFEN, W. L., TORSVIK, T. H. & SAEED, A. 2005. Timing of Late Neoproterozoic glaciation on Baltica constrained by detrital zircon geochronology in the Hedmark Group, south-east Norway. *Terra Nova*, **17**, 250–258.

BODISELITCH, B., KOEBERL, C., MASTER, S. & REIMOLD, W. U. 2005. Estimating duration and intensity of Neoproterozoic Snowball Glaciations from Ir anomalies. *Science*, **308**, 239–242.

BRASIER, M. D. & SHIELDS, G. 2000. Neoproterozoic chemostratigraphy and correlation of the Port Askaig glaciation, Dalradian Supergroup of Scotland. *Journal of the Geological Society, London*, **157**, 909–914.

BRASIER, M. D., CORFIELD, R. M., DERRY, L. A., ROZANOV, A. Y. & ZHURAVLEV, A. Y. 1994. Multiple delta ^{13}C excursions spanning the Cambrian explosion to the Botomian crisis in Siberia. *Geology*, **22**, 455–458.

BUTTERFIELD, N. J. 2007. Macroevolution and macroecology through deep time. *Palaeontology*, **50**, 41–55.

BUTTERFIELD, N. J. & CHANDLER, F. W. 1992. Paleoenvironmental distribution of Proterozoic microfossils with an example from the Agu Bay Formation, Baffin Island. *Palaeontology*, **35**, 943–957.

CALVER, C. R. & WALTER, M. R. 2000. The late Neoproterozoic Grassy Group of King Island, Tasmania: correlation and palaeogeographic significance. *Precambrian Research*, **100**, 299–312.

CALVER, C. R., BLACK, L. P., EVERARD, J. L. & SEYMOUR, D. B. 2004. U–Pb zircon age constraints on late Neoproterozoic glaciation in Tasmania. *Geology*, **32**, 893–896.

CAWOOD, P. A. 2005. Terra Australis Orogen: rodinia breakup and development of the Pacific and Iapetus margins of Gondwana during the Neoproterozoic and Palaeozoic. *Earth-Science Reviews*, **69**, 246–279.

COLLINS, A. S. & PISAREVSKY, S. A. 2005. Amalgamating eastern Gondwana: the evolution of the Circum-Indian Orogens. *Earth-Science Reviews*, **71**, 229–270.

CONDON, D. J. & PRAVE, A. R. 2000. Two from Donegal: Neoproterozoic glacial episodes on the northeast margin of Laurentia. *Geology*, **28**, 951–954.

CONDON, D. J., PRAVE, A. R. & BENN, D. I. 2002. Neoproterozoic glacial-rainout intervals: Observations and implications. *Geology*, **30**, 35–38.

CONDON, D., ZHU, M., BOWRING, S., WANG, W., YANG, A. & JIN, Y. 2005. U–Pb Ages from the

Neoproterozoic Doushantuo Formation, China. *Science*, **308**, 95–98.

CORSETTI, F. A. & LORENTZ, J. 2006. On Neoproterozoic cap carbonates as chronostratigraphic markers. *In*: XIAO, S. & KAUFMAN, A. J. (eds) *Neoproterozoic Geobiology and Palaeobiology*. Springer, Berlin, 273–294.

CORSETTI, F. A., OLCOTT, A. N. & BAKERMANS, C. 2006. The biotic response to Neoproterozoic snowball Earth. *Palaeogeography, Palaeoclimatology, Palaeoecology*, **232**, 114–130.

CREASER, R. A. 2003. A review of the Rhenium–Osmium (Re–Os) isotope system with application to organic-rich sedimentary rocks. *In*: LENZ, D. R. (ed.) *Geochemistry of Sediments and Sedimentary Rocks: Evolutionary Considerations to Mineral Deposit-forming Environments*. Geological Association of Canada, St John's, Newfoundland, 79–83.

CREASER, R. A., SANNIGRAHI, P., CHACKO, T. & SELBY, D. 2002. Further evaluation of the Re–Os geochronometer in organic-rich sedimentary rocks: a test of hydrocarbon maturation effects in the Exshaw Formation, Western Canada Sedimentary Basin. *Geochimica et Cosmochimica Acta*, **66**, 3441–3452.

DE ALVARENGA, C. J. S., SANTOS, R. V., DARDENNE, M. A., DANTAS, E. L., BROD, E. R. & GIOIA, S. M. C. L. 2003. C, O and Sr isotope evidence of Sturtian and Marinoan glaciations in Brazil. *In*: FRIMMEL, H. E. (ed.) *III International Colloquium Vendian–Cambrian of W-Gondwana, Programme and Extended Abstracts*, 6–9.

DEMPSTER, T. J., ROGERS, J. *ET AL*. 2002. Timing of deposition, orogenesis and glaciation within the Dalradian rocks of Scotland: constraints from U–Pb zircon ages. *Journal of the Geological Society, London*, **159**, 83–94.

DUNN, P. R., THOMSON, B. P. & RANKAMA, K. 1971. Late Pre-Cambrian glaciation in Australia as a stratigraphic boundary. *Nature*, **231**, 498–502.

DYSON, I. A. & VON DER BORCH, C. C. 1994. Sequence stratigraphy of an incised-valley fill: the Neoproterozoic Seacliff Sandstone, Adelaide Geosyncline, South Australia. *In*: DALRYMPLE, R. W., BOYD, R. & ZAITLIN, B. A. (eds) *Incised-valley Systems; Origin and Sedimentary Sequences*. SEPM, Special Publications, **51**, 209–222.

ETIENNE, J. L., ALLEN, P., LE GUERROUE, E. & RIEU, R. 2006. *Snowball Earth 2006*. ETH, Zurich.

ETIENNE, J. L., ALLEN, P. A., RIEU, R. & LE GUERROUE, E. 2007. Neoproterozoic glaciated basins: a critical review of the Snowball Earth hypothesis by comparison with Phanerozoic glaciations. *In*: HAMBREY, M. J., CHRISTOFFERSEN, P., GLASSER, N. F. & HUBBARD, B. (eds) *Glacial Processes and Products*. International Association of Sedimentologists, Special Publications, **39**, 343–399.

EVANS, D. A. D. 2000. Stratigraphic, geochronological, and palaeomagnetic constraints upon the Neoproterozoic climatic paradox. *American Journal of Science*, **300**, 347–433.

EYLES, C. H. 2005. Frozen in time: concepts of global glaciation from die Eiszeit (1837) to Snowball Earth (1998). *Geoscience Canada*, **31**, 157–166.

EYLES, N. & JANUSZCZAK, N. 2004. 'Zipper-rift': a tectonic model for Neoproterozoic glaciations during the breakup of Rodinia after 750 Ma. *Earth-Science Reviews*, **65**, 1–73.

EYLES, C. H., EYLES, N. & GREY, K. 2007. Palaeoclimate implications from deep drilling of Neoproterozoic strata in the Officer Basin and Adelaide Rift Complex of Australia; a marine record of wet-based glaciers. *Palaeogeography, Palaeoclimatology, Palaeoecology*, **248**, 291–312.

FAIRCHILD, I. J. & KENNEDY, M. J. 2007. Neoproterozoic glaciation in the Earth System. *Journal of the Geological Society, London*, **164**, 1–27.

FANNING, C. M. & LINK, P. K. 2004. U–Pb SHRIMP ages of Neoproterozoic (Sturtian) glaciogenic Pocatello Formation, southeastern Idaho. *Geology*, **32**, 881–884.

FRIMMEL, H. E., KLOTZLI, U. S. & SIEGFRIED, P. R. 1996. New Pb–Pb single zircon age constraints on the timing of Neoproterozoic glaciation and continental break-up in Namibia. *Journal of Geology*, **104**, 459–469.

GRADSTEIN, F. M., OGG, J. G. & SMITH, A. G. 2004. *A Geologic Time Scale 2004*. Cambridge University Press, Cambridge.

GREY, K. 2005. *Ediacaran Palynology of Australia*. Memoirs of the Association of Australasian Palaeontologists, **31**.

GREY, K. & CORKERON, M. 1998. Late Neoproterozoic stromatolites in glacigenic successions of the Kimberley region, Western Australia: evidence for a younger Marinoan glaciation. *Precambrian Research*, **92**, 65–87.

GROTZINGER, J. P. & KNOLL, A. H. 1995. Anomalous carbonate precipitates: is the Precambrian the key to the Permian? *Palaios*, **10**, 578–596.

GUERROT, C. & PEUCAT, J. J. 1990. U–Pb Geochronology of the Upper Proterozoic Cadomian Orogeny in the northern Armorican Massif, France. *In*: D'LEMOS, R. S., STRACHAN, R. A. & TOPLEY, C. G. (eds) *The Cadomian Orogeny*. Geological Society, London, Special Publications, **51**, 13–26.

HALLIDAY, A. N., GRAHAM, C. M., AFTALION, M. & DYMOKE, P. 1989. The depositional age of the Dalradian Supergroup: U–Pb and Sm–Nd isotopic studies of the Tayvallich Volcanics, Scotland. *Journal of the Geological Society, London*, **146**, 3–6.

HALVERSON, G. P. 2006. A Neoproterozoic chronology. *In*: XIAO, S. & KAUFMAN, K. J. (eds) *Neoproterozoic Geobiology*. Kluwer Academic, Dordrecht, 231–271.

HALVERSON, G. P., MALOOF, A. C. & HOFFMAN, P. F. 2004. The Marinoan glaciation (Neoproterozoic) in northeast Svalbard. *Basin Research*, **16**, 297–324.

HALVERSON, G. P., HOFFMAN, P. F., SCHRAG, D. P., MALOOF, A. C. & RICE, A. H. N. 2005. Toward a Neoproterozoic composite carbon-isotope record. *Geological Society of America Bulletin*, **117**, 1181–1207.

HALVERSON, G. P., DUDÁS, F. Ö., MALOOF, A. C. & BOWRING, S. A. 2007. Evolution of the $^{87}Sr/^{86}Sr$ composition of Neoproterozoic seawater. *Palaeogeography, Palaeoclimatology, Palaeoecology*, **256**, 103–129.

HAMBREY, M. J. & HARLAND, W. B. 1981. *Earth's Pre-Pleistocene Glacial Record*. Cambridge University Press, Cambridge.

HARLAND, W. B. 1964. Critical evidence for a great infra-Cambrian ice age. *Geologisches Rundschau*, **54**, 45–61.

HOFFMAN, P. F. 2005. On Cryogenian (Neoproterozoic) ice-sheet dynamics and the limitations of the glacial sedimentary record 28th DeBeers Alex. Du Toit Memorial Lecture, 2004. *South African Journal of Geology*, **108**, 557–577.

HOFFMAN, P. F. & SCHRAG, D. P. 2002. The snowball Earth hypothesis: testing the limits of global change. *Terra Nova*, **14**, 129–155.

HOFFMAN, P. F., KAUFMAN, A. J., HALVERSON, G. P. & SCHRAG, D. P. 1998. A Neoproterozoic snowball Earth. *Science*, **281**, 1342–1346.

HOFFMAN, P. F., HALVERSON, G. P., DOMACK, E. W., HUSSON, J. M., HIGGINS, J. A. & SCHRAG, D. P. 2007. Are basal Ediacaran (635 Ma) post-glacial 'cap dolostones' diachronous? *Earth and Planetary Science Letters*, **258**, 114–131.

HOFFMANN, K. H. & PRAVE, A. R. 1996. A preliminary note on a revised subdivision and regional correlation of the Otavi Group based on glaciogenic diamictites and associated cap dolostones. *Communications of the Geological Survey of Namibia*, **11**, 77–82.

HOFFMANN, K. H., CONDON, D. J., BOWRING, S. A. & CROWLEY, J. L. 2004. U–Pb zircon date from the Neoproterozoic Ghaub Formation, Namibia: constraints on Marinoan glaciation. *Geology*, **32**, 817–820.

HOLLAND, H. D. 2003. The early history of seawater. *In*: ELDERFIELD, H., HOLLAND, H. D. & TUREKIAN, K. K. (eds) *The Oceans and Marine Geochemistry*. Treatise on Geochemistry, **6**, 625.

HUNTLEY, J. W., XIAO, S. & KOWALEWSKI, M. 2006. 1.3 Billion years of acritarch history: an empirical morphospace approach. *Precambrian Research*, **144**, 52–68.

JAFFRÉS, J. B. D., SHIELDS, G. A. & WALLMANN, K. 2007. The oxygen isotope evolution of seawater: a critical review of a long-standing controversy and an improved geological water cycle model for the past 3.4 billion years. *Earth-Science Reviews*, **83**, 83–122.

JAMES, N. P., NARBONNE, G. M. & KYSER, T. K. 2001. Late Neoproterozoic cap carbonates: mackenzie Mountains northwestern Canada: precipitation and global glaciation. *Canadian Journal of Earth Sciences*, **38**, 1229–1262.

JIANG, J., KENNEDY, M. J. & CHRISTIE-BLICK, N. 2003. Stable isotopic evidence for methane seeps in Neoproterozoic postglacial cap carbonates. *Nature*, **426**, 822–826.

KELLER, B. M. 1979. Precambrian stratigraphic scale of the USSR. *Geological Magazine*, **116**, 419–429.

KENDALL, B. S., CREASER, R. A., ROSS, G. M. & SELBY, D. 2004. Constraints on the timing of Marinoan 'Snowball Earth' glaciation by ^{187}Re–^{187}Os dating of a Neoproterozoic, post-glacial black shale in Western Canada. *Earth and Planetary Science Letters*, **222**, 729–740.

KENDALL, B., CREASER, R. A. & SELBY, D. 2006. Re–Os geochronology of postglacial black shales in Australia: constraints on the timing of 'Sturtian' glaciation. *Geology*, **34**, 729–732.

KENNEDY, M. J. 1996. Stratigraphy, sedimentology, and isotopic geochemistry of Australian Neoproterozoic postglacial cap dolostones: deglaciation, δ^{13}C excursions, and carbonate precipitation. *Journal of Sedimentary Research*, **66**, 1050–1064.

KENNEDY, M. J., RUNNEGAR, B., PRAVE, A. R., HOFFMANN, K. H. & ARTHUR, M. 1998. Two or four Neoproterozoic glaciations? *Geology*, **26**, 1059–1063.

KENNEDY, M. J., CHRISTIE-BLICK, N. & SOHL, L. E. 2001. Are Proterozoic cap carbonates and isotopic excursions a record of gas hydrate destabilization following Earth's coldest intervals? *Geology*, **29**, 443–446.

KHOMENTOVSKII, V. V. & KARLOVA, G. A. 2005. The Tommotian Stage Base as the Cambrian lower boundary in Siberia. [Translated from *Stratigrafiya, Geologicheskaya Korrelyatsiya*.] *Stratigraphy and Geological Correlation*, **13**, 21–34.

KILNER, B., MAC NIOCAILL, C. & BRASIER, M. 2005. Low-latitude glaciation in the Neoproterozoic of Oman. *Geology*, **33**, 413–416.

KIRSCHVINK, J. L. 1992. Late Proterozoic low-latitude global glaciation: the Snowball Earth. *In*: SCHOPF, J. W. & KLEIN, C. (eds) *The Proterozoic Biosphere: A Multidisciplinary Study*. Cambridge University Press, Cambridge, 51–52.

KNOLL, A. H. & WALTER, M. R. 1992. Latest Proterozoic stratigraphy and Earth history. *Nature*, **356**, 673–678.

KNOLL, A. H., HAYES, J. M., KAUFMAN, A., SWETT, K. & LAMBERT, I. B. 1986. Secular variation in carbon isotope ratios from Upper Proterozoic successions of Svalbard and East Greenland. *Nature*, **321**, 832–838.

KNOLL, A. H., WALTER, M. R., NARBONNE, G. M. & CHRISTIE-BLICK, N. 2004. A new period for the geologic time scale. *Science*, **305**, 621–622.

KNOLL, A. H., WALTER, M. R., NARBONNE, G. M. & CHRISTIE-BLICK, N. 2006. The Ediacaran Period: a new addition to the geologic time scale. *Lethaia*, **39**, 13–30.

LANDING, E. L. 1994. Precambrian–Cambrian boundary global stratotype ratified and a new perspective of Cambrian time. *Geology*, **22**, 179–182.

LI, Z. X., LI, X. H., KINNY, P. D., WANG, J., ZHANG, S. & ZHOU, H. 2003. Geochronology of Neoproterozoic syn-rift magmatism in the Yangtze Craton, South China and correlations with other continents: evidence for a mantle superplume that broke up Rodinia. *Precambrian Research*, **122**, 85–109.

LONERGAN, L., MAIDMENT, S. C. R. & COLLIER, J. S. 2006. Pleistocene subglacial tunnel valleys in the central North Sea basin: 3-D morphology and evolution. *Journal of Quaternary Science*, **21**, 891–903.

LUND, K., ALEINIKOFF, J. N., EVANS, K. V. & FANNING, C. M. 2003. SHRIMP U–Pb geochronology of neoproterozoic Windermere Supergroup, central Idaho: implications for rifting of western Laurentia and synchroneity of Sturtian glacial deposits. *Geological Society of America Bulletin*, **115**, 349–372.

MACGABHANN, B. A. 2005. Age constraints on Precambrian glaciations and the subdivision of Neoproterozoic time. *Annual Report of IGCP Project 512*, 13.

MALOOF, A. C., HALVERSON, G. P., KIRSCHVINK, J. L., SCHRAG, D. P., WEISS, B. P. & HOFFMAN, P. F. 2006. Combined palaeomagnetic, isotopic, and stratigraphic evidence for true polar wander from the Neoproterozoic Akademikerbreen Group, Svalbard, Norway. *Geological Society of America Bulletin*, **118**, 1099–1124.

McArthur, J. M., Mutterlose, J., Price, G. D., Rawson, P. F., Ruffell, A. & Thirlwall, M. F. 2004. Belemnites of Valanginian, Hauterivian and Barremian age: Sr-isotope stratigraphy, composition ($^{87}Sr/^{86}Sr$, $\delta^{13}C$, $\delta^{18}O$, Na, Sr, Mg,), and palaeo-oceanography. *Palaeogeography, Palaeoclimatology, Palaeoecology*, **202**, 253–272.

McCall, G. J. H. 2006. The Vendian (Ediacaran) in the geological record: enigmas in geology's prelude to the Cambrian explosion. *Earth-Science Reviews*, **77**, 1–229.

McCay, G. A., Prave, A. R., Alsop, G. I. & Fallick, A. E. 2006. Glacial trinity: neoproterozoic Earth history within the British–Irish Caledonides. *Geology*, **34**, 909–912.

Melezhik, V. A., Gorokhov, I. M., Kuznetsov, A. B. & Fallick, A. E. 2001. Chemostratigraphy of Neoproterozoic carbonates: implications for 'blind dating'. *Terra Nova*, **13**, 1–11.

Menchikoff, N. 1949. Quelques traits de l'histoire geologique du Sahara occidentail. *Annales Hebert et Haug*, **7**, 303–325.

Myrow, P. M. & Kaufman, A. J. 1999. A newly discovered cap carbonate above Varanger-age glacial deposits in Newfoundland, Canada. *Journal of Sedimentary Research*, **69**, 784–793.

Narbonne, G. M. 1998. The Ediacaran biota: a terminal Proterozoic experiment in the evolution of life. *GSA Today*, **8**, 1–7.

Narbonne, G. M. 2005. The Ediacara biota; Neoproterozoic origin of animals and their ecosystems. *Annual Review of Earth and Planetary Sciences*, **33**, 421–442.

Narbonne, G. M. & Gehling, J. G. 2003. Life after snowball: the oldest complex Ediacaran fossils. *Geology*, **31**, 27–30.

Noble, S. R., Hyslop, E. K. & Highton, A. J. 1996. High-precision U–Pb monazite geochronology of the *c*. 806 Ma Grampian Shear Zone and the implications for the evolution of the Central Highlands of Scotland. *Journal of the Geological Society, London*, **153**, 511–514.

Page, A., Zalasiewicz, J., Williams, M. & Popov, L. 2007. Were transgressive black shales a negative feedback modulating glacioeustasy in the Early Palaeozoic greenhouse? *In*: Williams, M., Haywood, A. M., Gregory, F. J. & Schmidt, D. N. (eds) *Deep-Time Perspectives on Climate Change: Marrying the Signal From Computer Models and Proxies*. The Micropalaeontological Society, Special Publications. Geological Society, London, 123–156.

Park, J. K. 1997. Palaeomagnetic evidence for low-latitude glaciation during deposition of the Neoproterozoic Rapitan Group, Mackenzie Mountains, N.W.T., Canada. *Canadian Journal of Earth Sciences*, **34**, 34–49.

Plumb, K. A. 1991. New Precambrian time scale. *Episodes*, **14**, 139–140.

Preiss, W. V. 2000. The Adelaide geosyncline of South Australia and its significance in Neoproterozoic continental reconstruction. *Precambrian Research*, **100**, 21–63.

Pruvost, P. 1951. L'Infracambrien. *Bulletin de la Société Belge Geologie Palaeontologie et Hydrologie*, **60**, 43–65.

Pyle, L. J., Narbonne, G. M., James, N. P., Dalrymple, R. W. & Kaufmann, A. J. 2004. Integrated Ediacaran chronostratigraphy Wernecke Mountains, northwestern Canada. *Precambrian Research*, **132**, 1–27.

Ravizza, G. E. & Zachos, J. C. 2003. Records of Cenozoic ocean chemistry. *In*: Holland, H. D. & Turekian, K. K. (eds) *The Oceans and Marine Geochemistry*. Treatise on Geochemistry, **6**, 551–581.

Robb, L. J., Knoll, A. H., Plumb, K. A., Shields, G. A., Strauss, H. & Veiser, J. 2004. The Precambrian. *In*: Gradstein, F. M., Ogg, J. G. & Smith, A. G. (eds) *A Geologic Timescale 2004*. Cambridge University Press, Cambridge, 129–140.

Ross, G. M. & Villeneuve, M. E. 1997. U–Pb geochronology of stranger stones in Neoproterozoic diamictites, Canadian Cordillera: implications for provenance and ages of deposition. *Geological Survey of Canada Current Research*, **1997-F**, 141–155.

Ross, G. M., Bloch, J. D. & Krouse, H. R. 1995. Neoproterozoic strata of the Southern Canadian Cordillera and the isotopic evolution of seawater sulfate. *Precambrian Research*, **73**, 71–99.

Selby, D. & Creaser, R. A. 2005. Direct radiometric dating of hydrocarbon deposits using rhenium–osmium isotopes. *Science*, **308**, 1293–1295.

Shergold, J. H. & Cooper, R. A. 2004. The Cambrian period. *In*: Gradstein, F., Ogg, J. & Smith, A. G. (eds) *A Geologic Time Scale 2004*. Cambridge University Press, Cambridge, 147–164.

Shields, G. A. 2005. Neoproterozoic cap carbonates: a critical appraisal of existing models and the plumeworld hypothesis. *Terra Nova*, **17**, 299–310.

Smith, A. G. & Pickering, K. T. 2003. Oceanic gateways as a critical factor to initiate icehouse Earth. *Journal of the Geological Society, London*, **160**, 337–340.

Sokolov, B. S. & Fedonkin, M. A. 1990. *The Vendian System: Volume 2. Regional Geology* [translation of 1985 publication in Russian]. Springer, Berlin.

Sprigg, R. C. 1947. Early Cambrian (?) jellyfishes from the Flinders Ranges, South Australia. *Transactions of the Royal Society of South Australia*, **71**, 212–224.

Summons, R. E., Jahnke, I. L., Hope, J. M. & Logan, G. A. 1999. 2-Methylhopanoids as biomarkers for cyanobacterial oxygenic photosynthesis. *Nature*, **400**, 554–557.

Thompson, M. D. & Bowring, S. A. 2000. Age of the Squantum 'tillite', Boston Basin, Massachusetts: U–Pb zircon constraints on terminal Neoproterozoic glaciation. *American Journal of Science*, **300**, 630–655.

Tollo, R. P. & Hutson, F. E. 1996. 700 Ma rift event in the Blue Ridge province of Virginia: a unique time constraint on pre-Iapetan rifling of Laurentia. *Geology*, **24**, 59–62.

Trindade, R. I. F., Font, E., D'Agrella-Filho, A. S., Nogueira, A. C. R. & Riccomini, C. 2003.

Low-latitude and multiple geomagnetic reversals in the Neoproterozoic Puga cap carbonate, Amazon craton. *Terra Nova*, **15**, 441–446.

TRIPATI, A., BACKMAN, J., ELDERFIELD, H. & FERRETTI, P. 2005. Eocene bipolar glaciation associated with global carbon cycle changes. *Nature*, **436**, 341–346.

TROMPETTE, R., DE ALVARENGA, C. J. S. & WALDE, D. 1998. Geological evolution of the Neoproterozoic Corumba graben system (Brazil). Depositional context of the stratified Fe and Mn ores of the Jacadigo Group. *Journal of South American Earth Sciences*, **11**, 587–597.

XIAO, S. & KAUFMAN, A. J. 2007. *Neoproterozoic Geobiology and Palaeobiology*. Topics in Geobiology, **27**. Springer, New York.

XIAO, S. H., BAO, H. *ET AL*. 2004. The Neoproterozoic Quruqtagh Group in eastern Chinese Tianshan: evidence for a post-Marinoan glaciation. *Precambrian Research*, **130**, 1–26.

XU, B., JIAN, P., ZHENG, H., ZOU, H., ZHANG, L. & LIU, D. 2005. U–Pb zircon geochronology and geochemistry of Neoproterozoic volcanic rocks in the Tarim Block of northwest China: implications for the breakup of Rodinia supercontinent and Neoproterozoic glaciations. *Precambrian Research*, **136**, 107–123.

YE, M., KELER, B. M., SOKOLOV, B. S., SOLOMNTSOV, L. F. & SHUL'GA, P. L. 1980. A general scheme for upper Precambrian stratigraphy of the Russian platform. *International Geology Review*, **22**, 444–458.

ZHOU, C. M., TUCKER, R., XIAO, S., PENG, Z., YUAN, X. & CHEN, Z. 2004. New constraints on the ages of Neoproterozoic glaciations in south China. *Geology*, **32**, 437–440.

Macroevolutionary turnover through the Ediacaran transition: ecological and biogeochemical implications

NICHOLAS J. BUTTERFIELD

Department of Earth Sciences, University of Cambridge, Downing Street, Cambridge CB2 3EQ, UK
(e-mail: njb1005@esc.cam.ac.uk)

Abstract: Ecological and evolutionary principles are often context-dependent, particularly where the context is biologically defined. Organ-grade animals (eumetazoans) are particularly powerful contextual agents, with a unique capacity to drive escalatory co-evolution and build multi-tiered food-webs. The evolution of eumetazoans through the Ediacaran and early Cambrian fundamentally altered macroecological and macroevolutionary dynamics, including the structure and function of the marine carbon cycle. Pelagic eumetazoans can be held responsible for driving the evolution of relatively large eukaryotic phytoplankton, thereby shifting the system from a turbid, stratified, cyanobacteria-dominated stable state to the clear-water, well-oxygenated, algae-dominated condition typical of the Phanerozoic. Intermittent return to the pre-Ediacaran state during Phanerozoic extinctions and oceanic anoxic events suggests that the widespread anoxia detected in pre-Ediacaran deep-marine sequences may be a consequence of this alternate biological pump rather than a reflection of fundamentally lower levels of atmospheric oxygen. The transition between the pre- and post-Ediacaran states is also associated with the oldest commercially exploitable hydrocarbons, a possible by-product of invading animals and their top-down impact on the biological pump.

One of the first principles of the Earth sciences is uniformitarianism – that phenomena observed in the geological record should be interpreted in terms of modern, empirical process. But this sort of Lyellian reasoning has its limits. Not only is the timescale of many geological phenomena far too large for direct observation, but also the processes themselves have often changed over time. Nowhere does the principle fail quite so dramatically as in palaeobiology, where the long-term record reveals patterns that are not in any way inferable from modern population biology. Larger-scale pattern and process can, of course, be drawn from the geological record, but even these are likely to be time- and context-dependent. In this paper I will examine the structure, function and 'evolution' of marine ecosystems prior to the early Cambrian entrenchment of animal biology, and the degree to which Phanerozoic-based 'rules' can be extended back into the Proterozoic. Recognition of key differences promises to illuminate many of the broader issues relating to the Proterozoic–Phanerozoic transition, including the context-dependent nature of biological evolution, ecosystem function and the global carbon cycle.

The Tommotian–Recent record

The present is undoubtedly the key to the relatively recent past. Thus, the rich record of diatoms, coccolithophores and dinoflagellates over the past 50–100 Ma reveals a recognizably modern source of primary production in the marine realm, from which it is safe to infer a modern-style structure and function for marine ecology in general. By the same token, absence of such forms in the Palaeozoic reflects an unfamiliar, relatively non-uniformitarian biosphere. Even so, Palaeozoic ecosystems still contain a readily recognizable suite of metazoan guilds and tiered trophic structures extending back to the early (but not the earliest) Cambrian, suggesting a broad continuity of process. Indeed, it was the Tommotian (*c.* 530 Ma) radiation of small skeletal fossils, archaeocyaths, phytoplankton and sedimentary trace fossils – colloquially known as the Cambrian explosion – that appears to have set the overall context for the Phanerozoic (Butterfield 2007).

The pre-Tommotian record

Sub-Tommotian unconformities may have locally exaggerated the abruptness of the Cambrian explosion (Knoll *et al.* 1995), but there is little doubt that the immediately preceding Nemakit Daldynian stage (*c.* 543–530 Ma) presents a fundamentally different state of affairs. Body-fossil and trace-fossil diversity/abundance is conspicuously lower and represents a fundamentally more limited repertoire of behaviours than found in the remaining Phanerozoic (Zhuravlev 2001; Droser *et al.* 2002). There is,

From: CRAIG, J., THUROW, J., THUSU, B., WHITHAM, A. & ABUTARRUMA, Y. (eds) *Global Neoproterozoic Petroleum Systems: The Emerging Potential in North Africa.* Geological Society, London, Special Publications, **326**, 55–66.
DOI: 10.1144/SP326.3 0305-8719/09/$15.00

for example, no direct evidence of pre-Tommotian trophic structures extending beyond a single level of primary predation, no evidence of deposit feeding (McIlroy & Logan 1999), and a dearth of morphologically differentiated phytoplankton points to the absence of significant mesozooplankton and accompanying pelagic food-web (Butterfield 1997, 2001*b*, 2003, 2007; Peterson & Butterfield 2005). Indeed, there is good case for linking the Cambrian–Tommotian explosion to the expansion of animals into the pelagic realm.

The base of the Nemakit Daldynian nominally coincides with the base of the Cambrian, which has been formally placed in SE Newfoundland. However, there is little evidence for a significant macroecological or macroevolutionary break at this time. *Treptichnus pedum*, the trace fossil originally used to define the boundary, is now known to extend into Precambrian strata (Gehling *et al.* 2001), whereas Ediacaran-type macrofossils and microbial mat fabrics are increasingly recognized in the early Cambrian (Jensen *et al.* 1998; Hagadorn *et al.* 2000; Dornbos *et al.* 2004). Likewise, biomineralization, SSF (small shelly fossil) diversity, and trace fossil evidence for infaunal, epifaunal and predatory activity (Wood *et al.* 2002; Droser *et al.* 2002; Hua *et al.* 2003) are all broadly continuous through the Late Ediacaran–Early Cambrian transition.

Ediacaran biotas do become progressively less Phanerozoic-like with age, such that metazoan trace fossils and biologically controlled biomineralization have disappeared from the record by 555 Ma (Martin *et al.* 2000), and Ediacaran-type macrofossils by about 575 Ma (Narbonne 2005). Even so, there is still a strong case for the presence of tissue-/organ-grade animals (eumetazoans) through the whole of the Ediacaran (*c.* 635–543 Ma). The Early Ediacaran Doushantuo Formation (*c.* 635–551 Ma), for example, preserves phosphatized embryos (Xiao *et al.* 1998; Yin *et al.* 2007), as well as branched septate tubes (Liu *et al.* 2008) and spirally constructed macrofossils (Tang *et al.* 2008), all of which are reasonably interpreted as metazoan. More generally, the diverse morphologies and measurable evolutionary turnover of early Ediacaran acritarchs (e.g. Knoll 1994; Grey 2005; Huntley *et al.* 2006) serve as convincing co-evolutionary proxies for the presence of early Ediacaran eumetazoans (Peterson & Butterfield 2005).

The base of the Ediacaran Period has been formally placed at the base of the Nuccaleena Formation 'cap carbonate' immediately overlying glacial diamictites in South Australia – the first chronostratigraphic boundary not to be defined on biostratigraphic criteria (Knoll *et al.* 2006). The reasons for abandoning the fossil record at this point were largely practical, a combination of an increasingly cryptic palaeontological signal and the availability of a more conspicuous, apparently global, event horizon. Even so, there is also evidence for a fundamental shift in ecological/evolutionary mode at or around the base of the Ediacaran.

Body fossils of early Ediacaran and pre-Ediacaran age are almost entirely microscopic, dominated by simple sphaeromorphic acritarchs and a background of microbial mat biotas, neither of which offer any useful stratigraphic or evolutionary resolution. Under certain circumstances, however, these are accompanied by large acanthomorphic acritarchs that do. In the early Ediacaran, for example, Grey (2005) has proposed a five-part biozonation based on an integrated study of acritarchs, carbon isotopes and event beds in the Centralian Superbasin of Australia. There are problems with this scheme – not least, substantial inconsistencies with broadly correlative units in South China – but it does appear that most of the described acanthomorphs are unique to the early Ediacaran, possibly to the approximately 55 Ma interval separating the Marinoan and Gaskiers glaciations (Zhou *et al.* 2007).

Acanthomorphic acritarchs also occur in pre-Ediacaran strata, but these older forms express an entirely different character. In addition to the marked asymmetry of their processes, they are fundamentally less diverse and disparate than their early Ediacaran counterparts (Knoll 1994; Butterfield 2004, 2007; Huntley *et al.* 2006; Knoll *et al.* 2006). Most significantly, they reveal a pattern of profound morphological stasis, with age ranges typically of several hundred million to more than a billion years (Knoll 1994; Peterson & Butterfield 2005; Butterfield 2007). Such a pattern clearly limits the potential for biostratigraphic partitioning of pre-Ediacaran time, but it does have important theoretical implications (cf. Gould & Eldredge 1993). The pervasive morphological stasis of pre-Ediacaran life reflects a background of pervasively monotonous, non-escalatory environments, and provides compelling *positive* evidence for the absence of eumetazoans – even those too small or ephemeral to be preserved as body fossils (Peterson & Butterfield 2005).

The macroecological and macroevolutionary impact of animals

So what is it that makes eumetazoans and a eumetazoan biosphere so different? At the most basic level, it is a matter of multicellular sophistication and accompanying behaviour. Eumetazoans are heterotrophic eukaryotes with an integrated, usually motile, food-capturing and processing system based on a differentiated gut and complex motor-neuron control. The key distinction from unicellular heterotrophs (and sub-tissue-grade multicellular heterotrophs such as sponges and fungi) is the presence of differentiated tissues and organs. By providing functionally specialized components at a multicellular grade of organization, such structures open the door to an essentially inexhaustible range of morphology and ecology based on the ingestion of larger particles.

Tissues and organs give eumetazoans a unique ability to drive reciprocal, co-evolutionary arms races. Whereas sponges feed almost exclusively on bacterioplankton, eumetazoan heterotrophy typically involves the capture and ingestion of eukaryotic organisms – and, unlike bacteria, eukaryotes are capable of responding to predation pressure via morphological adaptation (Peterson & Butterfield 2005). At a unicellular level, the eukaryotic cytoskeleton and associated control systems allows the differentiation of cell shape/ornamentation and cell composition/biomineralization, both of which reduce the likelihood of metazoan capture (Butterfield 1997; Hamm *et al.* 2003). Likewise, multicellular eukaryotes readily adapt to metazoan predation through morphological co-evolution, exploiting their capacity for large size and anatomical/behavioural differentiation; indeed, metazoan predation has been shown to induce multicelluarity in otherwise unicellular algae (e.g. Lürling 1999). Such responses, of course, will be countered by eumetazoan predators equally capable of morphological/behavioural adaptation, leading to cycles of escalatory, reciprocal, co-evolution (Vermeij 1994). Thus, the first appearance of eumetazoans is likely to be accompanied by significant increases in evolutionary turnover, as well as organismal size, morphology, biomineralization and behaviour. To an important degree, the evolution of eumetazoans introduced the 'progressive' biological environments that characterize the Phanerozoic.

Eumetazoans also prey on other animals, thereby constructing multi-tiered trophic hierarchies, adding a unique vertical dimension to morphological co-evolution. Analogous structures do occur in microbial ecosystems (e.g. Diehl & Feissel 2001), but the unicellular grade of organization sets a fundamental constraint on their evolutionary potential. In aquatic ecosystems, metazoan predators are typically much larger than their prey and this, in combination with expanded trophic tiering (Link 2002), gives rise to body sizes ranging over 20 orders of magnitude (Brown *et al.* 2004). Except for the (mostly) microscopic primary productivity/detritus at their base, these trophic structures are built entirely of eumetazoans and account for the vast majority of organismal size and diversity in the modern/Phanerozoic oceans (Fig. 1).

Metazoan food-webs express a host of novel qualities and emergent characters that impart a unique dynamic to Phanerozoic ecosystems (Butterfield 2007). As a consequence of their compounding effects on organismal diversity and size, for example, they account for an enormous repertoire of activities and secondary effects that otherwise would not exist – from ambush predation to faecal pellet formation, bioturbation, deposit feeding and biomineralization. Moreover, the 'big fish eat little fish' vertical structure of aquatic food-webs, combined with the three-quarters scaling relationship between metabolic rate and body size, means that most of the biomass in the modern oceans is likely to be represented by animals, especially larger animals (Sheldon *et al.* 1972; Kerr & Dickie 2001; Brown *et al.* 2004), analogous to the large, long-lived, slowly metabolizing biomass represented by trees on land.

Body-size correlates inversely with population size (Sheldon *et al.* 1972; Irwin *et al.* 2006), such that larger-bodied species are subject to distinct modes of genomic evolution (Lynch & Conery 2003) and biogeographic partitioning (Finlay 2002). By the same token, larger, less populous species are at an increased risk of extinction, which can give rise to secondary and cascading extinctions. Simple food chains with relatively low connectance, for example, are particularly prone to trophic cascades, whereas greater diversity and higher web-like connectance impart a dynamic stability (McCann 2000; Duffy *et al.* 2007), if not always in a straightforward fashion (Petchey *et al.* 2008).

There is a whole science focused on the relationship between diversity and stability in modern ecosystems (e.g. May 1973; McCann 2000; Worm & Duffy 2003; Tilman *et al.* 2006), at least some of which may be applicable to understanding the patterns of Phanerozoic macroevolution. Thus, high rates of extinction in the early Palaeozoic might be related to the limited diversity and connectance of Cambrian food-webs (Butterfield 2001*b*), while the overall decline in extinction rates through the Phanerozoic correlates with major increases in diversity, ecological specialization and connectance. Even the most diversified portfolios, however, are subject to perturbation, and tiered superstructures are especially vulnerable to cascading collapse. On ecological timescales, such perturbations can induce major regime shifts that are associated with extensive local extinction and the establishment of alternate stable states (Scheffer *et al.* 2001). On macroevolutionary scales, the extinctions can be global and the consequences profound. Phanerozoic mass extinctions are notable for the preferential loss of interdependent ecologies, including the large predators and ecological

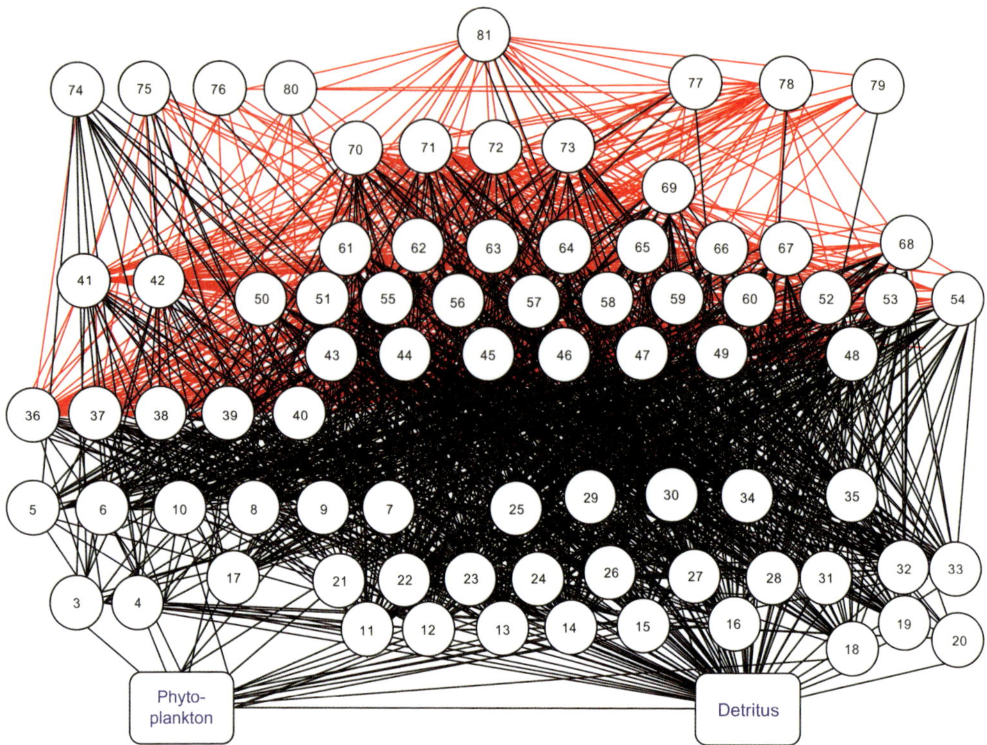

Fig. 1. Schematic depiction of trophic structure in the modern NW Atlantic Ocean, probably comparable to marine food-webs throughout most of the Phanerozoic – and fundamentally distinct from (eumetazoan-free) pre-Ediacaran systems. Primary productivity and detritus is at the base, with both trophic level and organism size increasing towards the top. Each node represents a separate species or trophic group (see Link 2002 for details). The left-hand side of the figure is represented primarily by pelagic organisms, whereas the middle and right-hand portions are dominated by benthic/demersal forms. Red lines indicate predation on fish. Modified from Link (2002).

specialists derived from progressively complexifying food-webs. To this extent, it is the particular evolvability and diversified interdependence of eumetazoan food-webs that accounts for the overall stability of Phanerozoic ecosystems, but also their intermittent, commonly catastrophic, collapse.

Pre-Tommotian macroecology

At some point in the past there were no eumetazoans and no eumetazoan food-webs. Such a biosphere would have been dominated by small–microscopic organisms characterized by conspicuously low diversity, enormous population sizes, rapid generation times, little if any biogeographic partitioning and low standing biomass. Moreover, background evolutionary turnover would have lacked the impetus of an escalatory 'organ-grade' biological environment, while the absence of a vertically developed trophic structure would have precluded cascading Phanerozoic-type mass extinctions.

The profoundly non-uniformitarian aspect of this pre-eumetazoan biosphere should make it easy to recognize in the geological record, despite the vagaries of fossil preservation. And, indeed, it is. As discussed earlier, the profound stasis, low diversity and microbial aspect of pre-Ediacaran biotas are an unambiguous signature for the global absence of eumetazoans. Conversely, the rich morphological diversity of post-Nemakit Daldynian biotas, combined with their fundamentally more rapid evolutionary turnover (see Knoll 1994), reflect the enormous ecological and evolutionary impact of eumetazoans. It is this positive, qualitative difference in biospheric expression – as opposed to the mere absence of early fossils – that provides the definitive test for the presence of Eumetazoa (Peterson & Butterfield 2005). The premise is that eumetazoans do not live in isolation, and cannot help but induce escalatory co-evolution in their trophically interlinked environment.

Phanerozoic-style ecosystems did not arise instantaneously with the appearance of eumetazoans, and it is clear that the approximately 100 Ma interval comprising the Ediacaran and Nemakit Daldynian represents the transition from a non-uniformitarian 'microbial world' to the more or less modern marine ecosystems of the Phanerozoic (Butterfield 2007). Thus, the early Ediacaran first-appearance of eumetazoans can be recognized by the comprehensive turnover in acritarch biotas and order-of-magnitude increase in evolutionary rates. But this is not to suggest that the biosphere became immediately familiar. Indeed, it was not until the middle Ediacaran (*c.* 575 Ma) that bona fide macrofossils put in an appearance (Narbonne 2005; Tang *et al.* 2008) – possibly representing a size-refuge response to the evolution of bilaterians (Peterson & Butterfield 2005) – and it is not until the Tommotian that there was a significant diversity of macroscopic life, accompanied by major increases in biomineralization, bioturbation and phytoplankton diversity (Zhuravlev 2001; Droser *et al.* 2002), and a further order-of-magnitude increase in evolutionary turnover (Knoll 1994). The fact that demonstrably planktic acritarchs radiate in concert with the Cambrian–Tommotian explosion – v. the conspicuously larger, benthic, pre-Cambrian acanthomorphs (Butterfield & Chandler 1992; Butterfield 1997, 2001*a*) – suggests that it was the expansion of complex food-webs into the pelagic realm that finally established the familiar uniformitarian world of the Phanerozoic.

Pre-Tommotian primary productivity

Animals clearly play a defining role in the macroscopic component of modern marine ecosystems, but does this bear significantly on the underlying, mostly microscopic, primary production that fuels them? Certainly there has been continuous phytoplankton export over the past 3.5 billion years (Ba, $\times 10^9$ years), as revealed both by the long-term record of organic-rich sediments and the long-term continuity of ^{13}C signatures – but there are also pronounced differences in the quality and quantity of sedimentary carbon through time.

Primary production in the modern oceans is dominated by unicellular phytoplankton, principally cyanobacteria and eukaryotic 'algae'. Cyanobacteria appear to make the larger contribution in terms of abundance and overall production, but because of their small, mostly picoplanktic cell size (0.2–2.0 μm) they tend to be rapidly recycled in a surface-water 'microbial loop.' By contrast, the relatively larger eukaryotic nanoplankton (2–20 μm) and microplankton (20–200 μm) preferentially sink out of the photic zone (= export production) to form the bulk of sedimentary organic carbon and its geological derivatives. Analysis of microfossils and fossil biomarker molecules shows that this surface-to-sediment biological pump has been dominated by eukaryotic algae through most of the Phanerozoic, despite conspicuous changes in the contributing taxa (Moldowan *et al.* 1996; Sinninghe Damsté *et al.* 2004; Knoll *et al.* 2007). In the Proterozoic, however, the situation appears to be reversed, with elevated levels of 2-methylhopanoid biomarkers indicating that cyanobacteria (and/or other bacterial photoautotrophs; Rashby *et al.* 2007) were the principal contributors to export production (Summons *et al.* 1999). Eukaryote-derived steranes were present throughout the Meso- and Neoproterozoic, but appear to have contributed little to export production (Knoll *et al.* 2007).

The terminal Proterozoic Ediacaran Period represents an interesting transitional phase in the expression of sedimentary organic carbon. In addition to hosting the oldest commercially exploited petroleum deposits – in Oman and eastern Siberia – the accompanying oils are notable for their high levels of 2-methylhopanoids, along with a conspicuous presence of C_{29} steranes, problematic C_{20+} mid-chain methyl alkanes and exceptionally light ^{13}C values (Grantham 1986; Fowler & Douglas 1987; Höld *et al.* 1999; Summons *et al.* 1999; Knoll *et al.* 2007). Along with elevated sulphur contents and depressed pristane/phytane ratios, these features impart a unique, effectively diagnostic signature to Ediacaran hydrocarbons, but one that is also becoming increasingly familiar in terms of eukaryotic input.

Logan *et al.* (1995, 1997) have also noted qualitative differences in the ^{13}C isotopic expression of Phanerozoic v. Proterozoic hydrocarbons: whereas *n*-alkanes extracted from Phanerozoic sediments tend to be lighter than their associated kerogens, those of Proterozoic age are curiously heavier. The most likely explanation lies in the isotopic enrichment that accompanies heterotrophic reworking in the context of a non-uniformitarian biological pump. In the Phanerozoic oceans, export is dominated by relatively large-celled eukaryotic algae delivered rapidly to the sediment via faecal pellets, mass accumulation and marine snow (Turner 2002), such that both *n*-alkanes and kerogens express a light, phytoplankton-derived ^{13}C signature. By contrast, the small, unpackaged, mostly cyanobacterial export of the Proterozoic would have experienced much slower rates of sinking leading to extensive degradation within the water column and thereby a fundamentally greater contribution of (isotopically enriched) heterotrophic biomass to soluble sedimentary hydrocarbons (Logan *et al.* 1995).

But why were the Proterozoic oceans so dominated by cyanobacteria in the first place? One possibility is that it was simply the belated evolutionary appearance of crown-group eukaryotes (e.g. Brasier & Lindsay 1998; Cavalier-Smith 2006), but this is inconsistent with the much earlier appearance of crown-group algae (e.g. Butterfield *et al.* 1994; Butterfield 2000, 2004; Javaux *et al.* 2001; Knoll *et al.* 2007). More intriguingly, Anbar & Knoll (2002) have argued that eukaryotic phytoplankton may have been constrained by severe N-limitation in the presence of a widely euxinic Proterozoic ocean: under such conditions dissolved molybdenum might be sequestered as insoluble sulphides, thereby reducing its availability for Mo-nitrogenases and imparting a selective advantage to N-fixing cyanobacteria. It turns out, however, that neither N-fixation nor growth rate in cyanobacteria is substantially depressed under modelled Mesoproterozoic Mo/Fe concentrations (Zerkle *et al.* 2006); which, in any event, fail to take into account potentially significant aeolian inputs of Mo/Fe from non-vegetated (non-uniformitarian) Proterozoic continents. An alternative model notes that the transition to an oxygenated ocean would have been accompanied by a major N-crisis as upwelling-reduced N became denitrified at an expanding anoxic–oxic chemocline (Fennel *et al.* 2005); although evidence from analagous Cretaceous oceanic anoxic events (OAEs) suggests that cyanobacterial N-fixation can more than make up any N shortfall (Kuypers *et al.* 2004).

The implication of these N-limitation hypotheses is that the pre-Cambrian/pre-Ediacaran oceans would have been profoundly oligotrophic, and that cyanobacteria out-competed eukaryotic algae by fixing their own nitrogen and exploiting a purportedly greater capacity to scavenge and assimilate limited NO_3^- (Anbar & Knoll 2002). Such conditions, however, would also have been accompanied by an overwhelming predominance of picoplankton and a microbial-loop style of recycling primary production within the water column, which is difficult to reconcile with the regular recurrence of organic-rich Proterozoic sediments (see Summons *et al.* 1999) and ^{13}C values reflecting the substantial long-term burial of organic carbon (*c.* 20% of total carbon).

An alternative model

There is certainly more to phytoplankton biology than nutrient acquisition. At least, in the Phanerozoic, there is also a significant, possibly overarching, link to pelagic metazoans by way of small grazing mesozooplankton (Verity & Smetacek 1996; Butterfield 1997, 2001*b*, 2007; Irigoien *et al.* 2004), with recent mesocosm experiments demonstrating order-of-magnitude changes in microbial functional groups accompanying mesozooplankton-induced trophic cascades (Zöllner *et al.* 2009). Not only do phytoplankton radiate in concert with the Cambrian–Tommotian explosion and, more broadly, with the Mesozoic marine revolution (Vermeij 1977; Falkowski *et al.* 2004), but they also crash in concert with metazoan extinctions (D'Hondt *et al.* 1998), sometimes to the extent of reverting to a Proterozoic-like system dominated by cyanobacterial export (e.g. Kuypers *et al.* 2004; Xie *et al.* 2005).

A conventional view of marine mass extinctions, most notably the K–T (Cretaceous–Cenozoic), is that they propagate through food-webs from the bottom-up, following collapse of planktic productivity (Vermeij 2004). This is, undoubtedly, the case for trophically balanced metazoans; however, it does not obviously account for the simultaneous declines in phytoplankton diversity, or the relatively drawn-out post-extinction recoveries. Like cyanobacteria, most eukaryotic phytoplankton are simply too abundant and too widely distributed to be eliminated by physical perturbation, and their rapid generation times ensure near instantaneous recovery from relict populations (see Lynch & Conery 2003). The fact that, for example, coccolithophores survived the K–T extinction, but lost most of their morphologically differentiated taxa (Bown 2005), suggests that there were additional, diversity-maintaining, factors that disappeared at the same time. Given the general collapse of pelagic food-webs at the end of the Cretaceous, there is little doubt that phytoplankton-grazing mesozooplankton were also decimated, such that the complex, possibly even chaotic (Benincá *et al.* 2008), top-down selective pressures on primary producers were released, allowing them to revert to simpler, less specialized modes of life. By the same token, the recovery of phytoplankton diversity – and its accompanying contribution to the biological pump (D'Hondt *et al.* 1998) – would require the reassembly of comparably complex food-webs built over evolutionary timescales (cf. Kirchner & Weil 2000).

Size lies at the heart of pelagic ecology, especially the export production that contributes to the long-term carbon cycle (Wassmann 1998). The inverse relationship between phytoplankton size and nutrient absorption/light-harvesting ability, for example, suggests that picoplankton should dominate marine photosynthesis under almost all circumstances (Raven 1998). The fact that they do not, at least in the Phanerozoic, is convincingly ascribed to top-down forcing by mesozooplankton (Butterfield 1997, 2001*b*, 2007; Irigoien *et al.* 2004; Jiang *et al.* 2005). By exploiting their ability to modify cell morphology, eukaryotic phytoplankton can evade predation by increasing their size, despite the negative effects on nutrient and light acquisition – an ecological and evolutionary trade-off.

And size changes everything. In the modern oceans, larger-celled phytoplankton form the vast majority of export production and accumulating sedimentary organic carbon because of their disproportionately high rates of mass sedimentation and preferential incorporation into rapidly sinking marine snow and faecal pellets (Butterfield 1997; Smetacek 2000; Turner 2002). Prior to the appearance of eumetazoan grazing pressures, however, all phytoplankton are likely to have converged on picoplanktic dimensions, leading to exceptionally high cell densities and high levels of oxygen-depleting heterotrophy within the water column (cf. Logan *et al.* 1995). Thus, although picoplankton are too small to sink on their own, at some elevated concentration they presumably aggregate and fall out of the water column. This is certainly a biological pump, but it differs fundamentally from the familiar Phanerozoic one (Fig. 2).

A picoplankton-dominated water column (under non-oligotrophic conditions) would also have been conspicuously turbid, leading to reduced light penetration and a strong selective advantage for shade-tolerant cyanobacteria. Because cyanobacteria are also the principal cause of such turbidity, they reinforce continued cyanobacterial dominance (see Scheffer *et al.* 1997). Exit from this system can only be effected by a major regime shift, whereby the water column is cleared and picoplankton replaced by larger, more export-prone, eukaryotes conducive to high light penetration (Fig. 2). In modern lacustrine systems there is pronounced hysteresis between these two alternative stable states, typically driven by metazoan activity (e.g. Gragnani *et al.* 1999; Sarnelle 2005; Ibelings *et al.* 2007). In Phanerozoic oceans the predominance of metazoan activity has generally kept the system in a clear-water, algal state, although perturbations associated with mass extinctions and OAEs (e.g. Joachimski *et al.* 2001; Kuypers *et al.* 2004; Xie *et al.* 2005) have induced intermittent returns to the turbid, cyanobacteria-dominated, anoxic–euxinic state typical of the Proterozoic. Intriguingly, even high-energy marine ecosystems such as kelp forests have been shown to be susceptible to the intense light attenuation of plankton blooms, with a potential for fundamental, hysteretic shifts in community ecology (Kavanaugh *et al.* 2009).

In the pre-eumetazoan Proterozoic oceans the potential for a clear-water state would have been limited. Severe nutrient crises may have cleared the water to something like the modern mid-Pacific gyres, but only in concert with exclusively non-exporting picoplankton. McFadden *et al.* (2008), Shen *et al.* (2008) and others have argued for a pulsed oxidation of the deep ocean through the Ediacaran, but this does not offer any obvious exit from the turbid, cyanobacteria-dominated state of the Proterozoic ocean: neither atmospheric oxidation nor deep-water sulphate reduction are likely to have induced the evolution of larger, more exportable phytoplankton. A hard 'snowball Earth' (Hoffman *et al.* 1998) could conceivably have given eukaryotic algae a localized advantage (see Boyle *et al.* 2007), but once again there are no obvious selective pressures for 'suboptimally' large phytoplankton.

The more useful explanations for the Ediacaran regime shift are biological. Logan *et al.* (1995), for example, have proposed that the early Cambrian water column could have been cleared of suspended phytoplankton by the packing effects of newly introduced zooplankton. The problem here is that most export production, at least in the modern oceans, does not travel via faecal pellets. Rather, it is transported to the sediment by means of aggregation, mass sedimentation and/or marine snow (Butterfield 1997, 2001*a*; Smetacek 2000; Turner 2002). Animals, specifically mesozooplankton, are nevertheless the key to this export, because they provide the ecological insult (i.e. grazing) necessary to drive the evolution of large-celled phytoplankton.

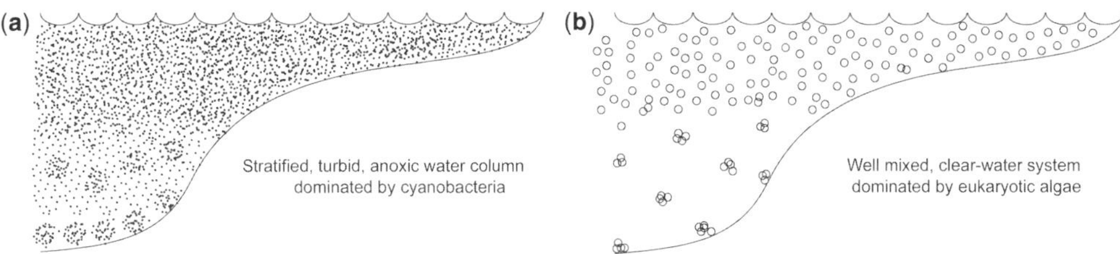

Fig. 2. Schematic diagrams of the two alternate stable states for the biological pump. (**a**) Dominance of shade-tolerant and shade-inducing cyanobacterial picoplankton in the absence of top-down selection for larger eukaryotic phytoplankton; the buoyancy of such forms gives rise to a stratified, turbid, anoxic water-column typical of pre-Tommotian oceans and Phanerozoic OAEs. (**b**) Dominance of relatively large eukaryotic phytoplankton, probably induced by the Early Cambrian introduction of grazing mesozooplankton. The exponentially higher rates of sinking experienced by larger phytoplankton would have switched the system to a non-stratified, clear-water, eukaryote-dominated condition, typical of Phanerozoic oceans.

In the process they invented an alternate, stable, clear-water state for pelagic ecosystems.

Sterane biomarkers point to an increasing contribution of eukaryotic phytoplankton to export production through the Ediacaran (Knoll *et al.* 2007), but it is not until the early Cambrian–Tommotian that sedimentary organic matter takes on its (more or less) modern aspect. The transition may have been facilitated by Ediacaran expansion of suspension-feeding sponges and rangeomorphs (Sperling *et al.* 2007), although the restriction of such forms to benthic habitats and bacteria-sized diets would have limited their overall impact. A larger effect is likely to have come from the invasion of pelagic environments by eumetazoans, and their incremental 'engineering' of novel clear-water environments – including fundamental shifts in nutrient cycling (Hall *et al.* 2007; Hernández-León *et al.* 2008), biomass distribution, life-history trade-offs, biogeography, planktic–benthic coupling and macroevolutionary dynamics (Butterfield 2007). By extension, much of the biogeochemical perturbation associated with the Ediacaran transition (e.g. Fike *et al.* 2006; Canfield *et al.* 2007; McFadden *et al.* 2008; Shen *et al.* 2008) should probably be viewed as consequences, rather than causes, of the associated evolutionary events.

Discussion

Given their relatively isotropic physical environment, the high diversity of phytoplankton in the modern oceans has long been considered a paradox (Hutchinson 1961). The absence of any comparable Proterozoic diversity, followed by a dramatic Tommotian radiation, convincingly links phytoplankton morphology to the top-down effects of small planktic metazoans (cf. Zöllner *et al.* 2009), even if this was subsequently maintained by more complex, even chaotic, dynamics (e.g. Benincá *et al.* 2008). The forcing of eukaryotic phytoplankton to co-evolve into diverse size- and morphology-based refugia would also have sequestered increasingly high proportions of limiting abiotic resources, tilting the overall budget away from cyanobacteria (whose response to grazing was limited largely to chemical unpalatability; see Jang *et al.* 2007). At the same time, increased cell size allowed eukaryotic phytoplankton to adopt further novel habits, such as nutrient storage under fluctuating conditions, physiologically controlled buoyancy (Irwin *et al.* 2006) and the formation of sinkable aestivating cysts.

A turbid, cyanobacteria-dominated ocean also has important implications for tracking the oxygenation of the Proterozoic atmosphere and oceans. Geochemical signatures record extensive deep-water anoxia–euxinia throughout all but the latest Proterozoic, which, along with the belated appearance of eumetazoans, has been widely interpreted as evidence for low levels of atmospheric oxygen (e.g. Anbar & Knoll 2002; Arnold *et al.* 2004; Canfield *et al.* 2007; McFadden *et al.* 2008; Shen *et al.* 2008) – somewhere between 1 and 40% pre-industrial atmospheric levels (PALs; Canfield 2005). Whether 40 or 20 or 10% PAL would have impinged significantly on the early evolution of eumetazoans remains to be demonstrated (see Catling *et al.* 2005; Budd 2008), but it is also worth appreciating that even these broad estimates are based on uniformitarian assumptions of ocean–atmosphere exchange. In the absence of a eukaryote-dominated biological pump, the pre-Ediacaran deep oceans are expected to have been anoxic under all but the most oligotrophic conditions simply because of the high buoyancy – and, therefore, high level of oxygen consumption within the water column (Logan *et al.* 1995) – of the cyanobacterial stable state. Under such conditions, geochemical proxies for marine oxidation do not translate directly to atmospheric composition, just as the widespread anoxic–euxinic conditions of (cyanobacteria-dominated) Devonian, Permo-Triassic and Mesozoic OAEs bear little relationship to contemporaneous atmospheric oxygen (see Sinninghe Damsté & Köster 1998; Joachimski *et al.* 2001; Kuypers *et al.* 2004; Grice *et al.* 2005; Xie *et al.* 2005; Pearce *et al.* 2008). In this light, it is at least plausible that Meso-/Neoproterozoic atmospheric oxygen was comparable to levels in the Phanerozoic. The absence of anoxia–euxinia in shallow-water Mesoproterozoic sequences (Scott *et al.* 2008), along with ongoing deep-water anoxia in the earliest Cambrian (Goldberg *et al.* 2007; Schröder & Grotzinger 2007; Wille *et al.* 2008) further attests to the non-uniformitarian nature of vertical mixing.

Oxygen on its own does not provide any selective pressure for larger cell size and does not alter the quality of the water column or the nature of the biological pump. The most likely reason for the return to anoxic–euxinic oceans during Phanerozoic OAEs is the loss of top-down selective pressures for larger cell size following disruption of pelagic food-webs. Likewise, the most likely reason for the Ediacaran–Cambrian transition to eukaryote-dominated export was the evolution and ecological expansion of pelagic metazoans. Oxygen on its own does not provide any selective pressure for zooplankton, but a diversified zooplankton, driven by both predators and prey, will oxygenate–ventilate the deep ocean.

The transition from a cyanobacteria-dominated to an algae-dominated biological pump also has economic implications. The oldest commercially exploited hydrocarbons are of Ediacaran age – specifically late–latest Ediacaran age – that correlate with a marked, if fluctuating, increase in the abundance of eukaryotic biomarkers (McKirdy *et al.* 2006; Knoll *et al.* 2007; Kodner *et al.* 2008). It is possible that the absence of older, pre-Ediacaran oil fields is simply a matter of geological history (e.g. Buick *et al.* 1998), but the distinctive, apparently less oil-prone biochemistry of cyanobacteria (Logan *et al.* 1997) suggests that the hydrocarbon potential of these earlier systems was limited from the outset. As on other fronts, what works for the Phanerozoic cannot necessarily be extrapolated back into the very different, non-uniformitarian world of the Proterozoic.

I thank the organizers of the Jammu conference, especially G. M. Bhat and B. Thusu, for the invitation to contribute to these proceedings. J. Link kindly contributed the food-web depicted in Figure 1. G. Logan and an anonymous referee made a number of useful suggestions on an earlier draft. Cambridge Earth Science contribution 1097.

References

ANBAR, A. D. & KNOLL, A. H. 2002. Proterozoic ocean chemistry and evolution: a bioinorganic bridge? *Science*, **297**, 1137–1142.

ARNOLD, G. L., ANBAR, A. D., BARLING, J. & LYONS, T. W. 2004. Molybdenum isotope evidence for widespread anoxia in mid-Proterozoic oceans. *Science*, **304**, 87–90.

BENINCÁ, E., HEERKLOSS, R. *ET AL.* 2008. Chaos in a long-term experiment with a plankton community. *Nature*, **451**, 822–826.

BOWN, P. 2005. Selective calcareous nannoplankton survivorship at the Cretaceous–Tertiary boundary. *Geology*, **33**, 653–656.

BOYLE, R. A., LENTON, T. M. & WILLIAMS, H. T. P. 2007. Neoproterozoic 'snowball Earth' glaciations and the evolution of altruism. *Geobiology*, **5**, 337–349.

BRASIER, M. D. & LINDSAY, J. F. 1998. A billion years of environmental stability and the emergence of eukaryotes: New data from northern Australia. *Geology*, **26**, 555–558.

BROWN, J. H., GILLOOLY, J. F., ALLEN, A. P., SAVAGE, V. M. & WEST, G. B. 2004. Toward a metabolic theory of ecology. *Ecology*, **85**, 1771–1789.

BUDD, G. E. 2008. The earliest fossil record of the animals and its significance. *Philosophical Transactions of the Royal Society B*, **363**, 1425–1434.

BUICK, R., RASMUSSEN, B. & KRAPEZ, B. 1998. Archean oil: evidence for extensive hydrocarbon generation and migration 2.5–3.5 Ga. *AAPG Bulletin*, **82**, 50–69.

BUTTERFIELD, N. J. 1997. Plankton ecology and the Proterozoic–Phanerozoic transition. *Paleobiology*, **23**, 247–262.

BUTTERFIELD, N. J. 2000. *Bangiomorpha pubescens* n. gen., n. sp.: implications for the evolution of sex, multicellularity and the Mesoproterozoic–Neoproterozoic radiation of eukaryotes. *Paleobiology*, **26**, 386–404.

BUTTERFIELD, N. J. 2001*a*. Cambrian food webs. *In*: BRIGGS, D. E. G. & CROWTHER, P. R. (eds) *Palaeobiology II, A Synthesis*. Blackwell Scientific, Oxford, 40–43.

BUTTERFIELD, N. J. 2001*b*. Ecology and evolution of the Cambrian plankton. *In*: ZHURAVLEV, A. YU. & RIDING, R. (eds) *Ecology of the Cambrian Radiation*. Columbia University Press, New York, 200–216.

BUTTERFIELD, N. J. 2003. Exceptional fossil preservation and the Cambrian explosion. *Integrative and Comparative Biology*, **43**, 166–177.

BUTTERFIELD, N. J. 2004. A vaucheriacean alga from the middle Neoproterozoic of Spitsbergen: implications for the evolution of Proterozoic eukaryotes and the Cambrian explosion. *Paleobiology*, **30**, 231–252.

BUTTERFIELD, N. J. 2007. Macroecovolution and macroecology through deep time. *Palaeontology*, **50**, 41–55.

BUTTERFIELD, N. J. & CHANDLER, F. W. 1992. Paleoenvironmental distribution of Proterozoic microfossils, with an example from the Agu Bay Formation, Baffin Island. *Palaeontology*, **35**, 943–957.

BUTTERFIELD, N. J., KNOLL, A. H. & SWETT, K. 1994. Paleobiology of the Neoproterozoic Svanbergfjellet Formation, Spitsbergen. *Fossils & Strata*, 34.

CANFIELD, D. E. 2005. The early history of atmospheric oxygen: homage to Robert M. Garrels. *Annual Review of Earth and Planetary Science*, **33**, 1–36.

CANFIELD, D. E., POULTON, S. W. & NARBONNE, G. M. 2007. Late-Neoproterozoic deep-ocean oxygenation and the rise of animal life. *Science*, **315**, 92–95.

CATLING, D. C., GLEIN, C. R., ZAHNLE, K. J. & MCKAY, C. P. 2005. Why O_2 is required by complex life on habitable planets and the concept of planetary 'oxygenation time'. *Astrobiology*, **5**, 414–438.

CAVALIER-SMITH, T. 2006. Cell evolution and Earth history: stasis and revolution. *Philosophical Transactions of the Royal Society B*, **361**, 969–1006.

D'HONDT, S., DONAGHAY, P., ZACHOS, J. C., LUTTENBERG, D. & LINDINGER, M. 1998. Organic carbon fluxes and ecological recovery from the Cretaceous–Tertiary mass extinction. *Science*, **282**, 276–279.

DIEHL, S. & FEISSEL, M. 2001. Intraguild prey suffer from enrichment of their resources: a microcosm experiment with ciliates. *Ecology*, **82**, 2977–2983.

DORNBOS, S. Q., BOTTJER, D. J. & CHEN, J.-Y. 2004. Evidence of seafloor microbial mats and associated metazoan lifestyles in Lower Cambrian phosphorites of southwest China. *Lethaia*, **37**, 127–137.

DROSER, M. L., JENSEN, S. & GEHLING, J. G. 2002. Trace fossils and substrates of the terminal Proterozoic–Cambrian transition: implications for the record of early bilaterians and sediment mixing. *Proceedings of the National Academy of Sciences, USA*, **99**, 12572–12576.

DUFFY, J. E., CARDINALE, B. J., FRANCE, K. E., MCINTYRE, P. B., THÉBAULT, E. & LOREAU, M. 2007. The functional role of biodiversity in ecosystems: incorporating trophic complexity. *Ecology Letters*, **10**, 522–538.

FALKOWSKI, P. G., KATZ, M. E., KNOLL, A. H., QUIGG, A., RAVEN, J. A., SCHOFIELD, O. & TAYLOR, F. J. R. 2004. The evolution of modern eukaryotic phytoplankton. *Science*, **305**, 354–360.

FENNEL, K., FOLLOWS, M. & FALKOWSKI, P. G. 2005. The co-evolution of the nitrogen, carbon and oxygen cycles in the Proterozoic ocean. *American Journal of Science*, **305**, 526–545.

FIKE, D., GROTZINGER, J., PRATT, L. & SUMMONS, R. 2006. Oxidation of the Ediacaran ocean. *Nature*, **444**, 744–747.

FINLAY, B. J. 2002. Global dispersal of free-living microbial eukaryote species. *Science*, **296**, 1061–1063.

FOWLER, M. G. & DOUGLAS, A. G. 1987. Saturated hydrocarbon biomarkers in oils of Late Precambrian age from Eastern Siberia. *Organic Geochemistry*, **11**, 201–213.

GEHLING, J. G., JENSEN, S., DROSER, M. L., MYROW, P. M. & NARBONNE, G. M. 2001. Burrowing below the basal Cambrian GSSP, Fortune Head, Newfoundland. *Geological Magazine*, **138**, 213–218.

GOLDBERG, T., STRAUSS, H., GUO, Q. & LIU, C. 2007. Reconstructing marine redox conditions for the early Cambrian Yangtze Platform: evidence from biogenic sulphur and organic carbon isotopes. *Palaeogeography, Palaeoclimatology, Palaeoecology*, **254**, 175–193.

GOULD, S. J. & ELDREDGE, N. 1993. Punctuated equilibrium comes of age. *Nature*, **366**, 223–227.

GRAGNANI, A., SCHEFFER, M. & RINALDI, S. 1999. Top-down control of cyanobacteria: a theoretical analysis. *American Naturalist*, **153**, 59–72.

GRANTHAM, P. J. 1986. The occurrence of unusual C_{27} and C_{29} sterane predominances in two types of Oman crude oil. *Organic Geochemistry*, **9**, 1–10.

GREY, K. 2005. *Ediacaran Palynology of Australia.* Association of Australasian Palaeontologists Memoirs, **31**.

GRICE, K., CAO, C. *ET AL.* 2005. Photic zone euxinia during the Permian–Triassic superanoxic event. *Science*, **307**, 706–709.

HAGADORN, J. W., FEDO, C. M. & WAGGONER, B. M. 2000. Early Cambrian Ediacaran-type fossils from California. *Journal of Paleontology*, **74**, 731–740.

HALL, S. R., LEIBOLD, M. A., LYTLE, D. A. & SMITH, V. H. 2007. Grazers, producer stoichiometry, and the light: nutrient hypothesis revisited. *Ecology*, **88**, 1142–1152.

HAMM, C. E., MERKEL, R., SPRINGER, O., JURKOJC, P., MAIER, C., PRECHTEL, K. & SMETACEK, V. 2003. Architecture and material properties of diatom shells provide effective mechanical protection. *Nature*, **421**, 841–843.

HERNÁNDEZ-LEÓN, S., FRAGA, C. & IKEDA, T. 2008. A global estimation of mesozooplankton ammonium excretion in the open ocean. *Journal of Plankton Research*, **30**, 577–585.

HOFFMAN, P. F., KAUFMAN, A. J., HALVERSON, G. P. & SCHRAG, D. P. 1998. A Neoproterozoic snowball earth. *Science*, **281**, 1342–1346.

HÖLD, I. M., SCHOUTEN, S., JELLEMA, J. & SINNINGHE DAMSTÉ, J. S. 1999. Origin of free and bound mid-chain methyl alkanes in oils, bitumens and kerogens of the marine, Infracambrian Huqf Formation (Oman). *Organic Geochemistry*, **30**, 1411–1428.

HUA, H., PRATT, B. R. & ZHANG, L.-Y. 2003. Borings in Cloudina shells: complex predator–prey dynamics in the terminal Neoproterozoic. *Palaios*, **18**, 454–459.

HUNTLEY, J. W., XIAO, S. & KOWALEWSKI, M. 2006. 1.3 Billion years of acritarch history: an empirical morphospace approach. *Precambrian Research*, **144**, 52–68.

HUTCHINSON, G. E. 1961. The paradox of the plankton. *American Naturalist*, **95**, 137–145.

IRIGOIEN, X., HUISMAN, J. & HARRIS, R. P. 2004. Global biodiversity patterns of marine phytoplankton and zooplankton. *Nature*, **429**, 863–867.

IBELINGS, B. W., PORTIELJE, R., LAMMENS, E. H. R. R., NOORDHUIS, R., VAN DEN BERG, M. S., JOOSSE, W. & MEIJER, M. L. 2007. Resilience of alternative stable states during the recovery of shallow lakes from eutrophication: lake Veluwe as a case study. *Ecosystems*, **10**, 4–16.

IRWIN, A. J., FINKEL, Z. V., SCHOFIELD, O. M. E. & FALKOWSKI, P. G. 2006. Scaling-up from nutrient physiology to the size-structure of phytoplankton communities. *Journal of Plankton Research*, **28**, 459–471.

JACKSON, G. A. 1990. A model of the formation of marine algal flocs by physical coagulation processes. *Deep Sea Research*, **37**, 1197–1211.

JANG, M.-H., JUNG, J.-M. & TAKAMURA, N. 2007. Changes in microcystin production in cyanobacteria exposed to zooplankton at different population densities and infochemical concentrations. *Limnology and Oceanography*, **52**, 1454–1466.

JAVAUX, E. J., KNOLL, A. H. & WALTER, M. R. 2001. Morphological and ecological complexity in early eukaryotic ecosystems. *Nature*, **412**, 66–69.

JENSEN, S., GEHLING, J. G. & DROSER, M. L. 1998. Ediacara-type fossils in Cambrian sediments. *Nature*, **393**, 567–569.

JIANG, L., SCHOFIELD, O. M. E. & FALKOWSKI, P. G. 2005. Adaptive evolution of phytoplankton cell size. *American Naturalist*, **166**, 496–505.

JOACHIMSKI, M. M., OSTERTAG-HENNING, C. *ET AL.* 2001. Water column anoxia, enhanced productivity and concomitant changes in $\delta^{13}C$ and $\delta^{34}S$ across the Frasnian–Famennian boundary (Kowala–Holy Cross Mountains/Poland). *Chemical Geology*, **175**, 109–131.

KAVANAUGH, M. T., NIELSEN, K. J., CHAN, F. T., MENGE, B. A., LETELIER, R. M. & GOODRICH, L. M. 2009. Experimental assessment of the effects of shade on an intertidal kelp: do phytoplankton blooms inhibit growth of open-coast macroalgae? *Limnology and Oceanography*, **54**, 276–288.

KERR, S. R. & DICKIE, L. M. 2001. *The Biomass Spectrum. A Predator–Prey Theory of Aquatic Production.* Columbia University Press, New York.

KIRCHNER, J. W. & WEIL, A. 2000. Delayed biological recovery from extinctions throughout the fossil record. *Nature*, **404**, 177–180.

KNOLL, A. H. 1994. Proterozoic and Early Cambrian protists: evidence for accelerating evolutionary tempo. *Proceedings of the National Academy of Sciences, USA*, **91**, 6743–6750.

KNOLL, A. H., SEMIKHATOV, M. A., GROTZINGER, J. P. & ADAMS, W. 1995. Sizing up the sub-Tommotian unconformity in Siberia. *Geology*, **23**, 1139–1143.

KNOLL, A. H., SUMMONS, R. E., WALDBAUER, J. R. & ZUMBERGE, J. E. 2007. The geological succession of primary producers in the oceans. *In*: FALKOWSKI, P. G. & KNOLL, A. H. (eds) *Evolution of Primary Producers in the Sea*. Elsevier Academic, Burlington, MA, 133–163.

KNOLL, A. H., WALTER, M. R., NARBONNE, G. M. & CHRISTIE-BLICK, N. 2006. The Ediacaran Period: a new addition to the geological time scale. *Lethaia*, **39**, 13–30.

KODNER, R., PEARSON, A., SUMMONS, R. E. & KNOLL, A. H. 2008. Sterols in red and green algae: quantification, phylogeny and relevance for the interpretation of geologic steranes. *Geobiology*, **6**, 411–420.

KUYPERS, M. M. M., VAN BREUGEL, Y., SCHOUTEN, S., ERBA, E. & SINNINGHE DAMSTE, J. S. 2004. N_2-fixing cyanobacteria supplied nutrient N for Cretaceous oceanic anoxic events. *Geology*, **32**, 853–856.

LINK, J. 2002. Does food web theory work for marine ecosystems? *Marine Ecology Progress Series*, **230**, 1–9.

LIU, P., XIAO, S., YIN, C., ZHOU, C., GAO, L. & TANG, F. 2008. Systematic description and phylogenetic affinity of tubular microfossils from the Ediacaran Doushantuo Formation at Weng'an, South China. *Palaeontology*, **51**, 339–356.

LOGAN, G. A., HAYES, J. M., HIESHIMA, G. B. & SUMMONS, R. E. 1995. Terminal Proterozoic reorganization of biogeochemical cycles. *Nature*, **376**, 53–56.

LOGAN, G. A., SUMMONS, R. E. & HAYES, J. M. 1997. An isotopic biogeochemical study of Neoproterozoic and Early Cambrian sediments from the Centralian Superbasin, Australia. *Geochimica et Cosmochimica Acta*, **61**, 5391–5409.

LÜRLING, M. 1999. Grazer-induced coenobial formation in clonal cultures of Scenendesmus obliquus (Chlorococcales, Chlorophyceae). *Journal of Phycology*, **35**, 19–23.

LYNCH, M. & CONERY, J. S. 2003. The origins of genome complexity. *Science*, **302**, 1401–1404.

MARTIN, M. W., GRAZHDANKIN, D. V., BOWRING, S. A., EVANS, D. A. D., FEDONKIN, M. A. & KIRSCHVINK, J. L. 2000. Age of Neoproterozoic bilatarian body and trace fossils, White Sea, Russia: implications for metazoan evolution. *Science*, **288**, 841–845.

MAURER, B. A. 2003. Adaptive diversification of body size: the roles of physical constraint, energetics and natural selection. *In*: BLACKBURN, T. M. & GASTON, K. J. (eds) *Macroecology: Concepts and Consequences*. Blackwell, Oxford, 174–191.

MAY, R. M. 1973. *Stability and Complexity in Model Ecosystems*. Princeton University Press, Princeton, NJ.

MCCANN, K. 2000. The diversity–stability debate. *Nature*, **405**, 228–233.

MCFADDEN, K. A., HUANG, J. ET AL. 2008. Pulsed oxidation and biological evolution in the Ediacaran Doushantuo Formation. *Proceedings of the National Academy of Sciences, USA*, **105**, 3197–3202.

MCILROY, D. & LOGAN, G. A. 1999. The impact of bioturbation on infaunal ecology and evolution during the Proterozoic–Cambrian transition. *Palaios*, **14**, 58–72.

MCKIRDY, D. M., WEBSTER, L. J., AROURI, K. R., GREY, K. & GOSTIN, V. A. 2006. Contrasting sterane signatures in Neoproterozoic marine rocks of Australia before and after the Acraman asteroid impact. *Organic Geochemistry*, **37**, 189–207.

MOLDOWAN, J. M., DAHL, J. ET AL. 1996. Chemostratigraphic reconstruction of biofacies: molecular evidence linking cyst-forming dinoflagellates with pre-Triassic ancestors. *Geology*, **24**, 159–162.

NARBONNE, G. 2005. The Ediacara biota: neoproterozoic origin of animals and their ecosystems. *Annual Review of Earth and Planetary Science*, **33**, 421–442.

PEARCE, C. R., COHEN, A. S., COE, A. L. & BURTON, K. W. 2008. Molybdenum isotope evidence for global ocean anoxia coupled with perturbations to the carbon cycle during the Early Jurassic. *Geology*, **36**, 231–234.

PETCHEY, O. L., EKLÖF, A., BORRVALL, C. & EBENMAN, B. 2008. Trophically unique species are vulnerable to cascading extinction. *American Naturalist*, **171**, 568–579.

PETERSON, K. J. & BUTTERFIELD, N. J. 2005. Origin of the Eumetazoa: testing ecological predictions of molecular clocks against the Proterozoic fossil record. *Proceedings of the National Academy of Sciences, USA*, **102**, 9547–9552.

RASHBY, S. E., SESSIONS, A. L., SUMMONS, R. E. & NEWMAN, D. K. 2007. Biosynthesis of 2-methylbacteriohopanepolyols by an anoxygenic phototroph. *Proceedings of the National Academy of Sciences, USA*, **104**, 15099–15104.

RAVEN, J. A. 1998. The twelfth Tansley Lecture. Small is beautiful: the picophytoplankton. *Functional Ecology*, **12**, 503–513.

SARNELLE, O. 2005. Daphnia as keystone predators: effects on phytoplankton diversity and grazing resistance. *Journal of Plankton Research*, **27**, 1229–1238.

SCHEFFER, M., CARPENTER, S., FOLEY, J. A., FOLKE, C. & WALKERK, B. 2001. Catastrophic shifts in ecosystems. *Nature*, **413**, 591–596.

SCHEFFER, M., RINALDI, S., GRAGNANI, A., MUR, L. R. & VAN NES, E. H. 1997. On the dominance of filamentous cyanobacteria in shallow, turbid lakes. *Ecology*, **78**, 272–282.

SCHRÖDER, S. & GROTZINGER, J. P. 2007. Evidence for anoxia at the Ediacaran–Cambrian boundary: the record of redox-sensitive trace elements and rare earth elements in Oman. *Journal of the Geological Society, London*, **164**, 175–187.

SCOTT, C., LYONS, T. W., BEKKER, A., SHEN, Y., POULTON, S. W., CHU, X. & ANBAR, A. D. 2008. Tracing the stepwise oxygenation of the Proterozoic ocean. *Nature*, **452**, 456–459.

SHELDON, R. W., PRAKASH, A. & SUTCLIFFE, W. H. 1972. The size distribution of particles in the ocean. *Limnology and Oceanography*, **17**, 327–340.

SHEN, Y., ZHANG, T. & HOFFMAN, P. F. 2008. On the coevolution of Ediacaran oceans and animals. *Proceedings of the National Academy of Sciences, USA*, **105**, 7376–7381.

SINNINGHE DAMSTÉ, J. S., MUYZER, G. ET AL. 2004. The rise of the rhizosolenid diatoms. *Science*, **304**, 584–587.

SINNINGHE DAMSTÉ, J. S. & KÖSTER, J. 1998. A euxinic southern North Atlantic Ocean during the

Cenomanian/Turonian oceanic anoxic event. *Earth and Planetary Science Letters*, **158**, 165–173.

SMETACEK, V. S. 2000. The giant diatom dump. *Nature*, **406**, 574–575.

SPERLING, E. A., PISANI, D. & PETERSON, K. J. 2007. Poriferan paraphyly and its implications for Precambrian palaeobiology. *In*: VICKERS-RICH, P. & KOMAROWER, P. (eds) *The Rise and Fall of the Ediacaran Biota*. Geological Society, London, Special Publications, **286**, 355–368.

SUMMONS, R. E., JAHNKE, L. L., HOPE, J. M. & LOGAN, G. A. 1999. 2-Methylhopanoids as biomarkers for cyanobacterial oxygenic photosynthesis. *Nature*, **400**, 554–557.

TANG, F., YIN, C., BENGTSON, S., LIU, P., WANG, Z. & GAO, L. 2008. Octoradiate spiral organisms in the Ediacaran of South China. *Acta Geologica Sinica*, **82**, 27–34.

TILMAN, D., REICH, P. B. & KNOPS, J. M. H. 2006. Biodiversity and ecosystem stability in a decadelong grassland experiment. *Nature*, **441**, 629–632.

TURNER, J. T. 2002. Zooplankton fecal pellets, marine snow and sinking phytoplankton blooms. *Aquatic Microbial Ecology*, **27**, 57–102.

VERITY, P. G. & SMETACEK, V. 1996. Organism life cycles, predation, and the structure of marine pelagic ecosystems. *Marine Ecology Progress Series*, **130**, 277–293.

VERMEIJ, G. J. 1977. The Mesozoic marine revolution; evidence from snails, predators and grazers. *Paleobiology*, **3**, 245–258.

VERMEIJ, G. J. 1994. The evolutionary interaction among species: selection, escalation, and coevolution. *Annual Review of Ecology and Systematics*, **25**, 219–236.

VERMEIJ, G. J. 2004. Ecological avalanches and the two kinds of extinction. *Evolutionary Ecology Research*, **6**, 315–337.

WASSMANN, P. 1998. Retention versus export food chains: processes controlling sinking loss from marine pelagic systems. *Hydrobiologia*, **363**, 29–57.

WILLE, M., NÄGLER, T. F., LEHMANN, B., SCHRÖDER, S. & KRAMERS, D. J. 2008. Hydrogen sulphide release to surface waters at the Precambrian/Cambrian boundary. *Nature*, **453**, 767–769.

WOOD, R. A., GROTZINGER, J. P. & DICKSON, J. A. D. 2002. Proterozoic modular biomineralized metazoan from the Nama Group, Namibia. *Science*, **296**, 2383–2386.

WORM, B. & DUFFY, J. E. 2003. Biodiversity, productivity and stability in real food webs. *Trends in Ecology and Evolution*, **18**, 628–632.

XIAO, S., ZHANG, Y. & KNOLL, A. H. 1998. Three-dimensional preservation of algae and animal embryos in a Neoproterozoic phosphorite. *Nature*, **391**, 553–558.

XIE, S., PANCOST, R. D., YIN, H., WANG, H. & EVERSHED, R. P. 2005. Two episodes of microbial change coupled with Permo/Triassic faunal mass extinction. *Nature*, **434**, 494–497.

YIN, L., ZHU, M., KNOLL, A. H., YUAN, X., ZHANG, J. & HU, J. 2007. Doushantuo embryos preserved inside diapause egg cysts. *Nature*, **446**, 661–663.

ZERKLE, A. L., HOUSE, C. H., COX, R. P. & CANFIELD, D. E. 2006. Metal limitation of cyanobacterial N_2 fixation and implications for the Precambrian nitrogen cycle. *Geobiology*, **4**, 285–297.

ZHURAVLEV, A. YU. 2001. Biotic diversity and structure during the Neoproterozoic–Ordovician transition. *In*: ZHURAVLEV, A.YU. & RIDING, R. (eds) *Ecology of the Cambrian Radiation*. Columbia University Press, New York, 173–199.

ZHOU, C., XIE, G., MCFADDEN, K., XIAO, S. & YUAN, X. 2007. The diversification and extinction of Doushantuo–Pertatataka acritarchs in South China: causes and biostratigraphic significance. *Geological Journal*, **42**, 229–262.

ZÖLLNER, E., HOPPE, H.-G., SOMMER, U. & JÜRGENS, K. 2009. Effect of zooplankton-mediated trophic cascades on marine microbial food web components (bacteria, nanoflagellates, ciliates). *Limnology and Oceanography*, **54**, 262–275.

Late Proterozoic plate tectonics and palaeogeography: a tale of two supercontinents, Rodinia and Pannotia

CHRISTOPHER R. SCOTESE

Department of Earth and Environmental Sciences, University of Texas at Arlington, Arlington, TX 76019, USA
(e-mail: chris@scotese.com)

Abstract: The plate tectonic and palaeogeographic history of the late Proterozoic is a tale of two supercontinents: Rodinia and Pannotia. Rodinia formed during the Grenville Event (*c.* 1100 Ma) and remained intact until its collision with the Congo continent (800–750 Ma). This collision closed the southern part of the Mozambique Seaway, and triggered the break-up of Rodinia. The Panthalassic Ocean opened as the supercontinent of Rodinia split into a northern half (East Gondwana, Cathyasia and Cimmeria) and a southern half (Laurentia, Amazonia–NW Africa, Baltica, and Siberia). Over the next 150 Ma, North Rodinia rotated counter-clockwise over the North Pole, while South Rodinia rotated clockwise across the South Pole. In the latest Precambrian (650–550 Ma), the three Neoproterozoic continents – North Rodinia, South Rodinia and the Congo continents – collided during the Pan-Africa Event forming the second Neoproterozoic supercontinent, Pannotia (Greater Gondwanaland). Pan-African mountain building and the fall in sea level associated with the assembly of Pannotia may have triggered the extreme Ice House conditions that characterize the middle and late Neoproterozoic. Although the palaeogeographic maps presented here do not prohibit a Snowball Earth, the mapped extent of Neoproterozoic ice sheets favour a bipolar Ice House World with a broad expanse of ocean at the equator. Soon after it was assembled (*c.* 560 Ma), Pannotia broke apart into the four principal Palaeozoic continents: Laurentia (North America), Baltica (northern Europe), Siberia and Gondwana. The amalgamation and subsequent break-up of Pannotia may have triggered the 'Cambrian Explosion'. The first economically important accumulations of hydrocarbons are from Neoproterozoic sources. The two major source rocks of this age (Nepa of Siberia and Huqf of Oman) occur in association with massive Neoproterozoic evaporite deposits and in the warm equatorial–subtropical belt, within 30° of the equator.

In this report we tell the tale of two late Precambrian supercontinents: Rodinia and Pannotia. We describe the configuration of continents that comprised Rodinia and Pannotia, outline a plausible history of their formation and break-up, reconstruct long-gone mountain ranges and seaways, and infer the location of ancient plate boundaries. The Rodinia–Pannotia model that we construct can be used as a framework to understand better the depositional environments of late Precambrian oil source rocks and the history of late Precambrian global climate change.

Time interval of interest: late Mesoproterozoic and Neoproterozoic

This study describes the plate tectonic events of the late Mesoproterozoic (*c.* 1200 Ma) and the entire Neoproterozoic (1000–542 Ma). Four palaeogeographic maps have been constructed for the mid-Cryogenian (750 Ma), late Cryogenian (690 Ma), middle Ediacaran (600 Ma) and earliest Cambrian (540 Ma) (see figures later in this chapter). The absolute ages assigned to these maps are based on the ICS International Time Scale (Gradstein *et al.* 2004). These palaeogeographic maps show the ancient configuration of mountains, lowlands, shallow seas and deep ocean basins, as well as the likely location of mid-ocean ridges, subduction zones and regions of continent–continent collision.

A very difficult task

Producing plate tectonic and palaeogeographic maps for the late Proterozoic is a difficult task. Some of the reasons are obvious: less than 3% of the world's mapped outcrop is Neoproterozoic in age (Lowe 1992); much of the rock record is deformed or highly metamorphosed; within this broad time interval (spanning over 500 Ma), it is difficult to correlate widely separated geological units; and absolute age dates are sparse in some areas, have error margins in tens of millions of years and often give conflicting results.

For these and other reasons, it would be impossible to make global plate tectonic and palaeogeographic maps for the Late Proterozoic based solely on the available geological and geophysical data. There are too many missing pieces. Fortunately, other tools and techniques are available that make an impossible task merely very difficult.

From: Craig, J., Thurow, J., Thusu, B., Whitham, A. & Abutarruma, Y. (eds) *Global Neoproterozoic Petroleum Systems: The Emerging Potential in North Africa.* Geological Society, London, Special Publications, **326**, 67–83.
DOI: 10.1144/SP326.4 0305-8719/09/$15.00

Plate tectonic reconstruction methodology and assumptions

The Plate Tectonic Reconstruction Methodology, or PALEOMAP Method, is the technique used by the author to build a plate tectonic framework upon which the available geological and geophysical data can be assembled.

The PALEOMAP Method

(1) Define the lithospheric plates (also called tectonic elements or plate polygons) that have had an independent history of motion.
(2) Compile geological and tectonic evidence to constrain the timing of important plate tectonic events such as continental break-up, the location and persistence of active subduction or continent–continent collision.
(3) Use available palaeomagnetic, linear magnetic anomaly and hot spot data to orient these plates with respect to each other and with respect to the spin axis or hot spot reference frame. (For the Neoproterozoic, only palaeomagnetic information is available.)
(4) Once the tectonic history is 'roughed out', finite rotations are then used to build a hierarchical plate model (plate circuit) that describes the motion of the plates with respect to the spin axis or hot spot reference frame (Ross & Scotese 1988).

Underlying the PALEOMAP Method are several hypotheses concerning behaviour of the plate tectonic system and the tempo and mode of plate motions. These plate tectonic hypotheses provide a guiding framework that permits interpolation and extrapolation when solid evidence is otherwise lacking. These hypotheses have played a particularly important role in the production of the late Precambrian reconstructions presented in this paper.

Plate tectonic hypotheses

(1) The negative buoyancy of the subducting slab (slab pull) is the dominant driving force of plate tectonics (Forsyth & Uyeda 1975; Conrad & Lithgow-Bertelloni 2002). Two corollaries of this hypothesis are: (a) that plates are pulled towards the subduction zone in a direction approximately normal to the orientation of the trench axis (Scotese & Rowley 1985); and (b) that mid-ocean ridges tend to be aligned parallel to major subduction zones.
(2) Large continental plates move slowly (*c*. 2 cm year^{-1}) (e.g. Eurasia during the Mesozoic and Cenozoic). At this rate, a supercontinent will move only 40° (4400 km) in 200 Ma.
(3) There has been no significant True Polar Wander.
(4) Plate tectonics is a catastrophic system. Long periods of slow and steady plate motions are interrupted by relatively brief periods of global plate tectonic reorganization. The two most likely causes of global plate tectonic reorganizations are continent–continent collision (i.e. the loss of a subduction zone) and the subduction of a mid-ocean ridge (e.g. the subduction of the Tethyan mid-ocean ridge triggered the break-up of Pangaea: Scotese 1991).
(5) Finally, we can take a uniformitarian approach to plate tectonics. In other words, when modelling plate motions during the late Precambrian, we can expect the same elegant and simple pattern of plate motions that characterized the Mesozoic and Cenozoic.

Reconstruction of Rodinia and Pannotia

In order to produce the plate tectonic and palaeogeographic reconstructions presented here, we have made another important assumption, namely, that during the mid and Late Neoproterozoic there were two Pangaea-sized supercontinents: Rodinia (Fig. 1) and Pannotia (Fig. 2).

The name *Rodinia*, which figuratively means 'mother of all continents', was proposed by McMenamin & McMenamin (1990), and was adopted by Dalziel (1991), Moores (1991) and Hoffman (1991). *Rodinia* initially referred to the presumed late Precambrian supercontinent that broke apart at the start of the Palaeozoic. It now is used to describe the first of two Neoproterozoic Pangaeas. Rodinia formed during the Grenville Event (*c*. 1100 Ma) and broke apart about 750 Ma ago. It remained intact for approximately 300 Ma, making it a long-lived supercontinent.

The second late Neoproterozoic supercontinent, *Pannotia* (which means 'all southern land': Powell 1995) formed at the very end of the Precambrian. Other names for Pannotia are 'Greater Gondwanaland' (Stern 1994, 2002), the 'Vendian supercontinent' (Meert & Torsvik 2004) or the 'Pan-African supercontinent'. As the name 'Greater Gondwanaland' implies, the Palaeozoic supercontinent of Gondwana(land) formed the core of Pannotia. Pannotia was a short-lived supercontinent. It assembled 650–500 Ma ago during the Pan-African Event, and had begun to break apart approximately 560 Ma ago with the opening of the Iapetus Ocean (Cawood *et al.* 2001, 2007).

Reconstruction of Rodinia

Figure 1 is the prototypical reconstruction of Rodinia. Although there are various reconstructions

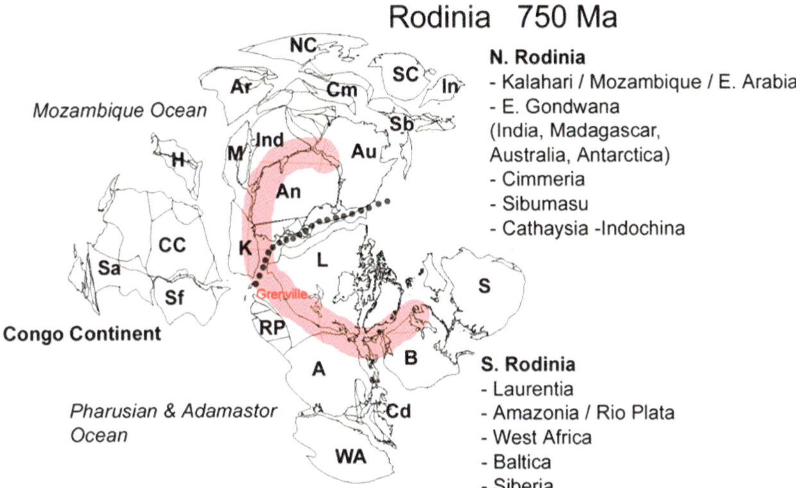

Fig. 1. Rodinia. A, Amazon Craton; An, Antarctica; Ar, Arabia; Au, Australia; B, Baltica; CC, Congo continent; Cd, Cadomian–Avalonian arc; Cm, Cimmeria; In, Indochina; H, Hijaz arc; Ind, India; K, Kalahari Craton; L, Laurentia; M, Madagascar; NC, North China; RP, Rio Plata; S, Siberia; Sa, Saharan shield; SC, South China; Sf, Sao Francisco; WA, West African Craton; the dotted line is the location of the Panthalassic Rift.

of Rodinia (Hoffman 1991; Powell *et al.* 1993; Dalziel 1997; Weil *et al.* 1998; Karlstrom *et al.* 1999; Pisarevsky *et al.* 2003; Li *et al.* 2007), they all share these five elements: (1) North America, or more properly the Laurentian shield (Laurentia), lies near the core of Rodinia; (2) the west coast of North America is adjacent, in some fashion, to the eastern margins of Antarctica and Australia (East Gondwana); (3) the Amazon Craton lies to the east of Laurentia; (4) Siberia (the Aldan and Anabar shields) generally lies to the north of Arctic Canada or Baltica; and (5) Baltica (northern Europe), rotated by various degrees, is adjacent to eastern Greenland. If any of these relationships

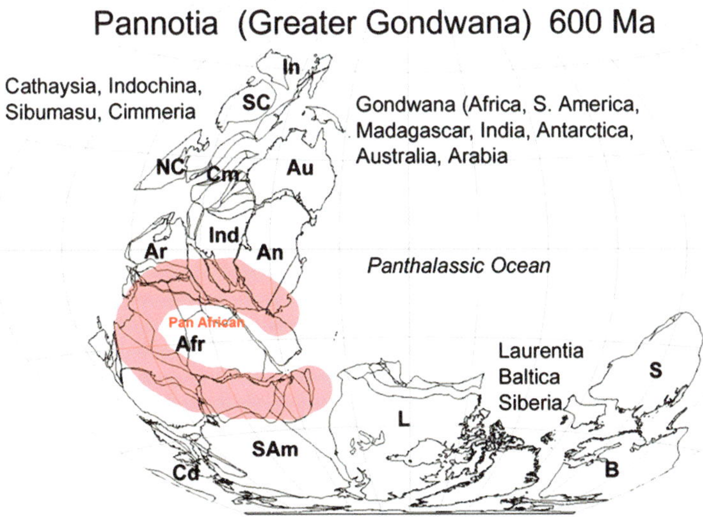

Fig. 2. Pannotia. Afr, Africa; An, Antarctica; Ar, Arabia; Au, Australia; B, Baltica; Cd, Cadomian–Avalonian arc; Cm, Cimmeria; In, Indochina; Ind, India; L, Laurentia; NC, North China; S, Siberia; Sam, South America; SC, South China.

can be shown to be incorrect, then the concept of a unified Rodinia supercontinent would be invalidated (Hoffman 1999).

Although all reconstructions of Rodinia share these five elements, there is considerable variability regarding the exact arrangement of these elements.

- *SWEAT–AUSWUS fit*: the exact placement of Antarctica and Australia along the western margin of North America varies. The classic SWEAT fit of Dalziel (1991), Hoffman (1991) and Moores (1991) is used in our reconstruction of Rodinia (Fig. 1) (see also Young 1992). In the AUSWUS version of the Rodinia reconstruction (Karlstrom *et al.* 1999; Burrett & Berry 2000) Antarctica and Australia are displaced southwards relative to North America so that eastern Australia, not Antarctica, is adjacent to the SW USA. Burrett & Berry (2002), using multivariate statistics, determined that the AUSWUS fit did a better job of matching the adjacent geological provinces than the SWEAT fit. Recently, other authors have proposed variations of the SWEAT or AUSWUS fit (Wingate *et al.* 2002; Meert 2003; Cawood *et al.* 2007; Li *et al.* 2007). Given the uncertainties in the palaeomagnetic data, however, all of the Laurentia/Australia–Antarctica reconstructions described above are permissible.
- *Siberia*: the location of Siberia in the reconstruction of Rodinia is also variable. Sometimes it is placed next to Baltica, as shown here. However, most of the time it is adjacent to the Arctic Islands (Dalziel 1997; Hoffman 1991; Pelechaty 1996). In the most radical reconstruction of Rodinia (Sears & Price 2000), Siberia is placed along the western margin of North America replacing Antarctica and Australia. This alternate fit of Siberia is based on strong stratigraphic ties and matching Neoproterozoic facies belts.
- *Baltica*: although located next to Greenland in most Rodinia reconstructions, it is shown rotated by different degrees. In the original Rodinia reconstruction, the margin of Norway was adjacent to SE Greenland (Hoffman 1991), so that Grenville-age rocks in southern Scandinavia matched up with the Grenville Front in eastern North America. Other reconstructions of Rodinia place the Norwegian margin of Baltica in the same orientation that it occupied in the late Palaeozoic (Pangaea) (Dalziel 1997). This fit is implausible because it would require that the early Palaeozoic Iapetus Ocean opened and closed in exactly the same location. In the Greenland–Baltica fit shown here, Baltica is rotated 90° clockwise, so that the Tornquist margin lies adjacent to eastern Greenland. This fit of Baltica and Laurentia agrees better with Cambrian orientations based on a more complete palaeomagnetic data set (Torsvik *et al.* 1991, 1992; Torsvik & Rehnstrom 2001; Cocks & Torsvik 2002). The nearly coincident age of rifting along the Tornquist Line and along the eastern margin of Laurentia (620–530 Ma: Cawood *et al.* 2007) suggests that they may have been conjugate rift margins. This configuration of Baltica also satisfactorily aligns the trend of the Grenville Front from Laurentia to Baltica.

The positions of the other continental blocks around the periphery of Rodinia are less certain, but some arguments concerning their configuration can be made.

Amazon Craton. A key element to the Rodinia reconstruction is the fit between eastern Laurentia and western South America (Amazon Craton and Rio Plata). Bond *et al.* (1984) were among the first authors to suggest that there was a Precambrian Pangaea that rifted apart during the latest Precambrian (625–555 Ma). They noted that North America (Laurentia) was surrounded by passive margins with thick post-rift sedimentary accumulations that formed during the latest Precambrian (*c.* 600 Ma). They suggested that the western margin of South America was the conjugate rift margin of eastern Laurentia.

West African Craton. The West African Craton of NW Africa is simply an extension of the Amazon Craton. The same Precambrian basement trends and older zones of deformation can be traced across the South Atlantic from NE Brazil to the Ivory Coast (e.g. Sergipe–Oubanguides fold belts, the Trans-Brazilian shear zone Dahomeyides and the '4 degree 50 minute' shear zone of the Hoggar region: Trompette 1994, 1997). The Amazon Craton and the West African Craton were together throughout the Late Proterozoic and Palaeozoic, and separated when the South Atlantic opened during the early Cretaceous.

East Gondwana. In this reconstruction of Rodinia, East Gondwana is a single block composed of Antarctica, Australia, India, Madagascar, Somalia, Mozambique and South Africa (Kalahari Craton). This is a simplification. It is likely that some of the East Gondwana terranes were separated by middle and late Proterozoic oceans (Bozhko 1986; Boger *et al.* 2001; Meert 2003). However, the location, extent and timing of closure of these oceans are not well known and deserve further study.

North and South China. Similarly, the Cathaysian terranes (North China, South China and Indochina) are placed adjacent to the Indo-Australian margin of East Gondwana. Palaeomagnetic and biogeographic arguments can be made (Burrett *et al.* 1990; Scotese & McKerrow 1990) that this is a reasonable configuration for the early Palaeozoic and, by inference, the late Neoproterozoic. In addition, this location for South China places it near the North Pole for most of the late Neoproterozoic, a location consistent with the presence of Sinian tillites and glacial deposits (Wang *et al.* 1981). In an alternate reconstruction of Rodinia (Li & Powell 1996), South China lies along the NW coast of Laurentia, adjacent to the Yukon and Northwest Territories; North China is in the vicinity of Siberia (Li *et al.* 2007).

Congo continent. The most problematic continental block in any reassembly of Rodinia is the Congo continent. The Congo continent is composed of the Congo continent plus the Saharan shield minus the West African Craton and the Hijaz island arcs of Sudan, Egypt and Arabia (Johnson & Woldehaimanot 2003). The Mozambique Belt suture (East African Orogen: Stern 1994) separates the Congo continent from East Gondwana and the Damara/Lufilian–Zambezi suture (Watters 1976; Daly 1986; John *et al.* 2003; Johnson *et al.* 2005) separates the Congo continent from the Kalahari Craton to the south. The Sao Francisco Craton of eastern Brazil (Chemale *et al.* 1993), which lies to the east of the Pan-African/Brazilide orogenic belt of central Brazil (Trompette 1994, 1997), was originally part of the Congo continent. It became part of Brazil when, in the early Cretaceous, the South Atlantic Ocean opened to the west of this branch of the Pan-African suture.

Although the original Rodinia reconstruction places the Congo continent adjacent to the other Rodinia continental blocks (Hoffman 1991; Dalziel 1997), we believe that during the late Mesoproterozoic (1100 Ma) the Congo continent was separated from Rodinia by a wide ocean (>10 000 km). By the early Neoproterozoic (900 Ma), the Mozambique Seaway separated the Hijaz island arcs (Egypt and Sudan: Stern 1994) and Congo continent from the northern half of Rodinia (East Gondwana and Cathaysia). The Pharusian Ocean separated the Congo continent from the southern half of Rodinia (Laurentia, Baltica, Siberia and the West African–Amazon Craton). The collision of the Congo continent with Rodinia during the middle Neoproterozoic (Cryogenian, *c.* 750 Ma) triggered the break-up of Rodinia into two halves (see Fig. 3). This collision and the subsequent break-up of northern and southern Rodinia is the key to the tectonic model presented here, and will be discussed in more detail in the following section.

Table 1 is a list of the finite poles of rotation used to reassemble Rodinia at 750 Ma.

Reconstruction of Pannotia

Although the name 'Rodinia' is better known than 'Pannotia', we know much more about the supercontinent of Pannotia than we know about Rodinia. In many respects 'Greater Gondwanaland' would be a better name for this latest Precambrian Pangaea, because at the core of Pannotia is the Palaeozoic supercontinent of Gondwanaland (also simply called Gondwana). Gondwana, and hence Pannotia, was assembled when the Congo continent was caught between the northern and southern halves of Rodinia. This series of collisions, which is known as the Pan-African Event, began approximately 750 Ma ago (initial collision along the Mozambique belt: Stern 1994) and may have extended into the early Cambrian (530 Ma: Meert 2003). The peak in Pan-African orogenesis dates from 610 to 640 Ma (Meert 2003). As can be seen in the reconstructions for 750, 700 and 600 Ma (Fig. 3), Pannotia was formed as a result of Rodinia 'turning itself inside-out' (Hoffman 1991).

Reviewing the configuration of continents that make up Pannotia (Fig. 2), we see that Africa is at its centre. The other Gondwanan continents of Arabia, Madagascar, India, Antarctica, and Australia and South America surround Africa. The fit of the continents that make up Gondwana is well established (Lawver & Scotese 1987; Scotese *et al.* 1999; Smith 1999); there is no debate as to the relative positions of these continental blocks.

Pannotia also includes the early Palaeozoic continents of Laurentia (North America), Baltica (northern Europe) and Siberia. These continental blocks have the same relative positions in both the Rodinia and the Pannotia reconstructions.

The Cathaysian terranes (North China, South China and Indochina), together with the Cimmerian terranes (Sibumasu, Qiang Tang, Lhasa, Afghanistan, Iran and Turkey) were adjacent to the Arabian–Indian–Australian margin of Pannotia (Audley-Charles 1988; Sengor *et al.* 1988; Sengor & Natal'in 1996; Metcalfe 1993, 1999). Along the northern coasts of Africa and South America were several smaller terranes (central Europe, Iberia, Britain, New England and Maritime Canada, Florida and Yucatan) that made up the Avalonian–Cadomian active margin (Murphy & Nance 1989, 1991; Nance *et al.* 1991; Nance & Murphy 1994; McNamara *et al.* 2001). After the consolidation of Pannotia, the Avalonian–Cadomian active margin may have extended northwards along the Uralian margin of the Baltic Shield (Timanide Orogen: Gee & Pease 2004; Scarrow *et al.* 2001).

A final note concerning the term 'Pannotia', in some of the papers discussing Precambrian supercontinents the term 'Rodinia' rather than 'Pannotia' is used to describe the Pangaea-like configuration

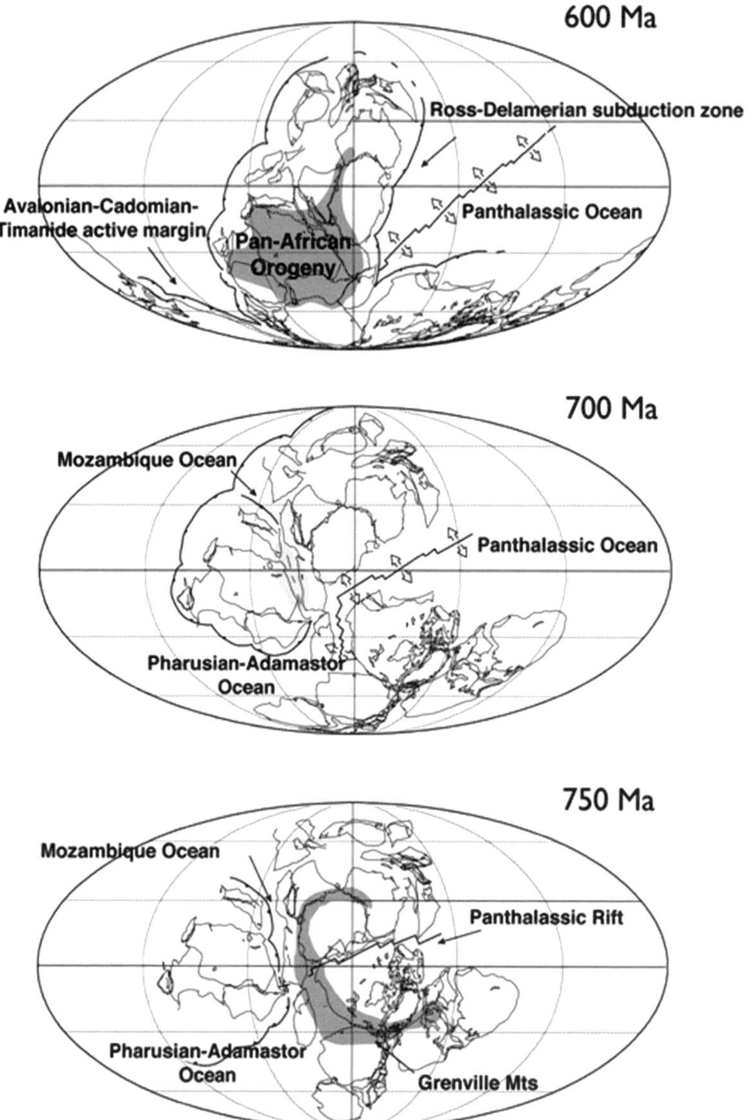

Fig. 3. Transition from Rodinia to Pannotia (750–600 Ma); after Scotese *et al.* (1999).

that existed at the end of the Precambrian (600–540 Ma ago). This is incorrect. Rodinia did not exist 600 Ma ago. The term Pannotia should be used for Precambrian *pangaeas* younger than 700 Ma.

Table 2 is a list of the finite poles of rotation used to reassemble Pannotia at 600 Ma.

Neoproterozoic plate tectonic and palaeogeographic maps

This section outlines a plate tectonic model that describes the assembly of Rodinia and its subsequent break-up to form Pannotia. Figure 3 illustrates the motions of Rodinia, the Congo continent and Pannotia, and the plate boundaries (mid-ocean ridges, subduction zones, island arcs, Andean margins and collision zones) that were active during the Neoproterozoic. Also presented are a set of full-colour palaeogeographic reconstructions for 750, 690, 600 and 540 Ma ago (Figs 4–9). These maps illustrate: the collision of the Congo continent with Rodinia at approximately 750 Ma (Fig. 4); the formation of the Panthalassic Rift that split Rodinia in two (North and South Rodinia)

Table 1. *Finite rotations used to assemble Rodinia at 750 Ma*

Plate polygon	Latitude	Longitude	Angle (°)
Amazonia (A)	−8	48	118
Arabia (Ar)	65	40	102
Australia (Au)	32	30	94
Baltica (B)	−31	92	−171
Congo (CC)	44	67	115
East Antarctica (An)	27	43	119
Hijaz (H)	58	64	107
India (Ind)	74	−20	69
Indochina (In)	−47	−27	156
Kalhari (K)	−58	−120	−119
Laurentia (L)	−40	12	138
Madagascar (Md)	56	37	109
Mozambique (M)	−58	−120	−119
North China (NC)	42	42	44
Rio Plata (RP)	−8	48	118
Saõ Francisco (SF)	32	37	109
Siberia (S)	−25	17	148
South China (SC)	52	147	−164
West Africa (WA)	−3	78	127

Note: All finite rotations follow the right-hand rule. Negative latitude values denote the southern hemisphere. Negative longitude values denote the western hemisphere.

Table 2. *Finite rotations used to assemble Pannotia at 600 Ma*

Plate polygon	Latitude	Longitude	Angle (°)
Amazonia (A)	38	66	138
Arabia (Ar)	49	109	119
Australia (Au)	43	74	74
Baltica (B)	−4	113	129
Congo (CC)	46	106	122
East Antarctica (An)	24	75	95
Hijaz (H)	46	106	122
India (Ind)	69	162	91
Indochina (In)	−28	−39	121
Kalhari (K)	46	106	122
Laurentia (L)	0	41	144
Madagascar (Md)	52	93	107
Mozambique (M)	46	106	122
North China (NC)	48	143	52
Rio Plata (RP)	38	66	138
Sao Francisco (SF)	38	66	138
Siberia (S)	11	47	167
South China (SC)	32	136	−133
West Africa (WA)	46	106	122

Note: All finite rotations follow the right-hand rule. Negative latitude values denote the southern hemisphere. Negative longitude values denote the western hemisphere.

(Figs 4 & 8); the gradual closing of the Pharusian Ocean–Mozambique Seaway and the Pan-African (Adamastor) Ocean to form Pannotia (Figs 4–7); the uplift of the U-shaped Pan-African collisional mountain ranges (Figs 5–7); and the subsequent break-up of Pannotia into the Palaeozoic continents: Gondwana, Laurentia, Baltica and Siberia (Figs 6 & 9).

In addition to showing the active plate boundaries, these reconstructions attempt to portray the approximate distribution of deep ocean, shallow seas, lowlands and highlands. These palaeotopographic and palaeobathymetric interpretations are based on the database of Neoproterozoic depositional environments, lithologies and palaeoclimate indicators assembled by Metz (2001), supplemented by the recent compilation of Stewart (2007). Our estimate of the extent of the Neoproterozoic ice sheets is shown in Figure 10. A more detailed description of this Neoproterozoic geological

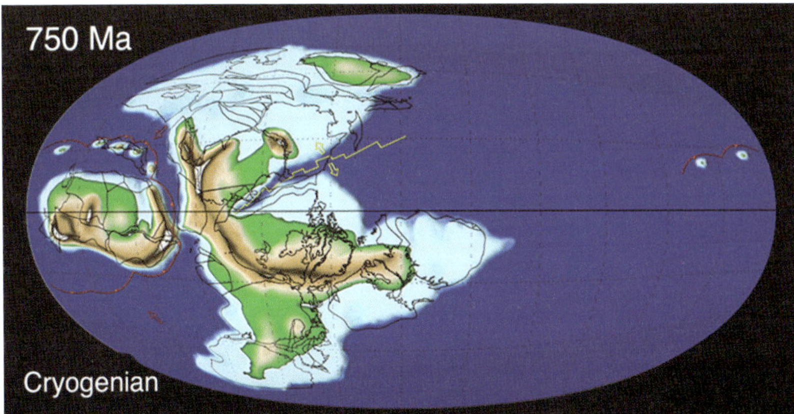

Fig. 4. Middle Neoproterozoic reconstruction (Cryogenian; 750 Ma; oval projection). Blue, deep ocean; light blue, shallow sea; green, lowlands; brown, uplands; white, high mountains; yellow line, mid ocean ridge; red line, subduction zone; red X's, collision zone.

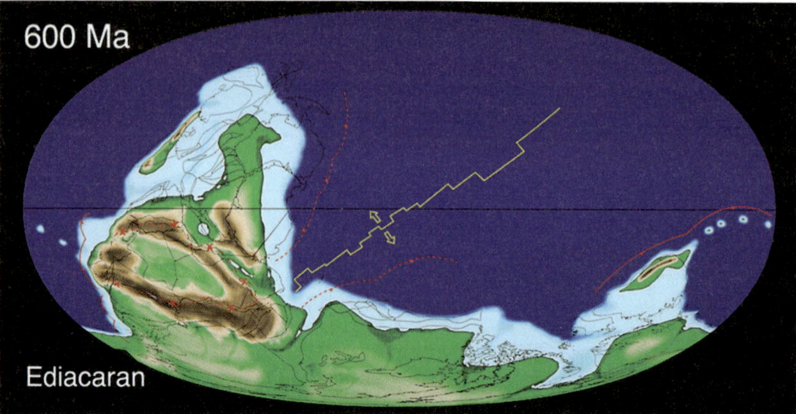

Fig. 5. Late Neoproterozoic reconstruction (Ediacaran, 600 Ma, oval projection). For an explanation of colours see Figure 4.

database, and a palaeoclimatic test of the Rodinia and Pannotia reconstructions will be the subject of a companion paper.

From a map-making point of view, it is not possible to produce global plate tectonic and palaeogeographic reconstructions for any time prior to the middle Mesoproterozoic (*c.* 1300 Ma). There is insufficient geological and palaeomagnetic information to constrain continental positions. It is not possible to make precise stratigraphic correlations so we cannot know whether tectonic events in widely separated areas were synchronous. Finally, prior to the assembly of Rodinia, the continental terranes were too numerous and too widely dispersed to reassemble in any meaningful fashion. Despite these difficulties, Rodgers (1996) has made an imaginative attempt to map plate motions back to 3 Ga (Ga – giga-annum – is 10^9 years).

The building of Rodinia (1200–1050 Ma)

The core of the supercontinent of Rodinia was assembled during the Grenville Event, 1200–1050 Ma ago (Rivers 1997). A minimum of two, large, unnamed continents collided along the Grenville suture, which can be traced from the Sveconorwegian belt in southern Norway (Gower & Owen 1984; Cosca 1998), across the Atlantic Ocean to East Greenland (Kalsbeek *et al.* 2000), to Labrador, through Quebec and Ontario, beneath the Palaeozoic strata of the North American mid-continent, emerging briefly in the Llano uplift and Marathon mountains of west Texas. Similar age deformation is found in the Namaqua–Natal orogenic belt of the Kalahari Craton (Dalziel *et al.* 2000), along the Dronning Maud Land sector of East Antarctic (Dirks & Wilson 1995; Bauer *et al.*

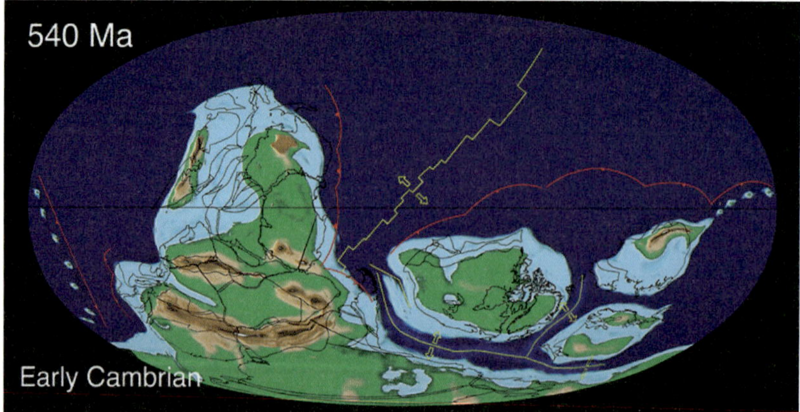

Fig. 6. Early Cambrian Reconstruction (540 Ma, oval projection). For an explanation of colours see Figure 4.

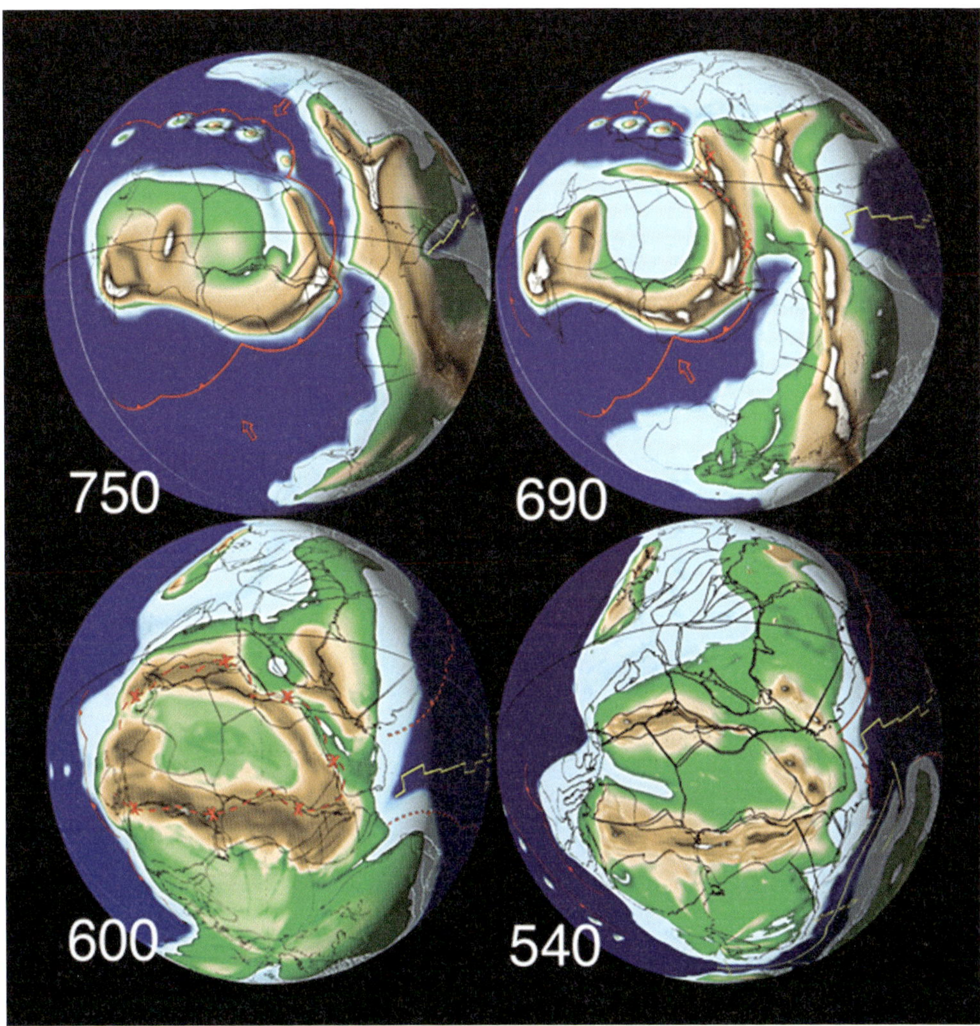

Fig. 7. Pan-African Orogeny (750, 690, 600 and 540 Ma, orthographic projection). For an explanation of colours see Figure 4.

2003; Jacobs *et al.* 2003) and extending into central Australia (Albany–Frasierbelt and Musgrave block: Boger *et al.* 2001). The opposite side of this collision belt runs parallel to the Tornquist Line, continues along the western edge South America and divides East Gondwana into unequal halves (Fig. 1). Grenville-age deformation (1.2 Ga) in western Brazil (Rondonia) supports the argument that the Amazon Craton collided with Laurentia during the Grenville Event (Tohver *et al.* 2002).

The break-up of Rodinia (800–700 Ma)

After its assembly approximately 1000 Ma ago, Rodinia remained intact for another 300 Ma. During the Cryogenian (*c.* 750 Ma), Rodinia split into two large continents (North Rodinia and South Rodinia) (Figs 3 & 8). North Rodinia was made up of East Gondwana, Cathaysia and the Cimmerian terranes. South Rodinia was composed of Laurentia, Baltica, Siberia, the Amazon–West African Craton and, possibly, the Rio de la Plata Craton. As illustrated in Figures 3 and 8, a new continental rift separated western Laurentia from East Antarctica (Rowell *et al.* 1993; Goodge 2002) and eastern Australia. Eventually, the rift developed into the central mid-oceanic ridge of the Panthalassic Ocean (Fig. 5).

Why did Rodinia break apart? We believe that redirected slab-pull forces triggered by the initial

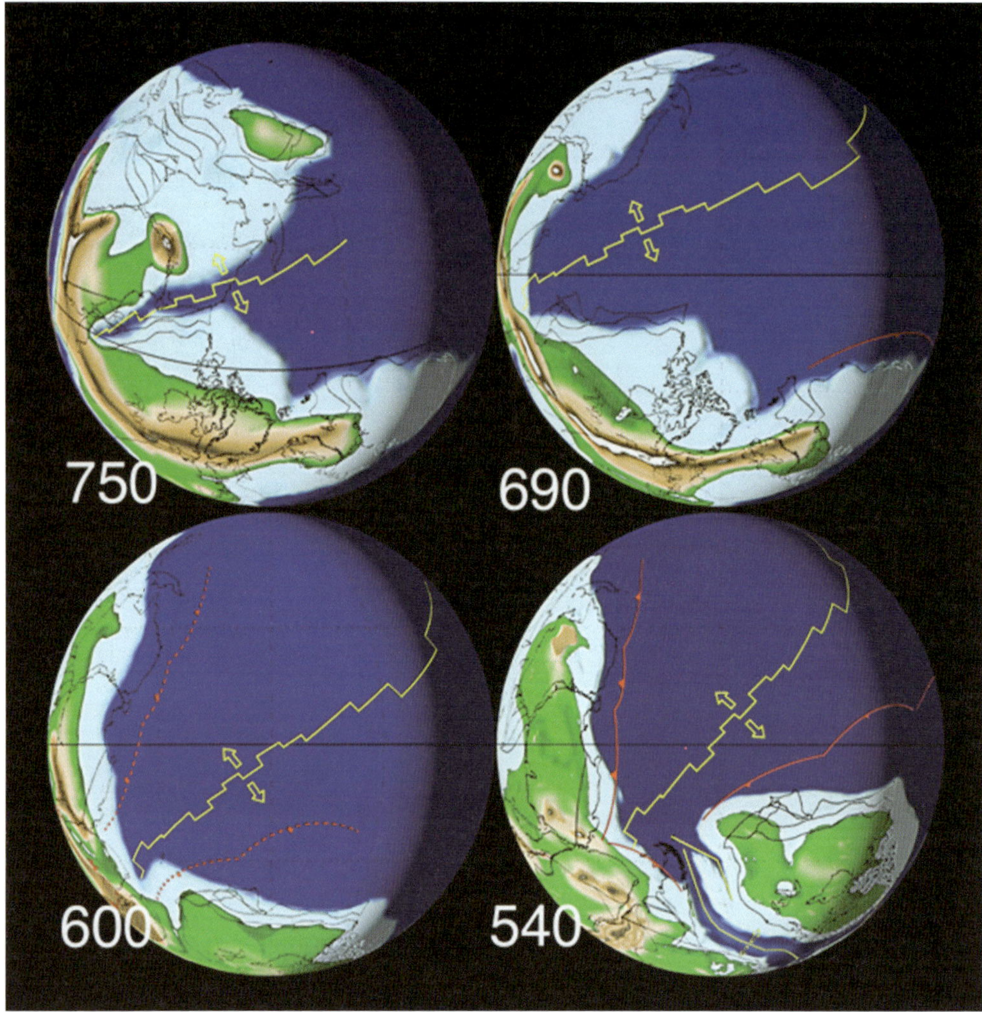

Fig. 8. Opening of the Panthalassic Ocean (750, 690, 600 and 540 Ma, orthographic projection). For an explanation of colours see Figure 4.

collision of the Congo continent (*c*. 750 Ma) played an important role in the break-up of Rodinia. In the following scenario we emphasize the important role that slab pull may have played.

Prior to 750 Ma, the Mozambique Belt subduction zone and the Pharusian–Adamastor subduction zone were connected and surrounded the Congo continent (Fig. 4). At that time, the vast region of oceanic lithosphere attached to the western part of Rodinia was subducted to the west beneath the Congo continent. As a consequence, Rodinia was drawn inexorably closer to the Congo continent (Fig. 7).

At about 750 Ma the eastern edge of the Congo continent collided with Rodinia in the vicinity of Namaqua–Natal orogenic belt of the Kalahari Craton. Westwards subduction at the point of collision came to a halt. However, vigorous south-directed subduction continued beneath the Hijaz island arcs along the northern rim of the Congo continent, and north-directed subduction continued beneath the back-arc basins and Andean margin along the southern edge of the Congo continent (Fig. 7).

The once continuous circum-Congo subduction zone was now divided into two parts. Each subduction zone began to tug on Rodinia in a different direction. North Rodinia began to rotate counter-clockwise as it was drawn by slab pull into the Mozambique–Hijaz subduction zone. Conversely, South Rodinia began to rotate clockwise, also drawn by slab pull, towards the Pharusian–Adamastor subduction zone. We propose that these differential plate tectonic stresses tore Rodinia in half

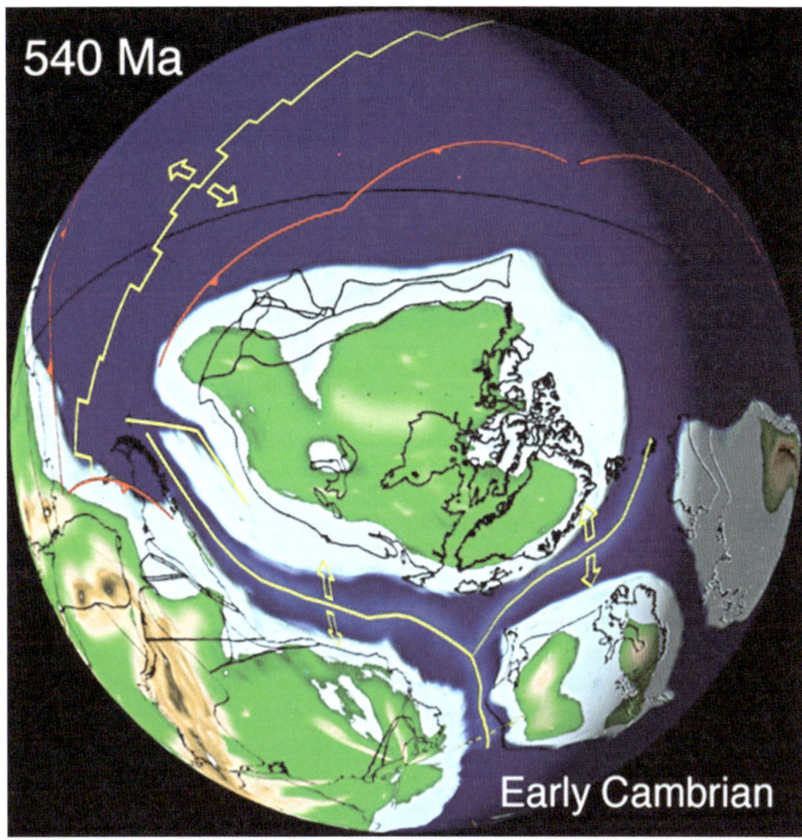

Fig. 9. Iapetus Ocean (Early Cambrian, orthographic projection). For an explanation of colours see Figure 4.

(Fig. 8). In the next 100 Ma Rodinia would be turned inside out.

The timing of the initial collision of the Congo continent with Rodinia and the subsequent rifting of Rodinia has been debated. The best estimate is about 750 Ma. The key features are: (1) the deep crustal burial and metamorphism along the southern portion of the Mozambique Belt (Stern 1994);

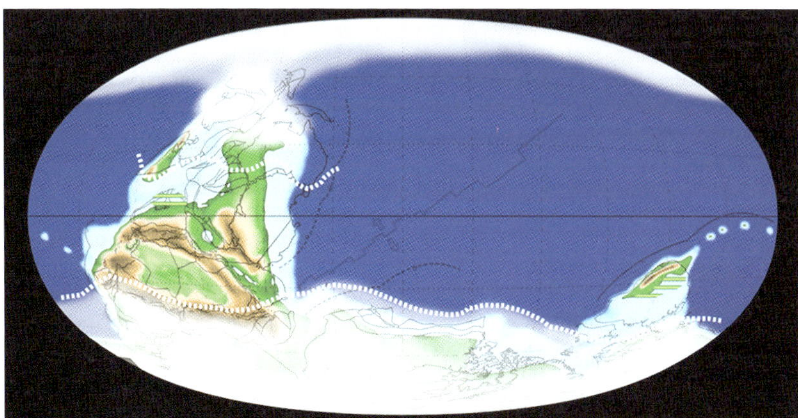

Fig. 10. Neoproterozoic Ice House World (dashed white line, maximum glacial advance; horizontal green lines, oil source rocks). For an explanation of colours see Figure 4.

(2) the formation of passive margin sequences in western North America (Bond *et al.* 1984), East Antarctica and Australia; and (3) the eruption of rift-related basalts in Antarctica (Goodge 2002) and North America that have been dated at about 750 Ma.

An additional estimate of the timing of rifting of Rodinia can be made based on the rates of plate motion. Looking forwards from the time of this rifting event (*c.* 750 Ma), we know that eventually the two halves of Rodinia collided with the Congo continent to form the core of Pannotia. This collision (Pan-African Orogeny) took place about 600 Ma ago (Figs 5 & 7). From its initial rift position (SWEAT fit) to its final docking configuration, North Rodinia and South Rodinia each approximately travelled 5500 km. Large continental plates move at rates of less than 2 cm year^{-1}. At these rates it would have taken approximately 180 Ma to complete this journey. This requires that the break-up of Rodinia began approximately 180 Ma earlier, or at about 780 Ma ($\pm$30 Ma).

Building Pannotia: the Pan-African Event

Continental collisions are notoriously messy affairs lasting tens of millions of years. This is particularly true when assembling supercontinents from multiple continental blocks. In this regard, it should come as no surprise that Pannotia took more than 100 Ma to assemble (Pan-African Event) (Rogers 1995). The collisions that closed the ocean basins separating the Congo continent from North Rodinia and South Rodinia were diachronous. The oldest collisions took place in the southernmost arm of the Mozambique Seaway (800–750 Ma: Stern 1994). The Mozambique–Hijaz Ocean closed sequentially from south (Mozambique) to north (Egypt and Sudan). The last phase of collision between North Rodinia and the Congo continent collapsed the Hijaz island arcs between Arabia and Egypt–Sudan (*c.* 600 Ma: Stern 1994).

The Pharusian–Adamastor Ocean along the southern margin of the Congo continent was wider (>5500 km) and took a longer time to close. Final collision and suturing of the Congo continent with South Rodinia may have extended into the earliest Cambrian (Trompette 1994).

The break-up of Pannotia

Unlike Rodinia, which was a long-lived supercontinent (*c.* 300 Ma), Pannotia appears to have rifted apart soon after it was assembled. In the latest Precambrian (*c.* 560 Ma), the Palaeozoic continents of Laurentia and Baltica rifted away from Pannotia forming the Iapetus Ocean and Tornquist Sea (Cocks & Torsvik 2002). Siberia is also shown rifting away from Pannotia in the early Cambrian (Fig. 9). By the early Cambrian, Pannotia had broken up into four principal Palaeozoic continents: Gondwana, Laurentia, Baltic and Siberia. The Cadomian–Avalonian terrane would rift away from northern Gondwana in the early Ordovician (Scotese & McKerrow 1990; Cocks & Torsvik 2002). The Cathaysian terranes would separate from equatorial Gondwana by the middle Palaeozoic (Scotese & McKerrow 1990; Cocks & Torsvik 2002), and the Cimmerian terranes would follow the Cathyasian terranes northward into Tethys by the late Palaeozoic (Audley-Charles 1988; Sengor *et al.* 1988; Sengor & Natal'in 1996; Metcalfe 1999).

The closure of the ocean basins between North Rodinia, South Rodinia and the Congo continent also eliminated several major subduction zones. In all, more than 40 000 linear kilometres of subduction zone were destroyed. This is comparable to the entire length of Circum-Pacific Ring of Fire. As sometimes happens after continental collision, new subductions were initiated (Ross–Delamerian subduction zone: Ireland *et al.* 1998; Veevers 2000; Boger & Miller 2004) or became more active (Cadomian–Avalonian–Timanide active margin: Murphy & Nance 1989, 1991; Nance *et al.* 1991; Nance & Murphy 1994; Scarrow *et al.* 2001; Gee & Pease 2004).

Palaeomagnetic underpinnings

Kirschvink (1992*b*) deserves special recognition for producing the first set of palaeomagnetically derived plate tectonic reconstructions for the Neoproterozoic. Li *et al.* (2007) have published a set of 12 plate tectonic reconstructions for the late Mesoproterozoic, Neoproterozoic and early Cambrian (1100–530 Ma) based on a comprehensive compilation of Neoproterozoic palaeomagnetic and tectonic data. Other reviews of Neoproterozoic palaeomagnetic data include: Weil *et al.* (1998), Smith (2001), Pisarevsky *et al.* (2003) and Meert & Torsvik (2004). Condie (2003) has published an excellent summary of the Neoproterozoic tectonic events related to the assembly and break-up of Rodinia and Pannotia.

A minimal amount of palaeomagnetic data was used to produce the reconstructions shown in Figures 4–10. In the reconstructions presented here, the Rodinia supercontinent was oriented with respect to the spin axis at 750 Ma using the Laurentian and Australian palaeomagnetic poles of Powell *et al.* (1993). The orientation of Pannotia at 600 Ma is based on results from the Catoctin volcanic province and nine other North American sites (Meert & Van der Voo 1994) that place Laurentia near the South Pole, and the well-established cluster in northwesternmost South America of more than a dozen late Neoproterozoic poles for Gondwana (Li & Powell 1993; Powell *et al.* 1993). The movement of the continents between 750 and 600 Ma was interpolated from these two palaeomagnetic end points.

Summary and conclusions

Summary of plate tectonic events

The most important plate tectonic events during the late Mesoproterozoic (1100 Ma) and the Neoproterozoic (1000–540 Ma) were as follows.

- The assembly of the supercontinent of Rodinia (Fig. 1) during the Grenville Event (*c.* 1100 Ma).
- The collision of the Congo continent with Rodinia (*c.* 800–750 Ma) closed the southern part of the Mozambique Seaway and triggered the break-up of Rodinia.
- The Panthalassic Ocean opened as the supercontinent of Rodinia split into a northern half (East Gondwana, Cathyasia and Cimmeria) and a southern half (Laurentia, Amazonia-West African Craton, Baltica and Siberia).
- Over the next 150 Ma North Rodinia rotated counter-clockwise over the North Pole, while South Rodinia rotated clockwise across the South Pole.
- During the late Neoproterozoic (700–550 Ma), the three Neoproterozoic continents – North Rodinia, South Rodinia and the Congo continent – collided (Pan-Africa Event) forming the second Neoproterozoic supercontinent, Pannotia (or Greater Gondwanaland).
- The collisions that built Pannotia were diachronous. In East Africa, collision and suturing began first along the southern arm of the Mozambique Seaway (800–750 Ma), and then progressed northwards towards the Hijaz arcs of Arabia, Egypt and Sudan (600 Ma). In West Africa and Brazil, collision and suturing began approximately 600 Ma and lasted until the end of the Precambrian (*c.* 550 Ma).
- Pan-African mountain building and the fall in sea level associated with the assembly of Pannotia, may have triggered the extreme Ice House conditions that characterize the middle and late Neoproterozoic.
- Soon after it was assembled (*c.* 560 Ma), Pannotia broke apart into the four principal Palaeozoic continents: Laurentia (North America), Baltica (northern Europe), Siberia and Gondwana.
- New subductions were initiated (Ross–Delamerian Orogen and Alexander arc) or became more active (Avalonian–Cadomian–Timanide active margin).
- The amalgamation and subsequent break-up of Pannotia in the early Cambrian was accompanied by a rise in sea level that flooded the continents and may have triggered the 'Cambrian Explosion' (Brasier & Lindsay 2001).

Application of plate tectonic hypotheses to Neoproterozoic reconstructions

The plate tectonic hypotheses outlined at the beginning of this paper have been applied in the following manner to produce the Neoproterozoic reconstructions presented in this paper.

- Subduction zones were mapped so that after the break-up of Rodinia, the two halves, North and South Rodinia, moved towards the subduction zones. The Panthalassic Rift was oriented approximately parallel to the Pharusian–Adamastor and Hijaz–Mozambique subduction zones.
- Because Rodinia and Pannotia were large continental plates, their plate velocities were kept to 2–3 cm $year^{-1}$.
- 'Outlier' palaeomagnetic poles that could only be explained by episodes of True Polar Wander were excluded from this analysis. For arguments in favour of Neoproterozoic–Cambrian True Polar Wander see Kirschvink (1992*b*), Kirschvink *et al.* (1997), Evans (1998), Hoffman (1999) and Li *et al.* (2004).
- The break-up of Rodinia was triggered by a global plate tectonic reorganization resulting from the collision of Rodinia with the Congo continent.
- The overall motions of the three major Neoproterozoic plates (Congo continent, North Rodinia and South Rodinia) were co-ordinated, simple and elegant. There were no spurious rotations or unusual accelerations in plate motion.

Snowball or no Snowball?

The Snowball Earth hypothesis (Kirschvink 1992*a*; Hoffman *et al.* 1998) proposes that, at times, during the Neoproterozoic the continents were covered by snow and ice and that the oceans were frozen solid. To test this hypothesis, the distribution of middle and late Neoproterozoic glacial deposits (Hambrey & Harland 1981; Scotese *et al.* 1999) were plotted on the late Neoproterozoic palaeogeographic reconstruction (600 Ma) (Fig. 10). The equatorward edge of both the north and south polar ice caps was mapped (dashed white line). The following conclusions may be drawn.

- Both the northern and southern hemispheres were extensively glaciated.

- Like the Permo-Carboniferous Ice House World, the southern hemisphere was mostly land and, thus, supported a larger ice cap.
- In the southern hemisphere, the ice sheets extended to within 25° of the equator.
- In the northern hemisphere, the ice sheets extended to within 5° of the equator.
- Much of the subtropical and equatorial region was not glaciated.
- Cratonic Siberia was not glaciated.

Although the palaeogeographic maps presented here do not prohibit a Snowball Earth, the extent of ice sheets favour an extensive, bipolar Ice House World.

Habitat of Neoproterozoic oil

The first economically important accumulations of hydrocarbons are from Neoproterozoic sources. The two major source rocks of this age (Nepa of Siberia and Huqf of Oman) occur in association with massive Neoproterozoic evaporite deposits. The locations of these fields are plotted in Figures 7 and 10. On both maps the Neoproterozoic source rocks occur in the warm equatorial–subtropical belt, within 30° of the equator. If one were to explore for additional Neoproterozoic source rocks, the most likely place to look would be the warm shallow seas of East Gondwana, the Canadian Arctic and cratonic Siberia.

The author would like to thank the Geological Society of London for the opportunity to present this work at the Conference on Global Infracambrian Hydrocarbon Systems, London, 29–30 November 2006. J. Craig deserves credit and thanks for encouraging the author to finish this contribution. In addition, the author also would like to thank A. Smith, A. Collins and Z. X. Li for their constructive comments regarding the manuscript. The author greatly benefited from discussions regarding Neoproterozoic plate tectonics with D. Nance, W. S. McKerrow, J. Goodge, P. Hoffman, J. Meert, R. Hanson, M. Brookfield and R. Stern. Much of the original research concerning the location of Pan-African sutures and timing of plate tectonic events was carried out by D. Nance and W. S. McKerrow. The research presented here was made possible by the companies that support the plate tectonic, palaeogeographic and palaeoclimatic research of the PALEOMAP Project. All of the figures presented in this paper are used with the permission of the PALEOMAP Project. This paper is dedicated to the memory of C. Powell, who was an inspirational pioneer in the field of 'impossible' Neoproterozoic plate tectonics.

References

AUDLEY-CHARLES, M. 1988. Evolution of the southern margin of Tethys (North Australian region) from early Permian to late Cretaceous. *In*: AUDLEY-CHARLES, M. & HALLAM, A. (eds) *Gondwana and Tethys*. Geological Society, London, Special Publications, **37**, 79–100.

BAUER, W., THOMAS, R. J. & JACOBS, J. 2003. Proterozoic–Cambrian history of Dronning Maud Land in the context of Gondwana Assembly. *In*: YOSHIDA, M., WINDLEY, B. F. & DASGUPTA, S. (eds) *Proterozoic East Gondwana: Supercontinent Assembly and Breakup*. Geological Society, London, special Publications, **206**, 247–269.

BOGER, S. D. & MILLER, J. MCL. 2004. Terminal suturing of Gondwana and the onset of the Ross–Delamerian Orogeny: the cause and effect of an Early Cambrian reconfiguration of plate motions. *Earth and Planetary Science Letters*, **219**, 35–48.

BOGER, S. D., WILSON, C. J. L. & FANNING, C. M. 2001. Early Palaeozoic tectonism within the East Antarctic craton: the final suture between East and West Gondwana? *Geology*, **29**, 463–466.

BOND, G. C., NICKERSON, P. A. & KOMINZ, M. A. 1984. Breakup of a supercontinent between 625 and 555 Ma: new evidence and implications for continental histories. *Earth and Planetary Science Letters*, **70**, 325–345.

BOZHKO, N. A. 1986. The evolution of the mobile zones of Gondwana and Laurasia in the Late Precambrian. *Tectonophysics*, **126**, 125–135.

BRASIER, M. D. & LINDSAY, J. F. 2001. Did supercontinent amalgamation trigger the Cambrian Explosion? *In*: ZHURAVLEV, A. Y. & RIDING, R. (eds) *The Ecology of the Cambrian Radiation. Perspectives in Palaeobiology and Earth History*. Columbia University Press, New York, 69–89.

BURRETT, C. & BERRY, R. 2000. Proterozoic Australia–Western United States (AUSWUS) fit between Laurentia and Australia. *Geology*, **28**, 103–116.

BURRETT, C. & BERRY, R. 2002. A statistical approach to defining Proterozoic crustal provinces and testing continental reconstructions of Australia and Laurentia – SWEAT or AUSWUS? *Gondwana Research*, **5**, 109–122.

BURRETT, C., LONG, J. & STAIT, B. 1990. Early–Middle Palaeozoic biogeography of Asia terranes derived from Gondwana. *In*: MCKERROW, W. S. & SCOTESE, C. R. (eds) *Palaeozoic Palaeogeography and Biogeography*. Geological Society, London, Memoirs, **12**, 163–174.

CAWOOD, P. A., MCCAUSLAND, P. J. A. & DUNNING, G. R. 2001. Opening Iapetus: constraints from the Laurentian margin in Newfoundland. *Geological Society of America Bulletin*, **113**, 443–453.

CAWOOD, P. A., NEMCHIN, A. A., STRACHAN, R., PRAVE, T. & KRABBENDAM, M. 2007. Sedimentary basin and detrital zircon record along East Laurentia and Baltica during the assembly and breakup of Rodinia. *Journal of the Geological Society, London*, **164**, 257–275.

CHEMALE, F., ALKMIM, F. F. & ENDO, I. 1993. Late Proterozoic tectonism in the interior of the Sao Francisco craton. *In*: FINDLAY, R. H., UNRUG, R., BANKS, M. R. & VEEVERS, J. J. (eds) *Gondwana Eight: Assembly, Evolution and Dispersal*. Balkema, Rotterdam, 29–42.

COCKS, L. R. M. & TORSVIK, T. H. 2002. Earth geography from 500 to 400 million years ago: a faunal and palaeomagnetic review. *Journal of the Geological Society, London*, **159**, 631–644.

CONDIE, K. C. 2003. Supercontinents, superplumes and continental growth: the Neopropterozoic record. *In*: YOSHIDA, M., WINDLEY, B. F. & DASGUPTA, S. (eds) *Proterozoic East Gondwana: Supercontinent Assembly and Breakup*. Geological Society, London, Special Publications, **206**, 1–21.

CONRAD, C. P. & LITHGOW-BERTELLONI, C. 2002. How mantle slabs drive plate tectonics. *Science*, **298**, 207–209.

COSCA, M. A. 1998. The Baltica–Laurentia connection: Sveconorwegian (Grenvillian) metamorphism, cooling, and unroofing in the Bamble sector, Norway. *Journal of Geology*, **106**, 549–552.

DALY, M. C. 1986. Crustal shear zones and thrust belts: their geometry and continuity in Central Africa. *Philosophical Transactions of the Royal Society of London*, **A317**, 111–128.

DALZIEL, I. W. D. 1991. Pacific margins of Laurentia and East Antarctica as a conjugate rift pair: evidence and implications for an Eocambrian supercontinent. *Geology*, **19**, 598–601.

DALZIEL, I. W. D. 1997. Neoproterozoic–Palaeozoic geography and tectonics: review, hypothesis and environmental speculation. *Geological Society of America Bulletin*, **109**, 16–42.

DALZIEL, I. W. D., MOSHER, S. & GAHAGAN, L. M. 2000. Laurentia–Kalahari collision and the assembly of Rodinia. *Journal of Geology*, **108**, 499–513.

DIRKS, P. H. G. M. & WILSON, C. J. L. 1995. Crustal evolution of the East Antarctic mobile belt in Prydz Bay: continental collision at 500 Ma? *Precambrian Research*, **75**, 189–207.

EVANS, D. A. 1998. True polar wander, a supercontinental legacy. *Earth and Planetary Science Letters*, **157**, 1–8.

GEE, D. G. & PEASE, V. 2004. *The Neoproterozoic Timanide Orogen of Eastern Baltica*. Geological Society, London, Memoirs, **30**, 255.

FORSYTH, D. & UYEDA, S. 1975. On the relative importance of driving forces of plate motion. *Geophysical Journal of the Royal Astronomical Society*, **43**, 163–189.

GOODGE, J. W. 2002. From Rodinia to Gondwana: supercontinent evolution in the Transantarctic Mountains. *In*: GAMBLE, J. & SKINNER, D. A. (eds) *Antarctica at the Close of a Millenium. Proceedings of the 8th International Symposium on Antarctic Earth Science*. Royal Society of New Zealand Bulletin, **35**, 61–74.

GOWER, C. F. & OWEN, V. 1984. Pre-Grenvillilan and Grenvillian lithotectonic regions in eastern Labrador – correlations with the Sveconorwegian orogenic belt in Sweden. *Canadian Journal of Earth Sciences*, **21**, 678–693.

GRADSTEIN, F. M., OGG, J. G. & SMITH, A. G. 2004. *A Geologic Time Scale 2004*. Cambridge University Press, Cambridge.

HAMBREY, M. J. & HARLAND, W. B. 1981. *Earth's Pre-Pleistocene Glacial Record*. Cambridge Univeristy Press, Cambridge.

HOFFMAN, P. F. 1991. Did the breakout of Laurentia turn Gondwanaland inside out? *Science*, **252**, 1409–1412.

HOFFMAN, P. F. 1999. The break-up of Rodinia, birth of Gondwana, true polar wander and the snowball Earth. *In*: DEWIT, M. J. (ed.) *Gondwana 10: Event Stratigraphy of Gondwana – Keynote Presentations. Journal of African Earth Sciences*, **28**, 17–33.

HOFFMAN, P. F., KAUFMAN, A. J., HALVERSON, G. P. & SCHRAG, D. P. 1998. A Neoproterozoic snowball Earth. *Science*, **281**, 1342–1346.

IRELAND, T. R., FLOETTMANN, C. M., FANNING, C. M., GIBSON, G. M. & PREISS, W. V. 1998. Development of the Early Palaeozoic Pacific margin of Gondwana from detrital-zircon ages across the Delamerian Orogen. *Geology*, **26**, 243–246.

JACOBS, J., KLEMD, R., FANNING, C. M., BAUER, W. & COLOMBO, F. 2003. Extensional collapse of the late Neoproterozoic–early Palaeozoic East Africa–Antarctic orogen in central Dronning Maud Land, East Antarctica. *In*: YOSHIDA, M., WINDLEY, B. F. & DASGUPTA, S. (eds) *Proterozoic East Gondwana: Supercontinent Assembly and Breakup*. Geological Society, London, Special Publications, **206**, 271–287.

JOHN, T., SCHENK, V., HAUSE, K., SCHEERER, E. & TEMBO, F. 2003. Evidence for a Neoproterozoic ocean in south-central Africa from mid-ocean-ridge-type geochemical signatures and pressure-temperature estimates of Zambian eclogites. *Geology*, **31**, 243–246.

JOHNSON, P. R. & WOLDEHAIMANOT, B. 2003. Development of the Arabian–Nubian shield: perspectives on accretion and deformation in the northern East African Orogen and the assembly of Gondwana. *In*: YOSHIDA, M., WINDLEY, B. F. & DASGUPTA, S. (eds) *Proterozoic East Gondwana: Supercontinent Assembly and Breakup*. Geological Society, London, Special Publications, **206**, 289–325.

JOHNSON, S. P., RIVERS, T. & DE WAELE, B. 2005. A review of the Mesoproterozoic to early Palaeozoic magmatic and tectonothermal history of south-central Africa: implications for Rodinia and Gondwana. *Journal of the Geological Society, London*, **162**, 433–450.

KALSBEEK, F., THRANE, K., NUTMAN, A. P. & JEPSEN, H. F. 2000. Late Mesoproterozoic to early Neoproterozoic history of East Greenland Caledonides: evidence for Grenvillian orogenisis? *Journal of the Geological Society, London*, **157**, 1215–1225.

KARLSTROM, K. E., HARLAN, S. S., WILLIAMS, M. L., MCCLELLAND, J., GEISSMAN, J. W. & AHALL, K.-I. 1999. Refining Rodinia: geologic evidence for the Australia – western US connection in the Proterozoic. *GSA Today*, **9**, 1–7.

KIRSCHVINK, J. L. 1992*a*. Late Proterozoic low-latitude global glaciation: the Snowball Earth. *In*: SCHOPF, J. W. & KLEIN, C. (eds) *The Proterozoic Biosphere*. Cambridge University Press, Cambridge, 51–52.

KIRSCHVINK, J. L. 1992*b*. A palaeogeographic model for Vendian and Cambrian time. *In*: SCHOPF, J. W. & KLEIN, C. (eds) *The Proterozoic Biosphere*. Cambridge University Press, Cambridge, 569–581.

KIRSCHVINK, J. L., RIPPERDAN, R. L. & EVANS, D. A. 1997. Evidence for large-scale reorganization of Early Cambrian continental masses by inertial interchange true polar wander. *Science*, **277**, 541–545.

LAWVER, L. A. & SCOTESE, C. R. 1987. A revised reconstruction of Gondwanaland. *In*: MCKENZIE, G. D. (ed.) *Gondwana Six: Structure, Tectonics and*

Geophysics. American Geophysical Union, Geophysical Monograph, **40**, 17–24.

LI, Z. X. & POWELL, C. MCA. 1993. Late Proterozoic to early Palaeozoic palaeomagnetism and the formation of Gondwanaland. *In*: FINDLAY, R. H., UNRUG, R., BANKS, M. R. & VEEVERS, J. J. (eds) *Gondwana Eight: Assembly, Evolution and Dispersal*. Balkema, Rotterdam, 9–21.

LI, Z. X., ZHANG, L. & POWELL, C. MCA. 1996. Positions of the East Asian cratons in the Proterozoic supercontinent Rodinia. *Australian Journal of Earth Sciences*, **43**, 593–604.

LI, Z. X., EVANS, D. A. D. & ZHANG, S. 2004. A 90 degree spin on Rodinia: possible causal links between the Neoproterozoic supercontinent, superplume, true polar wander and low-latitude glaciation. *Earth and Planetary Science Letters*, **220**, 409–421.

LI, Z. X., BOGDANOVA, S. V. *ET AL*. 2007. Assembly, configuration, and break-up history of Rodinia: a synthesis. *Precambrian Research*, **160**, 179–210.

LOWE, D. R. 1992. Major events in the geological events in the geological development of the Precambrian Earth. *In*: SCHOPF, J. W. & KLEIN, C. (eds) *The Proterozoic Biosphere*. Cambridge University Press, Cambridge, 67–75.

MCMENAMIN, M. A. S. & MCMENAMIN, D. L. S. 1990. *The Emergence of Animals: The Cambrian Breakthrough*. Columbia University Press, New York.

MCNAMARA, A. K., MAC NIOCAILL, C., VAN DER PLUIJM, B. A. & VAN DER VOO, R. 2001. West African proximity of the Avalon terrane in the latest Precambrian. *Geological Society of America Bulletin*, **1131**, 1161–1170.

MEERT, J. G. 2003. A synopsis of events related to the assembly of eastern Gondwana. *Tectonophysics*, **362**, 1–40.

MEERT, J. G. & TORSVIK, T. H. 2004. Palaeomagnetic constraints on Neoproterozoic 'Snowball Earth' continental reconstructions. *In*: JENKINS, G. S., MCMENAMIN, M. A., MCKAY, C. P. & SOHL, L. (eds) *The Extreme Proterozoic: Geology, Geochemistry and Climate*. American Geophysical Union, Geophysical Monograph, **146**, 5–12.

MEERT, J. G. & VAN DER VOO, R. 1994. Palaeomagnetism of the Catoctin volcanic province: a new Vendian–Cambrian apparent polar wander path for North America. *Journal of Geophysical Research*, **99**, 4625–4641.

METCALFE, I. 1993. Southeast Asian terranes: Gondwanaland origins and evolution. *In*: FINDLAY, R. H., UNRUG, R., BANKS, M. R. & VEEVERS, J. J. (eds) *Gondwana Eight: Assembly, Evolution and Dispersal*. Balkema, Rotterdam, 181–200.

METCALFE, I. 1999. Gondwana dispersion and Asian accretion: an overview. *In*: METCALFE, I., JISHUN, R., CHARVET, J. & HADA, S. (eds) *Gondwana Dispersion and Asia Accretion*. Balkema, Rotterdam, 9–28.

METZ, K. S. 2001. *The Palaeogeography of the Proterozoic*. Masters Thesis, Department of Earth and Environmental Sciences, University of Texas at Arlington, Arlington, TX.

MOORES, E. M. 1991. Southwest U.S.–East Antarctica (SWEAT) connection: a hypothesis. *Geology*, **19**, 425–428.

MURPHY, J. B. & NANCE, R. D. 1989. Model for the evolution of the Avalonian–Cadomian belt. *Geology*, **17**, 735–738.

MURPHY, J. B. & NANCE, R. D. 1991. Supercontinent model for the contrasting character of Late Proterozoic orogenic belts. *Geology*, **19**, 469–472.

NANCE, R. D. & MURPHY, J. B. 1994. Contrasting basement isotopic signatures and the palinspastic restoration of peripheral orogens: example from the Neoproterozoic Avalonian–Cadomian belt. *Geology*, **22**, 617–620.

NANCE, R. D., MURPHY, J. B., STRACHAN, R. A., D'LEMOS, R. S. & TAYLOR, G. K. 1991. Late Proterozoic tectonostratigraphic evolution of the Avalonian and Cadomian terranes. *Precambrian Research*, **53**, 41–78.

PELECHATY, S. M. 1996. Stratigraphic evidence for Siberia–Laurentia connection and Early Cambrian Rifting. *Geology*, **24**, 719–722.

PISAREVSKY, S. A., WINGATE, M. T. D., POWELL, C. MCA., JOHNSON, S. & EVANS, D. A. D. 2003. Models of Rodinia assembly and fragmentation. *In*: YOSHIDA, M., WINDLEY, B. F. & DASGUPTA, S. (eds) *Proterozoic East Gondwana: Supercontinent Assembly and Breakup*. Geological Society, London, Special Publications, **206**, 35–55.

POWELL, C. MCA., MCELHINNY, M. W., LI, Z. X., MEERT, J. G. & PARK, J. K. 1993. Palaeomagnetic constraints on the timing of the Neoproterozoic breakup of Rodinia and the Cambrian formation of Gondwana. *Geology*, **21**, 889–892.

POWELL, C. MCA. 1995. Are Neoproterozoic glacial deposits preserved on the margins of Laurentia related to the fragmentation of two supercontinents? Comment. *Geology*, **23**, 1053–1054.

RIVERS, T. 1997. Lithotectonic elements of the Grenville Province: review and tectonic implications. *Precambrian Research*, **86**, 117–154.

ROGERS, J. J. W. 1995. Tectonic assembly of Gondwana. *Journals of Geodynamics*, **19**, 1–34.

ROSS, M. I. & SCOTESE, C. R. 1988. A hierarchical tectonic model of the Gulf of Mexico and Caribbean Region. *In*: SCOTESE, C. R. & SAGER, W. W. (eds) *Mesozoic and Cenozoic Plate Reconstructions. Tectonophysics*, **155**, 139–168.

ROWELL, A. J., REES, M. N., DUEBENDORFER, E. M., WALLIN, E. T., VAN SCHMUS, W. R. & SMITH, E. I. 1993. An active Neoproterozoic margin: evidence from the Skelton Glacier area, Transantarctic Mountains. *Journal of the Geological Society, London*, **150**, 677–682.

SCARROW, J. H., PEASE, V., FLEUTELOT, C. & DUSHIN, V. 2001. The late Neoproterozoic Enganepe ophiolite, Polar Urals Russia: an extension of the Cadomian arc? *Precambrian Research*, **110**, 255–275.

SCOTESE, C. R. 1991. Jurassic and Cretaceous plate tectonic reconstructions. *Palaeogeography, Palaeoecology, Palaeoclimatology*, **87**, 493–501.

SCOTESE, C. R. 1998. A tale of two supercontinents: the assembly of Rodinia; its break-up, and the formation of Pannoia during the Pan-African event. *In*: ALMOND, J., ANDERSON, J. *ET AL*. (eds) *Gondwana 10: Event Stratigraphy of Gondwana* (Special Abstracts Issue). *Journal of African Earth Sciences*, **27**, 171.

SCOTESE, C. R. & MCKERROW, W. S. 1990. Revised world maps and Introduction. *In*: MCKERROW, W. S. & SCOTESE, C. R. (eds) *Palaeozoic Palaeogeography and Biogeography*. Geological Society, London, Memoirs, **12**, 1–21.

SCOTESE, C. R. & ROWLEY, D. B. 1985. The orthogonality of subduction: an empirical rule? *Tectonophysics*, **116**, 173–187.

SCOTESE, C. R., BOUCOT, A. J. & MCKERROW, W. S. 1999. Gondwanan palaeogeography and palaeoclimatology. *In*: DEWIT, M. J. (ed.) *Gondwana 10: Event Stratigraphy of Gondwana – Keynote Presentations. Journal of African Earth Sciences*, **28**, 99–114.

SEARS, J. W. & PRICE, R. A. 2000. New look at the Siberian connection: no SWEAT. *Geology*, **28**, 423–426.

SENGOR, A. M. C., ALTINER, D., CIN, A., USTAOMER, T. & HSU, K. 1988. Origin and assembly of the Tethyside orogenic collage at the expense of Gondwana Land. *In*: AUDLEY-CHARLES, M. & HALLAM, A. (eds) *Gondwana and Tethys*. Geological Society, London, Special Publications, **37**, 119–181.

SENGOR, A. M. C. & NATAL'IN, B. A. 1996. Palaeotectonics of Asia: fragments of a synthesis. *In*: YIN, A. & HARRISON, M. (eds) *The Tectonic Evolution of Asia*. Cambridge University Press, Cambridge, 486–640.

SMITH, A. G. 1999. Gondwana: its shape, sise, position from Cambrian to Triassic times. *In*: DEWIT, M. J. (ed.) *Gondwana 10: Event Stratigraphy of Gondwana – Keynote Presentations. Journal of African Earth Sciences*, **28**, 71–98.

SMITH, A. G. 2001. Palaeomagnetically and tectonically based global maps for the Vendian to mid-Ordovician. *In*: ZHURAVLEV, A. YU. & RIDING, R. (eds) *The Ecology of the Cambrian Radiation. Perspectives in Palaeobiology and Earth History*. Columbia University Press, New York, 11–46.

STERN, R. J. 1994. Arc assembly and continental collision in the Neoproterozoic East Africa Orogen: implications for the consolidation of Gondwanaland. *Annual Reviews of Earth and Planetary Sciences*, **33**, 319–351.

STERN, R. J. 2002. Crustal evolution in the East African Orogen: a neodymium isotopic perspective. *Journal of African Earth Sciences*, **34**, 109–117.

STEWART, J. H. 2007. World Map showing surface and subsurface distribution, and lithologic character of Middle and Late Neoproterozoic Rocks. USGS Open-File Report 2007-1087, Menlo Park, 52.

TOHVER, E., VAN DER PLUIJM, B. A., VAN DER VOO, R., RIZZOTTO, G. & SCANDOLARA, J. E. 2002. Palaeogeography of the Amazon craton at 1.2 Ga: early Grenvillian collision with the Llano segment of Laurentia. *Earth and Planetary Science Letters*, **199**, 185–200.

TORSVIK, T. H. & REHNSTROM, E. F. 2001. Cambrian palaeomagnetic data from Baltica: implications for true polar wander and Cambrian palaeogeography. *Journal of the Geological Society, London*, **158**, 321–329.

TORSVIK, T. H., RYAN, P. D., TRENCH, A. & HARPER, D. A. T. 1991. Cambrian–Ordovician palaeogeography of Baltica. *Geology*, **19**, 7–10.

TORSVIK, T. H., SMETHHURST, M. A. *ET AL*. 1996. Continental breakup and collision in the Neoproterozoic and Palaeozoic – A tale of Baltica and Laurentia. *Earth Science Reviews*, **40**, 229–258.

TORSVIK, T. H., SMETHHURST, M. A., VAN DER VOO, R., TRENCH, A., ABRAHAMSEN, N. & HALVORSEN, E. 1992. Baltica: A synposis of Vendian–Permian palaeomagnetic data and their palaeotectonic implications. *Earth-Science Reviews*, **33**, 1–20.

TROMPETTE, R. 1994. *Geology of Western Gondwana (2000–500 Ma): Pan-African–Brasiliano Aggregation of South America and Africa* (transl. by CAROZZI, A. V.). Balkema, Rotterdam, 350.

TROMPETTE, R. 1997. Neoproterozoic (~600 Ma) aggregation of Western Gondwana: a tentative scenario. *Precambrian Research*, **82**, 101–112.

VEEVERS, J. J. 2000. *Billion-year earth history of Australia and neighbors in Gondwanaland*. GEMOC Press, Sydney, Australia.

WANG, Y., LU, S., GAO, Z., LIN, W. & MA, G. 1981. Sinian tillites of China. *In*: HAMBREY, M. J. & HARLAND, W. B. (eds) *Earth's Pre-Pliestocene Glacial Record*. IGCP Project 38: Pre-Pleistocene Tillites. Cambridge Universiy Press, Cambridge, 386–401.

WATTERS, B. R. 1976. Possible late Precambrian subduction zone in South West Africa. *Nature*, **259**, 471–473.

WEIL, A. B., VAN DER VOO, R., MACNIOCALL, C. & MEERT, J. G. 1998. The Proterozoic supercontinent Rodinia, palaeomagnetically derived reconstructions for 1100–800 Ma. *Earth and Planetary Science Letters*, **154**, 13–24.

WINGATE, M. T. D., PISAREVSKY, S. A. & EVANS, D. A. 2002. Rodinia connections between Australia and Laurentia; no SWEAT, no AUSWUS? *Terra Nova*, **14**, 121–128.

YOUNG, G. M. 1992. Late Proterozoic stratigraphy and the Canada–Australia connection. *Geology*, **20**, 215–218.

^{187}Re–^{187}Os geochronology of Precambrian organic-rich sedimentary rocks

BRIAN KENDALL[1,2]*, ROBERT A. CREASER[2] & DAVID SELBY[3]

[1]*School of Earth and Space Exploration, Arizona State University, Tempe, AZ 85287-1404, USA*

[2]*Department of Earth and Atmospheric Sciences, University of Alberta, Edmonton, Alberta, Canada T6G 2E3*

[3]*Department of Earth Sciences, Science Labs, University of Durham, Durham DH1 3LE, UK*

**Corresponding author (e-mail: brian.kendall@asu.edu)*

Abstract: Global correlations of Precambrian stratigraphic successions can be hampered by the coarse resolution of biostratigraphic and chemostratigraphic records, and by the scarcity of reliable U–Pb zircon age constraints. The development of the ^{187}Re (rhenium)–^{187}Os (osmium) radioisotope system as an accurate deposition-age geochronometer for organic-rich sedimentary rocks (e.g. black shales) holds great potential for an improved radiometric calibration of the Precambrian rock record. Here, we review Re–Os isotope data obtained for Precambrian black shales and revisit the discrepancy in Re–Os ages for the Neoproterozoic Aralka Formation (central Australia). In addition, we introduce new Re–Os isotope data for the Late Neoproterozoic Doushantuo Formation (South China) that highlights the necessity of a rigorous sampling protocol for depositional age determinations. Improvements in sampling and analytical methodologies have permitted the determination of precise ages (<1%, 2σ) from Late Neoproterozoic to Late Archaean shales. Whole-rock digestion using a Cr^{VI}–H_2SO_4 solution minimizes the release of detrital Re and Os from shale matrices, and selectively attacks organic matter that hosts hydrogenous Re and Os. The Re–Os system in organic-rich sedimentary rocks appears to be robust during hydrocarbon maturation and up to the onset of lowermost greenschist facies metamorphism, but post-depositional hydrothermal fluid flow can result in scattered Re–Os isotope data. The Re–Os black shale geochronometer should find utility for constraining the age of a diverse range of Precambrian geological phenomena. In addition, the initial ^{187}Os/^{188}Os composition determined from Re–Os isochron regressions serves as a tracer for the Os isotope composition of Precambrian sea water.

Accurately determining the depositional ages of sedimentary rocks has proven extremely difficult to accomplish using the conventional long-lived radioisotope systems (e.g. Rb–Sr, Sm–Nd, U–Pb, K–Ar). The U–Pb SHRIMP (sensitive high-resolution ion microprobe) dating of detrital minerals such as zircon has proven useful for provenance studies and constraining the maximum depositional age (e.g. Bingen *et al.* 2005). However, authigenic minerals (e.g. apatite, glauconite, illite, K-feldspar, monazite) generally yield diagenetic ages that are variably younger than the depositional age of the host sedimentary rock. Diagenetic age determinations on authigenic minerals are also hampered by low closure temperatures of the applied radioisotope system, resulting in a high susceptibility to thermal resetting even during relatively low-temperature hydrothermal alteration or metamorphism (Dickin 2005). Diagenetic xenotime is found in a wide variety of siliciclastic and volcaniclastic rocks, and represents a robust U–Pb geochronometer with the potential for resolving complex geological histories within sedimentary basins (Rasmussen 2005). However, U–Pb xenotime dates also reflect the timing of diagenesis rather than deposition. In some cases, Pb/Pb ages from carbonates (e.g. Moorbath *et al.* 1987; Woodhead *et al.* 1998; Babinski *et al.* 2007) and phosphorites (Barfod *et al.* 2002; Chen *et al.* 2004) may yield depositional or early diagenetic ages using well-preserved material. However, Pb/Pb ages may be erroneously young or old owing to diagenetic or metamorphic recrystallization and detrital inheritance, respectively, or age information may be lost altogether due to post-depositional mobility of U and Pb (Rasmussen 2005).

Currently, the most reliable method of constraining the depositional age of sedimentary rocks is by U–Pb zircon dating of interbedded tuff horizons. In the case of Phanerozoic sedimentary basins, such U–Pb ages can be used to calibrate high-resolution Phanerozoic biostratigraphic, chemostratigraphic and magnetostratigraphic records, thereby facilitating regional and global correlations of stratigraphic successions (e.g. Gradstein *et al.* 2004). However, the generally coarse resolution of these

From: CRAIG, J., THUROW, J., THUSU, B., WHITHAM, A. & ABUTARRUMA, Y. (eds) *Global Neoproterozoic Petroleum Systems: The Emerging Potential in North Africa.* Geological Society, London, Special Publications, **326**, 85–107.
DOI: 10.1144/SP326.5 0305-8719/09/$15.00

chronostratigraphic methods for the Precambrian rock record does not currently permit this approach, and, in cases where datable ash beds are absent, the ages of Precambrian sedimentary rocks are generally only poorly constrained by radiometric dates from overlying and underlying volcanic or plutonic rocks, and/or cross-cutting plutonic rocks.

The development of the $^{187}Re-^{187}Os$ radioisotope system as a reliable deposition-age geochronometer for organic-rich sedimentary rocks (ORS; total organic carbon (TOC) $\geq$0.5%) like black shales (Ravizza & Turekian 1989; Cohen *et al.* 1999; Creaser *et al.* 2002; Selby & Creaser 2003; Kendall *et al.* 2004) has the potential to alleviate the problems associated with radiometric calibration of the Precambrian sedimentary rock record. Improvements in sampling and analytical methodologies, combined with the high precision of isotope dilution–negative thermal ionization mass spectrometry (ID-NTIMS: Creaser *et al.* 1991; Völkening *et al.* 1991; Walczyk *et al.* 1991), have made it possible to obtain a Re–Os age for black shale with a precision of better than $\pm$1% (2σ), with the absolute uncertainty comparable in some cases to the uncertainties on U–Pb ages derived from SHRIMP or laser ablation MC-ICP-MS (multicollector-inductively coupled plasma-mass spectrometry) analyses of zircons from tuffaceous beds (e.g. Kendall *et al.* 2004, 2006; Selby & Creaser 2005*a*; Anbar *et al.* 2007; Creaser & Stasiuk 2007).

Recently, the Re–Os ORS geochronometer has been successfully applied to studies regarding geological timescale calibration (Devonian–Mississippian boundary: Selby & Creaser 2005*a*), the timing of Proterozoic glaciation (Hannah *et al.* 2004; Kendall *et al.* 2004, 2006; Azmy *et al.* 2008), the Earth's early history of atmosphere and ocean oxygenation (Hannah *et al.* 2004; Anbar *et al.* 2007), and sedimentary basin analysis (Hannah *et al.* 2006; Creaser & Stasiuk 2007; Kendall *et al.* 2009*a*). In addition, the initial $^{187}Os/^{188}Os$ value (I_{Os}) from Re–Os isochron regressions has served as a tracer for the Os isotope composition of Phanerozoic sea water (Cohen *et al.* 1999, 2004; Creaser *et al.* 2002; Selby & Creaser 2003; Cohen 2004; Cohen & Coe 2007; Selby 2007). Here, we review recent applications of Re–Os geochronology to Precambrian ORS, and demonstrate how careful sampling and analytical methodologies are essential for precise and accurate depositional age determinations.

The $^{187}Re-^{187}Os$ system in ORS: a deposition-age geochronometer and sea-water Os isotope tracer

In addition to being siderophilic and chalcophilic (e.g. Shirey & Walker 1998), Re and Os are also organophilic and redox-sensitive (Ravizza *et al.* 1991; Ravizza & Turekian 1992; Colodner *et al.* 1993; Crusius *et al.* 1996; Levasseur *et al.* 1998; Selby & Creaser 2003, 2005*b*; Selby *et al.* 2005, 2007*a*). This geochemical behaviour of Re and Os in Earth surface environments enables the $^{187}Re-^{187}Os$ isotope system to be used as a deposition-age geochronometer for ORS and a tracer of palaeo-sea-water Os isotope composition.

Under oxidizing atmospheric conditions, Re is transported to the oceans primarily by rivers (as the highly soluble perrhenate anion ReO_4^-: Colodner *et al.* 1993). Rhenium is removed rapidly from reducing pore waters at centimetres to tens of centimetres below the sediment–water interface, and is sequestered into suboxic, anoxic and euxinic sediments (Colodner *et al.* 1993; Crusius *et al.* 1996; Morford & Emerson 1999; Nameroff *et al.* 2002; Sundby *et al.* 2004; Morford *et al.* 2005). Removal of Re from pore waters occurs by reductive capture ($Re^{VII}O_4^-$ is reduced to Re^{IV}: Colodner *et al.* 1993), the rate of which is controlled by slow precipitation kinetics (Crusius & Thomson 2000; Sundby *et al.* 2004).

The $^{187}Os/^{188}Os$ isotopic composition of present-day sea water is almost homogenous (*c.* 1.06: Sharma *et al.* 1997; Levasseur *et al.* 1998; Burton *et al.* 1999; Woodhouse *et al.* 1999; Peucker-Ehrenbrink & Ravizza 2000), consistent with an Os sea-water residence time of approximately 10^4 years (Oxburgh 1998, 2001; Levasseur *et al.* 1999). The dominant source of present-day sea-water Os (*c.* 70–80%) is from the weathering of the upper continental crust (McDaniel *et al.* 2004). Average $^{187}Os/^{188}Os$ for the currently eroding upper continental crust and riverine inputs are between 1.0 and 1.4 (Esser & Turekian 1993; Peucker-Ehrenbrink & Jahn 2001; Hattori *et al.* 2003) and approximately 1.5 (with an uncertainty of >20%: Levasseur *et al.* 1999), respectively. The remainder of the present-day marine Os budget is derived from extraterrestrial cosmic dust (Peucker-Ehrenbrink 1996) and the low- and high-temperature hydrothermal alteration of oceanic crust and peridotites (Ravizza *et al.* 1996; Sharma *et al.* 2000, 2007; Cave *et al.* 2003). Both sources contribute unradiogenic $^{187}Os/^{188}Os$ to sea water (0.126–0.130: Becker *et al.* 2001; Meisel *et al.* 1996, 2001; Walker *et al.* 2002*a*, *b*). Dissolved Os is probably present in sea water as an octavalent oxyanion (e.g. $HOsO_5^-$, $H_3OsO_6^-$). Osmium is removed into organic-rich sediments in direct association with organic matter (Levasseur *et al.* 1998) and/or is rapidly removed to organic-rich sediments first as Os (IV), and then is further reduced to Os (III) by organic complexation (Yamashita *et al.* 2007). Osmium removal from sea water into reducing sediments may occur below the depth of Re enrichment (Poirer 2006).

Rhenium and Os in ORS are predominantly hydrogenous, and are physically associated with organic matter (Ravizza & Turekian 1989, 1992; Ravizza *et al.* 1991; Ravizza & Esser 1993; Cohen *et al.* 1999; Martin *et al.* 2000; Peucker-Ehrenbrink & Hannigan 2000; Pierson-Wickmann *et al.* 2000, 2002; Creaser *et al.* 2002; Jaffe *et al.* 2002). Rhenium and Os are present in natural hydrocarbons (Poplavko *et al.* 1975; Barre *et al.* 1995; Woodland *et al.* 2001; Selby & Creaser 2005*b*; Selby *et al.* 2007*a*) and bitumens (Selby *et al.* 2005), and bulk organic matter is known to contain a large fraction of the Re and Os in shale source rocks (Ripley *et al.* 2001; Selby & Creaser 2003).

The fundamental assumptions behind the Re–Os ORS geochronometer include: (1) a hydrogenous origin for all Re and Os in ORS (i.e. negligible contributions from detrital and/or extraterrestrial particulates); (2) the closed-system behaviour of Re and Os in ORS following sediment deposition, such that Re–Os dates reflect the timing of deposition rather than diagenesis; (3) a homogenous I_{Os} derived from the contemporaneous sea water; and (4) the negligible post-depositional mobilization of Re and Os. Obtaining a precise Re–Os age by the isochron method requires a suitable range in initial ^{187}Re/^{188}Os such that β-decay of ^{187}Re to ^{187}Os over time generates a range in present-day ^{187}Os/^{188}Os. Although ORS may contain Re and Os concentrations ranging from average upper-crustal abundances (*c.* 0.2–2 ppb Re and 30–50 ppt Os: Esser & Turekian 1993; Peucker-Ehrenbrink & Jahn 2001; Hattori *et al.* 2003; Sun *et al.* 2003*a*, *b*) to hundreds of ppb Re and several ppb Os (e.g. Cohen *et al.* 1999), the high Re/Os ratios of ORS derived from sea water (present-day sea-water ^{187}Re/^{188}Os ratio is *c.* 3200–5300) means that only moderate enrichments in Re are required to generate radiogenic present-day ^{187}Os/^{188}Os ratios for Precambrian ORS (e.g. Kendall *et al.* 2004, 2006). One significant challenge of Re–Os ORS geochronology concerns the first assumption, which requires an analytical methodology to release hydrogenous Os from shales while minimizing release of detrital Re and Os that could otherwise affect the accuracy and precision of Re–Os age and I_{Os} determinations (e.g. Ravizza *et al.* 1991). Extraterrestrial material can be a negligible source of Os to ORS (e.g. <0.2% of total Os for ORS, with Os abundances >150 ppt, sediment accumulation rates >50 m Ma^{-1} and meteoritic fluxes comparable to Cenozoic fluxes) (Cohen *et al.* 1999; Kendall *et al.* 2004).

Development of the Re–Os geochronometer for ORS

The development of Re–Os ORS geochronology resulted from several studies on Phanerozoic shales. Using the NiS fire assay for pre-concentration of Re and Os (and acid dissolution for Re in some samples), and isotope dilution-secondary ion mass spectrometry (ID-SIMS), Ravizza & Turekian (1989) obtained Re–Os isotope data for the upper Mississippian Bakken Shale (North Dakota, USA). Regression of their data using Isoplot V. 3.0 (Ludwig 2003) and the value of 1.666×10^{-11} year^{-1} for λ^{187}Re (Smoliar *et al.* 1996; Selby *et al.* 2007*b*) yields a Re–Os age of 323 ± 110 Ma (2σ, Model 3, mean square of weighted deviates (MSWD) = 5.6) that is broadly consistent with known biostratigraphic age constraints (*c.* 360–352 Ma) for the Bakken Shale. The large uncertainty in the Re–Os date reflects in part the poor precision (3–10%) of ^{187}Os/^{186}Os ratios measured by SIMS (Luck & Allègre 1983). Ravizza & Turekian (1989) suggested that some of the scatter about the regression line resulted predominantly from disturbance of the Re–Os systematics by hydrocarbon maturation.

Using new analytical methodologies, including the Carius tube technique for whole-rock digestions in inverse aqua regia (Shirey & Walker 1995), improved chemical separation and purification protocols for Os (Cohen & Waters 1996) and ID-NTIMS analysis, Cohen *et al.* (1999) provided the first relatively precise Re–Os age determinations from hydrocarbon immature, high-TOC (>5%), Jurassic ORS from the UK. Cohen *et al.* (1999) obtained Re–Os ages of 207 ± 12 Ma (Model 3, MSWD = 88), 181 ± 13 Ma (Model 3, MSWD = 17) and 155.2 ± 4.3 Ma (Model 3, MSWD = 11) for Hettangian (Blue Lias, Dorset), Toarcian (Jet Rock, Yorkshire) and Kimmeridgian (Kimmeridge Clay, Dorset) shales. These ages agree within uncertainty to the respective stratigraphic ages of 198.0 ± 1.5, 179.3 ± 3.7 and 152.7 ± 1.9 Ma (Gradstein *et al.* 2004). Cohen *et al.* (1999) established that sampling of stratigraphic intervals representing short intervals of geological time were required to minimize I_{Os} heterogeneity and obtain precise Re–Os depositional ages. Cohen *et al.* (1999) also considered the selection of immature ORS, which have not undergone hydrocarbon maturation, as a necessary sampling protocol for precise and accurate Re–Os geochronology. However, Creaser *et al.* (2002) found that regression of hydrocarbon immature, mature and overmature samples gave a Re–Os age of 358 ± 10 Ma (Model 3, MSWD = 19) for the Late Devonian Exshaw Formation (Western Canada Sedimentary Basin) lower black shale member. The Re–Os age is consistent with a previous U–Pb monazite age (363.3 ± 0.4 Ma) from a tuff at the base of the lower black shale member (Richards *et al.* 2002). Individual regressions for immature and mature plus overmature samples yield statistically indistinguishable Re–Os dates (at 2σ uncertainties) and thus hydrocarbon maturation did not grossly perturb the Re–Os system in this ORS. Further, Creaser *et al.* (2002) showed that non-hydrogenous Re and Os can be liberated by inverse aqua regia whole-rock digestions, and thus negatively impact the precision and accuracy of Re–Os ages derived from low-TOC samples.

To limit the release of detrital Re and Os from ORS matrices, a new whole-rock digestion protocol (using a Cr^{VI}–H_2SO_4 solution) was developed by Selby & Creaser (2003). Using the same Exshaw Formation shale aliquots as Creaser *et al.* (2002), two subsets with distinctive $I_{Os(364\,Ma)}$ values were identified and separately regressed to yield Re–Os ages of 366.1 ± 9.6 Ma (Model 3, MSWD = 2.2; I_{Os} = 0.51 ± 0.06) and 363.4 ± 5.6 Ma (Model 1, MSWD = 1.6; I_{Os} = 0.41 ± 0.04). These Re–Os ages are more precise and better correlated than the Re–Os ages for the same subsets derived using the data from Creaser *et al.* (2002) for inverse aqua regia digestion (356 ± 23 Ma, MSWD = 11.1; and 356 ± 12 Ma, MSWD = 5.2). Selby & Creaser (2003) thus showed that the Cr^{VI}–H_2SO_4 medium represents a superior digestion protocol to inverse aqua regia for obtaining depositional ages and the $^{187}Os/^{188}Os$ isotope composition of contemporaneous sea water. Similarly, Kendall *et al.* (2004) obtained an imprecise Re–Os age of 634 ± 57 Ma (Model 3, MSWD = 65) with inverse aqua regia digestion, but a precise age of 607.8 ± 4.7 Ma (Model 1, MSWD = 1.2) was obtained using Cr^{VI}–H_2SO_4 digestion of the same shale powders of low-TOC (<1%) slates from the Late Neoproterozoic Old Fort Point Formation, Windermere Supergroup, Western Canada (Fig. 1). The 607.8 ± 4.7 Ma age agrees with existing Windermere Supergroup U–Pb age constraints that bracket the Old Fort Point Formation to between approximately 700–667 and 570 Ma (Colpron *et al.* 2002; Lund *et al.* 2003; Fanning & Link 2004, 2008). Neither chlorite-grade metamorphism nor low-TOC content precluded a precise Re–Os age determination (using Cr^{VI}–H_2SO_4 digestion) for the Old Fort Point Formation.

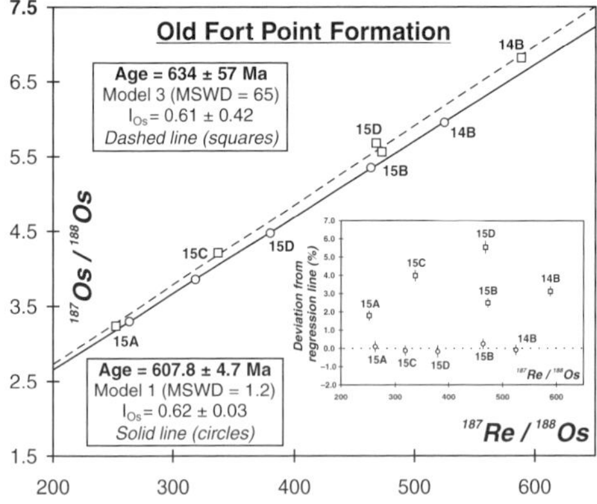

Fig. 1. Re–Os isochron diagram for the Neoproterozoic Old Fort Point Formation, Windermere Supergroup, Western Canada (Kendall *et al.* 2004). Regression of inverse aqua regia analyses (squares, dashed regression line) yields an imprecise Model 3 age, whereas regression of Cr^{VI}–H_2SO_4 digestion analyses (circles, solid regression line) yields a precise Model 1 age. The inset diagram shows the deviation of each point from the Cr^{VI}–H_2SO_4 best-fit regression line. Regression used measured 2σ uncertainties for both $^{187}Re/^{188}Os$ and $^{187}Os/^{188}Os$, and, for the first time, the error correlation function (ρ) (Ludwig 1980). Use of the latter is justified by generally significant error correlations between $^{187}Re/^{188}Os$ and $^{187}Os/^{188}Os$ (*c.* 0.4–0.6 in this case).

Accuracy of the Re–Os geochronometer for ORS and the ^{187}Re decay constant

An accurate and precise determination of the ^{187}Re decay constant, together with agreement between Re–Os black shale and U–Pb zircon ages from the same rock unit, are necessary prerequisites to qualify the Re–Os system as an accurate deposition-age geochronometer for ORS. The most widely used value of λ^{187}Re ($1.666 \pm 0.005 \times 10^{-11}$ year^{-1}), determined by Smoliar *et al.* (1996), was calculated using the slope of Re–Os data from group IIIA iron meteorites and the Pb/Pb age of 4557.8 ± 0.4 Ma determined for angrite meteorites (Lugmair & Galer 1992) that are assumed to form at the same time as group IIIA irons. This assumption is supported by ^{53}Mn–^{53}Cr ages for angrite and group IIIAB meteorites that constrain their formation to be within ±5 Ma of each other (Hutcheon & Olsen 1991; Hutcheon *et al.* 1992). Although the uncertainty in λ^{187}Re was calculated to be ±0.31% (2σ) by Smoliar *et al.* (1996), the ammonium hexachloro-osmate standard used by these researchers is only stoichiometric to within ±1.2% (Morgan *et al.* 1995) and thus the total uncertainty in λ^{187}Re would be approximately 1% (2σ) for spike solutions not calibrated against the particular Os standard used by Smoliar *et al.* (1996). Subsequent studies have yielded slightly different values for λ^{187}Re using either direct counting experiments or analyses from meteorites (Shukolyukov & Lugmair 1997; Birck & Allègre 1998; Shen *et al.* 1998). Recently, Selby *et al.* (2007*b*) intercalibrated the Re–Os molybdenite and U–Pb zircon chronometres using 11 magmatic–hydrothermal ore deposits spanning about 2700 Ma of Earth history. Calculated values for λ^{187}Re of $1.6668 \pm 0.0034 \times 10^{-11}$ year^{-1}

(using U decay constants from Jaffey *et al.* 1971) and $1.6689 \pm 0.0031 \times 10^{-11}$ year^{-1} (using $\lambda^{238}U$ from Jaffey *et al.* 1971; and $\lambda^{235}U$ from Schoene *et al.* 2006) are nominally higher (0.1–0.2%) than the value determined by Smoliar *et al.* (1996), but within calculated uncertainty.

To establish the utility of the Re–Os ORS geochronometer for geological timescale calibration, Selby & Creaser (2005*a*) studied black shales from the Devonian–Mississippian (D–M) boundary of the Exshaw Formation. Black shales obtained along a narrow 10 cm stratigraphic interval straddling the D–M boundary yield a Re–Os age of 361.3 ± 2.4 Ma (Model 1, including a $\lambda^{187}Re$ uncertainty of $\pm 0.35\%$, MSWD = 1.2) that agrees to within *c.* 0.2% of a recent calibration (360.7 ± 0.7 Ma) for the D–M boundary based on interpolation of U–Pb zircon dates from the lower Exshaw Formation and D–M boundary strata in New Brunswick, Canada, and Germany (Trapp *et al.* 2004). The excellent agreement between the Re–Os black shale and U–Pb zircon ages thus demonstrates the ability of the Re–Os ORS geochronometer for accurately constraining the timing, duration and rate of geological events and processes associated with sedimentary deposition.

Significance of appropriate sampling protocols for Re–Os geochronology

Another aspect of Re–Os ORS geochronology that deserves close scrutiny concerns the protocol for selecting and sampling appropriate material. This is not simply a matter of targeting ORS that have not been significantly affected by post-depositional processes such as weathering, metamorphism and hydrothermal fluid flow. For example, Selby & Creaser (2004) showed that, for a specific mineral separate, the required analysis aliquant size for obtaining reproducible and accurate Re–Os ages for molybdenite depends critically on the molybdenite grain size and age, and that small analysis aliquant sizes can result in inaccurate and/or non-reproducible Re–Os molybdenite ages. Similarly, it is necessary to determine carefully the optimal sampling strategies in order to avoid the generation of erroneous and/or imprecise Re–Os dates from ORS. The reliability of Re and Os isotope analyses can ideally be checked by replicate analyses of powder aliquots. Reproducible Re and Os abundance and isotope data for replicate analyses of a sample powder suggests a homogenous distribution of Re and Os within that powder. If Re and Os are distributed heterogeneously in a powder aliquot (incomplete powder homogenization during sample grinding), but have otherwise remained part of a closed system since deposition (negligible post-depositional mobility), then naturally coupled variations in $^{187}Re/^{188}Os$ and $^{187}Os/^{188}Os$ isotope ratios will result in reproducible sample I_{Os} with the added benefit of introducing additional variation along Re–Os isochrons (Creaser *et al.* 2002; Kendall *et al.* 2004). Significant post-depositional mobilization of Re and Os within shale samples may result in reproducible (if sample powders are homogeneous) or non-reproducible (if sample powders are heterogeneous) Re and Os abundance and isotope data for replicate analyses of powder aliquots, but in both cases the Re–Os isotope data will not yield a meaningful regression (e.g. Kendall *et al.* 2009*a*). However, it is important to distinguish between small-scale post-depositional diffusion and/or decoupling of Re and Os in otherwise pristine ORS, and large-scale element mobility resulting from some major post-depositional geological disturbance (e.g. weathering, metamorphism and hydrothermal fluid flow). Here, we demonstrate the importance of powder aliquot size for avoiding small-scale diffusion and/or elemental decoupling using ORS from the Neoproterozoic Doushantuo Formation, southern China.

The depositional age of the Doushantuo Formation is constrained by U–Pb zircon ages of 635.2 ± 0.6 and 551.1 ± 0.6 Ma from ash beds that occur near its base and top, respectively, in the Yangtze Gorges area (Fig. 2) (Condon *et al.* 2005). The latter U–Pb zircon age comes from an ash bed in the uppermost Miaohe Member (Jiuqunao section: see Zhou & Xiao 2007 for section locations) that comprises unmetamorphosed organic-rich shales with well-preserved carbonaceous compression fossils of prokaryotes, macroscopic multicellular algae and metazoans (cnidarians) (Xiao *et al.* 2002). Four samples of pyritic finely laminated black shale from the Miaohe Member (from stratigraphically lowest to highest over a *c.* 3.5 m stratigraphic interval: H_1O_{18}, H_1O_{19}, H_1O_{20}, H_1O_{21}, each comprising a volume of 40–80 cm^3) were subsampled into fractions of ≤ 10 g and ≥ 20 g of material. The samples are from the Yangtze Gorges area (Huajipo section that is less than 20 km from the Jiuqunao section: Zhou & Xiao 2007). Each subsample was ground to remove cutting marks and any weathered material, broken into small chips without metal contact and processed to a powder (*c.* 30 μm) in an automated agate mill. Rhenium and Os isotope analyses were carried out by ID-NTIMS following the methods outlined in Selby & Creaser (2001, 2003), Creaser *et al.* (2002) and Kendall *et al.* (2004), and references therein.

The Miaohe Member black shales are highly enriched in Re (30–566 ppb) and Os (1.14–8.24 ppb), and show a wide range in $^{187}Re/^{188}Os$

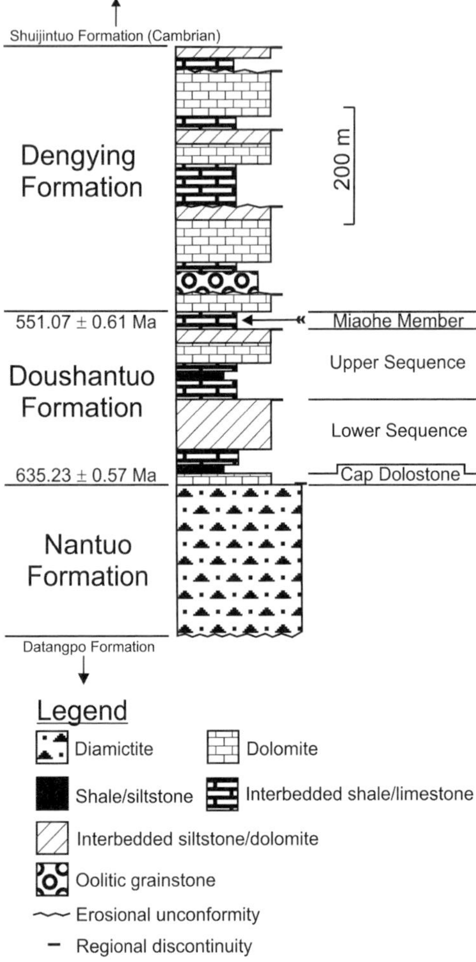

Fig. 2. Late Neoproterozoic stratigraphy, Yangtze Platform, South China. U–Pb zircon ages from ash beds are shown (Condon *et al.* 2005). The arrow indicates the sampled interval within the Miaohe Member of the uppermost Doushantuo Formation. Modified after Jiang *et al.* (2003).

(169–617) and $^{187}Os/^{188}Os$ (2.4–6.8) isotope ratios (Table 1). Duplicate analyses of $\geq$20 g aliquots of shale show very good reproducibility in Re and Os abundances (<3% variation), $^{187}Re/^{188}Os$ and $^{187}Os/^{188}Os$ isotope ratios (<2% variation), and $I_{Os\ (550\,Ma)}$ ($\leq$5% variation), suggesting that Re and Os are largely distributed homogenously in these aliquots. In contrast, for $\leq$10 g aliquots, larger variations are observed in Re and Os abundances (up to 4% for most samples), $^{187}Re/^{188}Os$ (up to 16%) and $^{187}Os/^{188}Os$ (up to 8%) isotope ratios, and $I_{Os(550\,Ma)}$ (>20 %). These observations are independent of sample powder aliquant size used for dissolution (0.2–0.8 g). Homogeneous distribution of Re and Os in a $\geq$20 g aliquot suggests that a $\leq$10 g aliquot derived from the same whole-rock hand sample also should have a homogeneous distribution of Re and Os because grinding conditions were the same in all samples. Our samples do not show secondary alteration from weathering or hydrothermal fluid flow (e.g. quartz/carbonate veinlets were not observed), and are unmetamorphosed. Thus, the contrast in reproducibility between the two Miaohe Member subsets most probably results from small-scale post-depositional diffusion and/or decoupling of Re and Os that is preserved within the $\leq$10 g aliquot subset. We interpret these observations to suggest that sample powders representing a larger rock volume ($\geq$20 g in this case) are required to remove the effect of small-scale diffusion and/or element decoupling.

Regression of all data for the Miaohe Member black shales yields a Model 3 Re–Os date of 598 $\pm$ 16 Ma (MSWD = 275; $I_{Os} = 0.65 \pm 0.13$) (Fig. 3a). Separate regressions of the $\leq$10 and $\geq$20 g aliquots yield Model 3 dates of 590 $\pm$ 20 (MSWD = 360; $I_{Os} = 0.69 \pm 0.16$) and 623 $\pm$ 13 Ma (MSWD = 16; $I_{Os} = 0.51 \pm 0.10$) (Fig. 3b, c). Although the $\geq$20 g aliquot sample subset yields a more precise date, all three Re–Os dates are significantly older than the U–Pb zircon age of 551.1 $\pm$ 0.6 Ma from the ash bed within the Miaohe Member. In addition, the 623 $\pm$ 13 Ma date is older than Pb/Pb phosphorite ages of 599.3 $\pm$ 4.2 (Barfod *et al.* 2002) and 576 $\pm$ 14 Ma (Chen *et al.* 2004) for the Upper Sequence of the Doushantuo Formation that stratigraphically underlies the Miaohe Member. We suggest below that this age discrepancy reflects I_{Os} heterogeneity related to temporal changes in sea-water Os isotope composition through the sampled stratigraphic interval of approximately 3.5 m.

In the Yangtze Gorges area, the Doushantuo Formation is about 250 m thick, but represents a duration of more than 80 Ma. Consequently, average sedimentation rates during Doushantuo time may have been quite slow, although at least two sequence boundaries are known to be present within the Doushantuo Formation (Xiao *et al.* 1998; Wang *et al.* 1998; Jiang *et al.* 2003; Condon *et al.* 2005; Zhou & Xiao 2007). Assuming relatively uniform sedimentation rates, the sampled stratigraphic interval of 3.5 m from the Miaohe Member could represent more than 1 Ma of elapsed time. Because of the relatively short sea-water residence time of Os (10^4 years), it is plausible that the Os isotope composition of sea water changed during the time period encompassed by the sampled stratigraphic interval. Of the $\geq$20 g aliquot subset, subsamples H_1O_{18}-3, H_1O_{19}-3, H_1O_{20}-4 and their replicate analyses have similar $I_{Os(550\,Ma)}$ (1.17–1.20) values and yield a Re–Os isochron

Table 1. *Re–Os abundances and isotope data for the Miaohe Member, upper Doushantuo Formation, South China*

Sample*	Aliquot mass (g)	Aliquant mass† (g)	Re (ppb)	Os (ppb)	^{192}Os (ppb)	$^{187}Re/^{188}Os$‡	$^{187}Os/^{188}Os$‡	ρ	I_{Os}§
H_1O_{18}-1	7.0	0.81	284.41	4.418	1.027	550.97 (2.30)	6.0788 (0.0110)	0.195	1.01
H_1O_{18}-1-rpt		0.80	275.38	4.510	1.049	522.45 (2.13)	6.0743 (0.0086)	0.140	1.27
H_1O_{18}-2	10.4	0.78	565.95	8.235	1.839	612.26 (2.57)	6.6343 (0.0125)	0.205	1.00
H_1O_{18}-3	39.8	0.78	402.35	6.225	1.414	566.22 (2.34)	6.3985 (0.0114)	0.166	1.19
H_1O_{18}-3-rpt		0.50	392.22	6.094	1.387	562.52 (2.32)	6.3655 (0.0120)	0.149	1.19
H_1O_{19}-1	7.8	0.80	388.52	5.971	1.388	556.92 (2.32)	6.0771 (0.0107)	0.193	0.95
H_1O_{19}-1-rpt		0.15	374.73	6.007	1.398	533.12 (2.20)	6.0583 (0.0096)	0.170	1.15
H_1O_{19}-2	6.6	0.80	392.16	5.654	1.264	617.10 (2.60)	6.6169 (0.0132)	0.212	0.94
H_1O_{19}-2-rpt		0.15	381.80	5.642	1.256	604.69 (2.88)	6.6769 (0.0258)	0.367	1.11
H_1O_{19}-2-rpt		0.79	384.57	5.715	1.265	604.89 (2.46)	6.7631 (0.0091)	0.137	1.19
H_1O_{19}-3	20.5	0.78	335.93	5.539	1.302	513.12 (2.12)	5.9227 (0.0109)	0.161	1.20
H_1O_{19}-3-rpt		0.58	338.91	5.578	1.312	513.92 (2.09)	5.9172 (0.0080)	0.129	1.19
H_1O_{20}-1	4.8	0.79	160.31	2.427	0.558	571.35 (2.42)	6.2219 (0.0127)	0.224	0.96
H_1O_{20}-1-rpt		0.26	154.73	2.490	0.579	531.48 (2.28)	6.0667 (0.0169)	0.201	1.17
H_1O_{20}-2	5.9	0.78	105.69	1.776	0.434	484.54 (2.06)	5.4170 (0.0118)	0.229	0.96
H_1O_{20}-2-rpt		0.30	102.42	1.834	0.461	441.61 (1.96)	5.0412 (0.0154)	0.272	0.98
H_1O_{20}-3	5.5	0.80	181.41	2.757	0.629	573.51 (2.41)	6.3271 (0.0128)	0.204	1.05
H_1O_{20}-3-rpt		0.25	178.81	2.809	0.647	549.62 (2.39)	6.1976 (0.0163)	0.256	1.14
H_1O_{20}-3-rpt		0.80	179.73	2.768	0.633	564.62 (2.30)	6.2948 (0.0083)	0.128	1.10
H_1O_{20}-4	22.0	0.82	167.29	2.661	0.615	541.20 (2.58)	6.1555 (0.0267)	0.331	1.17
H_1O_{20}-4-rpt		0.54	165.81	2.655	0.614	536.84 (2.20)	6.1346 (0.0086)	0.159	1.19
H_1O_{20}-5	32.8	0.78	117.44	1.945	0.464	503.58 (2.26)	5.7320 (0.0201)	0.270	1.10
H_1O_{20}-5-rpt		0.54	118.57	1.973	0.474	497.93 (2.06)	5.6414 (0.0093)	0.184	1.06
H_1O_{21}-1	8.9	0.67	33.76	1.185	0.368	182.28 (0.77)	2.6444 (0.0056)	0.180	0.97
H_1O_{21}-1-rpt		0.60	40.89	1.230	0.386	210.78 (1.01)	2.5529 (0.0103)	0.360	0.61
H_1O_{21}-1-rpt		0.75	39.09	1.206	0.377	206.27 (0.86)	2.5846 (0.0048)	0.192	0.69
H_1O_{21}-2	4.6	0.53	32.88	1.139	0.358	182.60 (0.78)	2.5224 (0.0051)	0.210	0.84
H_1O_{21}-2-rpt		0.62	34.52	1.175	0.373	184.00 (0.81)	2.4290 (0.0071)	0.246	0.74
H_1O_{21}-2-rpt		0.74	30.44	1.140	0.359	168.59 (0.70)	2.5037 (0.0043)	0.187	0.95
H_1O_{21}-3	32.2	0.49	37.09	1.176	0.368	200.68 (0.87)	2.5825 (0.0065)	0.279	0.74
H_1O_{21}-3-rpt		0.46	36.59	1.181	0.369	197.08 (0.89)	2.5864 (0.0082)	0.312	0.77

*rpt denotes a replicate analysis.
†Used for individual Re–Os analysis.
‡Numbers in parentheses denote measured 2σ uncertainty in the isotope ratio calculated by numerical error propagation.
§I_{Os}, initial $^{187}Os/^{188}Os$ isotope ratio calculated at 550 Ma.

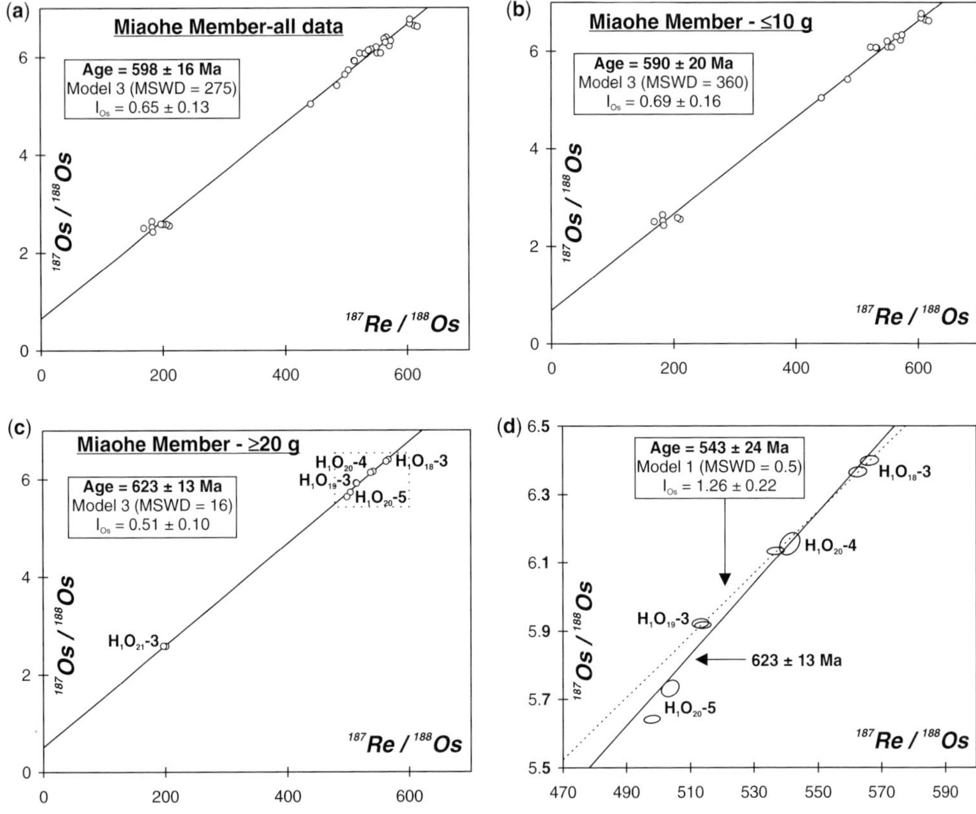

Fig. 3. Re–Os isochron diagrams for the Neoproterozoic Miaohe Member, uppermost Doushantuo Formation, South China. (**a**) All data. (**b**) Subset from $\leq$10 g sample aliquots. (**c**) Subset from $\geq$20 g aliquots. (**d**) Expanded view of the inset diagram in (c). Three samples (H_1O_{18}-3, H_1O_{19}-3 and H_1O_{20}-4) have similar $I_{Os\ (550\,Ma)}$ of 1.17–1.20 and plot on a 543 $\pm$24 Ma Model 1 isochron, but samples H_1O_{20}-5 and H_1O_{21}–3 (not shown) have less radiogenic $I_{Os\ (550\,Ma)}$ of 1.06–1.10 and 0.74–0.77, respectively, and plot below this isochron. Because sample H_1O_{21}-3 has much lower $^{187}Re/^{188}Os$ and $^{187}Os/^{188}Os$ relative to the other samples, it effectively controls the slope of an isochron defining an erroneously old Re–Os date of 623 $\pm$ 13 Ma. Regressions used measured 2σ uncertainties for $^{187}Re/^{188}Os$ and $^{187}Os/^{188}Os$, and the error correlation function (ρ) (Ludwig 1980; Kendall *et al.* 2004).

age of 543 $\pm$ 24 Ma (Model 1, MSWD = 0.5; I_{Os} = 1.26 $\pm$ 0.22) that is imprecise because of the restricted range in Re/Os, but overlaps the U–Pb zircon age of 551.1 $\pm$ 0.6 Ma for the Miaohe Member (Fig. 3d). However, both analyses of sample H_1O_{20}-5 do not plot on this regression, and have less radiogenic $I_{Os\ (550\,Ma)}$ values of 1.06–1.10 suggesting a possible change in sea water Os isotope composition is recorded within whole-rock H_1O_{20} (volume 80 cm^3). The stratigraphically highest subsample H_1O_{21}-3 has the lowest $I_{Os\ (550\,Ma)}$ values (0.74–0.77) and is characterized by a significantly less radiogenic Os isotope composition compared to the other $\geq$20 g aliquots. Thus, this one subsample exerted a large effect on the $\geq$20 g aliquot regression, which resulted in an erroneously old age estimate of 623 $\pm$ 13 Ma. The Miaohe Member Re–Os data demonstrate that a deceptively attractive date can be obtained for a sample suite with heterogeneous I_{Os}, especially when there is a large, but skewed, distribution in $^{187}Re/^{188}Os$ and $^{187}Os/^{188}Os$ isotope ratios.

Because of the potentially slow sedimentation rates for some condensed black shale sequences (e.g. <2 m Ma^{-1}: Arthur & Sageman 1994), a relatively short sea-water residence time for Os (probably $<10^4$ years in low-O_2 Proterozoic oceans) and potential for rapid variations in sea-water $^{187}Os/^{188}Os$, a thin stratigraphic sampling interval is critical to obtain homogenous I_{Os}. A thin stratigraphic sampling interval, together with a sufficiently large mass of black shale powder (e.g. >20 g), are thus key sampling protocols for precise and accurate Re–Os depositional age determinations. Recent studies have successfully produced precise Model 1 Re–Os ages for shales from stratigraphic intervals ranging from tens of centimetres to a few metres, and by using powder aliquots comprising >20 g of powdered shale (Kendall *et al.* 2004, 2006, 2009*a*, *b*; Selby & Creaser 2005*a*; Anbar *et al.* 2007; Creaser & Stasiuk 2007).

Discrepancy in Re–Os dates for the Neoproterozoic Aralka Formation, central Australia

Using the inverse aqua regia dissolution protocol, Schaefer & Burgess (2003) reported Re–Os dates of 592 ± 14 ($n = 3$, MSWD ≪1; $I_{Os} = 0.91 \pm 0.07$) and 623 ± 18 Ma ($n = 7$, MSWD = 5.2; $I_{Os} = 0.78 \pm 0.10$) for organic-rich dolomitic siltstones (TOC *c.* 0.5–1%) of the Aralka Formation (Amadeus Basin, central Australia). Because the seven samples defining the 623 ± 18 Ma date spanned a stratigraphic interval of about 10 m that could potentially be characterized by heterogeneous I_{Os}, Schaefer & Burgess (2003) suggested that the regression defined by the three samples in closest stratigraphic proximity (*c.* 1.6 m) represented the best estimate of the true depositional age. However, Kendall *et al.* (2006) subsequently obtained a significantly older Re–Os date of 657.2 ± 5.4 Ma (MSWD = 1.2; $I_{Os} = 0.82 \pm 0.03$) by Cr^{VI}–H_2SO_4 digestion of 18 samples (and two replicates) derived from *c.* 2 m stratigraphic interval within the larger 10 m interval sampled by Schaefer & Burgess (2003) (Fig. 4).

To determine whether the reason for this age discrepancy was simply the nature of the digestion mediums used (as suggested by Kendall *et al.* 2004), we reanalysed three of our Aralka Formation samples (BK-04-Wallara-1A, BK-04-Wallara-2A and BK-04-Wallara-5) using inverse aqua regia (Table 2). Our initial analyses used approximately 1.44–1.48 g of shale powder for Carius Tube digestion, and plot near the *c.* 623 Ma-isochron of Schaefer & Burgess (2003) (Fig. 5). However, replicate analyses for BK-04-Wallara-1A and BK-04-Wallara-5 using the same amount of sample powder were not reproducible in terms of Re and Os abundances, $^{187}Re/^{188}Os$ and $^{187}Os/^{188}Os$ isotope ratios, and $I_{Os\ (657\ Ma)}$. The large degree of scatter indicated to us that sample-spike equilibration had not occurred for these replicate analyses during Carius tube digestion. Subsequently, replicate analyses used 0.50–0.54 g of powder for inverse aqua regia digestion, and these data plot near the *c.* 657 Ma-isochron defined by the Cr^{VI}–H_2SO_4 analyses of Kendall *et al.* (2006). In addition, the *c.* 0.50–0.54 g inverse aqua regia analyses

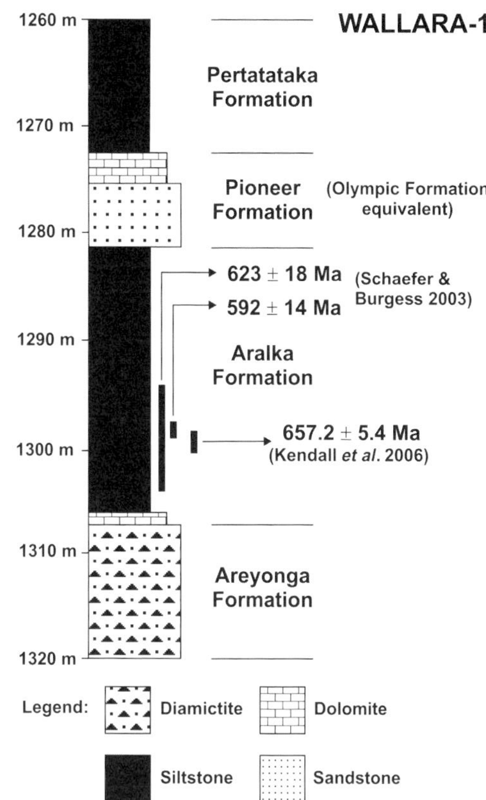

Fig. 4. Stratigraphic column for drillhole Wallara-1 (Amadeus Basin, central Australia) showing sampled intervals and Re–Os dates obtained for organic-rich siltstones of the Neoproterozoic Aralka Formation by Schaefer & Burgess (2003) and Kendall *et al.* (2006). Modified from Schaefer & Burgess (2003).

have broadly similar Re and Os abundances, $^{187}Re/^{188}Os$ and $^{187}Os/^{188}Os$ isotope ratios, and $I_{Os\ (657\ Ma)}$ as the Cr^{VI}–H_2SO_4 analyses from the same powder aliquots. Regression of our 0.50–0.54 g inverse aqua regia subset yields a Model 1 Re–Os date of 657 ± 15 Ma (MSWD = 0.7; $I_{Os} = 0.87 \pm 0.09$) that is statistically equivalent to the *c.* 657 Ma age obtained using Cr^{VI}–H_2SO_4 digestion. To obtain a direct estimate of the Re–Os systematics of the detrital component, we carried out an inverse aqua regia attack on the Carius tube residue left behind after Cr^{VI}–H_2SO_4 digestion of sample BK-04-Wallara-5. The Re–Os data for this residue have large uncertainty, but plot on the isochrons of both Schaefer & Burgess (2003) and Kendall *et al.* (2006) near the I_{Os} intercept. However, the residue has a calculated $I_{Os\ (657\ Ma)}$ of 0.85 similar to calculated $I_{Os\ (657\ Ma)}$ values of

Table 2. *Re–Os isotope data derived from inverse aqua regia and Cr^{VI}–H_2SO_4 digestions of three Aralka Formation samples*

Sample	Method*	Aliquant mass† (g)	Re (ppb)	Os (ppt)	^{192}Os (ppt)	$^{187}Re/^{188}Os$‡	$^{187}Os/^{188}Os$‡	ρ	I_{Os}§
BK-04-Wallara-1A	Cr^{VI}–H_2SO_4[a]	1.06	5.41	145.9	41.0	262.10 (1.23)	3.7170 (0.0125)	0.525	0.83
	Cr^{VI}–H_2SO_4-rpt[a]	1.48	5.41	145.3	40.9	263.41 (1.20)	3.7149 (0.0129)	0.436	0.82
	IAR[b]	1.45	6.01	145.0	40.8	293.02 (1.30)	3.7190 (0.0127)	0.331	0.49
	IAR-rpt[b]	1.46	5.36	148.1	41.8	255.03 (1.38)	3.6755 (0.0166)	0.642	0.87
	IAR-rpt[b]	0.54	5.33	146.7	41.4	256.18 (3.68)	3.6721 (0.0660)	0.747	0.85
BK-04-Wallara-2A	Cr^{VI}–H_2SO_4[a]	1.02	5.18	125.3	33.9	304.28 (1.54)	4.1659 (0.0164)	0.596	0.82
	IAR[b]	1.46	5.12	176.1	54.7	186.27 (0.91)	2.6518 (0.0097)	0.561	0.60
	IAR-rpt[b]	0.50	4.88	122.5	33.4	291.15 (4.14)	4.0802 (0.0611)	0.885	0.88
BK-04-Wallara-5	Cr^{VI}–H_2SO_4[a]	0.99	4.38	81.8	19.5	447.17 (3.28)	5.7469 (0.0421)	0.746	0.83
	Cr^{VI}–H_2SO_4-rpt[a]	1.49	4.36	81.8	19.5	443.99 (2.71)	5.7200 (0.0354)	0.625	0.83
	IAR[b]	1.48	4.67	87.5	22.0	421.86 (2.14)	5.0383 (0.0254)	0.421	0.40
	IAR-rpt[b]	1.44	4.34	85.5	19.7	437.89 (3.77)	6.2048 (0.0517)	0.843	1.39
	IAR-rpt[b]	0.52	4.44	84.3	20.2	438.58 (9.32)	5.7034 (0.1241)	0.946	0.88
	IAR-rpt[b]	0.51	4.18	79.8	19.2	434.01 (4.48)	5.6272 (0.0742)	0.677	0.85
	Cr^{VI}–H_2SO_4-res[b]	0.87	0.01	2.3	0.8	27.26 (6.45)	1.1466 (0.1803)	0.432	0.85

*rpt, IAR and res denote replicate, inverse aqua regia and residue analysis by IAR, respectively; [a]data from Kendall *et al.* (2006); [b]data from this study.
†Used for individual Re–Os analysis.
‡Numbers in parentheses denote measured 2σ uncertainty in the isotope ratio calculated by numerical error propagation.
§$I_{Os} = {}^{187}Os/{}^{188}Os$ isotope ratio calculated at 657 Ma (from Kendall *et al.* 2006).

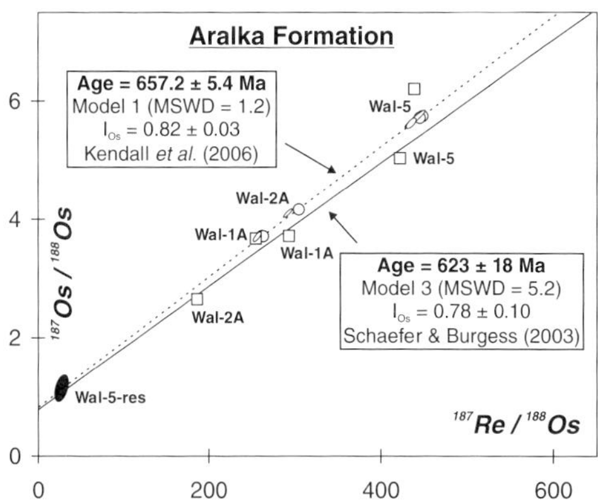

Fig. 5. Re–Os isochron diagram for the Aralka Formation showing the inverse aqua regia and Cr^{VI}–H_2SO_4 linear regressions from Schaefer & Burgess (2003) and Kendall *et al.* (2006), respectively. For clarity, only individual sample analyses of BK-04-Wallara-1A, BK-04-Wallara-2A and BK-04-Wallara-5 are shown. Circles, Cr^{VI}–H_2SO_4 analyses (Kendall *et al.* 2006); error ellipses, inverse aqua regia analyses using an aliquant size of *c.* 0.5 g (this study); squares, inverse aqua regia analyses using an aliquant size of *c.* 1.4–1.5 g (this study). Regression of the *c.* 0.5 g inverse aqua regia subset yields a Model 1 Re–Os date (not shown) of 657 ± 15 Ma (MSWD = 0.7; I_{Os} = 0.87 ± 0.09) that is equivalent within 2σ uncertainties to the Cr^{VI}–H_2SO_4 isochron age reported by Kendall *et al.* (2006). The inverse aqua regia analysis of the residue from Cr^{VI}–H_2SO_4 digestion of sample BK-04-Wallara-5 (Wal-5-res) provides an estimate of the detrital Re–Os systematics in the Aralka Formation and is shown for comparison.

0.81–0.84 for Cr^{VI}–H_2SO_4 analyses (Kendall *et al.* 2006), but is more radiogenic than calculated $I_{Os\,(657\,Ma)}$ values of 0.47–0.71 from the data of Schaefer & Burgess (2003). The similar ages derived here using the two dissolution mediums, together with the Re–Os data for the residue, suggests the detrital component in the Aralka Formation is minor and/or has similar I_{Os} as the hydrogenous Os fraction. Accordingly, the discrepancy between the Re–Os data of Schaefer & Burgess (2003) and Kendall *et al.* (2006) may relate to incomplete sample-spike equilibration during inverse aqua regia digestion of large (e.g. 1.4–1.5 g) aliquant fractions. In contrast, reproducible Re–Os data diagnostic of complete sample-spike equilibration during Cr^{VI}–H_2SO_4 digestions was obtained for the Aralka Formation using up to 1.5 g of sample (Kendall *et al.* 2006). Because sample details were not provided by Schaefer & Burgess (2003), further comparison with the data reported here and in Kendall *et al.* (2006) is not possible.

The $^{187}Os/^{188}Os$ isotope composition of Precambrian sea water

The value of I_{Os} from a Re–Os ORS isochron regression may record the $^{187}Os/^{188}Os$ isotope composition of the contemporaneous sea water at the time of sediment deposition if the hydrogenous Os fraction dominates the detrital/extraterrestrial Os fraction of ORS (Ravizza & Turekian 1989, 1992; Ravizza *et al.* 1991). This tenet is supported by the similar Os isotope composition of recent organic-rich sediments ($^{187}Os/^{188}Os$ = 0.98–1.07 based on data from three localities in the Pacific and Atlantic oceans) compared to present-day sea water (Ravizza & Turekian 1992). Similarly, Ravizza & Paquay (2008) show that Eocene–Oligocene organic-rich sediments yield initial Os isotope ratios (calculated using depositional ages derived from biostratigraphy) that agree well with Os isotope data from pelagic carbonates (which do not require correction for *in situ* decay of ^{187}Re to ^{187}Os). Because the present-day sea-water residence time of Os (10^4 years) is long compared to the ocean mixing time (2–3 ka), large global variations in the $^{187}Os/^{188}Os$ isotope composition of sea water are suggested to have occurred during the Cenozoic and Mesozoic. These variations result from major changes in the proportion of radiogenic Os (from oxidative weathering of upper continental crust) and unradiogenic Os (from dissolution of cosmic dust and hydrothermal alteration of oceanic crust and peridotites) delivered to the oceans (Peucker-Ehrenbrink *et al.* 1995; Pegram & Turekian 1999; Ravizza *et al.* 2001; Cohen & Coe 2002, 2007; Ravizza & Peucker-Ehrenbrink 2003; Cohen 2004; Cohen *et al.* 2004).

Determination of the Os isotope composition of Precambrian sea water is limited to a handful of I_{Os} determinations from precise Re–Os isochron regressions of Proterozoic and latest Archaean ORS (Hannah *et al.* 2004, 2006; Kendall *et al.* 2004, 2006, 2009*a*, *b*; Anbar *et al.* 2007; Creaser & Stasiuk 2007; Yang *et al.* 2009). The sea-water residence time of Os was probably low under conditions of a predominantly anoxic Archaean atmosphere and ocean prior to the 2.45–2.32 Ga Great Oxidation Event (Farquhar *et al.* 2000; Pavlov & Kasting 2002; Bekker *et al.* 2004; Goldblatt *et al.* 2006), or in a stratified Proterozoic ocean with oxidizing surface waters and suboxic, anoxic or euxinic deep waters after the Great Oxidation Event (Canfield 1998; Farquhar & Wing 2003; Slack *et al.* 2007). The deep ocean may not have become fully oxygenated until the Neoproterozoic (e.g. Fike *et al.* 2006; Kennedy *et al.* 2006; Canfield *et al.* 2007). On the modern Earth a restricted or semi-restricted (e.g. intracratonic) depositional environment can lead to preservation of I_{Os} in

ORS that reflects local sea water, but not the contemporaneous global sea water (Ravizza *et al.* 1991; Martin *et al.* 2000; Poirer 2006). Thus, a significantly shorter residence time of Os in Precambrian sea water would likely be accompanied by regional variations in sea-water Os isotope composition.

Nevertheless, some interesting insights can be gained from the limited information available for the Os isotope composition of Precambrian seawater. Weathering of ORS represents a major source of radiogenic Os to the Phanerozoic oceans (Peucker-Ehrenbrink & Hannigan 2000; Peucker-Ehrenbrink & Ravizza 2000; Jaffe *et al.* 2002; Pierson-Wickmann *et al.* 2002). However, for a predominantly anoxic Archaean atmosphere and oceans, riverine transport of soluble Re and radiogenic Os from weathering and erosion of crustal rocks would be negligible, resulting in deposition of Archaean shales with low Re abundances and $^{187}Re/^{188}Os$ and $^{187}Os/^{188}Os$ ratios (Yang & Holland 2002; Siebert *et al.* 2005; Wille *et al.* 2007). There may also have been a time lag between the Great Oxidation Event and appreciable riverine transport of radiogenic crustal Os to the oceans because of the low Re abundances in uplifted Archaean shales and/or insufficiently oxidizing conditions (Hannah *et al.* 2004). Chondritic or near-chondritic I_{Os} compositions may be typical of marine shales deposited under an anoxic weathering regime or under the conditions of a weakly oxygenated [e.g. $10^{-5}-10^{-2}$ PAL (present atmospheric level) for *c.* 2.45–2.00 Ga (Ga is 10^9 years): Farquhar & Wing 2003] or anoxic atmosphere with stratified oceans containing oxygenated shallow waters and anoxic or euxinic deep waters (Hannah *et al.* 2004, 2006; Siebert *et al.* 2005; Anbar *et al.* 2007; Wille *et al.* 2007). Chondritic I_{Os} values (e.g. *c.* 0.11) obtained from the 2.7 Ga Joy Lake Sequence, Superior Province, USA (Yang *et al.* 2009), the 2.50 Ga Mt McRae Shale, Hamersley Group, western Australia (Anbar *et al.* 2007) and 2.32 Ga Rooihoogte and Timeball Hill Formations, Transvaal Supergroup, South Africa (Hannah *et al.* 2004) are consistent with dominance of the late Archaean–early Palaeoproterozoic marine Os budget by extraterrestrial and magmatic/hydrothermal inputs (Fig. 6). If Earth's atmosphere and oceans were fully anoxic for most of the Archaean, depositional age determinations for most Archaean shales would not be feasible. In this case, the fundamental basis behind the Re–Os ORS geochronometer, namely the reductive capture of dissolved Re from an oxygenated water column into reducing ORS, breaks down. Such shales would be characterized by a subequal mixture comprising meteoritic and hydrogenous components with low Re/Os and chondritic $^{187}Os/^{188}Os$, and a detrital component with relatively higher Re/Os (radiogenic sulphides would persist as detrital phases during anoxic weathering and erosion: e.g. Fleet 1998; Rasmussen & Buick 1999; England *et al.* 2002) and elevated $^{187}Os/^{188}Os$.

Of particular interest is a precise determination of the time when pO_2 in the atmosphere rose sufficiently to allow oxidative weathering of continental crust and the development of a significant riverine flux of radiogenic Os to the oceans. The limited information available for the Os isotope composition of Precambrian sea water does not yet provide this information. A chondritic I_{Os} value of 0.133 ± 0.020 for *c.* 2.0 Ga sea water ('Productive Formation', Pechenga Greenstone Belt, Russia) may reflect a minimal riverine flux of radiogenic Os and/or dominance of mantle inputs in a rift-related tectonic setting (Hannah *et al.* 2006). A large time gap in the curve for Precambrian sea-water Os isotope compositions (see Fig. 6) exists until 1.54 Ga, where a I_{Os} determination of 0.51 ± 0.03 for the Douglas Formation (Athabasca Basin, western Canada) clearly indicates that crustal weathering and transport of radiogenic Os to sea water was possible during the Mesoproterozoic (Creaser & Stasiuk 2007). Two Re–Os isochron ages from the *c.* 1.4 Ga Velkerri Formation (McArthur Basin, northern Australia) yield I_{Os} values of 0.29 ± 0.18 and 0.06 ± 0.22 (Kendall *et al.* 2009*a*) that also permit some contribution of dissolved radiogenic crustal Os to *c.* 1.4 Ga oceans. By Late Neoproterozoic time, however, crustal Os fluxes from rivers exerted a significant control on the $^{187}Os/^{188}Os$ isotope composition of sea water. This is evidenced by moderate or radiogenic I_{Os} from the *c.* 657 Ma Aralka Formation, central Australia (0.82 ± 0.03), *c.* 643 Ma Tindelpina Shale Member, Tapley Hill Formation, southern Australia (0.95 ± 0.01), *c.* 640 Ma Black River Dolomite, northwestern Tasmania (1.00 ± 0.05), *c.* 608 Ma Old Fort Point Formation, western Canada (0.62 ± 0.03) and *c.* 551 Ma Miaohe Member, Doushantuo Formation (1.26 ± 0.22) (Kendall *et al.* 2004, 2006, 2009*b*; this study). Delineation of the broad features of the Proterozoic sea-water $^{187}Os/^{188}Os$ curve will depend upon further determination of I_{Os} values from precise Re–Os age determinations of ORS.

Re–Os geochronology of Precambrian ORS

With rigorous sampling and analytical protocols, the $^{187}Re-^{187}Os$ system can yield precise and accurate depositional ages for Precambrian ORS. In many Precambrian sedimentary basins (e.g.

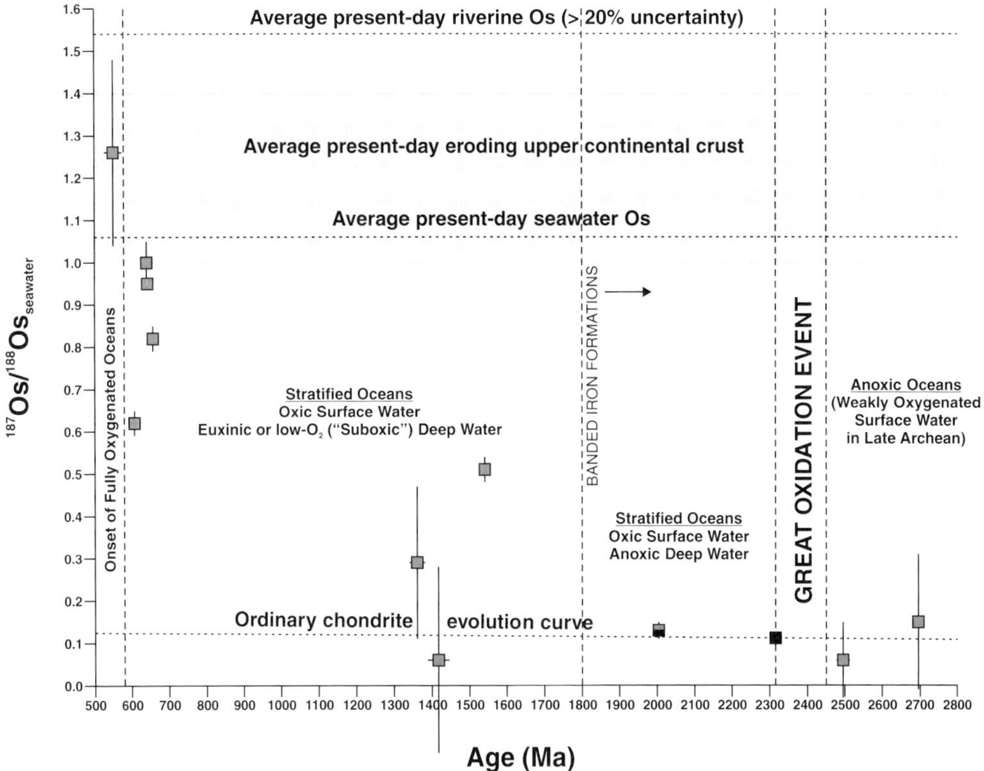

Fig. 6. Precambrian sea-water Os isotope compositions derived from black shale Re–Os isochron regressions of ORS. The suggested timeline for Precambrian Earth surface oxygenation is also shown for comparison. See text for discussion and sources of data. Grey squares, isochron regression I_{Os} derived from Cr^{VI}–H_2SO_4 digestion of ORS; black square, isochron regression I_{Os} derived from synsedimentary–early diagenetic pyrite in ORS; grey and black square, isochron regression I_{Os} derived using both methods. The 2.5 Ga Mt McRae Shale Re–Os age gave a subchondritic I_{Os} of 0.04 ± 0.06, but this age was derived from two sample subsets that were separated by more than 15 m of stratigraphy (Anbar *et al.* 2007). Because it is possible that the two sampled stratigraphic intervals may record distinctive sea-water Os isotope compositions, only the I_{Os} derived from Re–Os isochron regressions of the individual subsets ($I_{Os} = 0.86 \pm 0.86$ and 0.06 ± 0.09) may represent a robust estimate of the sea-water Os isotope composition. The more precise determination ($I_{Os} = 0.06 \pm 0.09$) is shown in the figure.

Neoproterozoic Windermere Supergroup, Mackenzie Mountains, Canada: Narbonne & Aitken 1995), ORS are typically more common than volcanic tuff horizons suitable for U–Pb zircon dating, so Re–Os ages from ORS may hold great promise for radiometric calibration of the Precambrian geological timescale. Nearly all precise Re–Os ages obtained from Precambrian ORS (described later) have been obtained using drill core, which avoids the issue of post-depositional mobility of Re and Os from ORS during oxidative surface weathering (Peucker-Ehrenbrink & Hannigan 2000; Jaffe *et al.* 2002; Pierson-Wickmann *et al.* 2002). Development of programmes for the acquisition of drill cores from Precambrian sedimentary successions (e.g. the Astrobiology Drilling Program of the NASA Astrobiology Institute) and preservation of drill core from petroleum exploration wells are thus important for successful Re–Os ORS geochronological studies and an improved Precambrian chronostratigraphy. Furthermore, organic-rich shales of Proterozoic age are increasingly being targeted as hydrocarbon source rocks for petroleum exploration. Several major oil-producing fields are known (e.g. Huqf Supergroup of Oman, Centralian Superbasin of Australia and the Siberian Craton: Peters *et al.* 2005) and exploration in Proterozoic basins is underway on most continents (e.g. Lottaroli & Craig 2006). Here, we review recent applications of the Re–Os system to Precambrian ORS that have constrained the timing of Proterozoic glaciations, Earth surface oxygenation and the depositional age of organic-rich shales with potential for generating hydrocarbons.

Timing of Neoproterozoic glaciation

Fine-grained siliciclastic sedimentary rocks such as siltstones and shales are commonly deposited during the post-glacial, eustatic sea-level rise that accompanies the end of major episodes of glaciation. If such mudrocks are organic-rich and were deposited under oxygen deficient, anoxic or suboxic conditions, they represent attractive targets for Re–Os geochronology because a Re–Os age would provide a minimum age constraint on the end of glaciation. This approach was one solution proposed to counter the large degree of uncertainty surrounding the number, timing, extent and duration of Neoproterozoic glaciations. These uncertainties result from a scarcity of suitable acid igneous rocks for U–Pb zircon dating of most Neoproterozoic glacial deposits (Schaefer & Burgess 2003; Kendall *et al.* 2004, 2006).

Kendall *et al.* (2004) obtained a Model 1 Re–Os age of 607.8 ± 4.7 Ma (see Fig. 1) for the Old Fort Point Formation (OFP), a widespread post-glacial marker horizon interpreted as the deep-water facies equivalent of cap carbonates overlying glacial deposits in the middle Windermere Supergroup (Ross *et al.* 1995). The significance of the Re–Os age for dating post-glacial sedimentation relied on lithostratigraphic correlations between the dated OFP interval from the Rocky Mountains, western Alberta (where no glacial deposits are preserved) and stratigraphic successions in northeastern British Columbia and the Mackenzie Mountains, N.W.T., that contain the glaciogenic Vreeland and Ice Brook formations (and their cap carbonates), respectively (Ross *et al.* 1995). Based on the geochronological database available for Neoproterozoic glaciation, Kendall *et al.* (2004) suggested that the Vreeland–Ice Brook glaciation represents part of a *c.* 620–600 Ma glaciation (e.g. represented, in part, by the Nantuo Formation, southern China: Barfod *et al.* 2002). Subsequently, new precise U–Pb zircon age constraints have been obtained for the Doushantuo Formation cap carbonate overlying the Nantuo Formation (635.23 ± 0.57 Ma: Condon *et al.* 2005) and the Ghaub Formation, Namibia (635.51 ± 0.54 Ma: Hoffmann *et al.* 2004). In light of these new age constraints, it is possible that the Vreeland–Ice Brook glaciation may have been wholly or partly younger than *c.* 635 Ma. Alternatively, if the Vreeland–Ice Brook glaciogenic diamictites represent part of a global glaciation that ended at *c.* 635 Ma (Condon *et al.* 2005), then the Re–Os age of 607.8 ± 4.7 Ma for the OFP may simply reflect a relatively condensed section (*c.* 80–120 m thick: Ross *et al.* 1995) and the underlying coarse-grained sandstones that are laterally equivalent to the Vreeland diamictites (Hein & McMechan 1994; McMechan 2000).

Prior to the study by Kendall *et al.* (2006), direct age constraints for the Sturtian and Elatina Formation glacial deposits of southern Australia were particularly poor, with the best age constraint being a U–Pb zircon age of 777 ± 7 Ma from volcanic rocks more than 6 km below Sturtian glacial deposits (Preiss 2000). Nevertheless, the Sturtian and Elatina Formation glacial deposits have served as marker horizons for global correlation schemes based largely on litho- and chemo-stratigraphy (Kennedy *et al.* 1998; Walter *et al.* 2000; Halverson *et al.* 2005). Kendall *et al.* (2006) tested proposed correlation schemes for a putative global 'Sturtian' ice age by obtaining Re–Os ages from post-glacial black shales immediately overlying Sturtian and Areyonga glaciogenic diamictites in drill cores from southern and central Australia, respectively. Two Re–Os ages of 647 ± 10 (Model 1, MSWD = 0.79) and 645.1 ± 4.8 Ma (Model 1, MSWD = 1.2) were obtained for the Tindelpina Shale Member (basal Tapley Hill Formation, Adelaide Rift Complex, southern Australia) from two separate drill cores (Fig. 7a). The sampled intervals from the two cores can be correlated on the basis of carbon isotope chemostratigraphy. Thus, combining the Re–Os data from both drill cores resulted in a depositional age of 643.0 ± 2.4 Ma (Model 1, MSWD = 1.1) for the Tindelpina Shale Member. A Re–Os age of 657.2 ± 5.4 Ma (Model 1, MSWD = 1.2) was also obtained for the Aralka Formation (Amadeus Basin, central Australia) from a single drill core. Because both diamictite–shale contacts appear conformable, the Re–Os ages were interpreted as tight minimum age constraints for the end of Sturtian and Areyonga glaciation. The new Re–Os ages suggested that the Sturtian and Areyonga glacial intervals were significantly younger than other radiometrically dated (*c.* 685–750 Ma) glacial intervals previously regarded as possible correlatives. A U–Pb SHRIMP zircon age of 659 ± 6 Ma from a tuffaceous bed within the glaciogenic Sturtian Wilyerpa Formation (Fanning & Link 2008) supports the interpretations of Kendall *et al.* (2006). Thus, the Sturtian ice age was diachronous and/or there were multiple episodes of 'Sturtian' glaciation between about 750 and 643 Ma, each of uncertain duration and extent.

Late Mesoproterozoic glaciation in Brazil?

The Vazante Group of the Sao Francisco Basin (east-central Brazil) represents another Precambrian sedimentary succession with glaciogenic

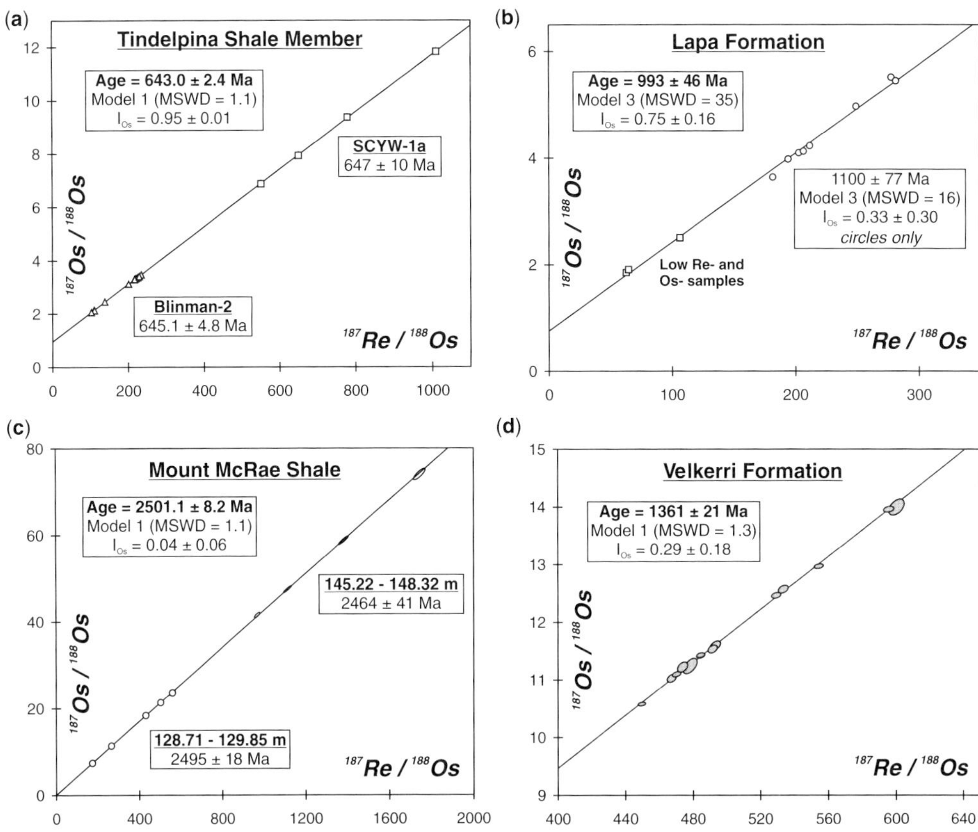

Fig. 7. Examples of Re–Os isochrons for Precambrian ORS. (**a**) Neoproterozoic Tindelpina Shale Member, basal Tapley Hill Formation, Adelaide Rift Complex, South Australia (Kendall *et al.* 2006). Triangles, Blinman-2 drillhole; squares, SCYW-1a drillhole. (**b**) Late Mesoproterozoic Lapa Formation, upper Vazante Group, Brasilia Fold Belt, Brazil (Azmy *et al.* 2008). Exclusion of the three low Re- and Os-samples (squares) yields a nominally older date of 1100 ± 77 Ma (Model 3, MSWD = 16; $I_{Os} = 0.33 \pm 0.30$; circles). (**c**) Late Archaean M McRae Shale, Hamersley Basin, western Australia (Anbar *et al.* 2007). Depth intervals refer to two stratigraphic horizons (represented by error ellipses and circles) sampled from the ABDP-9 drillhole. (**d**) Mesoproterozoic Velkerri Formation (upper organic-rich interval), McArthur Basin, northern Australia (Kendall *et al.* 2009*a*). All regressions used measured 2σ uncertainties for ^{187}Re/^{188}Os and ^{187}Os/^{188}Os, and the error correlation function (ρ).

diamictites and cap carbonates, and is characterized by a paucity of reliable U–Pb radiometric age constraints. Distinctive negative δ^{13}C isotope excursions in the cap carbonate, together with ^{87}Sr/^{86}Sr data (Azmy *et al.* 2001, 2006; Brody *et al.* 2004; Olcott *et al.* 2005), were used to correlate the upper Vazante Group diamictite unit and overlying post-glacial cap carbonate (Lapa Formation) with the $<746 \pm 2$ Ma (Hoffman *et al.* 1996) glaciogenic Chuos Formation and Rasthof Formation cap carbonate on the Congo Craton, Namibia (Azmy *et al.* 2001, 2006). However, a Re–Os date of 993 ± 46 Ma (Model 3, MSWD = 35; $I_{Os} = 0.75 \pm 0.16$) was obtained for ORS at the base of the Lapa Formation (Fig. 7b) (Azmy *et al.* 2008). Exclusion of three samples with low Re (0.4–0.9 ppb) and Os (33–54 ppt) abundances (e.g. possible significant detrital component) yields a nominally older Re–Os date of 1100 ± 77 Ma (Model 3, MSWD = 16; $I_{Os} = 0.33 \pm 0.30$). Elevated MSWD in these regressions may reflect post-depositional mobilization (e.g. chlorite-grade metamorphism or hydrothermal fluid flow), and/or heterogeneous sea-water I_{Os} (samples were derived from a *c.* 15 m stratigraphic interval to obtain a good range of Re/Os ratios). However, the Re–Os dates were suggested to represent a good estimate of the depositional age for the basal Lapa Formation on the basis of reasonable I_{Os} derived from the Re–Os isochron regressions, the absence of upper Vazante Group and basal Lapa Formation detrital zircons yielding concordant U–Pb ages younger than about 1000 Ma, and stromatolite biostratigraphic evidence (Cloud & Dardenne 1973) for a late Mesoproterozoic age for the upper Vazante Group. In addition, Geboy (2006) has obtained Model 3 Re–Os ages of 1353 ± 69 Ma and 1126 ± 47 Ma for ORS from the upper Vazante Group. Thus, the upper Vazante Group glaciogenic diamictites are not correlative with 'Sturtian' (i.e. 750–643 Ma) glacial deposits, but instead may record a late Mesoproterozoic glaciation.

Timing of Earth surface oxygenation

Anbar *et al.* (2007) presented a high-resolution chemostratigraphic profile for the late Archaean Mt McRae Shale (Hamersley Basin, western Australia) that showed an episode of enrichment of the redox-sensitive metals Mo and Re in black shales. The suggested mechanism for metal enrichment was oxidative weathering and dissolution of Mo and Re derived from crustal sulphide minerals, transportation within or into an ocean basin, and sequestration into organic-rich sediments. A Re–Os depositional age of 2501.1 ± 8.2 Ma (Model 1, MSWD = 1.1) (Fig. 7c) agrees well with a U–Pb SHRIMP zircon age of 2504 ± 5 Ma (Rasmussen *et al.* 2005) from a tuffaceous bed within the Mt McRae Shale. Thus, the episode of metal enrichment is a primary sedimentary feature and suggests that shallow-water ocean oxygenation commenced more than 50 Ma prior to the Great Oxidation Event (2.45–2.32 Ga; Bekker *et al.* 2004). Thus, Re–Os ORS geochronology, together with geological and geochemical redox proxies, holds great potential for radiometrically constraining the timing of stepwise increases in atmosphere and ocean oxygen abundances during the Precambrian Eon.

Application to Precambrian petroleum systems

An example of a Precambrian ORS unit that has been studied for its petroleum source rock potential is the Mesoproterozoic Velkerri Formation (Roper Group, McArthur Basin, northern Australia). Within the Velkerri Formation there are three intervals, tens of metres thick, that are highly organic-rich (TOC >5%) and represent oil source beds (Jackson & Raiswell 1991; Warren *et al.* 1998). The world's oldest live oil is from the Velkerri Formation and has been observed bubbling from drillhole BMR Urapunga-4 at multiple depths (e.g. Jackson *et al.* 1986; Crick *et al.* 1988). Organic matter in the Velkerri Formation is mature, except where affected by a dolerite sill intrusion (Crick *et al.* 1988; Crick 1992; Summons *et al.* 1994; George & Ahmed 2002). During and prior to late Mesoproterozoic–early Neoproterozoic McArthur Basin inversion, the Velkerri Formation was the most likely source rock for multiple episodes of hydrocarbon migration into a cross-cutting dolerite sill and the overlying Bessie Creek Sandstone. The oldest known hydrocarbon migration event occurred prior to 1280 Ma (minimum K–Ar age of a dolerite sill: McDougall *et al.* 1965) and is associated with dolerite sill intrusion, which resulted in flash pyrolysis of kerogen and generation of bitumen. At least one younger hydrocarbon migration event is associated with basin inversion, and is represented by non-biodegraded fluid inclusion oils (Dutkiewicz *et al.* 2003, 2004; Volk *et al.* 2005).

Two Re–Os ages of 1361 ± 21 (Model 1, MSWD = 1.3) and 1417 ± 29 Ma (Model 1, MSWD = 1.3) were obtained by Kendall *et al.* (2009*a*) for the upper and lower organic-rich intervals of the Velkerri Formation, respectively (Fig. 7d). Both ages are internally consistent with stratigraphic position and with a U–Pb SHRIMP zircon age of 1492 ± 4 Ma from tuff within the Wooden Duck Member of the Mainoru Formation (>700 m beneath the Velkerri Formation: Jackson *et al.* 1999). Thus, the multiple hydrocarbon migration events do not appear to have affected the Re–Os systematics in the Velkerri Formation source rock, consistent with previous studies suggesting the Re–Os ORS geochronometer is robust up to the onset of greenschist facies metamorphism (Creaser *et al.* 2002; Kendall *et al.* 2004; Selby & Creaser 2005*a*). The Re–Os ages constrain the age of biomarkers in the Velkerri Formation and, together with the U–Pb zircon age of 1492 ± 4 Ma for the Mainoru Formation, suggest a duration of more than 100 Ma for the acritarch and microfossil record of the Roper Group (Javaux *et al.* 2001). In addition, the hydrocarbon migration event associated with the dolerite sill intrusion can be bracketed between the minimum K–Ar age of 1280 Ma and the Re–Os age of 1361 ± 21 Ma.

The Re–Os systematics of ORS source rocks are not grossly affected by hydrocarbon maturation, suggesting minimal fractionation of Re from Os during hydrocarbon generation and/or domination of the Re and Os mass balance by the source rock (Creaser *et al.* 2002). However, Selby & Creaser (2005*b*) and Selby *et al.* (2005) have shown that the Re–Os systematics of maturated hydrocarbons may be reset, possibly by some organic mechanism, during the process of hydrocarbon migration to the reservoir rock. In addition, Selby *et al.* (2007*a*) observed a strong correlation between the Os isotope composition of whole oils with source rock age. Thus, the Re–Os system in hydrocarbon deposits (e.g. bitumen, crude oil and oil sands) may constrain the timing of hydrocarbon migration and entrapment, and serve as a tracer for fingerprinting hydrocarbon source rocks. We suggest that a joint application of the Re–Os system to Precambrian ORS source rocks and hydrocarbon deposits may ultimately prove valuable in understanding the age and nature of Precambrian petroleum systems. Such studies would complement more traditional organic geochemical methods such as source rock and hydrocarbon biomarker distributions, and thermal maturity indicators.

Summary and future directions

With rigorous sampling and analytical methodologies, we have shown that the ^{187}Re–^{187}Os isotope system represents a precise and accurate deposition-age geochronometer for ORS. However, we emphasize that some uncertainties still remain regarding the limitations of the Re–Os ORS geochronometer. For example, weathering (Peucker-Ehrenbrink & Hannigan 2000; Jaffe *et al.* 2002; Pierson-Wickmann *et al.* 2002) and hydrothermal fluid flow (Fig. 8) (Kendall *et al.* 2009*a*) are known to result in post-depositional mobilization of Re and Os. The robustness, or otherwise, of the Re–Os system in ORS at metamorphic grades above lowermost greenschist facies is not known. In addition, the mechanism controlling the range in initial ^{187}Re/^{188}Os of ORS is not fully understood, but may relate to the presence of multiple and/or separate organic complexes as host phases for Re and Os (Creaser *et al.* 2002; Selby & Creaser 2003). However, with further research, these questions should be resolved. If low abundances of Re in Archaean ORS are the norm (Yang & Holland 2002; Siebert *et al.* 2005; Wille *et al.* 2007), then the utility of Re–Os geochronology for ORS deposited prior to the Great Oxidation Event may be limited. Nevertheless, the Re–Os ORS geochronometer holds tremendous potential for constraining the rates, duration, timing and extent of geological phenomena during the Proterozoic Eon. Examples include timing of evolution and diversification of eukaryotes (including macroscopic metazoans), significant perturbations of biogeochemical cycles (e.g. positive and negative carbon isotope excursions), radiometric calibration of the Precambrian rock record and sedimentary basin analysis.

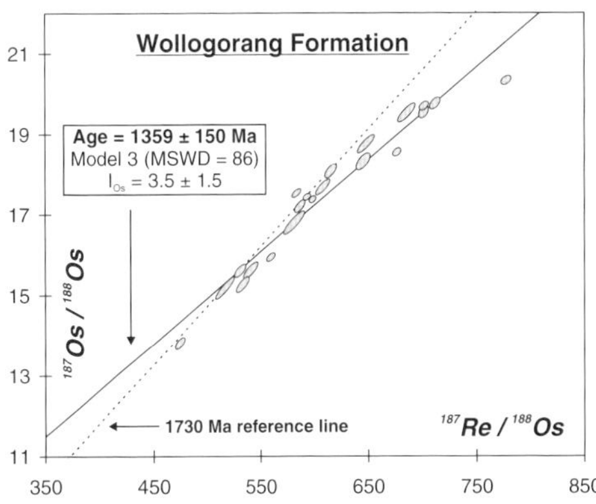

Fig. 8. Isochron diagram showing the scattered Re–Os isotope data for the Palaeoproterozoic Wollogorang Formation, McArthur Basin, northern Australia (Kendall *et al.* 2009*a*). The imprecise Re–Os date of 1359 ± 150 Ma is significantly younger than the U–Pb zircon ages of 1729 ± 4 and 1730 ± 3 Ma from tuffaceous beds within the Wollogorang Formation black shale unit that constrain the age of black shale deposition (Page *et al.* 2000). A *c.* 1730 Ma reference line (dashed line) with chondritic I_{Os} (0.115) is shown for comparison. Note the high I_{Os} value of 3.5 ± 1.5 from this regression that is significantly more radiogenic than currently eroding upper continental crust (e.g. ^{187}Os/^{188}Os *c.* 1.0–1.4: Esser & Turekian 1993; Peucker-Ehrenbrink & Jahn 2001; Hattori *et al.* 2003). Scattered Re–Os systematics are interpreted to result from post-depositional hydrothermal fluid flow, as evidenced by carbonate veinlets within the shale.

The constructive comments by G. Ravizza and A. Cohen improved the manuscript. This research was supported by a Natural Sciences and Engineering Research Council (NSERC) Discovery Grant to R. A. Creaser, and a NSERC Canada Graduate Scholarship, Alberta Ingenuity Fund PhD. Studentship and Geological Society of America Student Research Grant to B. Kendall. The Radiogenic Isotope Facility at the University of Alberta is supported in part by a NSERC Major Facilities Access Grant and Major Resources Support Grant. G. Jiang and M. Kennedy are thanked for Doushantuo Formation samples.

References

Anbar, A. D., Duan, Y. *et al.* 2007. A whiff of oxygen before the Great Oxidation Event. *Science*, **317**, 1903–1906.

Arthur, M. A. & Sageman, B. B. 1994. Marine black shales: depositional mechanisms and environments of ancient deposits. *Annual Review of Earth and Planetary Science*, **22**, 499–551.

Azmy, K., Kaufman, A. J., Misi, A. & Oliveira, T. F. 2006. Isotope stratigraphy of the Lapa Formation, São Francisco Basin, Brazil: implications for Late Neoproterozoic glacial events in South America. *Precambrian Research*, **149**, 231–248.

Azmy, K., Kendall, B., Creaser, R. A., Heaman, L., Misi, A. & de Oliveira, T. F. 2008. Global correlation of the Vazante Group, São Francisco Basin, Brazil: Re–Os and U–Pb radiometric age constraints. *Precambrian Research*, **164**, 160–172.

Azmy, K., Veiser, J., Misi, A., Oliveira, T. F., Sanches, A. L. & Dardenne, M. A. 2001.

Dolomitization and isotope stratigraphy of the Vazante Formation, São Francisco Basin, Brazil. *Precambrian Research*, **112**, 303–329.

BABINSKI, M., VIEIRA, L. C. & TRINDADE, R. I. F. 2007. Direct dating of the Sete Lagoas cap carbonate (Bambuí Group, Brazil) and implications for the Neoproterozoic glacial events. *Terra Nova*, **19**, 401–406.

BARFOD, G. H., ALBARÈDE, F., KNOLL, A. H., XIAO, S., TÉLOUK, P., FREI, R. & BAKER, J. 2002. New Lu–Hf and Pb–Pb age constraints on the earliest animal fossils. *Earth and Planetary Science Letters*, **201**, 203–212.

BARRE, A. B., PRINZHOFER, A. & ALLEGRE, C. J. 1995. Osmium isotopes in the organic matter of crude oil and asphaltenes. *Terra Abstracts*, **7**, 1999.

BECKER, H., MORGAN, J. W., WALKER, R. J., MACPHERSON, G. J. & GROSSMAN, J. N. 2001. Rhenium–osmium systematics of calcium–aluminum-rich inclusions in carbonaceous chondrites. *Geochimica et Cosmochimica Acta*, **65**, 3379–3390.

BEKKER, A. & HOLLAND, H. D. 2004. Dating the rise of atmospheric oxygen. *Nature*, **427**, 117–120.

BINGEN, B., GRIFFIN, W. L., TORSVIK, T. H. & SAEED, A. 2005. Timing of late Neoproterozoic glaciation on Baltica constrained by detrital zircon geochronology in the Hedmark Group, south-east Norway. *Terra Nova*, **17**, 250–258.

BIRCK, J.-L. & ALLÈGRE, C.-J. 1998. Rhenium 187–osmium 187 in iron meteorites and the strange origin of the Kodaikanal meteorite. *Meteoritic Planetary Science*, **33**, 647–653.

BRODY, K. B., KAUFMAN, A. J., EIGENBRODE, J. L. & CODY, G. D. 2004. Biomarker geochemistry of a post-glacial Neoproterozoic succession in Brazil. *Geological Society of America, Abstracts with Programs*, **36**(5), 477.

BURTON, K. W., BOURDON, B., BIRCK, J.-L., ALLÈGRE, C. J. & HEIN, J. R. 1999. Osmium isotope variations in the oceans recorded by Fe–Mn crusts. *Earth and Planetary Science Letters*, **171**, 185–197.

CANFIELD, D. E. 1998. A new model for Proterozoic ocean chemistry. *Nature*, **396**, 450–453.

CANFIELD, D. E., POULTON, S. W. & NARBONNE, G. M. 2007. Late-Neoproterozoic deep-ocean oxygenation and the rise of animal life. *Science*, **315**, 92–95.

CAVE, R. R., RAVIZZA, G. E., GERMAN, C. R., THOMSON, J. & NESBITT, R. W. 2003. Deposition of osmium and other platinum-group elements beneath the ultramafic-hosted Rainbow hydrothermal plume. *Earth and Planetary Science Letters*, **210**, 65–79.

CHEN, D. F., DONG, W. Q., ZHU, B. Q. & CHEN, X. P. 2004. Pb–Pb ages of Neoproterozoic Doushantuo phosphorites in South China: constraints on early metazoan evolution and glaciation events. *Precambrian Research*, **132**, 123–132.

CLOUD, P. E. & DARDENNE, M. A. 1973. Proterozoic age of the Bambuí Group in Brazil. *Geological Society of America Bulletin*, **84**, 1673–1676.

COHEN, A. S. 2004. The rhenium-osmium isotope system: applications to geochronological and palaeoenvironmental problems. *Journal of the Geological Society, London*, **161**, 729–734.

COHEN, A. S. & COE, A. L. 2002. New geochemical evidence for the onset of volcanism in the Central Atlantic magmatic province and environmental change at the Triassic–Jurassic boundary. *Geology*, **30**, 267–270.

COHEN, A. S. & COE, A. L. 2007. The impact of the Central Atlantic Magmatic Province on climate and on the Sr- and Os-isotope evolution of seawater. *Palaeogeography, Palaeoclimatology, Palaeoecology*, **244**, 374–390.

COHEN, A. S. & WATERS, F. G. 1996. Separation of osmium from geological materials by solvent extraction for analysis by thermal ionization mass spectrometry. *Analytical Chimica Acta*, **332**, 269–375.

COHEN, A. S., COE, A. L., BARTLETT, J. M. & HAWKESWORTH, C. J. 1999. Precise Re–Os ages of organic-rich mudrocks and the Os isotope composition of Jurassic seawater. *Earth and Planetary Science Letters*, **167**, 159–173.

COHEN, A. S., COE, A. L., HARDING, S. M. & SCHWARK, L. 2004. Osmium isotope evidence for the regulation of atmospheric CO_2 by continental weathering. *Geology*, **32**, 157–160.

COLODNER, D., SACHS, J., RAVIZZA, G., TUREKIAN, K., EDMOND, J. & BOYLE, E. 1993. The geochemical cycle of rhenium: a reconnaissance. *Earth and Planetary Science Letters*, **117**, 205–221.

COLPRON, M., LOGAN, J. M. & MORTENSEN, J. K. 2002. U–Pb zircon age constraint for late Neoproterozoic rifting and initiation of the lower Palaeozoic passive margin of western Laurentia. *Canadian Journal of Earth Sciences*, **39**, 133–143.

CONDON, D., ZHU, M., BOWRING, S., WANG, W., YANG, A. & JIN, Y. 2005. U–Pb ages from the Neoproterozoic Doushantuo Formation, China. *Science*, **308**, 95–98.

CREASER, R. A. & STASIUK, L. D. 2007. Depositional age of the Douglas Formation, northern Saskatchewan, determined by Re–Os geochronology. *In*: JEFFERSON, C. W. & DELANEY, G. (eds) *EXTECH IV: Geology and Uranium Exploration TECHnology of the Proterozoic Athabasca Basin, Saskatchewan and Alberta.* Geological Survey of Canada Bulletin, **588**, 341–346.

CREASER, R. A., PAPANASTASSIOU, D. A. & WASSERBURG, G. J. 1991. Negative thermal ion mass spectrometry of osmium, rhenium, and iridium. *Geochimica et Cosmochimica Acta*, **55**, 397–401.

CREASER, R. A., SANNIGRAHI, P., CHACKO, T. & SELBY, D. 2002. Further evaluation of the Re–Os geochronometer in organic-rich sedimentary rocks: a test of hydrocarbon maturation effects in the Exshaw Formation, Western Canada Sedimentary Basin. *Geochimica et Cosmochimica Acta*, **66**, 3441–3452.

CRICK, I. H. 1992. Petrological and maturation characteristics of organic matter from the Middle Proterozoic McArthur Basin, Australia. *Australian Journal of Earth Sciences*, **39**, 501–519.

CRICK, I. H., BOREHAM, C. J., COOK, A. C. & POWELL, T. G. 1988. Petroleum geology and geochemistry of Middle Proterozoic McArthur Basin, northern Australia II: Assessment of source rock potential. *AAPG Bulletin*, **72**, 1495–1514.

CRUSIUS, J. & THOMSON, J. 2000. Comparative behavior of authigenic Re, U, and Mo during reoxidation and subsequent long-term burial in marine sediments. *Geochimica et Cosmochimica Acta*, **64**, 2233–2242.

CRUSIUS, J., CALVERT, S., PEDERSEN, T. & SAGE, D. 1996. Rhenium and molybdenum enrichments in sediments as indicators of oxic, suboxic, and sulfidic conditions of deposition. *Earth and Planetary Science Letters*, **145**, 65–78.

DICKIN, A. P. 2005. *Radiogenic Isotope Geology*, 2nd edn. Cambridge University Press, Cambridge.

DUTKIEWICZ, A., VOLK, H., RIDLEY, J. & GEORGE, S. 2003. Biomarkers, brines, and oil in the Mesoproterozoic, Roper Superbasin, Australia. *Geology*, **31**, 981–984.

DUTKIEWICZ, A., VOLK, H., RIDLEY, J. & GEORGE, S. C. 2004. Geochemistry of oil in fluid inclusions in a middle Proterozoic igneous intrusion: implications for the source of hydrocarbons in crystalline rocks. *Organic Geochemistry*, **35**, 937–957.

ENGLAND, G. L., RASMUSSEN, B., KRAPEZ, B. & GROVES, D. I. 2002. Palaeoenvironmental significance of rounded pyrite in siliciclastic sequences of the Late Archaean Witwatersrand Basin: oxygen-deficient atmosphere or hydrothermal alteration? *Sedimentology*, **49**, 1133–1156.

ESSER, B. K. & TUREKIAN, K. K. 1993. The osmium isotopic composition of the continental crust. *Geochimica et Cosmochimica Acta*, **57**, 3093–3104.

FARQUHAR, J. & WING, B. A. 2003. Multiple sulfur isotopes and the evolution of the atmosphere. *Earth and Planetary Science Letters*, **213**, 1–13.

FARQUHAR, J., BAO, H. & THIEMENS, M. 2000. Atmospheric influence of Earth's earliest sulfur cycle. *Science*, **289**, 756–758.

FANNING, C. M. & LINK, P. K. 2004. U–Pb SHRIMP ages of Neoproterozoic (Sturtian) glaciogenic Pocatello Formation, southeastern Idaho. *Geology*, **32**, 881–884.

FANNING, C. M. & LINK, P. K. 2008. Age constraints for the Sturtian glaciation: data from the Adelaide Geosyncline, South Australia and Pocatello Formation, Idaho, USA. *In*: GALLAGHER, S. J. & WALLACE, M. W. (eds) *Neoproterozoic Extreme Climates and the Origin of Early Life*, Selwyn Symposium of the GSA Victoria Division. Geological Society of Australia Extended Abstracts, **91**, 57–62.

FIKE, D. A., GROTZINGER, J. P., PRATT, L. M. & SUMMONS, R. E. 2006. Oxidation of the Ediacaran ocean. *Nature*, **444**, 744–747.

FLEET, M. E. 1998. Detrital pyrite in Witwatersrand gold reefs: X-ray diffraction evidence and implications for atmospheric evolution. *Terra Nova*, **10**, 302–306.

GEBOY, N. J. 2006. *Rhenium-osmium age determinations of glaciogenic shales, Vazante Formation, Brazil*. Unpublished M.Sc. Thesis, University of Maryland, College Park.

GEORGE, S. C. & AHMED, M. 2002. Use of aromatic compound distributions to evaluate organic maturity of the Proterozoic Middle Velkerri Formation, McArthur Basin, Australia. *In*: KEEP, M. & MOSS, S. J. (eds) *The Sedimentary Basins of Western Australia, 3*. Proceedings of the Petroleum Exploration Society of Australia Symposium, 253–270.

GOLDBLATT, C., LENTON, T. M. & WATSON, A. J. 2006. Bistability of atmospheric oxygen and the Great Oxidation. *Nature*, **443**, 683–686.

GRADSTEIN, F. M., OGG, J. G. & SMITH, A. G. 2004. *A Geologic Time Scale 2004*. Cambridge University Press, Cambridge.

HALVERSON, G. P., HOFFMAN, P. F., SCHRAG, D. P., MALOOF, A. C. & RICE, A. H. N. 2005. Toward a Neoproterozoic composite carbon-isotope record. *Geological Society of America Bulletin*, **117**, 1181–1207.

HANNAH, J. L., BEKKER, A., STEIN, H. J., MARKEY, R. J. & HOLLAND, H. D. 2004. Primitive Os and 2316 Ma age for marine shale: implications for Palaeoproterozoic glacial events and the rise of atmospheric oxygen. *Earth and Planetary Science Letters*, **225**, 43–52.

HANNAH, J. L., STEIN, H. J., ZIMMERMAN, A., YANG, G., MARKEY, R. J. & MELEZHIK, V. A. 2006. Precise 2004 $\pm$ 9 Ma Re–Os age for Pechenga black shale: comparison of sulfides and organic material. *Geochimica et Cosmochimica Acta*, **70**(18; Supplement 1), A228.

HATTORI, Y., SUZUKI, K., HONDA, M. & SHIMIZU, H. 2003. Re–Os isotope systematics of the Taklimakan Desert sands, moraines and river sediments around the Taklimakan Desert, and of Tibetan soils. *Geochimica et Cosmochimica Acta*, **67**, 1195–1205.

HEIN, F. J. & MCMECHAN, M. E. 1994. Proterozoic–Lower Cambrian strata of the Western Canada Sedimentary Basin. *In*: MOSSOP, G. D. & SHETSEN, I. (eds) *Geological Atlas of the Western Canada Sedimentary Basin*. Canadian Society of Petroleum Geologists, Calgary, Alberta, 57–68.

HOFFMAN, P. F., HAWKINS, D. P., ISACHSEN, C. E. & BOWRING, S. A. 1996. Precise U–Pb zircon ages for early Damaran magmatism in the Summas Mountains and Welwitschia Inlier, northern Damara Belt, Namibia. *Communications of the Geological Survey of Namibia*, **11**, 47–52.

HOFFMANN, K.-H., CONDON, D. J., BOWRING, S. A. & CROWLEY, J. L. 2004. U–Pb zircon date from the Neoproterozoic Ghaub Formation, Namibia: constraints on Marinoan glaciation. *Geology*, **32**, 817–820.

HUTCHEON, I. D. & OLSEN, E. 1991. Cr isotopic composition of differentiated meteorites: A search for ^{53}Mn. *Lunar and Planetary Sciences*, **22**, 605–606.

HUTCHEON, I. D., OLSEN, E., ZIPFEL, J. & WASSERBURG, G. J. 1992. Cr isotopic composition of differentiated meteorites: Evidence for ^{53}Mn. *Lunar and Planetary Sciences*, **23**, 565–566.

JACKSON, M. J. & RAISWELL, R. 1991. Sedimentology and carbon–sulphur geochemistry of the Velkerri Formation, a Mid-Proterozoic potential oil source in northern Australia. *Precambrian Research*, **54**, 81–108.

JACKSON, M. J., POWELL, T. G., SUMMONS, R. E. & SWEET, I. P. 1986. Hydrocarbon shows and petroleum source rocks in sediments as old as 1.7×10^9 years. *Nature*, **322**, 727–729.

JACKSON, M. J., SWEET, I. P., PAGE, R. W. & BRADSHAW, B. E. 1999. The South Nicholson and Roper Groups: evidence for the early Mesoproterozoic Roper Superbasin. *In*: *Integrated Basin Analysis of the Isa Superbasin using Seismic, Well-log, and*

Geopotential Data: An Evaluation of the Economic Potential of the Northern Lawn Hill Platform. Australian Geological Survey Organisation Record 1999/19 (unpaginated).

JAFFE, L. A., PEUCKER-EHRENBRINK, B. & PETSCH, S. T. 2002. Mobility of rhenium, platinum group elements and organic carbon during black shale weathering. *Earth and Planetary Science Letters*, **198**, 339–353.

JAFFEY, A. H., FLYNN, K. F., GLENDENIN, L. E., BENTLEY, W. C. & ESSLING, A. M. 1971. Precision measurement of half-lives and specific activities of ^{235}U and ^{238}U. *Physics Reviews*, **4**, 1889–1906.

JAVAUX, E. J., KNOLL, A. H. & WALTER, M. R. 2001. Morphological and ecological complexity in early eukaryotic ecosystems. *Nature*, **412**, 66–69.

JIANG, G., SOHL, L. E. & CHRISTIE-BLICK, N. 2003. Neoproterozoic stratigraphic comparison of the Lesser Himalaya (India) and Yangtze block (south China): palaeogeographic implications. *Geology*, **31**, 917–920.

KENDALL, B. S., CREASER, R. A., ROSS, G. M. & SELBY, D. 2004. Constraints on the timing of Marinoan 'Snowball Earth' glaciation by ^{187}Re–^{187}Os dating of a Neoproterozoic, post-glacial black shale in Western Canada. *Earth and Planetary Science Letters*, **222**, 729–740.

KENDALL, B., CREASER, R. A. & SELBY, D. 2006. Re–Os geochronology of postglacial black shales in Australia: constraints on the timing of 'Sturtian' glaciation. *Geology*, **34**, 729–732.

KENDALL, B., CREASER, R. A., GORDON, G. W. & ANBAR, A. D. 2009*a*. Re–Os and Mo isotope systematics of black shales from the Middle Proterozoic Velkerri and Wollogorang Formations, McArthur Basin, northern Australia. *Geochimica et Cosmochimica Acta*, **73**, 2534–2558.

KENDALL, B., CREASER, R. A., CALVER, C. R., RAUB, T. D. & EVANS, D. A. D. 2009*b*. Correlation of Sturtian diamictite successions in southern Australia and northwestern Tasmania by Re–Os black shale geochronology and the ambiguity of "Sturtian"-type diamictite-cap carbonate pairs as chronostratigraphic marker horizons. *Precambrian Research*, **172**, 301–310.

KENNEDY, M., DROSER, M., MAYER, L. M., PEVEAR, D. & MROFKA, D. 2006. Late Precambrian oxygenation; inception of the clay mineral factory. *Science*, **311**, 1446–1449.

KENNEDY, M. J., RUNNEGAR, B., PRAVE, A. R., HOFFMANN, K.-H. & ARTHUR, M. A. 1998. Two or four Neoproterozoic glaciations? *Geology*, **26**, 1059–1063.

LEVASSEUR, S., BIRCK, J.-L. & ALLÈGRE, C. J. 1998. Direct measurement of femtomoles of osmium and the ^{187}Os/^{186}Os ratio in seawater. *Science*, **282**, 272–274.

LEVASSEUR, S., BIRCK, J.-L. & ALLÈGRE, C. J. 1999. The osmium riverine flux and the oceanic mass balance of osmium. *Earth and Planetary Science Letters*, **174**, 7–23.

LOTTAROLI, F., CRAIG, J. & THUSU, B. 2009. Neoproterozoic–Early Cambrian (Infracambrian) hydrocarbon prospectivity of North Africa: a synthesis. *In*: CRAIG, J., THUROW, J., THUSU, B., WHITHAM, A. & ABUTARRUMA, Y. (eds) *Global Neoproterozoic Petroleum Systems: The Emerging Potential in North Africa*. Geological Society, London, Special Publications, **326**, 137–156.

LUCK, J.-M. & ALLÈGRE, C. J. 1983. ^{187}Re–^{187}Os systematics in meteorites and cosmochemical consequences. *Nature*, **302**, 130–132.

LUDWIG, K. 1980. Calculation of uncertainties of U–Pb isotope data. *Earth and Planetary Science Letters*, **46**, 212–220.

LUDWIG, K. 2003. *Isoplot/Ex, Version 3: A Geochronological Toolkit for Microsoft Excel*. Berkely Geochronology Center, Berkeley, CA.

LUGMAIR, G. W. & GALER, S. J. G. 1992. Age and isotopic relationships among angrites Lewis Cliff 86010 and Angra dos Reis. *Geochimica et Cosmochimica Acta*, **56**, 1673–1694.

LUND, K., ALEINIKOFF, J. N., EVANS, K. V. & FANNING, C. M. 2003. SHRIMP U–Pb geochronology of Neoproterozoic Windermere Supergroup, central Idaho: Implications for rifting of western Laurentia and synchroneity of Sturtian glacial deposits. *Geological Society of America Bulletin*, **115**, 349–372.

MARTIN, C. E., PEUCKER-EHRENBRINK, B., BRUNSKILL, G. J. & SZYMCZAK, R. 2000. Sources and sinks of unradiogenic osmium runoff from Papua New Guinea. *Earth and Planetary Science Letters*, **183**, 261–274.

MCDANIEL, D. K., WALKER, R. J., HEMMING, S. R., HORAN, M. F., BECKER, H. & GRAUCH, R. I. 2004. Sources of osmium to the modern oceans: new evidence from the ^{190}Pt–^{186}Os system. *Geochimica et Cosmochimica Acta*, **68**, 1243–1252.

MCDOUGALL, I., DUNN, P. R., COMPSTON, W., WEBB, A. W., RICHARDS, J. R. & BOFINGER, V. M. 1965. Isotopic age determinations on Precambrian rocks of the Carpentaria region, Northern Territory, Australia. *Journal of the Geological Society of Australia*, **12**, 67–90.

MCMECHAN, M. E. 2000. Vreeland diamictites – Neoproterozoic glaciogenic slope deposits, Rocky Mountains, northeast British Columbia. *Bulletin of Canadian Petroleum Geology*, **48**, 246–261.

MEISEL, T., WALKER, R. J., IRVING, A. J. & LORAND, J.-P. 2001. Osmium isotopic compositions of mantle xenoliths: a global perspective. *Geochimica et Cosmochimica Acta*, **65**, 1311–1323.

MEISEL, T., WALKER, R. J. & MORGAN, J. W. 1996. The osmium isotopic composition of the Earth's primitive upper mantle. *Nature*, **383**, 517–520.

MOORBATH, S., TAYLOR, P. N., ORPEN, J. L., TRELOAR, P. & WILSON, J. F. 1987. First direct radiometric dating of Archaean stromatolitic limestone. *Nature*, **326**, 865–867.

MORFORD, J. L. & EMERSON, S. 1999. The geochemistry of redox sensitive trace metals in sediments. *Geochimica et Cosmochimica Acta*, **63**, 1735–1750.

MORFORD, J. L., EMERSON, S. R., BRECKEL, E. J. & KIM, S. H. 2005. Diagenesis of oxyanions (V, U, Re, and Mo) in pore waters and sediments from a continental margin. *Geochimica et Cosmochimica Acta*, **69**, 5021–5032.

MORGAN, J. W., HORAN, M. F., WALKER, R. J. & GROSSMAN, J. N. 1995. Rhenium–osmium concentration and isotope systematics in group IIAB iron meteorites. *Geochimica et Cosmochimica Acta*, **59**, 2331–2344.

NAMEROFF, T. J., BALISTRIERI, L. S. & MURRAY, J. W. 2002. Suboxic trace metal geochemistry in the eastern tropical North Pacific. *Geochimica et Cosmochimica Acta*, **66**, 1139–1158.

NARBONNE, G. M. & AITKEN, J. D. 1995. Neoproterozoic of the Mackenzie Mountains, northwestern Canada. *Precambrian Research*, **73**, 101–121.

OLCOTT, A. N., SESSIONS, A. L., CORSETTI, F. A., KAUFMAN, A. J. & OLIVIERA, T. F. 2005. Biomarker evidence for photosynthesis during Neoproterozoic glaciation. *Science*, **310**, 471–474.

OXBURGH, R. 1998. Variations in the osmium isotope composition of seawater over the past 200,000 years. *Earth and Planetary Science Letters*, **159**, 183–191.

OXBURGH, R. 2001. Residence time of osmium in the oceans. *Geochemistry, Geophysics, Geosystems*, **2**(6), 1018.

PAGE, R. W., JACKSON, M. J. & KRASSAY, A. A. 2000. Constraining sequence stratigraphy in north Australian basins: SHRIMP U–Pb zircon geochronology between Mt Isa and McArthur River. *Australian Journal of Earth Sciences*, **47**, 431–459.

PAVLOV, A. A. & KASTING, J. F. 2002. Mass-independent fractionation of sulfur isotopes in Archean sediments: strong evidence for an anoxic Archean atmosphere. *Astrobiology*, **2**, 27–41.

PEGRAM, W. J. & TUREKIAN, K. K. 1999. The osmium isotopic composition change of Cenozoic sea water as inferred from a deep-sea core corrected for meteoritic contributions. *Geochimica et Cosmochimica Acta*, **63**, 4053–4058.

PETERS, K. E., WALTERS, C. C. & MOLDOWAN, J. M. 2005. Petroleum systems through time, Chapter 18. *In*: *The Biomarker Guide, Volume 2: Biomarkers and Isotopes in Petroleum Exploration and Earth History*, 2nd edn. Cambridge University Press, Cambridge, 751–964.

PEUCKER-EHRENBRINK, B. 1996. Accretion of extraterrestrial matter during the last 80 million years and its effect on the marine osmium isotope record. *Geochimica et Cosmochimica Acta*, **60**, 3187–3196.

PEUCKER-EHRENBRINK, B. & HANNIGAN, R. E. 2000. Effects of black shale weathering on the mobility of rhenium and platinum group elements. *Geology*, **28**, 475–478.

PEUCKER-EHRENBRINK, B. & JAHN, B.-M. 2001. Rhenium–osmium isotope systematics and platinum group element concentrations: Loess and the upper continental crust. *Geochemistry Geophysics Geosystems*, **2**(10), 1061.

PEUCKER-EHRENBRINK, B. & RAVIZZA, G. 2000. The marine osmium isotope record. *Terra Nova*, **12**, 205–219.

PEUCKER-EHRENBRINK, B., RAVIZZA, G. & HOFMANN, A. W. 1995. The marine $^{187}Os/^{186}Os$ record of the past 80 million years. *Earth and Planetary Science Letters*, **130**, 155–167.

PIERSON-WICKMANN, A.-C., REISBERG, L. & FRANCE-LANORD, C. 2000. The Os isotopic composition of Himalayan river bedloads and bedrocks: importance of black shales. *Earth and Planetary Science Letters*, **176**, 203–218.

PIERSON-WICKMANN, A.-C., REISBERG, L. & FRANCE-LANORD, C. 2002. Behavior of Re and Os during low-temperature alteration: results from Himalayan soils and altered black shales. *Geochimica et Cosmochimica Acta*, **66**, 1539–1548.

POIRIER, A. 2006. Re–Os and Pb isotope systematics in reduced fjord sediments from Saanich Inlet (Western Canada). *Earth and Planetary Science Letters*, **249**, 119–131.

POPLAVKO, Y. M., IVANOV, V. V. *ET AL*. 1975. On the concentration of rhenium in petroleum, petroleum bitumens and oil shales. *Geochemistry International*, **11**, 969–972.

PREISS, W. V. 2000. The Adelaide Geosyncline of south Australia and its significance in Neoproterozoic continental reconstruction. *Precambrian Research*, **100**, 21–63.

RASMUSSEN, B. 2005. Radiometric dating of sedimentary rocks: the application of diagenetic xenotime geochronology. *Earth-Science Reviews*, **68**, 197–243.

RASMUSSEN, B. & BUICK, R. 1999. Redox state of the Archean atmosphere: evidence from detrital heavy minerals in ca. 3250–2750 Ma sandstones from the Pilbara Craton, Australia. *Geology*, **27**, 115–118.

RASMUSSEN, B., BLAKE, T. S. & FLETCHER, I. R. 2005. U–Pb zircon age constraints on the Hamersley spherule beds: Evidence for a single 2.63 Ga Jeerinah–Carawine impact ejecta layer. *Geology*, **33**, 725–728.

RAVIZZA, G. & ESSER, B. K. 1993. A possible link between the seawater osmium isotope record and weathering of ancient sedimentary organic matter. *Chemical Geology*, **107**, 255–258.

RAVIZZA, G. & PAQUAY, F. 2008. Os isotope chemostratigraphy applied to organic-rich marine sediments from the Eocene–Oligocene transition on the West African margin (ODP Site 959). *Palaeoceanography*, **23**, PA2204, doi:10.1029/2007PA001460.

RAVIZZA, G. & PEUCKER-EHRENBRINK, B. 2003. Chemostratigraphic evidence of Deccan volcanism from the marine osmium isotope record. *Science*, **302**, 1392–1395.

RAVIZZA, G. & TUREKIAN, K. K. 1989. Application of the $^{187}Re-^{187}Os$ system to black shale geochronometry. *Geochimica et Cosmochimica Acta*, **53**, 3257–3262.

RAVIZZA, G. & TUREKIAN, K. K. 1992. The osmium isotopic composition of organic-rich marine sediments. *Earth and Planetary Science Letters*, **110**, 1–6.

RAVIZZA, G., MARTIN, C. E., GERMAN, C. R. & THOMPSON, G. 1996. Os isotopes as tracers in seafloor hydrothermal systems: metalliferous deposits from the TAG hydrothermal area, 26°N Mid-Atlantic Ridge. *Earth and Planetary Science Letters*, **138**, 105–119.

RAVIZZA, G., NORRIS, R. N., BLUSZTAJN, J. & AUBRY, M. P. 2001. An osmium isotope excursion associated with the late Palaeocene thermal maximum: evidence of intensified chemical weathering. *Palaeoceanography*, **16**, 155–163.

RAVIZZA, G., TUREKIAN, K. K. & HAY, B. J. 1991. The geochemistry of rhenium and osmium in recent sediments from the Black Sea. *Geochimica et Cosmochimica Acta*, **55**, 3741–3752.

RICHARDS, B. C., ROSS, G. M. & UTTING, J. 2002. U–Pb geochronology, lithostratigraphy and biostratigraphy of tuff in the upper Famennian to Tournaisian Exshaw Formation: Evidence for a mid-Palaeozoic magmatic arc on the northwestern margin of North America. *In*: HILLS, L. V., HENDRSON, C. M. & BAMBER, E. W. (eds) *Carboniferous and Permian of the World: XIV ICCP Proceedings*. Canadian Society of Petroleum Geologists Memoir, **19**, 158–207.

RIPLEY, E. M., PARK, Y.-R., LAMBERT, D. D. & FRICK, L. R. 2001. Re–Os isotopic composition and PGE contents of Proterozoic carbonaceous argillites, Virginia Formation, Northeastern Minnesota. *Organic Geochemistry*, **32**, 857–866.

ROSS, G. M., BLOCH, J. D. & KROUSE, H. R. 1995. Neoproterozoic strata of the southern Canadian Cordillera and the isotopic evolution of seawater sulfate. *Precambrian Research*, **73**, 71–99.

SCHAEFER, B. F. & BURGESS, J. M. 2003. Re–Os isotopic age constraints on deposition in the Neoproterozoic Amadeus Basin: implications for the 'Snowball Earth'. *Journal of the Geological Society, London*, **160**, 825–828.

SCHOENE, B., CROWLEY, J. L., CONDON, D. J., SCHMITZ, M. D. & BOWRING, S. A. 2006. Reassessing the uranium decay constants for geochronology using ID-TIMS U–Pb data. *Geochimica et Cosmochimica Acta*, **70**, 426–445.

SELBY, D. 2007. Direct rhenium–osmium age of the Oxfordian–Kimmeridgian boundary, Staffin Bay, Isle of Skye, UK, and the Late Jurassic time scale. *Norwegian Journal of Geology*, **87**, 291–299.

SELBY, D. & CREASER, R. A. 2001. Re–Os geochronology and systematics in molybdenite from the Endako porphyry molybdenum deposit, British Columbia, Canada. *Economic Geology*, **96**, 197–204.

SELBY, D. & CREASER, R. A. 2003. Re–Os geochronology of organic-rich sediments: an evaluation of organic matter analysis methods. *Chemical Geology*, **200**, 225–240.

SELBY, D. & CREASER, R. A. 2004. Macroscale NTIMS and microscale LA-MC-ICP-MS Re–Os isotopic analysis of molybdenite: testing spatial restrictions for reliable Re–Os age determinations, and implications for the decoupling of Re and Os within molybdenite. *Geochimica et Cosmochimica Acta*, **68**, 3897–3908.

SELBY, D. & CREASER, R. A. 2005*a*. Direct radiometric dating of the Devonian–Mississippian time-scale boundary using the Re–Os black shale geochronometer. *Geology*, **33**, 545–548.

SELBY, D. & CREASER, R. A. 2005*b*. Direct radiometric dating of hydrocarbon deposits using rhenium–osmium isotopes. *Science*, **308**, 1293–1295.

SELBY, D., CREASER, R. A., DEWING, K. & FOWLER, M. 2005. Evaluation of bitumen as a ^{187}Re–^{187}Os geochronometer for hydrocarbon maturation and migration: a case study from the Polaris MVT deposit, Canada. *Earth and Planetary Science Letters*, **235**, 1–15.

SELBY, D., CREASER, R. A. & FOWLER, M. G. 2007*a*. Re–Os elemental and isotopic systematics in crude oils. *Geochimica et Cosmochimica Acta*, **71**, 378–386.

SELBY, D., CREASER, R. A., STEIN, H. J., MARKEY, R. J. & HANNAH, J. L. 2007*b*. Assessment of the ^{187}Re decay constant by cross calibration of Re–Os molybdenite and U–Pb zircon chronometres in magmatic ore systems. *Geochimica et Cosmochimica Acta*, **71**, 1999–2013.

SHARMA, M., PAPANASTASSIOU, D. A. & WASSERBURG, G. J. 1997. The concentration and isotopic composition of osmium in the oceans. *Geochimica et Cosmochimica Acta*, **61**, 3287–3299.

SHARMA, M., ROSENBERG, E. J. & BUTTERFIELD, D. A. 2007. Search for the proverbial mantle osmium sources to the oceans: hydrothermal alteration of mid-ocean ridge basalt. *Geochimica et Cosmochimica Acta*, **71**, 4655–4667.

SHARMA, M., WASSERBURG, G. J., HOFMANN, A. W. & BUTTERFIELD, D. A. 2000. Osmium isotopes in hydrothermal fluids from the Juan de Fuca Ridge. *Earth and Planetary Science Letters*, **179**, 139–152.

SHEN, J. J., PAPANASTASSIOU, D. A. & WASSERBURG, G. J. 1998. Re–Os systematics in pallasite and mesosiderite metal. *Geochimica et Cosmochimica Acta*, **62**, 2715–2723.

SHIREY, S. B. & WALKER, R. J. 1995. Carius tube digestion for low-blank rhenium–osmium analysis. *Analytical Chemistry*, **67**, 2136–2141.

SHIREY, S. B. & WALKER, R. J. 1998. The Re–Os isotope system in cosmochemistry and high temperature geochemistry. *Annual Reviews of Earth and Planetary Sciences*, **26**, 423–500.

SHUKOLYUKOV, A. & LUGMAIR, G. W. 1997. The ^{53}Mn–^{53}Cr isotope system in the Omolon pallasite and the half-life of ^{187}Re. *Lunar and Planetary Sciences Abstracts*, **28**, 1315–1316.

SIEBERT, C., KRAMERS, J. D., MEISEL, T., MOREL, P. & NÄGLER, T. F. 2005. PGE, Re–Os, and Mo isotope systematics in Archean and early Proterozoic sedimentary systems as proxies for redox conditions of the early Earth. *Geochimica et Cosmochimica Acta*, **69**, 1787–1801.

SLACK, J. F., GRENNE, T., BEKKER, A., ROUXEL, O. J. & LINDBERG, P. A. 2007. Suboxic deep seawater in the late Palaeoproterozoic: evidence from hematitic chert and iron formation related to seafloor-hydrothermal sulfide deposits, central Arizona, USA. *Earth and Planetary Science Letters*, **255**, 243–256.

SMOLIAR, M. I., WALKER, R. J. & MORGAN, J. W. 1996. Re–Os ages of Group IIA, IIIA, IVA, and IVB iron meteorites. *Science*, **271**, 1099–1102.

SUMMONS, R. E., TAYLOR, D. & BOREHAM, C. J. 1994. Geochemical tools for evaluating petroleum generation in Middle Proterozoic sediments of the McArthur Basin, Northern Territory, Australia. *Australian Petroleum Exploration Association Journal*, **34**, 692–706.

SUN, W., ARCULUS, R. J., BENNETT, V. C., EGGINS, S. M. & BINNS, R. A. 2003*a*. Evidence for rhenium enrichment in the mantle wedge from submarine arc-like volcanic glasses (Papua New Guinea). *Geology*, **31**, 845–848.

SUN, W., BENNETT, V. C., EGGINS, S. M., KAMENETSKY, V. S. & ARCULUS, R. J. 2003*b*. Enhanced mantle-to-crust rhenium transfer in undegassed arc magmas. *Nature*, **422**, 294–297.

SUNDBY, B., MARTINEZ, P. & GOBEIL, C. 2004. Comparative geochemistry of cadmium, rhenium, uranium, and molybdenum in continental margin sediments. *Geochimica et Cosmochimica Acta*, **68**, 2485–2493.

TRAPP, E., KAUFMANN, B., MEZGER, K., KORN, M. & WEYER, D. 2004. Numerical calibration of the Devonian–Mississippian boundary: Two new U–Pb isotope dilution-thermal ionization mass spectrometry single-zircon ages from Hasselbachtal (Sauerland, Germany). *Geology*, **32**, 857–860.

VOLK, H., GEORGE, S. C., DUTKIEWICZ, A. & RIDLEY, J. 2005. Characterisation of fluid inclusion oil in a Mid-Proterozoic sandstone and dolerite (Roper Superbasin, Australia). *Chemical Geology*, **223**, 109–135.

VÖLKENING, J., WALCZYK, T. & HEUMANN, K. G. 1991. Osmium isotope ratio determinations by negative thermal ionization mass spectrometry. *International Journal of Mass Spectrometry and Ion Processes*, **105**, 147–159.

WALCZYK, T., HEBEDA, E. H. & HEUMANN, K. G. 1991. Osmium isotope ratio measurements by negative thermal ionization mass spectrometry (NTI-MS): Improvement in precision and enhancement in emission by introducing oxygen or Freon into the ion source. *Fresenius' Journal of Analytical Chemistry*, **341**, 537–541.

WALKER, R. J., HORAN, M. F., MORGAN, J. W., BECKER, H., GROSSMAN, J. N. & RUBIN, A. E. 2002*a*. Comparative ^{187}Re–^{187}Os systematics of chondrites: implications regarding early solar system processes. *Geochimica et Cosmochimica Acta*, **66**, 4187–4201.

WALKER, R. J., PRICHARD, H. M., ISHIWATARI, A. & PIMENTEL, M. 2002*b*. The osmium isotopic composition of convecting upper mantle deduced from ophiolite chromites. *Geochimica et Cosmochimica Acta*, **66**, 329–345.

WALTER, M. R., VEEVERS, J. J., CALVER, C. R., GORJAN, P. & HILL, A. C. 2000. Dating the 840–544 Ma Neoproterozoic interval by isotopes of strontium, carbon, and sulfur in seawater, and some interpretative models. *Precambrian Research*, **100**, 371–433.

WANG, X., ERDTMANN, B.-D., CHEN, X. & MAO, X. 1998. Integrated sequence-, bio- and chemo-stratigraphy of the terminal Proterozoic to lowermost Cambrian 'black rock series' from central South China. *Episodes*, **21**, 178–189.

WARREN, J. K., GEORGE, S. C., HAMILTON, P. J. & TINGATE, P. 1998. Proterozoic source rocks: sedimentology and organic characteristics of the Velkerri Formation, Northern Territory, Australia. *AAPG Bulletin*, **82**, 442–463.

WILLE, M., KRAMERS, J. D. *ET AL*. 2007. Evidence for a gradual rise of oxygen between 2.6 and 2.5 Ga from Mo isotopes and Re-PGE signatures in shales. *Geochimica et Cosmochimica Acta*, **71**, 2417–2435.

WOODHEAD, J. D., HERGT, J. M. & SIMONSON, B. M. 1998. Isotopic dating of an Archean bolide impact horizon, Hamersley Basin, Western Australia. *Geology*, **26**, 47–50.

WOODHOUSE, O. B., RAVIZZA, G., FALKNER, K. K., STATHAM, P. J. & PEUCKER-EHRENBRINK, B. 1999. Osmium in seawater: vertical profiles of concentration and isotopic composition in the eastern Pacific Ocean. *Earth and Planetary Science Letters*, **173**, 223–233.

WOODLAND, S. J., OTTLEY, C. J., PEARSON, D. G. & SWARBRICK, R. E. 2001. Microwave digestion of oils for analysis of platinum group and rare earth elements by ICP-MS. *In*: HOLLAND, G. & BANDURA, D. (eds) *Plasma Source Mass Spectrometry. Current Trends and Future Developments*. Royal Society of Chemistry, Special Publications, **301**, 17–24.

XIAO, S., YUAN, X., STEINER, M. & KNOLL, A. H. 2002. Macroscopic carbonaceous compressions in a terminal Proterozoic shale: a systematic reassessment of the Miaohe biota, South China. *Journal of Palaeontology*, **76**, 347–376.

XIAO, S., ZHANG, Y. & KNOLL, A. H. 1998. Three-dimensional preservation of algae and animal embryos in a Neoproterozoic phosphorite. *Nature*, **391**, 553–558.

YAMASHITA, Y., TAKAHASHI, Y., HABA, H., ENOMOTO, S. & SHIMIZU, H. 2007. Comparison of reductive accumulation of Re and Os in seawater–sediment systems. *Geochimica et Cosmochimica Acta*, **71**, 3458–3475.

YANG, W. & HOLLAND, H. D. 2002. The redox-sensitive trace elements, Mo, U, and Re in Precambrian carbonaceous shales: indicators of the Great Oxidation Event. *Geological Society of America, Abstracts with Programs*, **34**, 381.

YANG, G., HANNAH, J. L., ZIMMERMAN, A., STEIN, H. J. & BEKKER, A. 2009. Re–Os depositional age for Archean carbonaceous slates from the southwestern Superior Province: challenges and insights. *Earth and Planetary Science Letters*, **280**, 83–92.

ZHOU, C. & XIAO, S. 2007. Ediacaran δ^{13}C chemostratigraphy of South China. *Chemical Geology*, **237**, 89–108.

Global Infracambrian petroleum systems: a review

K. A. R. GHORI[1]*, JONATHAN CRAIG[2], BINDRA THUSU[3], SEBASTIAN LÜNING[4,5] & MARKUS GEIGER[4,6]

[1]*Department of Industry and Resources, Perth, Australia*

[2]*Eni Exploration and Production Division, Via Emilia 1, 20097 San Donato Milanese, Milan, Italy*

[3]*MPRG (Maghreb Petroleum Research Group), University College London, Gower Street, London WC1E 6BT, UK*

[4]*Department of Geosciences, University of Bremen, 28334 Bremen, Germany*

[5]*Present address: RWE Dea Überseering 40, 22297 Hamburg, Germany*

[6]*StatoilHydro ASA, 4035 Stavanger, Norway*

**Corresponding author (e-mail: Ameed.GHORI@dmp.wa.gov.au)*

Abstract: This review covers global uppermost Neoproterozoic–Cambrian petroleum systems using published information and the results of studies undertaken by the Geological Survey of Western Australia (GSWA) on the Neoproterozoic Officer Basin. Both production and hydrocarbon (HC) shows sourced from, and reservoired in, uppermost Neoproterozoic–Cambrian successions occur worldwide, and these provide ample incentive for continuing exploration for these older petroleum systems. However, the risks of charge volume, timing of generation–migration v. trap formation and preservation of accumulation are significantly higher than in conventional Phanerozoic petroleum systems. Therefore, the location and assessment of preserved HC accumulations in such old petroleum systems presents a significant exploration challenge.

Organic-rich metamorphosed Proterozoic successions of SE Greenland, the Ukrainian Krivoy Roy Series, the Canadian Upper Huronian Series and the oil shales of the Russian Onega Basin are known as the world's oldest overmature petroleum source rocks. The oldest live oil has been recovered from the McArthur Basin of Australia (*c.* 1.4 Ga; Ga is 10^9 years), followed by the Nonesuch oil of Michigan. Numerous other petroleum shows have been reported from Australia, Canada, China, India, Morocco, Mauritania, Mali, Oman, Pakistan, Venezuela and the USA. These demonstrate that generation and migration of Proterozoic petroleum has occurred worldwide. The Siberian Lena–Tunguska province, the Russian Volga–Ural region and the Middle Eastern south Oman petroleum fields exemplify the productive potential of uppermost Neoproterozoic–Cambrian successions, where petroleum generation, migration and trapping were either late in the geological history (Palaeozoic–Mesozoic, Oman) or where accumulations have been preserved beneath highly effective super-seals (Lena–Tunguska). The total resource potential of the Lena–Tunguska petroleum province is estimated to be 2000 Mbbl (million barrels) oil and 83 Tcf (trillion cubic feet) gas. The equivalent proven and probable reserves derived from Neoproterozoic–Early Cambrian source rocks and trapped in Late Neoproterozoic (Ediacaran), Palaeozoic and Mesozoic reservoirs in Oman are at least 12 bbbl (billion barrels) of oil and an undetermined volume of gas.

The recovery of 12 Mcf (million cubic feet) of Precambrian gas from the Ooraminna-1 well in the Amadeus Basin in 1963, together with the occurrence of numerous HC shows within the Australian Centralian Superbasin, triggered the initial exploration for Proterozoic hydrocarbons in Australia. This included exploration in the Neoproterozoic Officer Basin, which is reviewed in this paper as a case study. Minor oil shows and numerous bitumen occurrences have been reported from the 24 petroleum exploration wells drilled in the Officer Basin to date, indicating the existence of a Neoproterozoic petroleum system. However, the potential of the Neoproterozoic petroleum system in the vast underexplored Officer Basin, with its sparse well control, remains unverified, but may be significant, as may that of many other 'Infracambrian' basins around the world.

The stratigraphic terminology and the position of key events used in this study are summarized in Figure 1, using numerical ages of the unit boundaries available from 2008 updated from International Commission on Stratigraphy web: http://www.stratigraphy.org/gssp.htm.

According to the presently known stratigraphic record, the conditions required for the deposition

From: CRAIG, J., THUROW, J., THUSU, B., WHITHAM, A. & ABUTARRUMA, Y. (eds) *Global Neoproterozoic Petroleum Systems: The Emerging Potential in North Africa.* Geological Society, London, Special Publications, **326**, 109–136.
DOI: 10.1144/SP326.6 0305-8719/09/$15.00

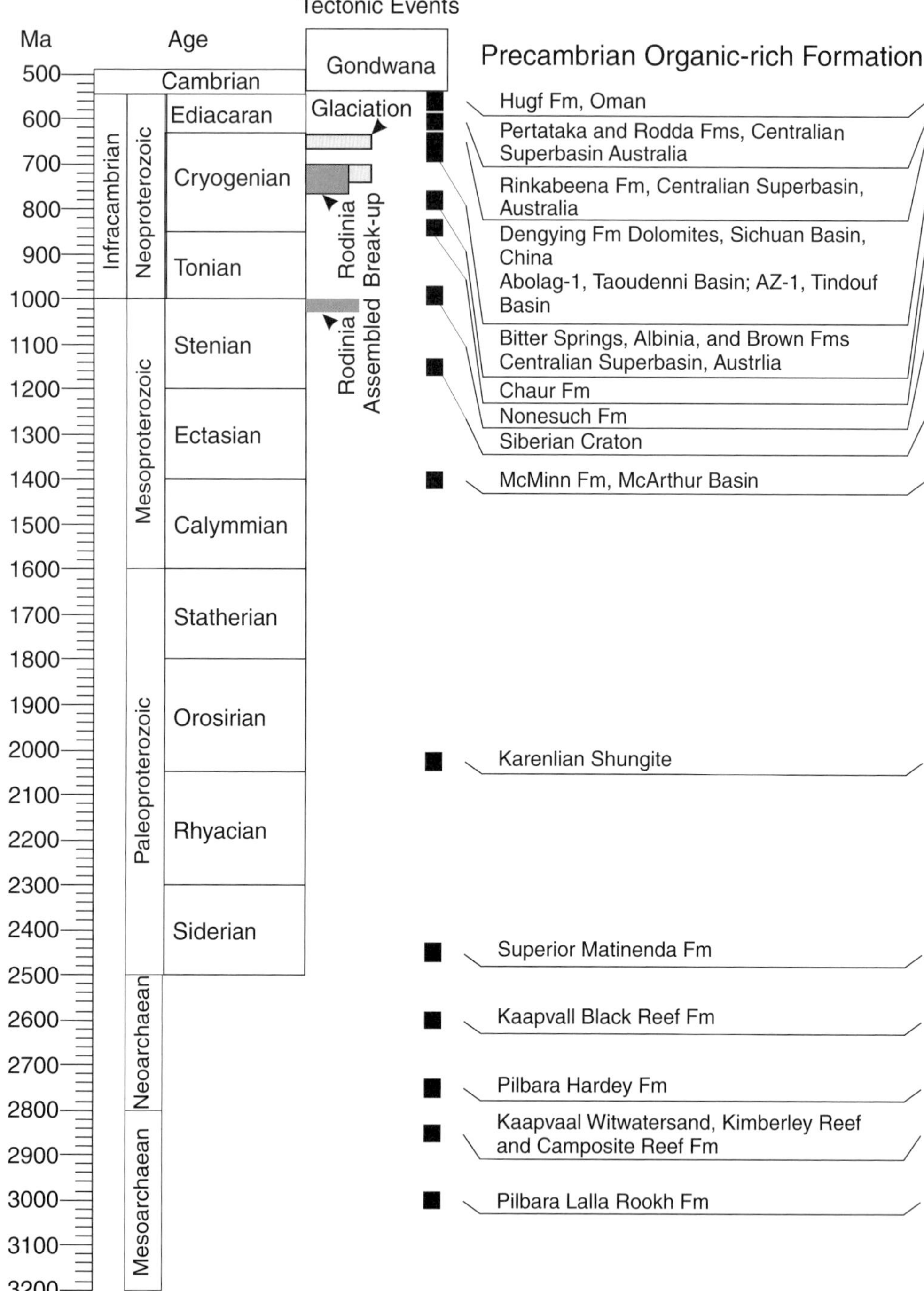

Fig. 1. Precambrian stratigraphic terminology and events used in this study.

of organic-rich sediments have occurred locally since Archaean time, but their frequency increased substantially during the Proterozoic and drastically during the Phanerozoic. The quality and richness of organic matter preserved in sediments increases in accordance with the evolution of the archea and eubacteria (prokaryotes) and the eukarya (eukaryotes or higher organisms) from the Archaean to the Present. The resulting organic-rich sediments are the source of most of the world's oil and gas reserves. These reserves are not evenly distributed either spatially or temporally. Globally, conditions for the deposition of source rocks were most conducive during the Mesozoic Era. This review provides a summary of the geographical and stratigraphic position of global Neoproterozoic–Early Cambrian petroleum systems from published information, followed by a summary of the results of studies undertaken by the Geological Survey of Western Australia (GSWA), on the Neoproterozoic Officer Basin, as a more detailed case study.

Petroleum system concept

Petroleum geochemistry has played the key role in developing the petroleum system concept since the early 1970s as discussed by Dow (1974), Demaison (1984), Meissner *et al.* (1984), Ulmishek (1986), Perrodon (1992), Bradshaw (1993), Magoon & Dow (1994), Magoon (1995), Magoon & Beaumont (1999) and Peters *et al.* (2005). The principal objective of the petroleum system approach is to define the three basic risk factors in evaluating effective petroleum systems: (1) magnitude of petroleum charge, which depends on source rock organic richness and facies, thickness and maturity; (2) trap, which depends on reservoir porosity, permeability, geometry, and seal quality; and (3) timing of charge v. trap formation. These three basic factors are essential to define the geographical boundaries of oil and gas occurrence, and to predict the location of areas that may have potential petroleum systems.

Precambrian petroleum systems

Precambrian sourced and reservoired oil and gas accumulations worldwide have been differentiated geochemically from equivalent Phanerozoic-sourced petroleum systems on the basis of their unusual biomarkers and light carbon isotopic signatures (McKirdy & Imbus 1992). The first prerequisite for petroleum formation is the deposition of organic-rich petroleum source rocks that are enriched in hydrogen and are exposed to sufficient heat that they generate and expel hydrocarbons. Most of the world's major petroleum source rocks have a total organic carbon (TOC) content of at least 2–3%, even after they have passed through the oil-generating stage. Such organic-rich sediments have been well documented from many Precambrian successions. The oldest of these potential or proven Precambrian source rocks include the metamorphosed Proterozoic succession of the Canadian Upper Huronian Series, broadly equivalent age successions in SE Greenland, the Ukrainian Krivoy Series and the oil shales of the Russian Onega Basin (Hunt 1996).

The commercial accumulations of indigenous Proterozoic-sourced petroleum in Siberia (Meyerhoff 1980), Oman (Al-Marjeby & Nash 1986; Alsharhan & Kendall 1986) and China (Korsch *et al.* 1991) demonstrate the economic viability of these 'old' petroleum systems. These commercial accumulations, which include several giant fields [>500 Mbbl (million barrels) of oil and/or 3 Tcf (trillion cubic feet) of gas] together with the widespread occurrence of hydrocarbon (HC) shows in Precambrian successions worldwide, have provided significant impetus for the exploration of other Proterozoic petroleum systems in recent years. However, a better understanding of the stratigraphy and structural history of Proterozoic sequences worldwide is required to encourage future exploration. Organic-rich rocks of Archaean–Palaeoproterozoic age (3.80–1.60 Ga; Ga is 10^9 years) have long been studied for their scientific interest, but organic-rich rocks of Mesoproterozoic–Neoproterozoic age (1.60–0.54 Ga) are also becoming of increasing interest from a commercial perspective.

Archaean–Palaeoproterozoic

The oldest recognizable petroleum source rocks were deposited in anoxic environment where methylotrophic bacteria played a key role in the production of oil-prone organic matter highly enriched in light ^{12}C. The oldest (*c.* 2 Ga) and organically richest rocks are reported from the Upper Zaonezhskaya Formation near Lake Onega, NW Russia. This includes a 600 m-thick sequence that contains shungite, a nearly pure carbonaceous material, averaging about 25% TOC (Melezhik *et al.* 1999; Peters *et al.* 2005). Relict Archaean petroleum systems are also recognized within the Pilbara Craton, Australia, and Kaapvaal Craton, South Africa, on the basis of pyrobitumen nodules of migrated oil and oil fluid inclusions (Buick *et al.* 1998; Dutkiewicz *et al.* 1998; Peters *et al.* 2005). Palaeoproterozoic rocks contain abundant microfossils and hydrocarbons. (Burlingame *et al.* 1965). The associated biomass, largely consisting of prokaryotes, was sufficient to produce organic-rich source rocks, while veins of pyrobitumen provide evidence of oil migration (Mancuso *et al.* 1989; Peters *et al.* 2005).

Mesoproterozoic–Neoproterozoic

Mesoproterozoic successions containing lacustrine and marine shales with source potential are reported from the Siberian Craton, central North America, and also from the McArthur Basin of Australia, where preserved Mesoproterozoic petroleum systems indicate the potential for commercial production of oil and gas (Peters *et al.* 2005). The Mesoproterozoic rocks of the McArthur Basin of Australia contain the world's oldest (*c.* 1.4 Ga) live oil, which flowed from black mudstone of the Velkerri Formation in BMR well Urapunga No. 4 (Muir *et al.* 1980). The Mesoproterozoic Nonesuch Formation (*c.* 1.05 Ga) of central North America contains oil seeps within the White Pine copper deposit, Michigan, USA (Imbus *et al.* 1988; Pratt *et al.* 1991; Mauk & Hieshima 1992; Price *et al.* 1996). The source of the eastern Siberian Precambrian oils is also believed to be organic-rich shales of Mesoproterozoic (Riphean) age (Ulmishek 2001). The Neoproterozoic was a period of Earth history during which one or more continental-scale, or even global-scale, glaciations affected the evolution of life (Hoffman *et al.* 1998; Hoffman & Schrag 2002: Halverson *et al.* 2005). Source rock deposition occurred between the pulses of glaciation during periods of prolonged oceanic anoxia and high rates of accumulation of organic matter in marine sediments, with TOCs locally as high as 20–30% (McKirdy & Imbus 1992). Rich Neoproterozoic source rocks and the world's oldest commercial oil and gas accumulations occur on the Siberian Platform (Lena–Tunguska region) and on the Arabian Shield (Oman). In the former Late Riphean and Vendian, siliciclastic and carbonate rocks are overlain by thick Lower Cambrian salt deposits that form an effective superseal to the petroleum system (McKirdy & Imbus 1992; Ulmishek *et al.* 2002). The giant Weiyuan gas field in the Sichuan Basin of eastern China produces from peritidal dolomitic carbonate rocks of the Sinian Period (Neoproterozoic), which include both basal argillaceous source rocks and the main reservoirs (Hao & Liu 1989). Countries currently producing Proterozoic oil and gas, and those that are being actively explored for Precambrian petroleum, are shown in Figure 2. Productive and potential Proterozoic petroleum systems are briefly reviewed later in this chapter.

Proterozoic petroleum production

The Sultanate of Oman salt basins, the Siberian Lena–Tunguska province of Russia and the Sichuan Basin of China all produce commercial volumes of oil and gas from the Meso-Neoproterozoic–Lower Palaeozoic petroleum systems, along with their Mesozoic–Cenozoic petroleum systems. The Meso-Neoproterozoic sourced petroleum systems of these areas are reviewed briefly here, together with their approximate stratigraphic position in relation to the latest International Geological Timescale (Gradstein *et al.* 2004; International Commission on Stratigraphy 2008).

Sultanate of Oman basins

The Sultanate of Oman is located on the southeastern margin of the Arabian plate. Figures 3 and 4

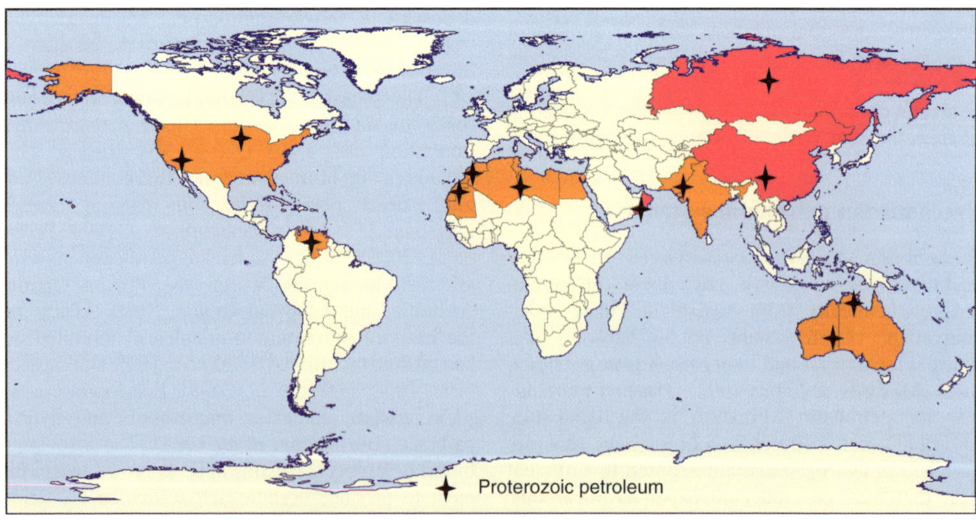

Fig. 2. Countries either producing (red) or potential to produce (orange) Proterozoic oil and gas.

summarize location, stratigraphy and petroleum systems of the Sultanate of Oman, which were compiled from various publications including: Pollastro (1999), Terken *et al.* (2001), Al-Lazki *et al.* (2002) and Al-Lazki (2003). The best estimate on the age of the Huqf Supergroup is 725–540 Ma, with the young constraint being the age of the Angudan unconformity (Allen 2007). The Huqf Supergroup is the oldest sedimentary sequence in Oman (Fig. 4), and consists of alternating clastic, carbonate and evaporitic units (Gorin *et al.* 1982; Pollastro 1999). It has been extensively studied for both commercial and scientific interests as it is a prolific petroleum producer, and provides critical information on the geological evolution of the Arabian–Persian Gulf region during the Neoproterozoic and constrains the Sturtian (740–700 Ma) and Marinoan (665–635 Ma) glacial episodes of the Neoproterozoic (Amthor 2006; Summons *et al.* 2006; Allen 2007).

The Huqf Supergroup contains several clastic and carbonate source rocks of exceptional quality, and these sourced the primary petroleum systems that produce hydrocarbons throughout Oman. The latest Neoproterozoic–earliest Cambrian Ara Group is a carbonate- and evaporate-dominated sequence containing salt deposits up to 1000 m thick (the Ghaba, South Oman and Fahud salt basins). The Ara evaporites were deposited in geographically restricted basins during periods of low relative sea level where stratified, anoxic conditions prevailed and organic-rich sediments and salt were deposited (Mattes & Conway-Morris 1990; Edgell 1991). Source rocks and reservoirs are widespread both geographically and stratigraphically in Oman (Fig. 4). The main source rocks have been identified in the Neoproterozoic Huqf Supergroup, the Silurian Safiq, the Upper Jurassic Sahtan (Diyab Formation), and the Cretaceous Kahmah (Bab Formation) and Wasia (Natih Formation) groups (Terken *et al.* 2001). Oil and gas derived from the Huqf Supergroup, which is believed to be the source of about 12 bbbl (billion barrels) of oil and an undetermined amount of gas, accumulated in Ediacaran and the overlying Palaeozoic and Mesozoic reservoirs (Grantham *et al.* 1987; Fritz 1989; Edgell 1991).

Five geochemical types of crude oil are recognized in Oman, including South and North Oman Huqf oils, 'Q' oils, Natih oils and Tuwaiq oils. These are correlated to four main source rocks. Of these, the South and North Oman Huqf oils and 'Q' oils are correlated to the Huqf Supergroup source rocks. The existence of prolific Ediacaran–Early Cambrian petroleum systems in Oman is one of the key drivers for the exploration of potentially equivalent petroleum systems within Precambrian successions elsewhere in the world.

Russian basins

The Lena–Tunguska Mesoproterozoic–Palaeozoic petroleum superprovince is located on the Siberian Craton in northern Russia between the Yenisey and Lena rivers (Fig. 5a). The region comprises three petroleum-bearing provinces: the Lena–Tunguska, Lena–Vilyuy and Yenisey–Anabar provinces. The Yurubchen–Tokhom Zone in Lena–Tunguska province is potentially the most important oil-producing area in the superprovince (Clarke 1985; Kontorovich *et al.* 1990; Kuznetsov 1997). The Siberian Craton rifted in the Riphean, and the rift basins were filled by thick synrift sedimentary sequences (Ulmishek *et al.* 2002). These are mainly overlain by Palaeozoic platform successions (Ulmishek *et al.* 2002). The petroleum-bearing sedimentary sequences are Riphean, Vendian, upper Vendian–Lower Cambrian, Cambrian, Ordovician–Devonian and Carboniferous–Triassic in age (Fig. 5b). The Riphean petroliferous sequences are particularly important in the Baikit and Predpatoma regions. The Vendian petroliferous sequences are commercially important in the Baikit, Katanga, Angara–Lena, Nepa, Botuoba and Predpatoma regions. The Vendian–Cambrian petroliferous sequences contain oil, oil and gas, and gas pools in the South Tunguska, Nepa–Botuoba, Baikit and Predpatoma regions. The Cambrian petroliferous sequences produce gas in the South Tunguska and Angara–Lena region and the Turukhan–Norilsk area (Kontorovich *et al.* 1990).

The Yurubchen–Tokhom productive area provides the most extensive geophysical and drilling database with over 100 wells. The first oil was tested in 1977. In 1982 it was proved to be light oil (42°–45° API) of low sulphur content (0.2–0.3%) and can be produced commercially. Major petroleum discoveries include the Jurubchenskoe, Kujumbinskoe, Talakanskoe, Chayadinskoe, Verkhne–Viluchanskoe, Kovyktinskoe and Verkhnevilyuy fields. Verkhnevilyuy is one of the largest fields in the region with proven plus probable reserves of 10.5 Tcf of gas and some 260 Mbbl of condensate (Meyerhoff 1980). The total resource potential of the Lena–Tunguska petroleum province is estimated to be 2.0 bbbl oil and 83 Tcf gas (Meyerhoff 1982). The principal source and reservoir rocks are within the Meso-Neoproterozoic succession. A thick Cambrian salt succession provides a regional super-seal that facilitated the preservation of hydrocarbons, which are considered to have been generated in pre-Devonian times.

The oldest petroleum systems on the Siberian Craton are unusual in that the kitchen areas were partially or fully destroyed by post-depositional tectonism in early Palaeozoic time (Ulmishek *et al.*

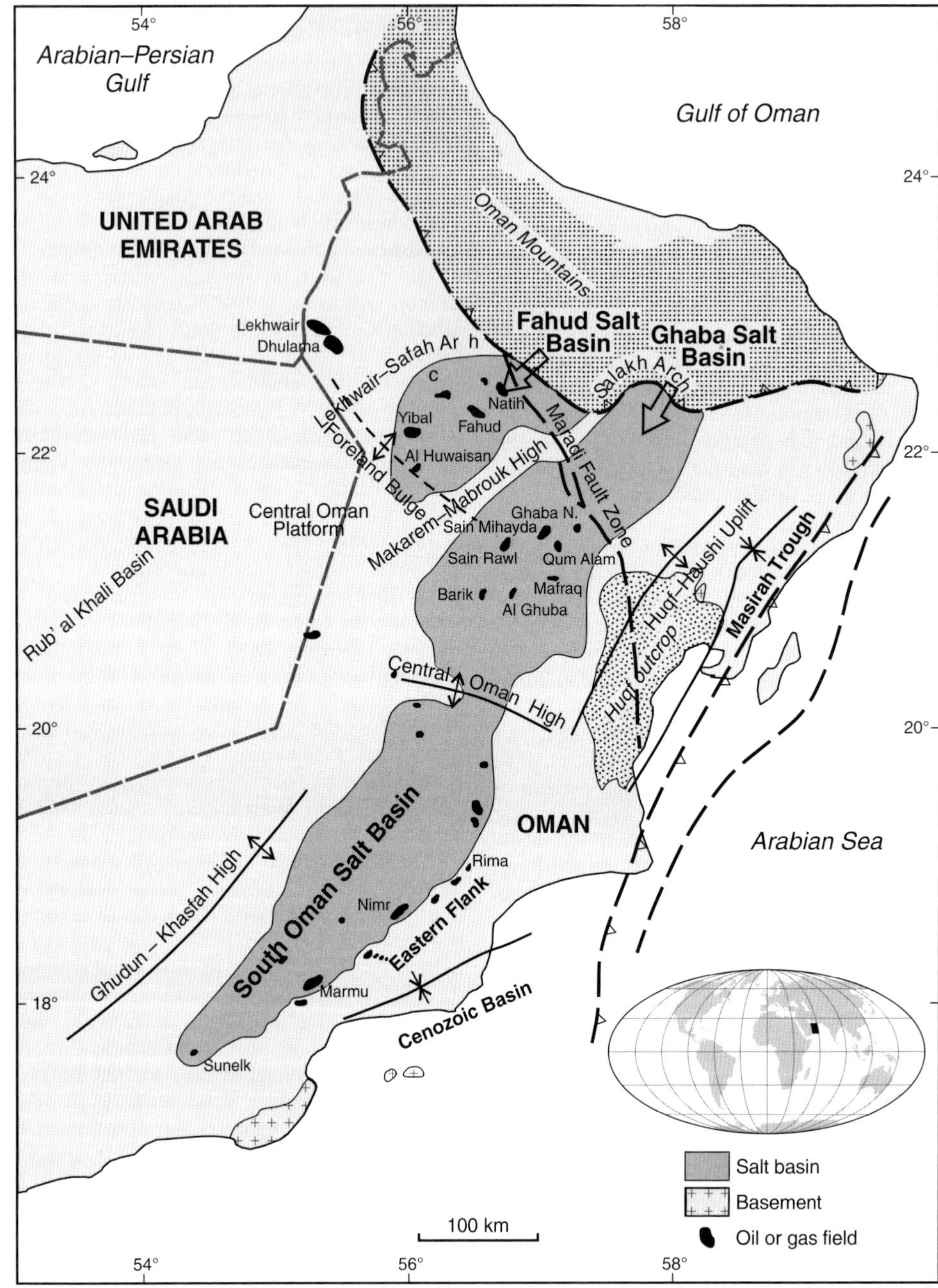

Fig. 3. Location of sedimentary basins of the Sultanate of Oman modified from Pollastro (1999).

2002). However, the source rocks reached maturity, and HCs were expelled and migrated into platform traps before the original HC kitchens were destroyed. The exceptionally long preservation of HCs was facilitated by the existence of the undeformed Cambrian salt super-seal. The undiscovered volumes of both oil and gas on the Siberian Craton, as assessed by United States Geological Survey (USGS), range from 2.8 bbbl of oil and 48.9 Tcf of gas (Masters *et al.* 1997) to 11.3 bbbl of oil and 175 Tcf of gas (Ulmishek 2001). The main reason for the difference in the assessments is the

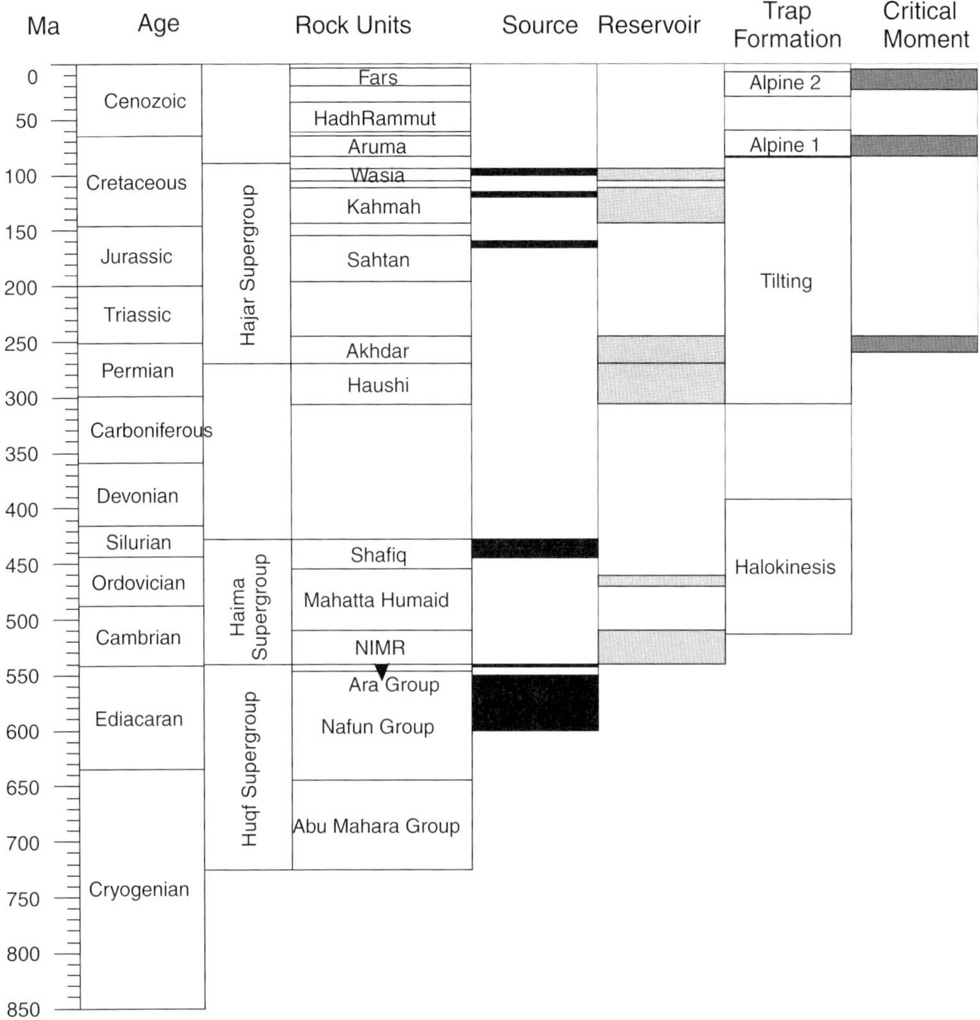

Fig. 4. Generalized stratigraphy and petroleum systems of the Sultanate of Oman.

assumption that much of undiscovered resources will be accounted for by reserve growth in two discovered giant fields, the Yurubchen–Takhoma oil field and the Kovykta gas field. In the 2000 assessment, field growth was considered separately (Ulmishek *et al.* 2002). There are more than 30 discoveries (mostly gas and condensate) in the Lena–Tunguska region. At least two of these are considered to be giant fields. Modelling of HC generation suggests that significant additional petroleum is yet to be discovered in the region. The principal HC prospective areas are related to zones of Riphean–Vendian reservoir rocks along palaeo-rift zones that contain thick source rock development in their central parts (Postnikov & Postnikova 2006).

Chinese basins

Sinian (Neoproterozoic) sourced and produced HCs occur in the Sichuan Basin of SW China (Fig. 6a), a large intracratonic sedimentary basin, covering an area of 180 000 km^2 in the western part of the Yangtze Craton. The relationship of the Yangtze Craton to the assembly and break-up of Rodinia and correlations with other continents are discussed by Li *et al.* (2003, 2005) and Greentree *et al.* (2006), The Yangtze Block sedimentary province is an important petroleum producing area with over 60 gas fields and 10 oil fields (Korsch *et al.* 1991). The basin contains up to 12 000 m of Neoproterozoic–Cenozoic sediments, deposited in a Neoproterozoic–middle Mesozoic

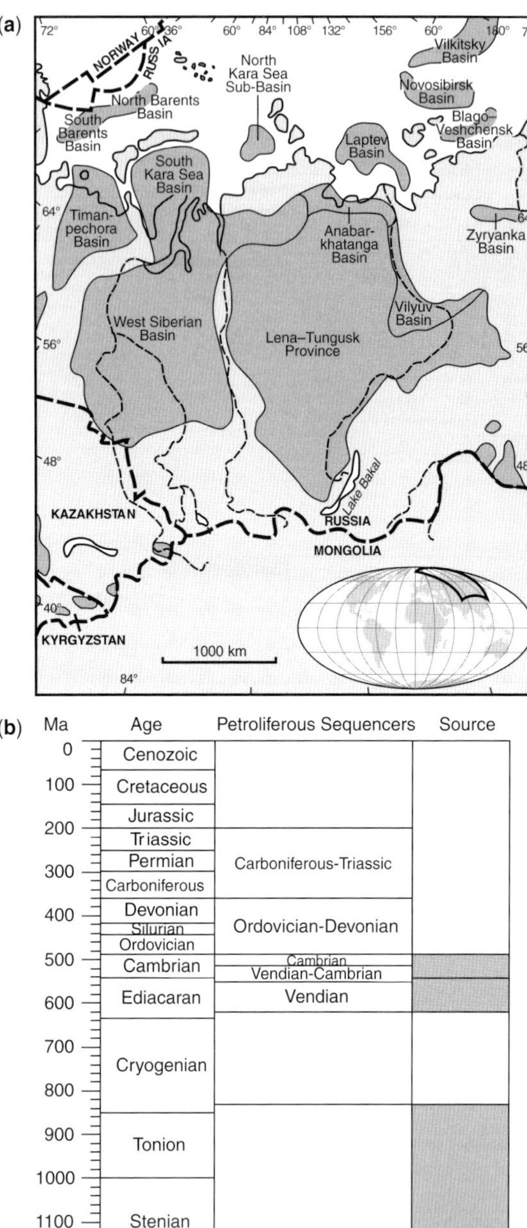

Fig. 5. Proterozoic of Russia: (**a**) location of sedimentary basins; and (**b**) petroliferous sequences and petroleum source rocks.

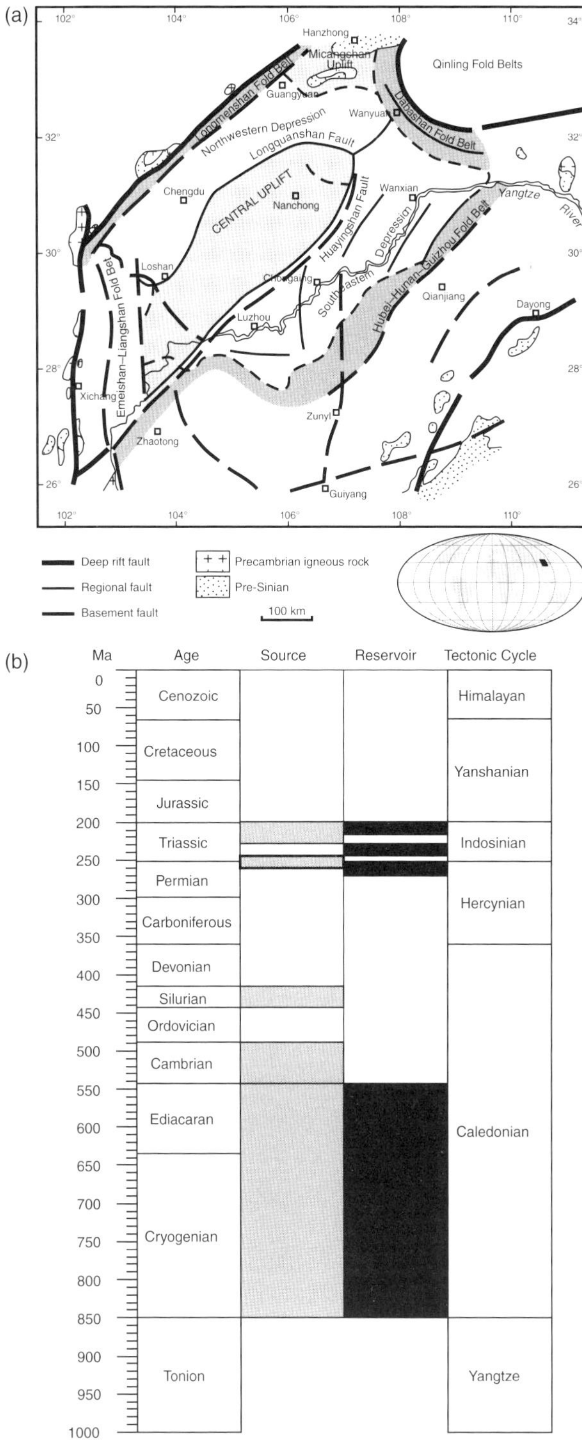

Fig. 6. The Sichuan Basin of China: (**a**) location; and (**b**) petroleum systems.

passive margin basin overlain by a late Mesozoic–Cenozoic foreland basin. The basin has three major subdivisions: the northwestern depression; the central uplift; and the southeastern depression. The central uplift is separated from the northwestern depression by the Longquanshan fault in the west and from the southeastern depression by the Huayingshan fault in the east (Ma *et al.* 2007).

The Sichuan Basin is a prolific petroleum province with an upside resource potential of 178 Tcf of gas and 26.18 bbbl of oil. Since 1950 over 245 gas and 15 oil pools had been discovered. The basin contains over 21 producing gas and oil reservoirs, distributed within the Sinian–Jurassic succession (Fig. 6b) at depths ranging from 7157 m (Guanji well) to a few hundreds of metres (Zhou 1996). The Weiyuan Gas Field within the Neoproterozoic–Palaeozoic succession of the Sichuan Basin is one of the largest gas fields in China, with estimated total reserves of up to 1.41 Tcf (Korsch *et al.* 1991). The gas is produced primarily from dolomite in the Sinian Dengying Formation and was generated chiefly from the microbial dolomite of the same Neoproterozoic sequence. Other major gas fields in the Sichuan Basin include the Wulunghou, Shiyougou, Luzhou, Shendengshan, Zigong, Huanchiachan, Tengchingkuan, Huangkuanshan and Yenkaohsi fields. Gas fields are located predominantly in the south of the basin, whereas oil fields occur predominantly in the centre and east. Other fields in Sichuan have an estimated ultimate recovery of 1.06 Tcf of gas. The oil and gas is predominantly reservoired in carbonate sequences deposited on a stable platform extending in time from the Neoproterozoic to the Jurassic.

The Bohai Bay Basin in northern China also has significant petroleum prospectivity within the Neoproterozoic succession. The Renqiu Oil Field in the Bohai Bay Basin is one of the largest oil fields in China. The oil in this field is produced primarily from dolomite in the Meso-Neoproterozoic Wumishan Formation, but was generated mainly from source rocks in the Oligocene Shahejie Formation. However, there is a possibility that the oil was also partly sourced from the Meso-Neoproterozoic rocks, since many oil and gas shows have been reported in the Meso-Neoproterozoic rocks of the Yanshan foldbelt at the northern margin of the Bohai Bay Basin. According to organic geochemical studies, the Meso-Neoproterozoic rocks in the area have good potential for HC generation, and conditions for the generation of HCs should exist in the Bohai Bay Basin and the nearby Yanshan Fold Belt. These Precambrian producing carbonate formations account for a considerable portion of the reserves and production in China (Hao & Liu 1989).

Proterozoic petroleum potential

Meso-Neoproterozoic sourced petroleum systems are reported from basins in North Africa, India, Pakistan, North America and Australia, but their commercial viability needs further systematic evaluation. These basins include: the Taoudenni Basin overlying the West African platform, the Bikaner–Nagaur Basin of India and the Punjab Platform of Pakistan, the Midcontinent Rift system and Grand Canyon areas of the United States, and the McArthur Basin and Centralian Superbasin of Australia. The potential of these Meso-Neoproterozoic sourced petroleum systems is briefly reviewed here.

North Africa basins

The Maghreb Petroleum Research Group (MPRG) at University College London, supported by the National Oil companies of Libya, Algeria and Morocco and by the Eni Exploration and Production Division, is actively assessing the potential for Mesoproterozoic and Neoproterozoic–Cambrian petroleum systems in North African basins. Their published and unpublished studies provided the basis for this review. The recovery of gas from 'Infracambrian' carbonates in the Taoudenni Basin and the success of Neoproterozoic–Early Cambrian plays globally provided the incentive to research and explore the prospectivity of Precambrian petroleum systems in North Africa, where Palaeozoic petroleum systems are already well established. The recovery of gas (480 million cubic feet day^{-1} (Mcfd)) from the Infracambrian succession of the Abolag-1 well within the Mauritanian part of the Taoudenni Basin in 1973 confirms the existence of a viable Infracambrian petroleum system in northwestern Africa. The Taoudenni Basin developed over part of the West African platform covering areas of present-day Mauritania, Mali and Algeria (Fig. 7). Deposition started at about 1000 Ma and continued until the end of the Carboniferous, producing a sedimentary succession with an average thickness of 3000 m. The Neoproterozoic–Carboniferous succession is exposed on the flanks of the basin, but is covered by a thin sequence of Mesozoic–Cenozoic rocks in centre of the basin.

The Infracambrian–earliest Palaeozoic petroleum potential of North African basins is summarized here based on detailed published field, laboratory and research reports, and unpublished data available from MPRG, including: Lüning *et al.* (2004, 2005), Geiger *et al.* (2004), Kolonic *et al.* (2004), Thusu *et al.* (2004) and Craig *et al.* (2008). According to these studies, polyphase evolution of the North African basins started during the Huqf Supergroup East African Orogeny

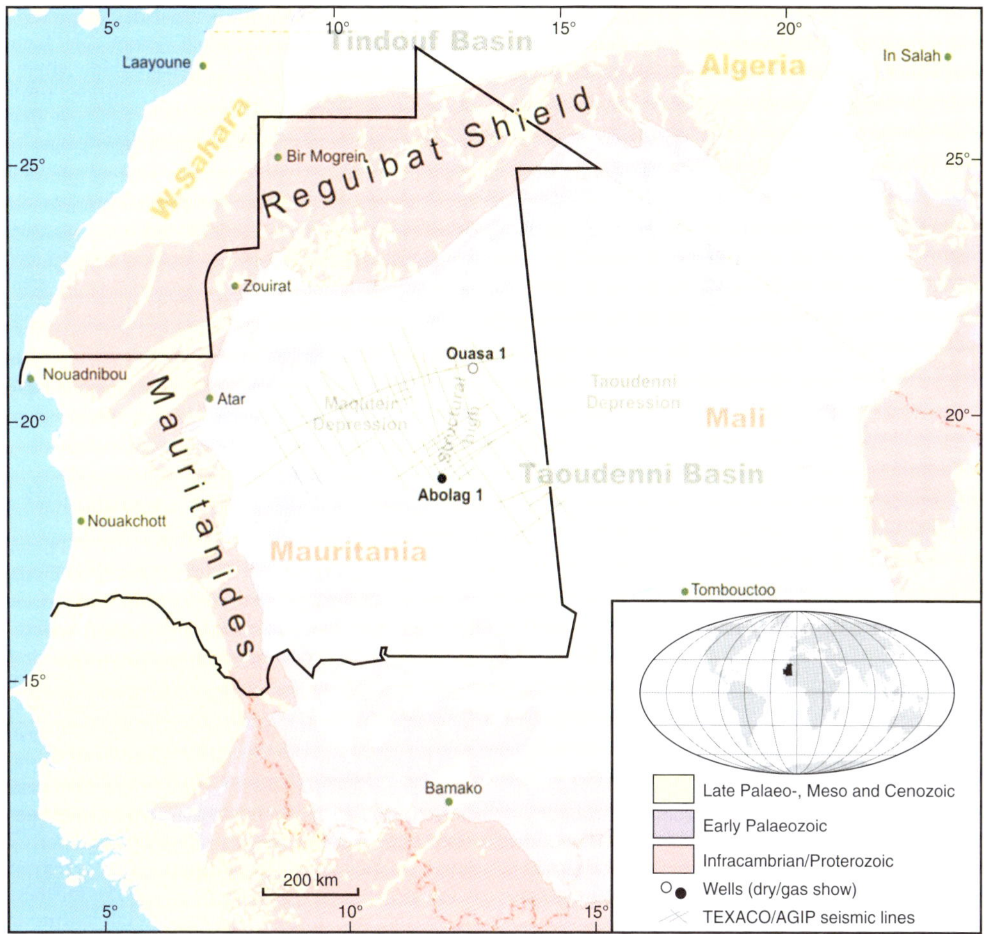

Fig. 7. Locations of outcrops, wells and seismic lines, and the extent of the Taoudenni Basin across Mauritania, Mali and Algeria.

(*c.* 650 Ma: Collins & Pisarevsky 2005) and involved at least seven major tectonic phases up to the final Oligocene–Miocene tectonic event associated with the development of the Red Sea, Gulf of Suez and Gulf of Aqaba rift systems. The effect of these different tectonic events varies greatly across the North African region (Craig *et al.* 2008).

Currently, Infracambrian organic-rich and/or black pyritic sediments in North Africa are known from the Taoudenni Basin, the Anti-Atlas and the Ahnet Basin. Field studies, laboratory analysis and synthesis indicate that Neoproterozoic and Lower Palaeozoic successions in the north-western African basins can be correlated through Morocco, Algeria, Libya and Egypt to Jordan and Oman. With the exception of the uppermost Neoproterozoic–Early Cambrian salt basins of Oman, the petroleum potential of the Neoproterozoic successions in the basins of North Africa and the Middle East is poorly known (Craig *et al.* 2008).

The Neoproterozoic–Early Cambrian plays in most of these basins are at the frontier stage of exploration with limited seismic data and very few wells. In Mauritania (Fig. 7), only two-dimensional (2D) seismic surveys, totalling 6268 km, have been acquired in the Mauritanian sector of the Taoudenni Basin, and only two wells have been drilled: Abolag-1 (Texaco: 1974) and Ouasa-1 (Agip: 1974). Texaco completed the Abolag-1 well in April 1974, and recorded gas shows at depths between 2300 and 3000 m. A drillstem test recorded the equivalent of 480 Mcfd from Infracambrian limestones at about 3000 m depth. In October 1974 Agip completed the Ouasa-1 well, but failed to reach the Infracambrian objective tested by Texaco. It is likely that Ouasa-1 did not test a valid closure. However, despite the Abolag-1 gas discovery, no further exploration has occurred since 1974 (Kolonic *et al.* 2004). New palynological and geochemical analyses on core and cutting samples from the Abolag-1 well indicate an age for the gas-bearing sequence ranging from Tonian to Early Cryogenian, with low organic richness and high maturity within the gas condensate window (Kolonic *et al.* 2004). A Tonian–Early Cryogenian age for this sequence is consistent with available radiometric dates of approximately 750–1000 Ma for the broadly equivalent carbonates of the Atar Group that crop out on the northern margin of the Taoudenni Basin (Deynoux 1980; Clauer & Deynoux 1987). In Morocco, gas shows in the AZ-1 well reported from a similar Infracambrian limestone succession on the northern margin of the Tindouf Basin, indicate the possible existence of another viable petroleum system in the Infracambrian Tata and Taroudant groups in this basin (Geiger *et al.* 2004). The Infracambrian–earliest Palaeozoic petroleum potential of North African basins is discussed in detail by Lüning *et al.* (2009).

Indian and Pakistani basins

The Bikaner–Nagaur Basin of India and its extension across the Punjab Platform of Pakistan contains a proven Infracambrian petroleum system (Fig. 8). Heavy asphaltic oil was first produced from fractured dolomite within the Infracambrian Salt Range Series at Karampur-1 on the Punjab Platform in 1959; and the first major heavy oil discovery was made at Baghewala-1 in the Bikaner–Nagaur Basin in 1991. Only some 7 bbl of 17.6° API gravity oil was originally produced from sandstone reservoirs (1103–1117 m) within the Infracambrian Jodhpur Formation in Baghewala-1. However, good shows were also reported from many other wells in the region including Bijnot-1 on the Punjab Platform, and Kalrewala-1 and Tavriwala-1 in the Bikaner–Nagaur Basin (Fig. 8). The geochemistry of the crude oil obtained from Baghewala-1 and Karampur-1 is very similar to that of oils and source rocks from the Neoproterozoic–Early Cambrian Huqf Supergroup of Oman, thus providing both strong evidence that viable Infracambrian sourced petroleum systems are present in the Bikaner–Nagaur Basin and Punjab Platform, and the rationale to further explore these basins (Peters *et al.* 1995; Sheikh *et al.* 2003).

The Baghewala-1 oil is non-biodegraded, heavy, sulphur-rich and considered to be sourced from organic-rich laminated dolomites within the Neoproterozoic–Early Cambrian Bilara Formation, based on the source and age-diagnostic biomarkers (Peters *et al.* 1995). Regionally, there are two different source rock types within the Neoproterozoic–Early Cambrian petroleum system. The 'laminated dolomitic beds' produce heavy, high sulphur oil (17°–25° API) during early maturation, while associated 'oil shales' produce low sulphur, light oil (42°–50° API), but require higher levels of maturation for oil expulsion. The light oil phase is more dominant in the Punjab and Potwar regions of Pakistan, but there are also unconfirmed rumours that light oil has been encountered in at least one well in the Indian portion of the Bikaner–Nagaur Basin. The reported reserves for the Baghewala Field are 628 Mbbl of oil; with the oil contained in four separate reservoirs. The oldest reservoir consists of sandstones within the Ediacaran Jodhpur Group, with a porosity of 16–25% and oil saturation ranging from 65 to 80%. The Early Cambrian Upper Carbonate Formation dolostone forms the youngest reservoir with porosities ranging from 7 to 15% (Peters *et al.* 1995; Sheikh *et al.* 2003).

The thick evaporite sequences of the Punjab Platform, the Bikaner–Nagaur Basin, and the Ghaba, Fahud and South Oman salt basins were deposited in a series of basins that formed during a period of latest Neoproterozoic–earliest Cambrian rifting and/or transtension (Pollastro 1999) from India and Pakistan across the Arabian Shield to central Iran. Plate-tectonic reconstruction suggests that these restricted marine-evaporite rift basins formed in close proximity on a broad carbonate shelf along the northern margin of the Gondwana supercontinent (Gorin *et al.* 1982; Lawyer & Scotese 1987; Husseini & Husseini 1990; McKerrow *et al.* 1992; Collins & Pisarevsky 2005). Of these salt basins, the Ghaba and Fahud Salt basins of Oman currently contain the most prolific Infracambrian petroleum systems. The Infracambrian plays on the Punjab Platform and in the Bikaner–Nagaur Basin are still very underexplored. However, the similarity of their depositional and tectonic history to the salt basins of Oman and the geochemical similarity of the oils encountered with those from the Neoproterozoic–Early Cambrian Huqf Supergroup, which is believed to be the source of about 12 bbbl of oil and an undetermined amount of gas in Neoproterozoic and overlying Palaeozoic and Mesozoic reservoirs in Oman (Grantham *et al.* 1987; Fritz 1989; Edgell 1991), provide a strong incentive for further systematic exploration for Neoproterozoic and Early Cambrian petroleum systems in the Infracambrian basins of India and Pakistan.

North American basins

Several areas in North America contain proven or potential Precambrian petroleum systems (Fig. 9). These include the Midcontinent Rift system, the Grand Canyon area in northern Arizona, the Unita

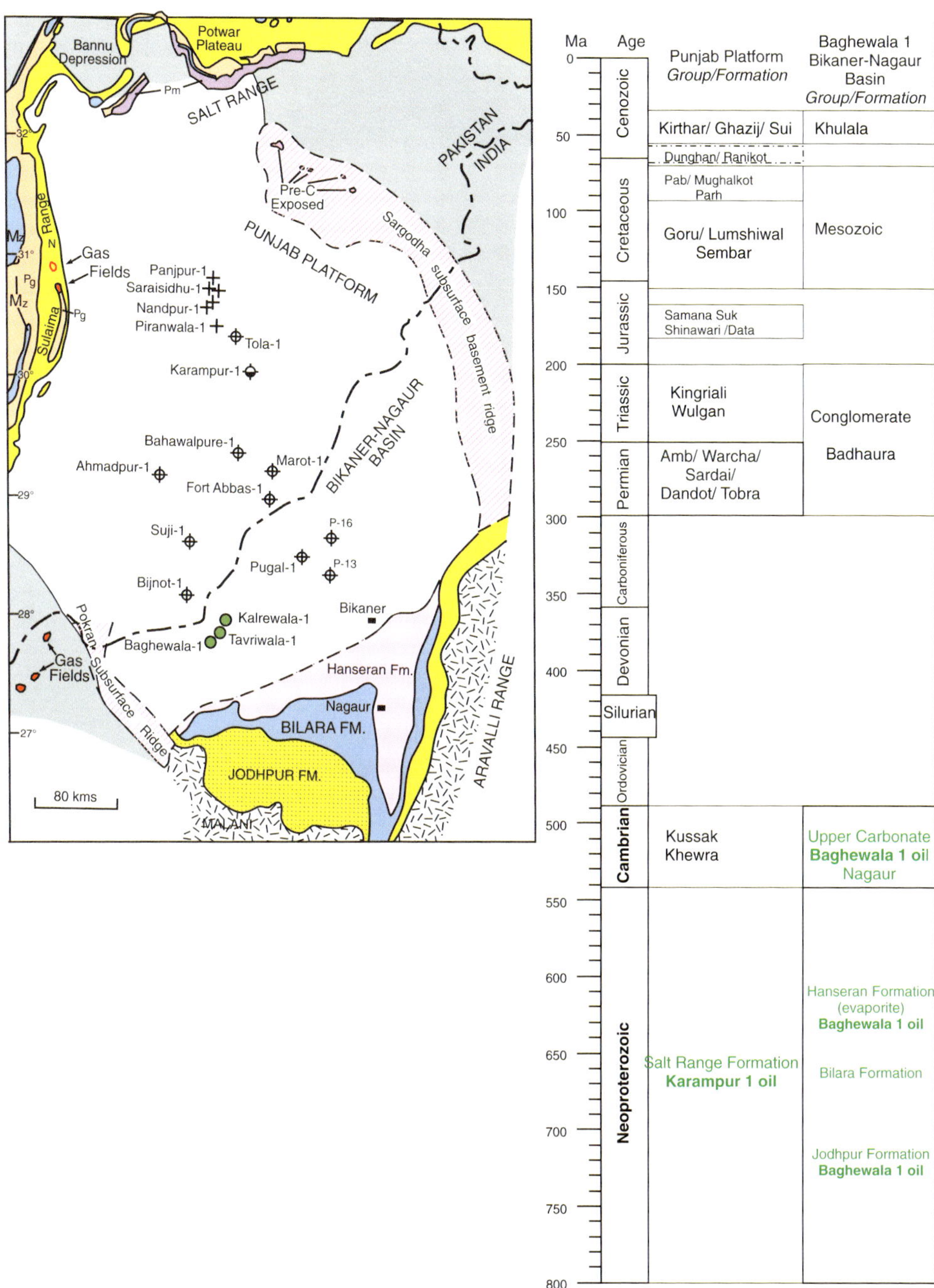

Fig. 8. Location and stratigraphy of the Bikaner–Nagaur Basin of India and the Punjab Platform of Pakistan.

Mountains and the Rocky Mountain overthrust belt of northwestern Montana. Of these, the best producing potential appears to be in the Meso-Neoproterozoic Oronto Group of the Midcontinent Rift system and in the Neoproterozoic Chuar Group of the Grand Canyon (Fig. 10). However, the potential of source rocks in these regions is poorly known, but they may have generated and expelled HCs that were subsequently trapped in Neoproterozoic–Palaeozoic reservoirs (Palacas 1997). The 1.1 Ga

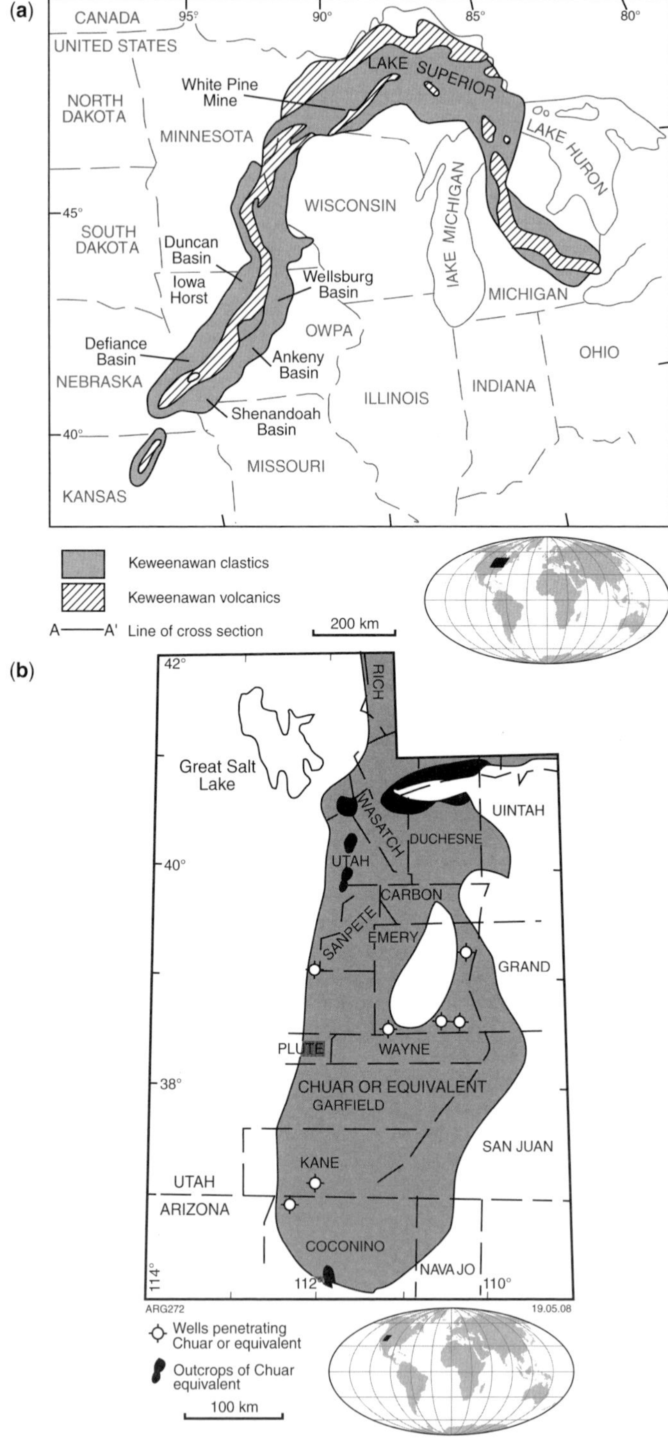

Fig. 9. Location of the Mid-Continent Rift and the Grand Canyon. Modified from Palacas (1997).

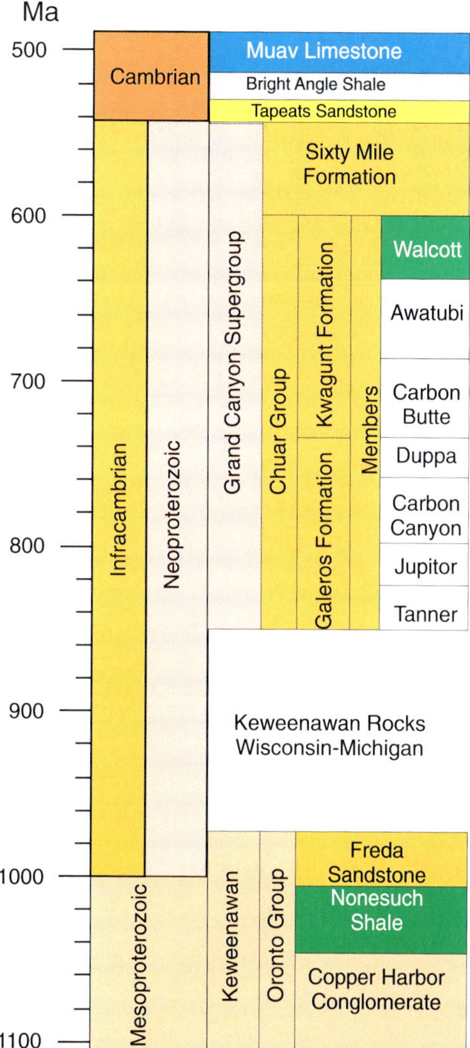

Fig. 10. Stratigraphic position of the Nonesuch Formation (Mid-Continent Rift) and the Chuar Group (Grand Canyon). Geographic positions are shown in Figure 9.

Midcontinent Rift system is a major structure in the North American Craton that is infilled with up to 15 km of dominantly mafic volcanic rocks, overlain by up to 10 km of clastic sedimentary rocks (Green 1983; Van Schmus & Hinze 1985; Mauk & Burruss 2002). These rocks crop out in the Lake Superior region of Michigan, northern Wisconsin and Minnesota, and extend in the subsurface through Minnesota, Iowa, Nebraska and into northeastern Kansas. The 1500 km-long Midcontinent Rift system is a failed rift, characterized by a series of asymmetric basins filled with a clastic succession up to 10 km thick (Anderson 1989) belonging to the Meso-Neoproterozoic Keweenawan Supergroup, which is subdivided into a lower Oronto Group and an upper Bayfield Group. The Copper Harbor Conglomerate and the Nonesuch and Freda formations together form the Oronto Group (Daniels 1982; Elmore *et al.* 1989).

In the Lake Superior area of the Midcontinent Rift system, indigenous oil seeps are present within thin intervals of silty shales in the Nonesuch Formation (1.05 Ga). These shales contain up to 3.0% TOC and are marginally mature to mature with respect to the zone of oil generation. Petroleum seeps and shows from the White Pine mine have been typed to the Nonesuch Formation (Pratt *et al.* 1991; Mauk & Hieshima 1992). The Neoproterozoic Chuar Group of the eastern Grand Canyon, Arizona, contains the organically richest, thermally mature source rocks with TOC contents of up to 10%. The generation potential of these source rocks is locally as high as 16 mg of HC per gram of rock and extractable organic matter contents of up to 4000 ppm. These organic-rich units of the Chuar Group may be the source of economic accumulations of gas and oil in Neoproterozoic–Palaeozoic reservoirs in northern Arizona and southern Utah (Palacas 1997).

The high risks of exploring Precambrian basins are well known, and hence careful evaluation of available geological, geophysical and geochemical data is essential in order to identify the most promising areas for the accumulation of HCs sourced from the Mesoproterozoic Nonesuch Formation of the Midcontinent Rift system and the Neoproterozoic Chuar Group of the Grand Canyon areas (Fig. 10).

Australian basins

Australia was the part of the Rodinia supercontinent until about 750 Ma when rifting occurred along both margins of the continent as Rodinia broke up. Australia then collided with India to form a major component of Gondwana as it finally amalgamated at around 530 Ma (Collins & Pisarevsky 2005). The assembling of Gondwana towards the end of the Precambrian and its Precambrian sedimentary record is similar to that of other Precambrian basins around the world. Three potential Proterozoic petroleum supersystems are recognized in Australia: the Palaeo-Mesoproterozoic McArthur, the Mesoproterozoic Urapungan and the Neoproterozoic Centralian systems (Bradshaw *et al.* 1994). As yet, there is no HC production from these Proterozoic sequences, but organic-rich rocks and significant oil and gas shows confirm the existence of Proterozoic petroleum systems. There is a significant gas accumulation at Dingo in the Amadeus Basin

(Ozimic *et al.* 1986; Bradshaw *et al.* 1994). The Palaeo-Mesoproterozoic McArthur Basin, the Neoproterozoic Centralian Superbasin and the Adelaide Rift Complex have been studied for scientific and commercial purposes since 1960 (Fig. 11). The Centralian Superbasin includes the Neoproterozoic fill (840–545 Ma) of the Amadeus, Georgina, Ngalia and Officer basins, which developed as a single depositional system but separated into different structural units mainly during the Petermann (600–540 Ma) and Alice Springs (400–300 Ma) orogenic events (Walter *et al.* 1992, 1995; Walter & Gorter 1993). The McArthur Basin covers an area of about 200 000 km^2 and contains four lithostratigraphic sequences (Tawallah, McArthur, Nathan and Roper groups) separated by regional unconformities (Plumb *et al.* 1980; Crick *et al.* 1988). The Tawallah group is mostly overmature. The McArthur and Nathan Groups consist mainly of evaporitic and stromatolitic cherty dolostones interbedded with dolomitic siltstone and shale that were deposited in a variety of marginal marine, lacustrine and fluvial environments. The Roper Group consists of quartz arenite, siltstone and shale deposited in a stable marine setting. The most organic-rich source rocks in the Palaeo-Mesoproterozoic McArthur Basin are reported from the lacustrine Barney Creek Formation (*c.* 1640 Ma) in the McArthur Group and from the marine Velkerri Formation (*c.* 1440 Ma) in the Roper Group (Jackson *et al.* 1986; Womer 1986; Rawlings 1999). Source rocks with comparable thickness and potential to Phanerozoic source rocks are present in these sequences with TOC of up to 7% containing type I and II kerogen, with thermal maturities ranging from overmature to marginal mature (Crick *et al.* 1988). Weeping oil and gas blowouts occurred in several shallow wells drilled in the McArthur Basin for lead–zinc exploration in the mid-1970s. Two different oil types have been observed: a heavily biodegraded oil containing associated galena, sphalerite and barite, which was probably generated and migrated during the phase of lead–zinc mineralization; and a 'golden honey colour', very volatile oil generated during the later tectonic events (Wilkins 2007). The Roper Group of McArthur Basin was one of the oldest sequences currently explored for HCs in Australia due to the

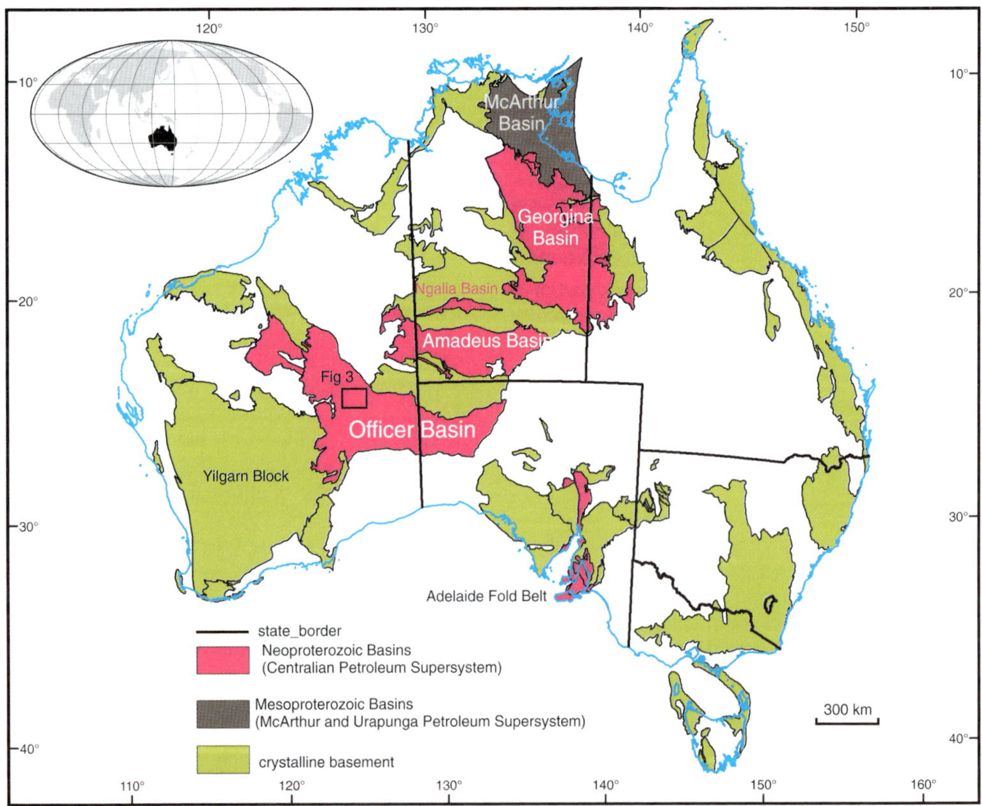

Fig. 11. Location of Australian basins containing prospective Mesoproterozoic and Neoproterozoic sequences.

presence of extensive oil and gas shows reported in stratigraphic and petroleum exploration wells drilled during the 1980s (Jackson *et al.* 1988).

The Neoproterozoic Centralian Superbasin contains four supersequences (Walter *et al.* 1995; Grey *et al.* 2005). Deposition of Supersequence 1 (early Cryogenian: Fig. 12) commenced with a thick sheet of sand, overlain by dolomites, limestones, evaporates and fine siliciclastics, which reach a total thickness of more than 3000 m in the Officer Basin (Yowalga area), as discussed later. This sequence is considered to be the most prospective for oil and gas exploration. The base of Supersequence 2 (mid to late Cryogenian: Fig. 12) is defined by the Sturtian glaciation deposits (Cryogenian), which are overlain by widespread igneous sills and by shales with interbedded carbonates and sandstones. This supersequence is mainly developed within the Amadeus Basin. The base of Supersequence 3 (late Cryogenian: Fig. 12) is defined by the Marinoan glaciation deposits, which are continent-wide (Preiss & Forbes 1981), and may be coeval with extensive tillites found over much of the globe (Knoll & Walter 1992; Hoffman & Schrag 2002) and the products of a period of global glaciation. Comparable glacigenic successions are reported from the Sinian of China (Yin 1985), the latest Proterozoic of Svalbard (Knoll 1992), the Vendian of Siberia (Moczydlowska *et al.* 1993) and the Cryogenian of Oman (Allen 2007).

The basal part of Supersequence 4 (Ediacaran: Fig. 12) contains an 'Ediacara fauna' and the upper part is Cambrian in age. This supersequence consists predominantly of sandstone, except within the Adelaide Rift complex where there are extensive shales and marls (the Brachina, Bunyeroo and Wonoka formations). A prominent central ridge existed in the Centralian Superbasin at this time (Paterson Province–Musgrave Block) and provided the main provenance for sand deposited on deltas building out into the northern basins. The northernmost part of the Centralian Superbasin remained marine at this time, while most of the southern region may have been emergent.

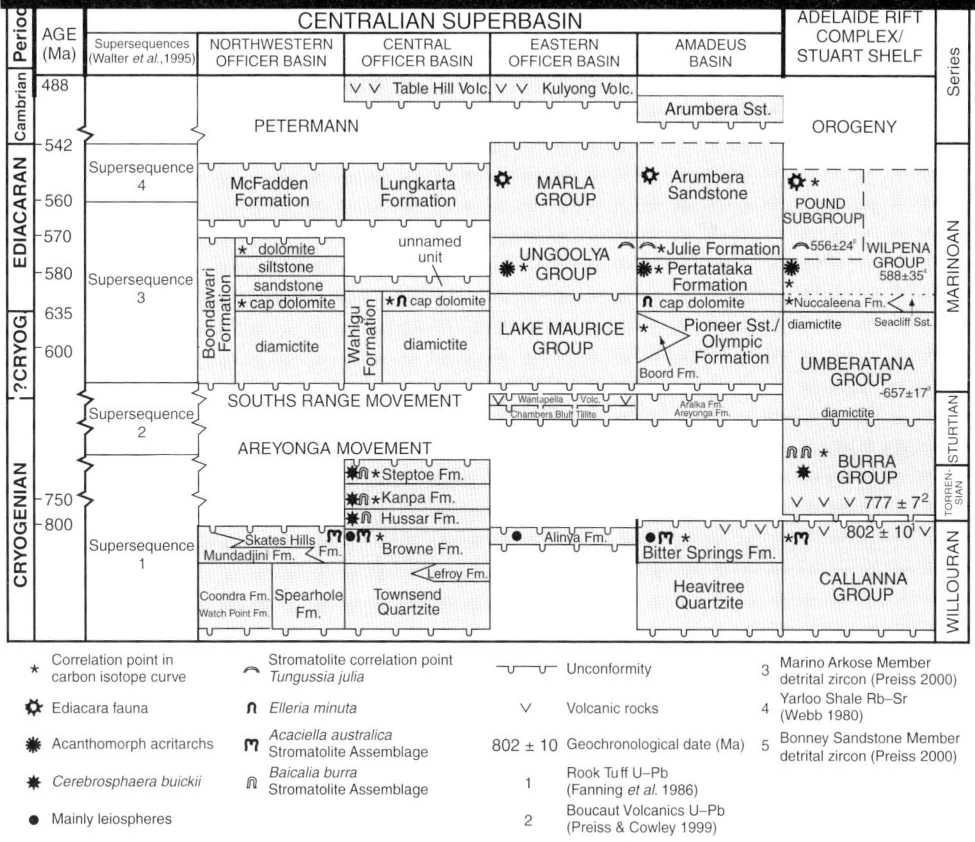

Fig. 12. Generalized stratigraphy of the Centralian Superbasin of Australia (after Grey *et al.* 2005).

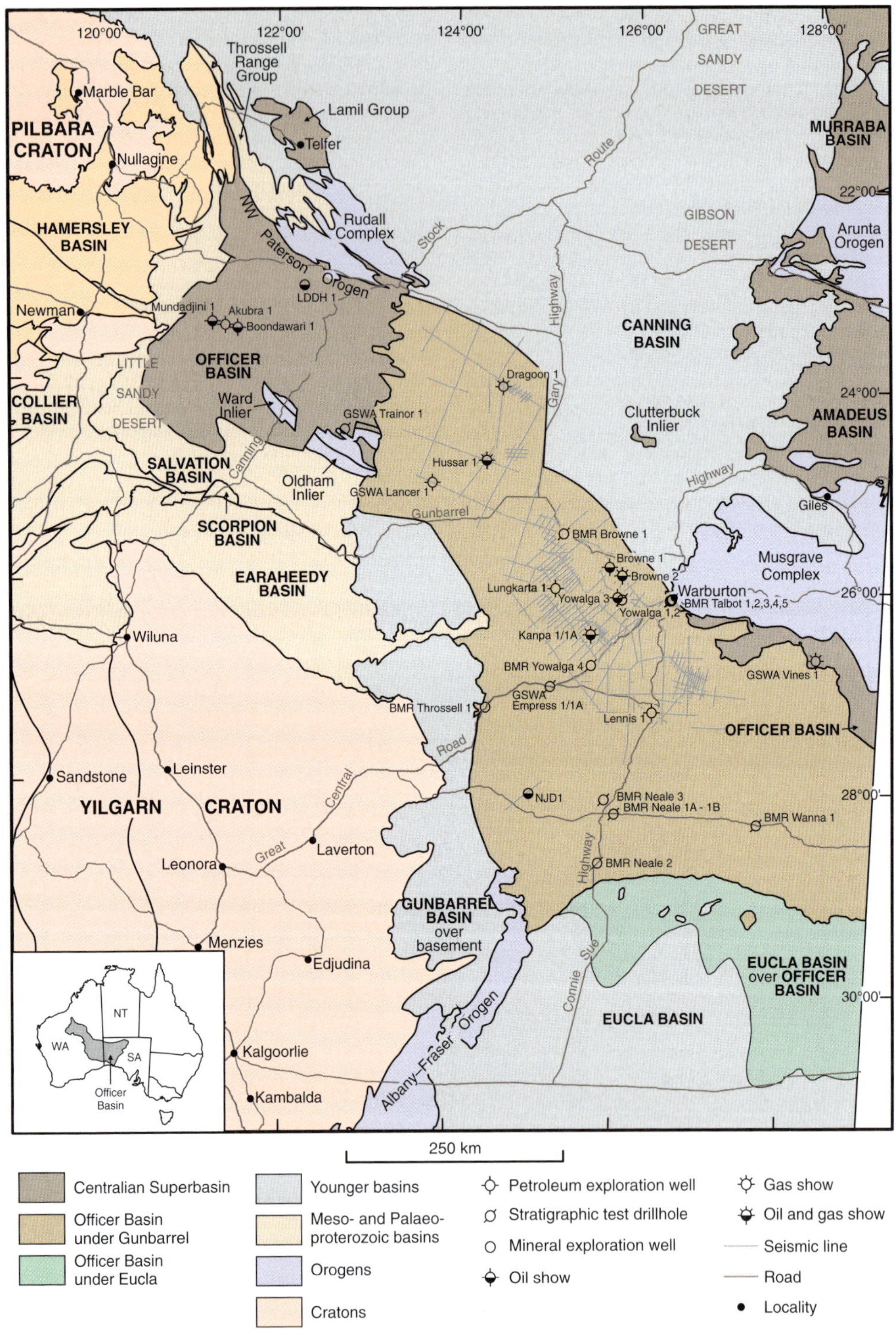

Fig. 13. Structural units, wells, seismic survey and petroleum shows in the Officer Basin, Western Australia.

Officer Basin of Western Australia petroleum geology

The petroleum systems of the Neoproterozoic Officer Basin of Western Australia are reviewed here as a case study. The Officer Basin is elongate with a NW–SE trend, and contains over 8000 m of Neoproterozoic strata, overlain by lower Palaeozoic rocks of the Gunbarrel Basin. Subsurface geochemical and geological data comprise 16 wells and 19 seismic surveys undertaken during three phases of petroleum exploration in the late 1960s (five wells) early 1980s (five wells) and the late 1990s (six wells) (Fig. 13). Yowalga-3 is the deepest

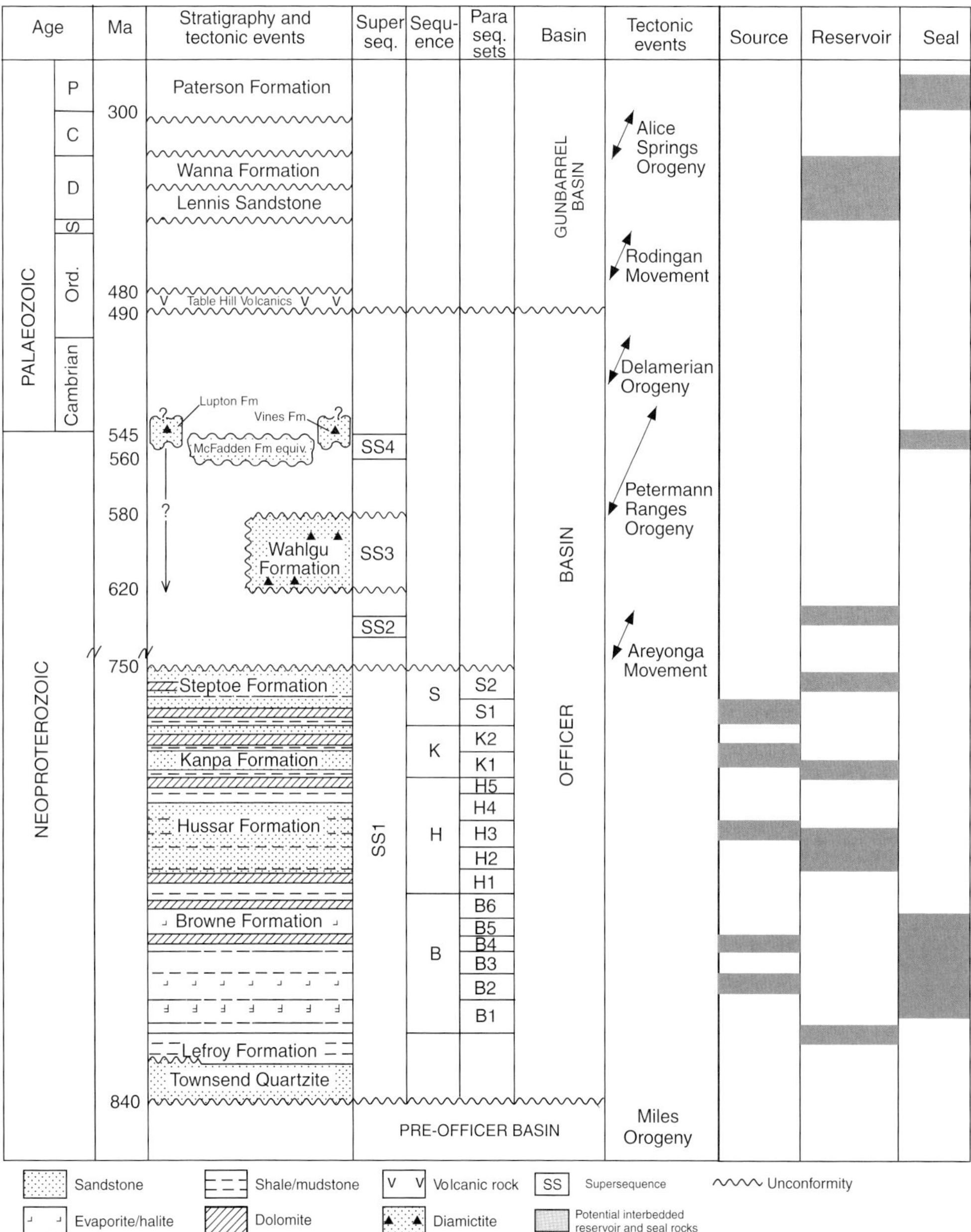

Fig. 14. Generalized time and seismic stratigraphy, tectonic events, source, reservoir and seal rocks for the Officer Basin, Western Australia.

well drilled in the basin. Chronostratigraphy, tectonic events, and the location of source, reservoir and seal rocks are summarized in Figure 14. Minor oil shows and numerous bitumen occurrences have been reported in many of the petroleum exploration wells drilled in the basin (Table 1), confirming the existence of a Neoproterozoic petroleum system. However, given the sparse well control, the ultimate petroleum potential of the vast under-explored Officer Basin remains unverified, but may be significant. Three unconformity bounded sedimentary successions exist throughout most of the Officer Basin in Western Australia (supersequences 1, 3 and 4) and these can be correlated with key tectonic episodes. The Areyonga Movement (Fig. 14) appears to be responsible for the larger structures in the Officer Basin, and separates Supersequence 1 from Supersequence 3. Structural and stratigraphic variations within the overlying supersequences 3 and 4 are attributed to later deformation related to the Petermann Orogeny (Wade *et al.* 2005). The Neoproterozoic traps are associated with faults, unconformities, facies changes and salt tectonics. Episodic salt movement may have resulted in the formation of traps within the younger successions. These younger traps occur throughout the basin, but remain untested. There is little seismic control on the distribution and potential facies of the lower units of Supersequence 1, namely the Townsend Quartzite and Lefroy Formation. However, the overlying units within Supersequence 1 are better understood, and consist of conformable and laterally correlative genetic parasequence units bounded by flooding surfaces. These parasequence units comprise the Browne (B 1–6), Hussar (H 1–5), Kanpa (K 1–2) and Steptoe formations (S 1–2). In most seismic lines Supersequence 1 is characterized by continuous parallel reflectors that are traceable across most of the basin, except where truncated by a younger unconformity (Fig. 15). The presence of mobile salt within the Officer Basin has resulted in a wide range of possible trap configurations. Warren (1989) defined many of the possible salt-related trap styles based on structure and porosity. Many of these traps are related to thrust faults that typically initiated within the Browne Formation salt units, and generate drag rollover structures within the overlying strata. Salt pillows also generate four-way dip closures in the suprasalt section. Unconformity-related traps may have developed where Supersequence 1 strata have been tilted and significantly eroded adjacent to salt injection features and along the basin margins, and extensive karst secondary porosity may have been created within carbonates through leaching of soluble components such as halite, anhydrite and carbonate. Stratigraphic traps may be related to differential subsidence that has resulted in downlap and onlap of units, creating pinch-outs. Additional traps related to facies changes may have developed where halite and anhydrite have been formed in desiccation zones, plugging the porosity of the sediments either during, or after, deposition. Such traps also formed early with respect to charge. Frequent emergence is well documented in Neoproterozoic successions in the Officer Basin, and channels filled with high-energy, reservoir-quality sediments, sealed by the subsequent transgressive shale, are identified as potential exploration targets. The development of structural traps of significant size in the Officer Basin of Western Australia is demonstrated in

Table 1. *Hydrocarbon shows, Officer Basin, Western Australia*

Well	Quality of show	Formation	Formation age
Boondawari-1	40% oil fluorescence in core	Spearhole Formation	Neoproterozoic
Browne-1	Gas cut mud, cut fluorescence, trace oil in core	Paterson Formation Browne Formation	Permian Neoproterozoic
Browne-2	Gas cut mud, cut fluorescence, trace oil in core	Paterson Formation	Permian
Dragoon-1	Mud gas to 10% methane equivalent, including hydrocarbons up to pentane	Browne Formation	Neoproterozoic
Hussar-1	Mud gas readings to 1000 ppm. Possible gas blow On air lift. Trip gas to 4.6% total gas. 72% oil saturation from log analysis	Kanpa Formation Hussar Formation	Neoproterozoic
Kanpa-1A	Dull yellow-orange sample fluorescence, light yellow-white cut fluorescence, brown oil stains in sandstone and dolomite cuttings	Kanpa Formation	Neoproterozoic
LDDH-1	Bitumen in core	Tarcunyah Group	Neoproterozoic
Mundadjini-1	10% oil fluorescence in core	Spearhole Formation	Neoproterozoic
NJD-1	Bleeding oil and bitumen in core	Neale Formation	Neoproterozoic
OD-23	Bleeding oil and bitumen in core	Scorpion Group	Mesoproterozoic
Vines-1	Total gas peaks 25 times background	?Supersequence 1	Neoproterozoic

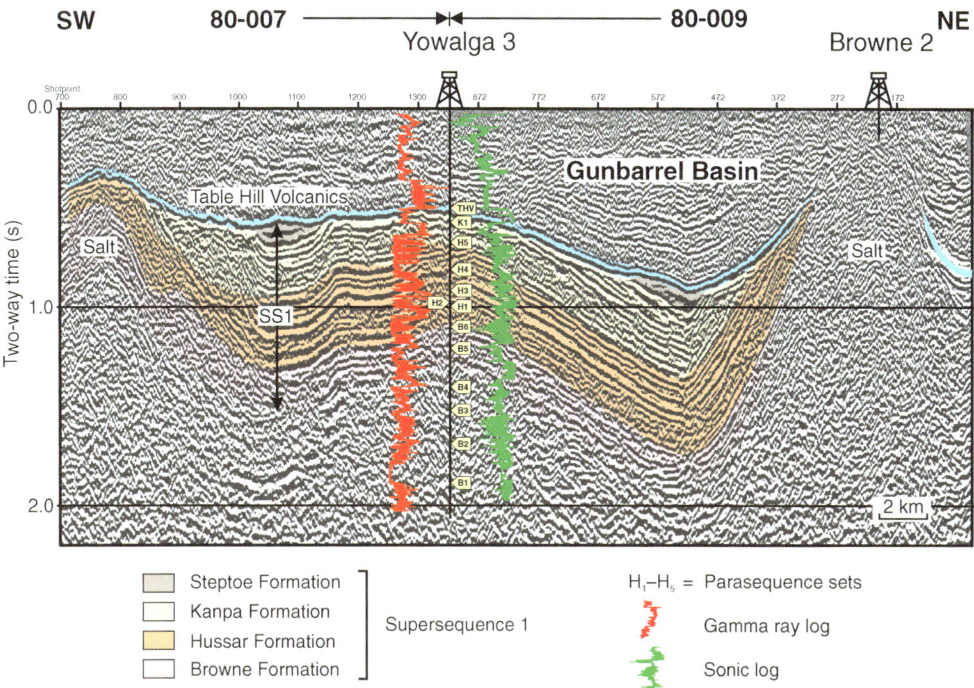

Fig. 15. Composite seismic section of Yowalga-3 and Browne-2 wells, and erosion of Supersequence 1 strata between salt emplacements in the Yowalga area.

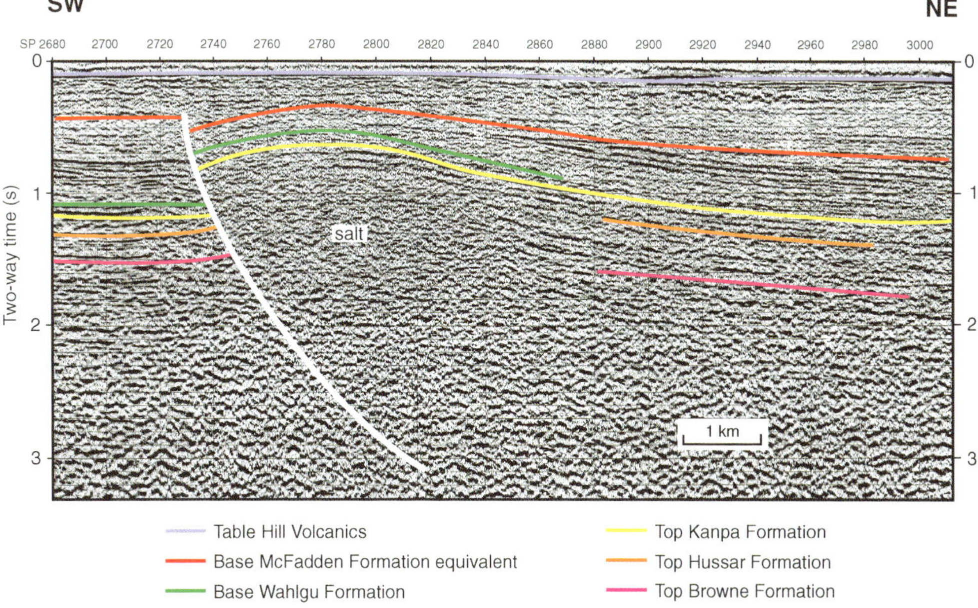

Fig. 16. Seismic line N 83-11 showing an angular unconformity between the Wahlgu and the McFadden Formation equivalent. The Wahlgu Formation is deeply eroded in this location. High-angle reverse faulting associated with salt emplacement resulted in minor displacement of the post-Supersequence 3 unconformity.

Table 2. *TOC and Rock-Eval data for samples with TOC content over 0.90%*

Well	Depth (m)	Sample	TOC (%)	T_{max} (°C)	S_1	S_2	S_3	$S_1 + S_2$	PI	HI	OI
Empress-1A	587.9	Core	1.38	418	0.43	6.20	0.11	6.63	0.06	449	8
Empress-1A	588.4	Core	1.52	413	0.46	6.44	0.15	6.90	0.07	424	10
Empress-1A	768.2	Core	0.93	421	0.66	4.02	0.44	4.68	0.14	432	47
Kanpa-1A	3412.2	SWC	1.41	436	1.78	4.83	0.75	6.61	0.27	343	53
NJD-1	327.5	Core	6.64	430	0.7	23.47	1.04	24.17	0.03	353	16
NJD-1	328.5	–	3.18	432	0.26	15.83	0.39	16.09	0.02	498	12
NJD-1	400.8	–	1.22	443	0.74	1.27	0.10	2.01	0.37	104	8
NJD-1	502.6	–	21.50	442*	32.11	104.00	1.94	136.11	0.24	484	9
NJD-1	502.7	–	1.10	430*	1.06	4.76	0.45	5.82	0.18	433	41
Normandy LDDH-1	222.8	Core	1.61	450	0.42	2.11	0.32	2.53	0.17	131	20
Normandy LDDH-1	529.6	Core	2.05	471	1.05	1.40	0.07	2.45	0.43	68	3
Throssell-1	184.9	Core	1.37	428	0.35	1.06	0.31	1.41	0.25	77	23
Yowalga-3	1484.0	Cuttings	1.23	421	0.49	3.00	0.47	3.49	0.014	244	38

Notes: TOC, total organic carbon; T_{max}, temperature of maximum pyrolytic yield (S_2); S_1, existing hydrocarbons (HC); S_2, pyrolytic yield (HC); S_3, organic carbon dioxide; $S_1 + S_2$, potential yield; PI, production index; HI, hydrogen index; OI, oxygen index; SWC, side wall core.
*Bitumen-rich samples.

Figure 16. Basin modelling studies indicate that HC generation occurred in the Officer Basin in three main phases, during the latest Neoproterozoic, the Cambrian and the Late Palaeozoic. These phases of oil generation correlate well with initial migration and trap formation during the Areyonga Movement, and late migration and trap formation during the later deformations (Apak *et al.* 2002; Ghori 2002).

Petroleum geochemistry

The available geochemical data indicate the presence of thin source rocks with fair–excellent HC generating potential within the Neoproterozoic successions in the Browne-1 and Browne-2, Empress-1/1A, Hussar-1, Kanpa-1A, LDDH-1, NJD-1, Throssell-1 and Yowalga-3 wells and mineral drill holes. The organic-rich source beds occur in the B2 and B4 parasequences of the Browne Formation, the H3 parasequence of the Hussar Formation, the K1 parasequence of the Kanpa Formation and the S1 parasequence of the Steptoe Formation. Pyrolysis gas chromatography and extract analyses indicate that most of the organic matter in the source beds is oil- and gas-generating type II kerogen (Table 2 and Fig. 17).

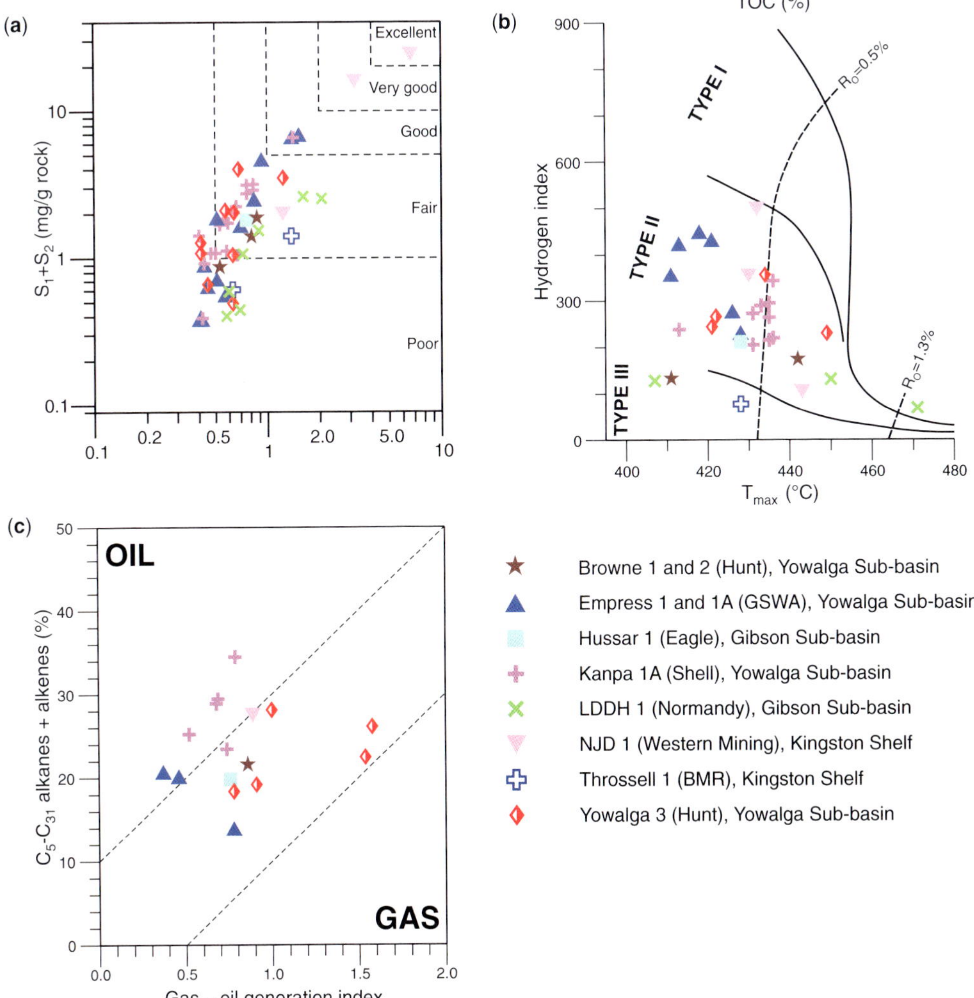

Fig. 17. Source-rock characterization of the Neoproterozoic Officer Basin: (**a**) petroleum-generating potential as a function of organic richness v. potential yield, for samples interpreted as reliable; (**b**) type of kerogen as a function of T_{max} v. hydrogen index, from Rock-Eval pyrolysis; and (**c**) type of kerogen as a function of oil proneness (C5–C31 alkanes + alkenes) v. gas–oil generation index (C1–C5)/C6+) from pyrolysis–gas chromatography.

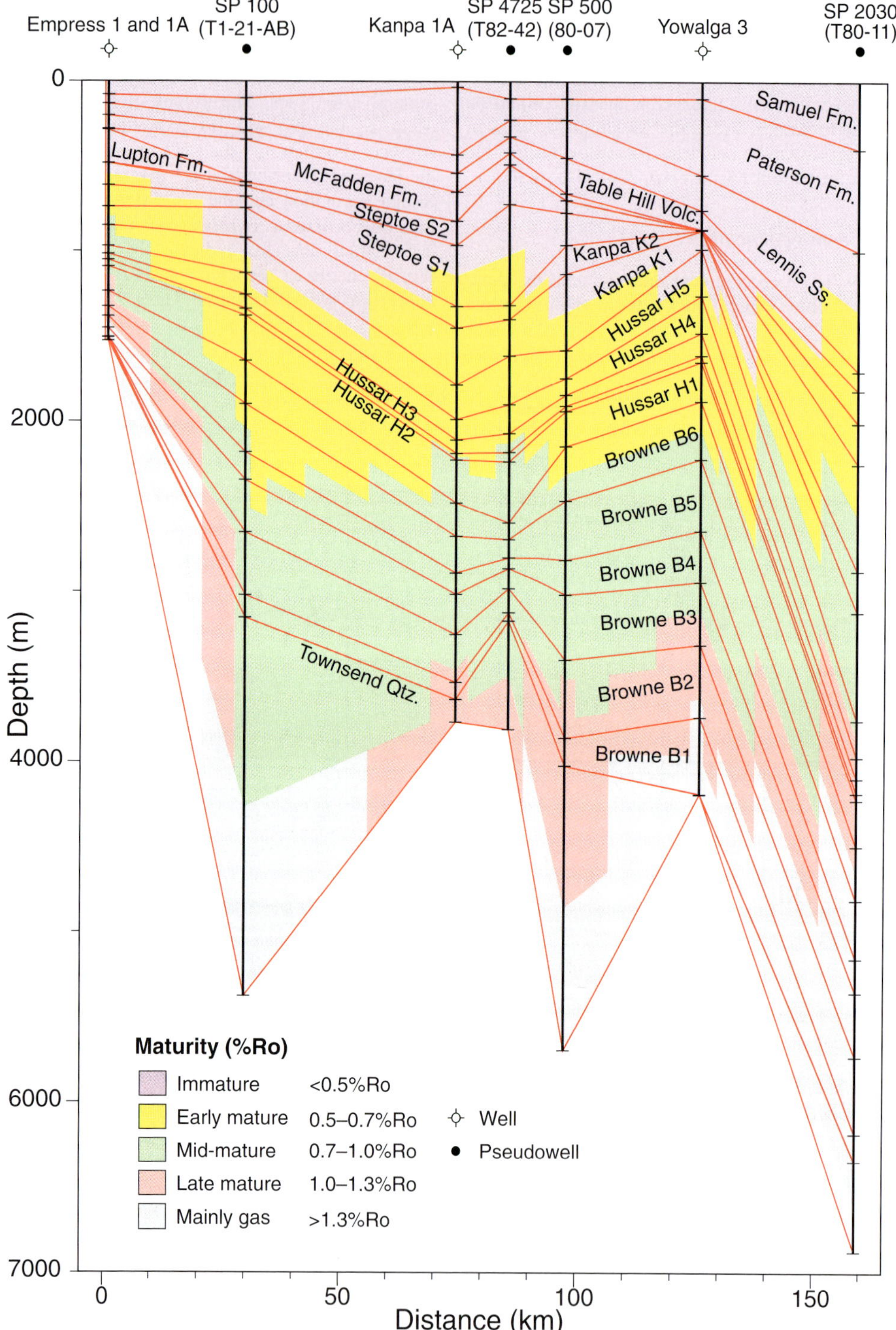

Fig. 18. Present-day maturity across the Yowalga area, from Empress-1/1A in the south to SP 2030 on seismic line T80-11 in the north, based on 2D modelling.

One- and two-dimensional basin modelling studies suggest that the optimum maturity for HC generation within the Browne Formation was reached early in the basin history and that most of the HC generation potential of this formation was exhausted during the Neoproterozoic. However, the Hussar, Kanpa and Steptoe formations were not buried so deeply, and the generation of HCs from these units extends into the Phanerozoic. In parts of the Gibson area (LDDH-1) and over the salt diapirs (Dragoon-1, Browne-1 and Browne-2) the oil-generating window is comparatively shallow, but the oil window is very deep in the Yowalga area (Yowalga-3). The extent and the effects of Mesozoic and Cenozoic tectonic events are poorly understood because the preserved post-Alice Spring Orogeny section is thin and irregularly distributed. Present-day maturity across the Yowalga area is shown in Figure 18. The vast area covered by the western Officer Basin is very poorly explored, and the sparse well control precludes a comprehensive assessment of the source rock potential of the Neoproterozoic succession. The existence of volumetrically significant source rocks of a commercially viable petroleum system cannot be verified with the available dataset. However, the fact that thin, but good-quality, source rocks have been identified in the Browne, Hussar, Kanpa and Steptoe formations, that a significant part of the Neoproterozoic section is presently within the oil window, and that the succession contains good-quality reservoirs and seals suggests that further exploration of this frontier region is warranted.

Conclusions

Oil and gas accumulations will continue to be found in Proterozoic rocks worldwide where organic-rich source rocks and good-quality reservoirs are present, and where the source rocks are not thermally overmature and/or the presence of effective super-seals allows preservation of early generated hydrocarbons. Many of the worlds 'traditional' proven petroliferous basins are approaching exploration maturity and much of the 'easy oil' has now been found. As a result, significant exploration effort is now being devoted to deep-water and harsh environments worldwide. This review suggests that there is also potential for new discoveries to be made in the Infracambrian (Neoproterozoic–earliest Cambrian age) basins worldwide, including those in Africa, Australia, India-Pakistan and the United States. With further concerted exploration these areas may yet join the already producing Neoproterozoic–early Cambrian basins of China, Russia and the Sultanate of Oman.

References

Al-Lazki, A. I. 2003. *Crustal and Upper Mantle Structure of Oman and the Northern Middle East.* Unpublished PhD thesis, Cornel University.

Al-Lazki, A. I., Seber, D. & Sandvol, E. 2002. A crustal transect across the Oman Mountains on the eastern margin of Arabia. *GeoArabia*, **7**, 47–78.

Allen, P. A. 2007. The Huqf Supergroup of Oman: basin development and context for Neoproterozoic glaciation. *Earth Science Reviews*, **84**, 139–185.

Al-Marjeby, A. & Nash, D. 1986. A summary of the geology and oil habitat of the Eastern Flank Hydrocarbon Province of south Oman. *Marine and Petroleum Geology*, **3**, 306–314.

Alsharhan, A. S. & Kendall, C. G. St. C. 1986. Precambrian to Jurassic rocks of Arabian Gulf and adjacent area: their facies, depositional setting, and hydrocarbon habitat. *AAPG Bulletin*, **70**, 997–1002.

Anderson, R. R. 1989. Gravity and magnetic modeling of central segment of Mid-Continent Rift in Iowa – New insights into its stratigraphy, structure, and geologic history abstract. *AAPG Bulletin*, **73**, 1043–1043.

Amthor, J. E. 2006. Precambrian hydrocarbon systems of Oman. *In*: *International Conference on Global Infracambrian Hydrocarbon Systems and the Emerging Potential in North Africa, 29–30 November 2006.* Geological Society, London, 9–10.

Apak, S. N., Ghori, K. A. R., Carlsen, G. M. & Stevens, M. K. 2002. Basin development with implications for petroleum trap styles of the Neoproterozoic Officer Basin, Western Australia. *In*: Keep, M. & Moss, S. J. (eds) *The Sedimentary Basins of Western Australia, 3. Proceedings of the West Australian Basins Symposium, Perth, Western Australia.* Petroleum Exploration Society of Australia, Perth, 913–927.

Bradshaw, M. T. 1993. Australian petroleum systems. *Petroleum Exploration Society of Australia Journal*, **21**, 43–53.

Bradshaw, M. T., Bradshaw, J. *et al.* 1994. Petroleum systems in West Australian basins. *In*: Purcell, R. G. & Purcell, R. R. (eds) *The Sedimentary Basins of Western Australia. Proceedings of the West Australian Basins Symposium, Perth, Western Australia.* Petroleum Exploration Society of Australia, Perth, 93–118.

Buick, R., Rasmussen, B. & Krapez, B. 1998. Archaean oil: evidence for extensive hydrocarbon generation and migration 2.5–3.5 Ga. *AAPG Bulletin*, **82**, 50–69.

Burlingame, A. I., Haug, P., Belsky, T & Calvin, M. 1965. Occurrence of biogenic steranes and pentacyclic triterpane in Eocene shale (52 million years) and in early Precambrian shale (2.7 billion years): a preliminary report. *Proceedings of the National Academy of Science*, **54**, 1406–1412.

Clarke, J. W. 1985. *Petroleum Geology of East Siberia.* USGS Open-file Report, **85–267**, 123.

Clauer, N. & Deynoux, M. 1987. New information on the probable isotopic age of the Late Proterozoic glaciation in West Africa. *Precambrian Research*, **37**, 89–94.

COLLINS, A. S. & PISAREVSKY, S. A. 2005. Amalgamating eastern Gondwana: the evolution of the Circum-Indian Orogens. *Earth Science Reviews*, **71**, 229–270.

CRAIG, J., RIZZI, C. *ET AL.* 2007. Structural styles and prospectivity in the Precambrian and Palaeozoic hydrocarbon systems of North Africa. *In*: SALEM, M. J., ABURAWI, R. M. & MISALLATI, A. A. (eds) *The Geology of East Libya, Volume IV, Sedimentary Basins of Libya, 3rd Symposium.* Earth Science Society of Libya, Tripoli, Libya, 51–122.

CRICK, I. H., BOREHAM, C. J., COOK, A. C. & POWELL, T. G. 1988. Petroleum geology and geochemistry of Middle Proterozoic McArthur Basin, Northern Australia II: assessment of source rock potential. *AAPG Bulletin*, **72**, 1495–1514.

ELMORE, R. D., MILAVEC, G. J., IMBUS, S. W. & ENGEL, M. H. 1989. The Precambrian Nonesuch Formation of the North American Mid-continent rift, sedimentology and organic geochemical aspects of lacustrine deposition. *Precambrian Research*, **43**, 191–213.

DANIELS, P. A., JR. 1982. Upper Precambrian sedimentary rocks: Oronto Group, Michigan–Wisconsin. *In*: WOLD, R. J. & HINZE, W. J. (eds) *Geology and Tectonics of the Lake Superior Basin*. Geological Society of America Memoir, **156**, 107–133.

DEMAISON, G. J. 1984. The generative basin concept. *In*: DEMAISON, G. J. & MURRIS, R. J. (eds) *Petroleum Geochemistry and Basin Evaluation*. American Association of Petroleum Geologists, Tulsa, OK, 1–14.

DEYNOUX, M. 1980. Les formations glaciaires du Précambrien terminal et de la fin de l'Ordovicien en afrique de l'ouest. *Travaux des laboratoires des sciences de la terre, série B*, **17**, 544.

DOW, W. G. 1974. Application of oil-correlation and source-rock data to exploration in Williston Basin. *AAPG Bulletin*, **58**, 1253–1262.

DUTKIEWICZ, A., RASMUSSEN, B. & BUICK, R. 1998. Oil preservation in fluid inclusions in Archaean sandstones. *Nature*, **395**, 885–888.

EDGELL, H. S. 1991. Proterozoic salt basins of the Persian Gulf area and their role in hydrocarbon generation. *Precambrian Research*, **54**, 1–14.

FRITZ, M. 1989. An old source surfaces in Oman. *Explorer.* American Association of Petroleum Geologists, **10**, 14–15.

GEIGER, M., LÜNING, S. & THUSU, B. 2004. *Infracambrian hydrocarbon potential of Morocco. Scouting Fieldtrip to the Anti-Atlas, September 2004.* Maghreb Petroleum Research Group (MPRG), University College London (unpublished).

GHORI, K. A. R. 2002. Modelling the hydrocarbon generative history of the Officer Basin, Western Australia. *Petroleum Exploration Society of Australia Journal*, **29**, 29–43.

GORIN, G. E., RACZ, L. G. & WALTER, M. R. 1982. Late Precambrian–Cambrian sediments of Huqf Group, Sultanate of Oman. *AAPG Bulletin*, **66**, 2609–2627.

GRADSTEIN, F. M., OGG, J. G., SMITH, A. G., BLEEKER, W. & LUCAS, L. J. 2004. A new geologic time scale, with special reference to Precambrian and Neogene. *Episodes*, **27**, 83–100.

GRANTHAM, P. J., LIJMBACH, G. W. M., POSTHUMA, J., HUGHES CLARK, M. W. & WILLINK, R. J. 1987. Origin of crude oils in Oman. *Journal of Petroleum Geology*, **11**, 61–80.

GREEN, J. C. 1983. Geologic and geochemical evidence for the nature and development of the Middle Proterozoic (Keweenawan) Midcontinent rift of North America. *Tectonophysics*, **94**, 413–437.

GREENTREE, M. R., LI, Z. X., LI, X. H. & WU, H. C. 2006. Late Mesoproterozoic to earliest Neoproterozoic basin record of the Sibao orogenesis in western South China and relationship to the assembly of Rodinia. *Precambrian Research*, **151**, 79–100.

GREY, K., HOCKING, R. M. *ET AL.* 2005. *Lithostratigraphic nomenclature of the Officer Basin and correlative parts of the Paterson Orogen, Western Australia.* Western Australia Geological Survey, Report, **93**, 89.

HALVERSON, G. P., HOFFMAN, P. F., SCHRAG, D. P., MALOOF, A. C. & RICE, A. H. N. 2005. Toward a Neoproterozoic composite carbon-isotope record. *Geological Society of America, Bulletin*, **117**, 1181–1207.

HAO, S. & LIU, G. 1989. Precambrian oil and gas in China. *AAPG Bulletin*, **73**, 412.

HOFFMAN, P. F. & SCHRAG, D. P. 2002. The snowball Earth hypothesis: testing the limits of global change. *Terra Nova*, **14**, 129–155.

HOFFMAN, P. F., KAUFMAN, A. J., HALVERSON, G. P. & SCHRAG, D. P. 1998. A Neoproterozoic snowball earth. *Science*, **281**, 1342–1346.

HUNT, J. M. 1996. *Petroleum Geochemistry and Geology*, 2nd edn. W. H. Freeman, New York, 16–18.

HUSSEINI, M. I. & HUSSEINI, S. I. 1990. Origin of the Infracambrian salt basins of the Middle East. *In*: BROOKS, J. (ed.) *Classic Petroleum Provinces.* Geological Society, London, Special Publications, **50**, 279–292.

IMBUS, S. W., ENGEL, M. H., ELMORE, R. D. & ZUMBERGE, J. E. 1988. The origin, distribution and hydrocarbon generation potential of the organic-rich facies in the Nonesuch Formation, Central North American Rift system: a regional study. *Organic Geochemistry*, **13**, 207–219.

INTERNATIONAL COMMISSION ON STRATIGRAPHY. 2008. *International Stratigraphic Chart*: web: http://www.stratigraphy.org/gssp.htm.

JACKSON, J. M., POWELL, T. G., SUMMONS, R. E. & SWEET, I. P. 1986. Hydrocarbon shows and petroleum source rocks in sediments as old as 1.7×10^9 years. *Nature*, **322**, 727–727.

JACKSON, J. M., SWEET, I. P. & POWELL, T. G. 1988. Studies on petroleum geology and geochemistry of the Middle Proterozoic McArthur Basin, northern Australia I: petroleum potential. *Australian Petroleum Exploration Association Journal*, **28**, 283–302.

KNOLL, A. H. 1992. Vendian microfossils in metasedimentary cherts of the Scotia Group, Prins Karls Forland. *Svalbard Palaeontology*, **35**, 751–774.

KNOLL, A. H. & WALTER, M. R. 1992. Latest Proterozoic stratigraphy and Earth history. *Nature*, **256**, 673–678.

KOLONIC, S., GEIGER, M., PETERS, H., THUSU, B. & LÜNING, S. 2004. *Infracambrian Hydrocarbon*

Potential of the Taoudenni Basin (Mauritania–Algeria–Mali). Maghreb Petroleum Research Group (MPRG), London–Bremen, **51** (unpublished).

KONTOROVICH, A. E., MANDEL'BAUM, M. M., SURKOV, V. S., TROFIMUK, A. A. & ZOLOTOV, A. N. 1990. Lena–Tunguska Upper Proterozoic petroleum superprovince. *In*: BROOKS, J. (ed.) *Classic Petroleum Provinces.* Geological Society, London, Special Publications, **50**, 473–489.

KORSCH, R. J., HUAZHAO, M., ZHAOCAI, S. & GORTER, J. D. 1991. The Sichuan Basin, southwest China: at Late Proterozoic (Sinian) petroleum province. *Precambrian Research*, **54**, 45–63.

KUZNETSOV, V. G. 1997. Riphean hydrocarbon reservoirs of the Yurubchen–Tokhom Zone, Lena-Tunguska Province, NE Russian. *Journal of Petroleum Geology*, **20**, 459–474.

LAWYER, L. A. & SCOTESE, C. R. 1987. A revised reconstruction of Gondwanaland. *In*: MCKENZIE, G. D. (ed.) *Gondwana Six – Structure, Tectonics, and Geophysics.* American Geophysical Union, Geophysical Monograph, **40**, 17–23.

LI, W. X., LI, X. H. & LI, Z. X. 2005. Neoproterozoic bimodal magmatism in the Cathaysia Block of South China and its tectonic significance. *Precambrian Research*, **136**, 51–66.

LI, Z. X., LI, X. H., KINNY, P. D., WANG, J., ZHANG, S. & ZHOU, H. 2003. Geochronology of Neoproterozoic syn-rift magmatism in the Yangtze Craton, South China and correlations with other continents: evidence for a mantle superplume that broke up Rodinia. *Precambrian Research*, **122**, 85–109.

LÜNING, S., GEIGER, M. & THUSU, B. 2005. *ENI North Africa Study, Infracambrian Prospectivity of North Africa, Summary and Recommendations.* Maghreb Petroleum Research Group (MPRG), London–Bremen (unpublished).

LÜNING, S., KOLONIC, S., GEIGER, M., THUSU, B., BELL, J. S. & CRAIG, J. 2009. Infracambrian hydrocarbon source rock potential and petroleum prospectivity of NW Africa. *In*: CRAIG, J., THUROW, J., THUSU, B., WHITHAM, A. & ABUTARRUMA, Y. (eds) *Global Neoproterozoic Petroleum Systems: The Emerging Potential in North Africa.* Geological Society, London, Special Publications, **326**, 157–180.

LÜNING, S., KOLONIC, S., GEIGER, M., THUSU, B. & CRAIG, J. 2004. *Infracambrian Hydrocarbon Source Rock Potential and Petroleum Prospectivity of North Africa.* Maghreb Petroleum Research Group (MPRG), London–Bremen (unpublished).

MAGOON, L. B. 1995. The play that complements the petroleum system – a new exploration equation. *Oil and Gas Journal*, **93**, 85–87.

MAGOON, L. B. & BEAUMONT, E. A. 1999. Petroleum systems. *In*: BEAUMONT, E. A. & FOSTER, N. H. (eds) *Handbook of Petroleum Geology: Exploring for Oil and Gas Traps.* American Association of Petroleum Geologists, Tulsa, OK, 31–34.

MAGOON, L. B. & DOW, W. G. 1994. The petroleum system. *In*: MAGOON, L. B. & DOW, W. G. (eds) *The Petroleum System – From Source to Trap.* American Association of Petroleum Geologists, Tulsa, OK, 3–24.

MANCUSO, J. J., KNELLER, W. A. & QUICK, J. C. 1989. Precambrian vein pyrobitumen: evidence for petroleum generation and migration 2 Ga ago. *Precambrian Research*, **44**, 137–146.

MA, Y., GUO, X., GUO, T., HUANG, R., CAI, X. & LI, G. 2007. The Puguang gas field: new giant discovery in the mature Sichuan Basin, southwest China. *AAPG Bulletin*, **91**, 627–643.

MASTERS, C. D., ROOT, D. H. & TURNER, R. M. 1997. World of resource statistics geared for electronic access. *Oil and Gas Journal*, **95**, 98–104.

MATTES, B. W. & CONWAY-MORRIS, S. 1990. Carbonate/evaporite deposition in the Late Precambrian–Early Cambrian Ara Formation of southern Oman. *In*: ROBERTSON, A. H. F., SEARLE, M. P. & RIES, A. C. (eds) *The Geology and Tectonics of the Oman Region.* Geological Society, London, Special Publications, **49**, 617–636.

MAUK, J. L. & BURRUSS, R. C. 2002. Water washing of Proterozoic oil in the Midcontinent rift system. *AAPG Bulletin*, **86**, 1113–1127.

MAUK, J. L. & HIESHIMA, G. B. 1992. Organic matter and copper mineralization at White Pine, Michigan. *Chemical Geology*, **99**, 189–211.

MCKERROW, W. S., SCOTESE, C. R. & BRASIER, M. D. 1992. Early Cambrian continental reconstruction. *Journal of the Geological Society, London*, **149**, 599–606.

MCKIRDY, D. M. & IMBUS, S. W. 1992. Precambrian petroleum: a decade of changing perceptions. *In*: SCHIDLOWSKI, M. (ed.) *Early Organic Evolution: Implications for Mineral and Energy Resources.* Springer, Berlin, 176–192.

MEISSNER, F. F., WOODWARD, J. & CLAYTON, J. L. 1984. Stratigraphic relationships and distribution of source rocks in the Greater Rocky Mountain Region. *In*: WOODWARD, J., MEISSNER, F. F. & CLAYTON, J. L. (eds) *Hydrocarbon Source Rocks of the Greater Rocky Mountain Region.* Rocky Mountain Association of Geologists, Denver, CO, 1–34.

MELEZHIK, V. A., FALLICK, A. E., FILIPPOV, M. M. & LARSEN, O. 1999. Karelian shungite – an indication of 2.0 Ga-old metamorphosed oil-shale generation of petroleum: geology, lithology and geochemistry. *Earth-Science Reviews*, **47**, 1–40.

MEYERHOFF, A. A. 1980. Geology and petroleum fields in Proterozoic and Lower Cambrian strata, Lena-Tunguska petroleum province, Eastern Siberia, USSR. *In*: HALBOUTY, M. T. (ed.) *Giant Oil and Gas Fields of the Decade 1968–78.* American Association of Petroleum Geologists, Memoir, **30**, 225–252.

MEYERHOFF, A. A. 1982. Hydrocarbon resources in arctic and sub-artic regions. *In*: EMBRY, A. F. & BALKWILL, H. R. (eds) *Arctic Geology and Geophysics.* Canadian Society of Petroleum Geologists, Memoir, **8**, 451–552.

MOCZYDLOWSKA, M., VIDAL, G. & RUDAVSKAYA, V. A. 1993. Neoproterozoic (Vendian) phytoplankton from the Siberian Platform, Yakutia. *Palaeontology*, **36**, 495–521.

MUIR, M. D., ARMSTRONG, K. J. & JACKSON, J. 1980. Precambrian hydrocarbon in the McArthur Basin, NT. *Journal of Australian Geology and Geophysics, Bureau of Mineral Resources*, **5**, 301–304.

OZIMIC, S., PASSMORE, V. L., PAIN, L. & LEVERING, I. H. 1986. *Australian Petroleum Accumulations Report 1, Amadeus Basin, Central Australia, Australia.* Bureau of Mineral Resources, Canberra, 64.

PALACAS, J. G. 1997. Source-rock potential of Precambrian rocks in selected basins of the United States. *United States Geological Survey Bulletin*, **2146-J**, 125–134.

PERRODON, A. 1992. Petroleum systems: models and applications. *Journal of Petroleum Geology*, **15**, 319–326.

PETERS, K. E., CLARK, M. E., GUPTA DAS, U., MCCAFFREY, M. A. & LEE, C. Y. 1995. Recognition of an Infracambrian source rock based on biomarkers in the Baghewala l Oil, India. *AAPG Bulletin*, **79**, 1481–1494.

PETERS, K. E., WATTERS, C. C. & MOLDOWAN, J. M. 2005. *The Biomarker Guide, Volume 2.* Cambridge University Press, Cambridge, 1155.

PLUMB, A. K., DERRICK, G. M. & WILSON, I. H. 1980. Precambrian geology of the McArthur River-Mount Isa region, northern Australia. *In*: HUNTER, D. R. (ed.) *Precambrian of the Southern Hemisphere.* Development in Precambrian Geology, 2, Elsevier, Amsterdam, 205–307.

POLLASTRO, R. M. 1999. Ghaba Salt Basin province and Fahud Salt Basin province – Oman: geological overview and total petroleum systems. *United States Geological Survey Bulletin*, **2167**, 1–41.

POSTNIKOV, A. V. & POSTNIKOVA, O. V. 2006. Vendian–Riphean deposits as main objective of hydrocarbon exploration on the Siberian Platform. *In*: *International Conference on Global Infracambrian Hydrocarbon Systems and the Emerging Potential in North Africa, 29–30 November 2006.* Geological Society, London, 39–40.

PRATT, L. M., SUMMONS, R. E. & HIESHIMA, G. B. 1991. Sterane and triterpane biomarkers in the Precambrian Nonesuch Formation, North American Midcontinent rift. *Geochimica et Cosmochimica Acta*, **55**, 911–916.

PREISS, W. V. & FORBES, B. G. 1981. Stratigraphy, correlation and sedimentary history Adelaidean (Latest Proterozoic) basin in Australia. *Precambrian Research*, **15**, 255–304.

PRICE, K. I., HUNTOON, J. E. & MCDOWELL, S. D. 1996. Thermal history of the 1.1 Ga Nonesuch Formation, North American Mid-Continent Rift, White Pine. Michigan. *AAPG Bulletin*, **80**, 1–15.

RAWLINGS, D. J. 1999. Stratigraphic resolution of a multiphase intracratonic basin system: the McArthur Basin, northern Australia. *Australian Journal of Earth Sciences*, **46**, 703–723.

SHEIKH, R. A., JAMIL, M. A., MCCANN, J. & SAQI, M. I. 2003. Distribution of Infracambrian reservoirs on Punjab Platform in central Indus Basin of Pakistan. *In*: *ATC 2003 Conference and Oil Show*, 3–5 October. Society of Petroleum Engineers (SPE) and Pakistan Association of Petroleum Geoscientists (PAPG), Islamabad, 1–17.

SUMMONS, R. E., LOVE, G. D. *ET AL.* 2006. New approaches to oilsource correlation in the Neoproterozoic–Cambrian, South Oman Salt Basin. *In*: *International Conference on Global Infracambrian Hydrocarbon Systems and the Emerging Potential in North Africa, 29–30 November 2006.* Geological Society, London, 53–55.

TERKEN, J. M. J., FREWIN, N. L. & INDRELID, S. L. 2001. Petroleum systems of Oman: charging timing and risks. *AAPG Bulletin*, **85**, 1817–1845.

THUSU, B., GEIGER, M. & LÜNING, S. 2004. *Infracambrian Stratigraphy and Hydrocarbon Potential of East Libya.* Maghreb Petroleum Research Group (MPRG), University College of London, London (unpublished).

ULMISHEK, G. E. 1986. Stratigraphic aspects of petroleum resource assessment. *In*: RICE, D. D. (ed.) *Oil and Gas Assessment – Method and Applications.* American Association of Petroleum Geologists, Tulsa, OK, 59–68.

ULMISHEK, G. F. 2001. Petroleum geology and resources of the Baykit High Province, East Siberia, Russia. *United States Geological Survey Bulletin*, **2201-F**, 1–18.

ULMISHEK, G. F., LINDQUIST, S. J. & SMITH-ROUCH, L. S. 2002. *Region 1 Former Soviet Union – Summary.* U. S. Geological Survey Digital Data Series, **60**. United States Geological Survey World Petroleum Assessment 2000.

VAN SCHMUS, W. R. & HINZE, W. J. 1985. The Midcontinent rift system. *Annual Review of Earth and Planetary Sciences*, **13**, 345–383.

WADE, B. P., HAND, M. & BAROVICH, K. M. 2005. Nd isotopic and geochemical constraints on provenance of sedimentary rocks in the eastern Officer Basin, Australia; implications for the duration of the intracratonic Petermann Orogeny. *Journal of the Geological Society, London*, **162**, 513–530.

WALTER, M. R. & GORTER, J. 1993. *Centralian Superbasin, Australia.* Petroconsultants Australasia, Sydney (unpublished).

WALTER, M. R., VEEVERS, C. R., CALVER, C. R., GREY, K. & HILYARD, D. 1992. *The Proterozoic Centralian Superbasin: A Frontier Petroleum Province.* 1992 American Association of Petroleum Geologists Conference, Abstracts, 77.

WALTER, M. R., VEEVERS, C. R., CALVER, C. R. & GREY, K. 1995. Neoproterozoic stratigraphy of the Centralian Superbasin, Australia. *Precambrian Research*, **73**, 173–195.

WARREN, J. K. 1989. *Evaporite Sedimentology: Its Importance in Hydrocarbon Accumulations.* Prentice-Hall, Englewood Cliffs. NJ.

WILKINS, N. 2007. Proterozoic evangelist tries to convert the unbelievers. *Oil & Gas Gazette*, December 2006–January 2007, 3–3.

WOMER, M. B. 1986. Hydrocarbon occurrence and digenetic history within Proterozoic sediments, McArthur River area, Northern Territory, Australia. *Australian Petroleum Exploration Association Journal*, **26**, 363–374.

YIN, L. 1985. Microfossils of the Doushantuo Formation in the Yangtze Gorge district, western Hebei, China. *Palaeontologica Cathayana*, **2**, 229–249.

ZHOU, J. G. 1996. Abstract evaluation of Sichuan Basin in China. *AAPG Bulletin*, **80**, 985–985.

Neoproterozoic–Early Cambrian (Infracambrian) hydrocarbon prospectivity of North Africa: a synthesis

FABIO LOTTAROLI[1]*, JONATHAN CRAIG[1] & BINDRA THUSU[2]

[1]*Eni Exploration and Production Division, Via Emilia 1, 20097 San Donato Milanese, Milan, Italy*

[2]*MPRG (Maghreb Petroleum Research Group), University College London, Gower Street, London WC1E 6BT, UK*

**Corresponding author (e-mail: fabio.lottaroli@eni.it)*

Abstract: Despite the existence of proven Neoproterozoic–Early Cambrian ('Infracambrian') hydrocarbon plays in many parts of the world, the Neoproterozoic Eon, from 1000 Ma to the base of the Cambrian at 542 Ma, is relatively poorly known from a petroleum perspective. The so-called 'Peri-Gondwanan Margin' is one region of the Neoproterozoic world that is exciting particular interest in the search for 'old' hydrocarbon plays, mainly due to exploration success in time-equivalent sequences of Oman. The 'Infracambrian' succession in North Africa is widely accessible, and is already emerging as a hydrocarbon exploration target with considerable potential and with proven petroleum systems in different areas. The Taoudenni Basin (Mauritania, Mali, Algeria) in western North Africa is an underexplored basin, despite the Abolag-1 well (Texaco 1974) gas discovery. New palynological data have recently provided the first definitive Late Riphean age dates for the stromatolitic limestone reservoir sequence in Abolag-1. The widespread presence of stromatolitic carbonate units of potential reservoir facies in many parts of North Africa has been confirmed by new fieldwork in the Taoudenni Basin, in the Anti-Atlas region of Morocco and in the Al Kufrah Basin of Libya. Similar biostratigraphic age constraints have also been obtained from subsurface sequences of the Cyrenaica Platform bordering the East Sirte Basin of Libya, many of which have been traditionally assigned an 'unconstrained' Cambro-Ordovician age on the basis of lithological characteristics. Besides the proven, producing, weathered-granite reservoir in East Sirte Basin, the hydrocarbon potential of Neoproterozoic–Early Cambrian sequences developed in structural troughs bordering the south Cyrenaica Platform is still being evalutated. Neoproterozoic–Early Cambrian organic-rich strata with hydrocarbon source rock potential are widespread along the Peri-Gondwanan Margin. Some of the black shales encountered on the West African Craton may be as old as 1000 Ma and predate the Pan-African orogenic event. The Late Ordovician–Early Silurian systems in North Africa and the Middle East may form a good analogue for post-glacial source rock depositional systems in the Neoproterozoic, where black shale deposition may also have been triggered by post-glacial sea-level rise.

In North Africa, the term 'Infracambrian' is generally used to denote the widely distributed successions of Neoproterozoic–earliest Cambrian age, deposited as pre-, syn- and post-Pan-African sequences, that often underlie the oldest dated Early Palaeozoic strata and overlie igneous or metamorphic basement.

The Neoproterozoic Eon (1000–542 Ma) is subdivided into Tonian (1000–850 Ma), Cryogenian (850–630 Ma) and Ediacaran (630–542 Ma) periods. It was a time of massive atmospheric, climatic and tectonic changes, dominated by the Cryogenian 'Snowball Earth' type glaciations (Hoffman & Schrag 2002) and by the assembly and break-up of two supercontinents: Rodinia and Greater Gondwana (Fig. 1).

The concept of the 'Snowball Earth', during which ice may have extended from the poles to cover the land masses that were clustered, at that time, around the equatorial zone, has stimulated considerable debate among Earth scientists (Harland & Rudwick 1964; Eyles 1993; Hoffman *et al.* 1998*a*, *b*; Evans 2000; Sankaran 2003; Eyles & Januszczak 2004) triggering a massive increase in knowledge about, and bibliography on, Late Neoproterozoic successions around the world. Recently, data have begun to emerge that also help to resolve the sequence of genetic and morphological innovations, environmental events and ecological interactions that were collectively responsible for the emergence of the first recognizable animal life around 600 Ma, during the Ediacaran Period, and, in turn, for the explosion of life forms that occurred during the latest Neoproterozoic and Early Cambrian (Knoll & Carroll 1999).

From: Craig, J., Thurow, J., Thusu, B., Whitham, A. & Abutarruma, Y. (eds) *Global Neoproterozoic Petroleum Systems: The Emerging Potential in North Africa*. Geological Society, London, Special Publications, **326**, 137–156.
DOI: 10.1144/SP326.7 0305-8719/09/$15.00

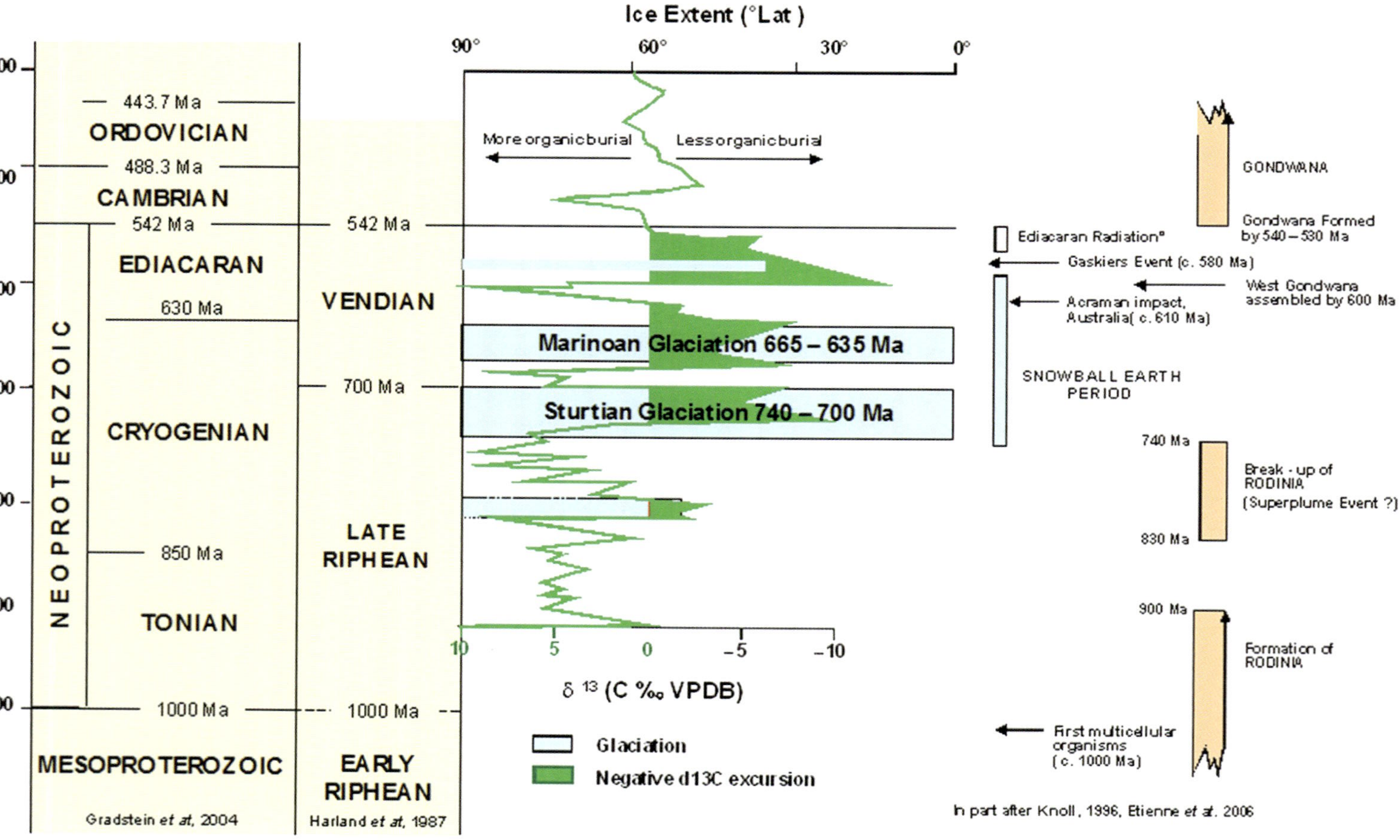

Fig. 1. Neoproterozoic timescale and key events.

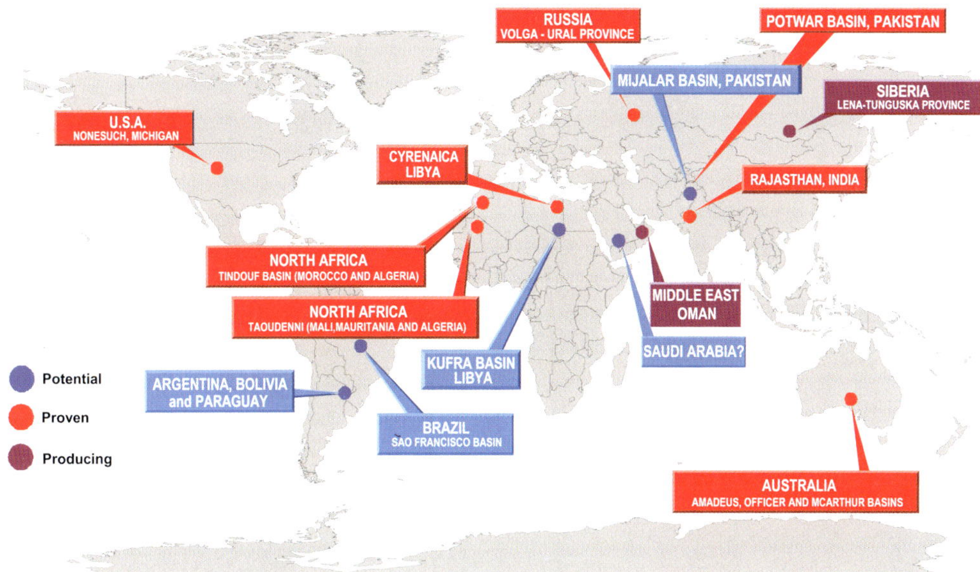

Fig. 2. Neoproterozoic (Infracambrian) petroleum systems of the world.

Proven and potential Neoproterozoic–Early Cambrian petroleum systems have been reported worldwide (Fig. 2). The Siberian Lena–Tunguska province (Volga–Ural region) and the south Oman petroleum fields exemplify the productive potential of these Neoproterozoic–Early Cambrian petroleum systems. Hydrocarbon shows (both oil and gas) are reported from Australia, Canada, China, India, Pakistan, North Africa, Venezuela and USA (Ghori *et al.* 2009), demonstrating that active petroleum systems are, indeed, widespread in the Neoproterozoic–Early Cambrian successions. Despite this global distribution, the 'Infracambrian' remains relatively poorly known from a petroleum perspective.

Neoproterozoic and Palaeozoic geology of the Peri-Gondwanan Margin and North Africa

The Rodinia supercontinent was assembled at the beginning of the Neoproterozoic Era and started to break up about 750 Ma, during the Mid Cryogenian. The break-up produced two major landmasses termed East Gondwana and West Gondwana, separated by the Panthalassic Ocean. During the Mid–Late Neoproterozoic (700–600 Ma), the closure of the oceans on both sides of the Congo Craton, along with a smaller ocean basin in East Gondwana (Fig. 3a), culminated in the Pan-African Orogeny (650–550 Ma) and the formation of the Greater Gondwana supercontinent (Fig. 3b, c). This second Neoproterozoic supercontinent was relatively short lived because, contemporary with the assemblage, rifting of peripheral plates resulted in the formation of the main Palaeozoic oceans (Iapetus, Palaeo-Tethys).

The so-called 'Peri- (or North) Gondwanan Margin', which extends from northern South America, across North Africa and the Middle East into Pakistan and NW India (Fig. 4), is the region of the Neoproterozoic world that has attracted most interest in recent years in the search for new Neoproterozoic–Early Cambrian hydrocarbon plays (Craig 2006). The interest is rooted in the greatly improved understanding of the prolific Late Neoproterozoic–Early Cambrian Huqf Supergroup of Oman, especially the intra-salt Ara 'Stringer' carbonate and Athel silicilyte plays in the South Oman Salt Basin (Amthor *et al.* 2005; Al-Siyabi 2005). Increasingly important older 'pre-salt' plays in South Oman, and karstified carbonates on Huqf highs in North Oman (Cozzi & Al-Siyabi 2004), contribute to the high level of attention given to these petroleum systems, which are estimated to be the source of more than 90% of Oman's oil. Similar plays are already being actively pursued in the age-equivalent Marwar Supergroup in Pakistan and in Rajasthan (Western India), further to the east. Recently, attention has also turned to the potential for age-equivalent and possibly older Neoproterozoic plays across the rest of the Middle East and North Africa.

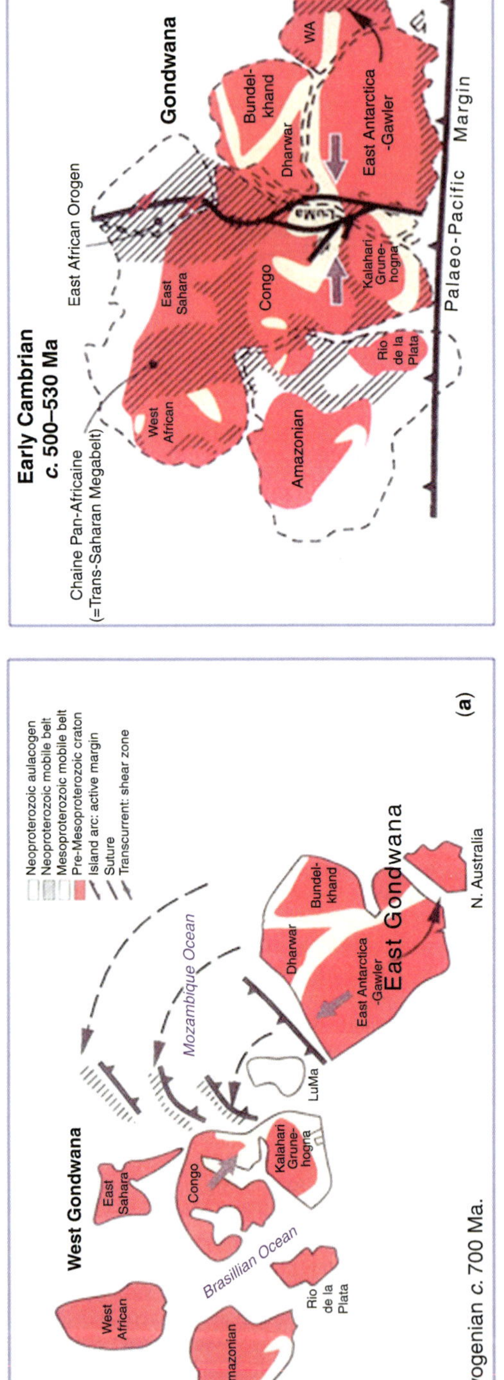

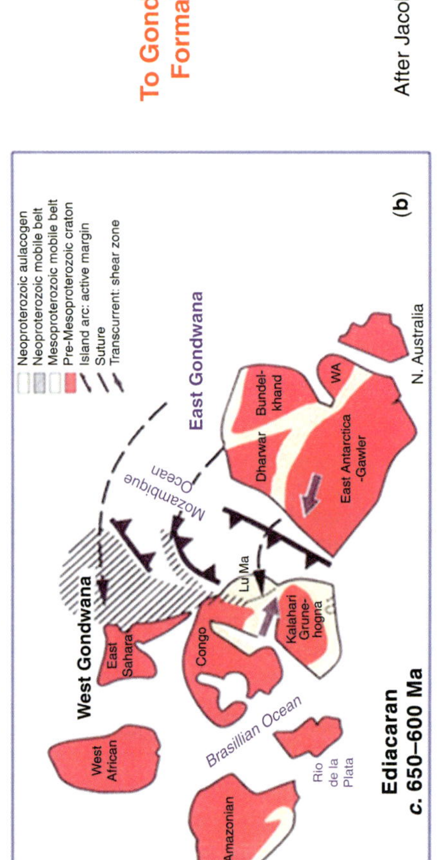

Fig. 3. Plate reconstruction of East and West Gondwana from the Middle Proterozoic to the Early Cambrian.

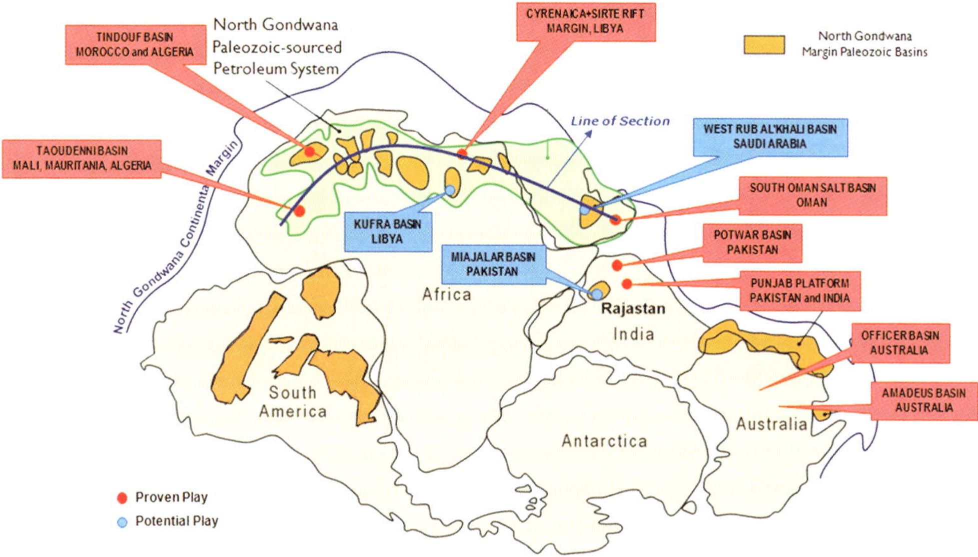

Fig. 4. Peri-Gondwanan Margin Palaeozoic and Neoproterozoic petroleum plays.

The Pan-African orogenic event shaped the continental crust of North Africa. It resulted in the development of two main orogenic belts, the East African Orogen and the Chaine Pan-Africaine (Fig. 3), that, together with remnant cratonic areas, now form the basement to Palaeozoic successions in North Africa and Arabia.

The East African Orogen, represented by the Arabian–Nubian Shield (Arabia, eastern Egypt, Sudan), formed by the gradual closure of the Mozambique Ocean from 900 Ma (Condie 2003). The Chaine Pan-Africaine developed between the West African and East Saharan cratons (Fig. 3c), and is exposed in different parts of North Africa, including the Anti-Atlas of Morocco and the Algerian Ougarta Ranges. Intramontane molasse basins are developed in both these Pan-African orogenic systems, as exemplified by the Murdama Basin in Arabia (Genna *et al.* 2002) and by Northwest Hoggar in Libya (Caby & Moniè 2003).

The timing of the start of Neoproterozoic sedimentation varies greatly across North Africa, with sediments older than 1000 Ma preserved on the West African Craton and generally younger Neoproterozoic sediments associated with Pan-African mobile belts (Lüning *et al.* 2009). Neoproterozoic–Early Cambrian sedimentation along the Peri-Gondwanan Margin, from Australia through Pakistan, Oman and North Africa, coincides with a major phase of crustal extension (Shandelmeier 1988; Husseini & Husseini 1990) during which basins of various different types were formed (e.g. half-graben, pull-apart basins, foreland basins, intramontane molasse basins).

Neoproterozoic deposits have been described from outcrops in many parts of North Africa, but their presence in the subsurface is only rarely proven by well penetrations because they are typically very deeply buried. Moreover, the paucity of suitable chronostratigraphic diagnostic criteria, both for relative (biostratigraphic) and absolute dating, is a consistent problem both in outcrops and the subsurface. Subsurface penetrations of the Neoproterozoic–Early Cambrian succession occur in the Taoudenni Basin (Mauritania, Mali, Algeria), together with extensive outcrops on the NW and SW basin margins (Moussine-Pouchine & Bertrand-Sarfati 1997; Deynoux *et al.* 2006). Infracambrian strata are also described at outcrop in the Anti-Atlas of Morocco, the Algerian Ougarta Range and in the Hoggar Massif and in the subsurface of the northern Tindouf Basin, and, questionably, in the Reggane and Ahnet basins of western Algeria.

The western portion of the North African Craton and surrounding mobile belts provides the best opportunity to examine the entire 'Infracambrian' sedimentary record, highlighting potential plays that are stratigraphically older than the proven Oman petroleum system (Fig. 5).

Few data have been published on the Infracambrian sequences that have been drilled in Cyrenaica (NE Libya), in the Murzuq Basin (SW Libya) and interpreted on seismic data in the Al Kufrah Basin (Craig *et al.* 2008) and/or their possible

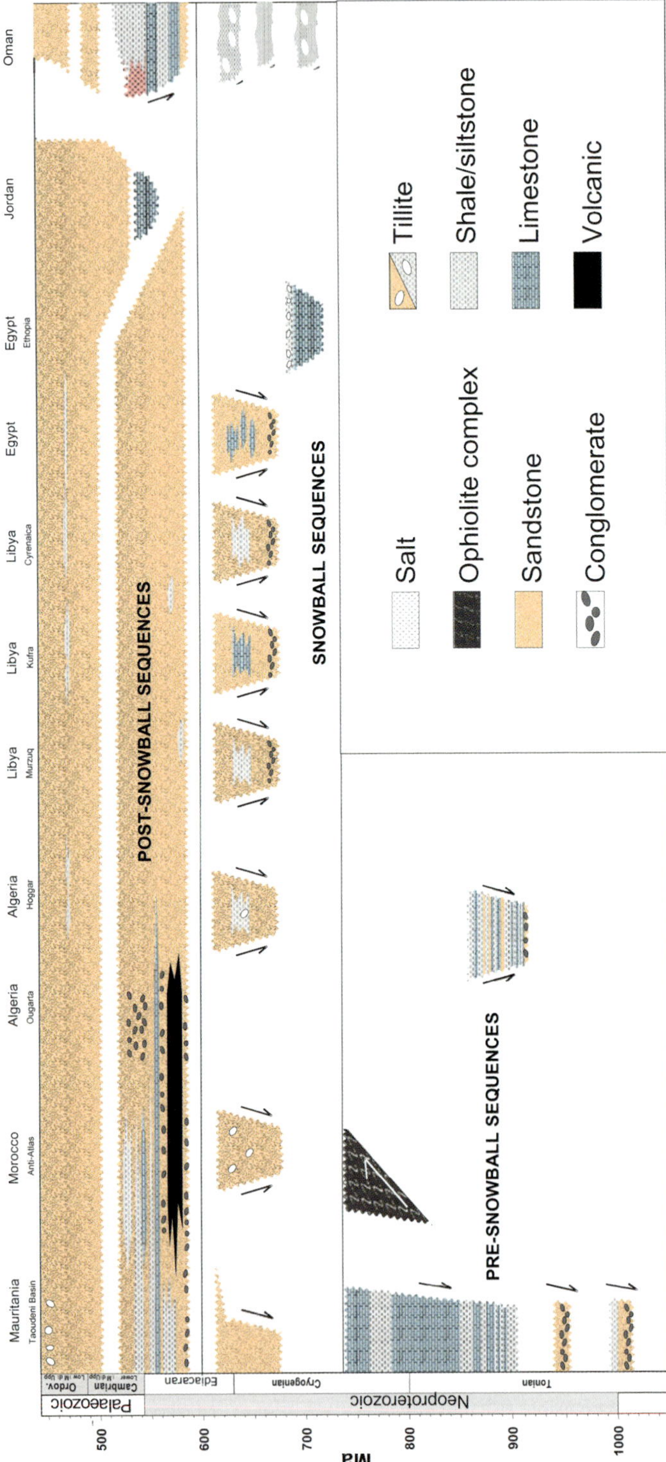

Fig. 5. Neoproterozoic chronostratigraphy: North Africa and Arabia.

extension in the Egyptian Western Desert (Fig. 5) (Selley 1971; Bellini & Massa 1980; Baudet 1988). In most of these areas, Neoproterozoic–Early Cambrian sequences have been considered to form the economic basement below Palaeozoic successions that are the main target of petroleum exploration activity. More than 50 Bboe (billion barrels of oil equivalent) of reserves are ascribed to the extremely productive Palaeozoic-sourced petroleoum system in North Africa (MacGregor 1996) (Figs 4 & 6). In addition, the Lower Silurian–Upper Ordovician petroleum system may constitute an attractive analogue for the Precambrian potential. Reservoirs deposited under glacially influenced conditions, linked to the Late Ordovician glaciation, are widespread in North Africa and contain at least 5 Bboe in more than 50 fields (Craig *et al.* 2008). The onset of anoxic conditions during the initial stages of the Early Silurian transgression following deglaciation led to the deposition of a 'world class' source rock (Basal Silurian 'Hot shales') that is the origin of most of the Palaeozoic-sourced hydrocarbons in North Africa (Lüning *et al.* 2000). The coincidence of glacigenic conditions and a proven petroleum system in the Lower Palaeozoic can be regarded as a potential analogue for the mid-Neoproterozoic, when major glacial episodes were also common.

Neoproterozoic petroleum systems of North Africa

Tonian–Cryogenian ('pre-Snowball'): Taoudenni Basin (Mauritania, Mali, Algeria)

The Taoudenni Basin is the largest basin on the West African Craton. It is a broad intra-cratonic sag extending over south and east Mauritania, western Mali and SW Algeria (Fig. 7). The basin contains a sedimentary sequence that is more than 5–6 km thick in the depocentre and mainly consists of Neoproterozoic–Early Palaeozoic megasequences. Despite its enormous size (1.8×10^6 km^2), the basin is virtually unexplored, with only two deep exploration wells, Abolag-1 (Texaco 1974) and Ouasa-1 (Agip 1974), that reached the Neoproterozoic sequence and a sparse regional two-dimensional (2D) seismic coverage. The Abolag-1 well tested gas at a rate of 48 000 scf (standard cubic feet) per day from Neoproterozoic stromatolitic limestones at around 3000 m of depth during a short duration open hole test. Ouasa-1 failed to reach the stromatolitic Limestone target. Despite the Abolag-1 gas discovery, no significant exploration activity has been carried out in the basin since 1974. Recently Eni SpA became the operator of a large tract of acreage (193 500 km^2) in the Mali portion of the basin, where Neoproterozoic sedimentary rocks represent one of the main exploration targets.

Neoproterozoic sequences exposed on the margins of the Taoudenni Basin have been the subject of detailed field studies (e.g. Moussine-Pouchkine & Bertrand-Sarfati 1997; Benan & Deynoux 1998; Deynoux *et al.* 2006). Moreover, new data presented in this paper have been collected during fieldwork in the Atar region, and new biostratigraphical and geochemical analyses of samples from the Abolag-1 well have been carried out by the Maghreb Petroleum Research Group, University College, London, within the framework of an Eni-funded project.

Major tectonic events affecting the Pan-African Orogenic Belt are reflected in the unconformities/disconformities defining the four tectono-sedimentary sequences that characterize the Neoproterozoic–Cambrian sedimentary cover of the Taoudenni Basin (Fig. 8).

The oldest sequence, broadly coincides with the Tonian–Mid Cryogenian' ('pre-Snowball') cycle and comprises mixed siliciclastic–carbonate rocks of the Duik–Char groups, overlain by stromatolitic limestones of the Hank–Atar Group, the latter deposited as aggradational cycles of sheet-like build-ups. This first megasequence records terrestrial–marginal marine deposition over a wide flat cratonic area, followed by the transgressive deposits of the lower Atar Group, dominated by stromatolitic limestones with a minor siliciclastic component, locally including black shales enriched in pyrite (Bertrand-Sarfati & Moussine-Pouchkine 1988).

The Upper Atar Group (Cryogenian 'Snowball' age), by contrast, typically consists of wedge-like sedimentary units deposited in fault-bounded extensional mini-basins that formed during a phase of pre-Pan-African crustal extension. Stromatolites are common in this megasequence (Dar Cheikh Group) and frequently pass downslope into black shales enriched in organic matter. Bioherms acted as barriers to water circulation, enhancing bottom-water anoxic conditions in local troughs (Moussine-Pouchkine & Bertrand-Sarfati 1997). The dominantly carbonate–shale sequence of the Atar Group in the north of the Taoudenni Basin is replaced by a sandstone–shale clastic sequence in the south.

The geological evolution during the deposition of the third and fourth megasequences (Ediacaran–Cambrian 'post-Snowball age') is strongly overprinted by the main phase of Pan-African deformation. Uplift and erosion were widespread on the craton along with formation of the Pan-African mobile belts, preceding a glacial event with the deposition of continental tillites and glacio-marine deposits (Moussine-Pouchkine &

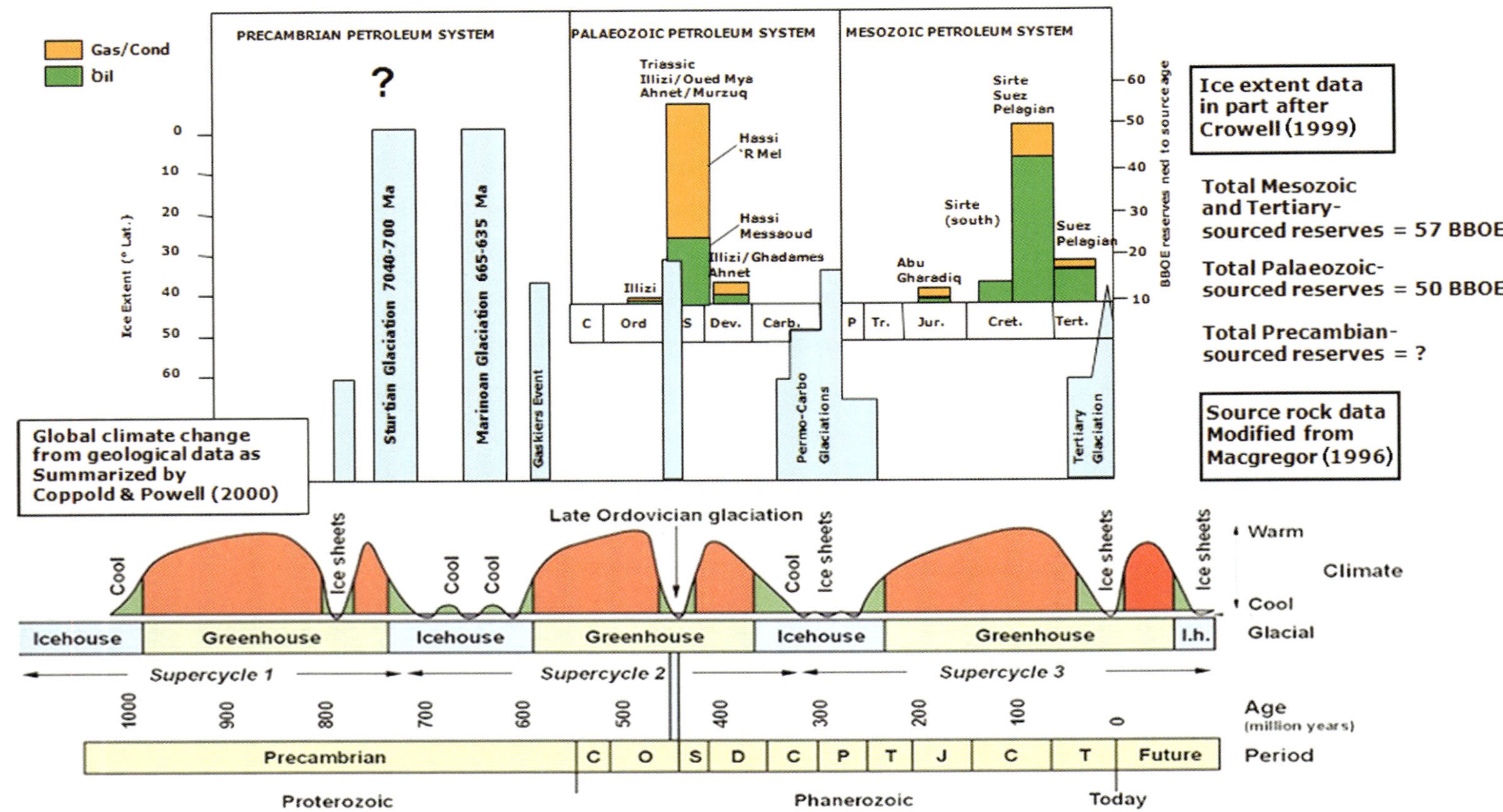

Fig. 6. Global climate, glaciations and source rocks in North Africa.

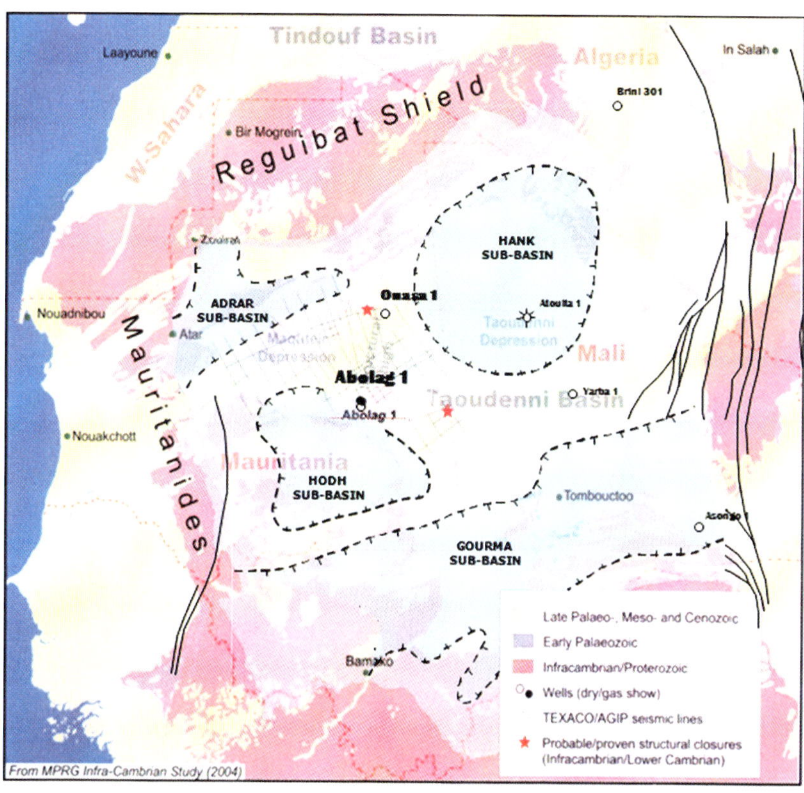

Fig. 7. Taoudenni Basin/Sub-basins (Mauritania, Mali and Algeria).

Bertrand-Sarfati 1997). Doubts still exist about the age of these glacial deposits, although they have been correlated with the Marinoan glaciation (Deynoux *et al.* 2006).

Potential stromatolitic reservoir facies and associated organic-rich shales of the Atar Group have a well-constrained Tonian–Mid Cryogenian (1000–770 Ma) radiometric age at outcrop (Fig. 9). In the subsurface, a rich and diverse palynoflora including acritarchs, cyanophytic filaments and amorphous organic matter (AOM) has recently been recovered from the Neoproterozoic sequence drilled by Abolag-1 and from a water well near Atar (Amrad 1986), in Mauritania. The assemblage represents a major biostratigraphic discovery that allows a Tonian–Cryogenian age to be assigned to the gas-bearing interval in the Abolag well. The first incoming downhole appearance at 1521 m of *Obruchevella parvissima*, *Siphnophycus inornatum*, *Leiosphaeridia asperata*, *L. crassa*, *Kildinella timanica*, *K. ripheica*, *Kildniosphaera chagrinata*, *K. Verrucata* and *Myxococcoides cantabrigiensis* identified in the assemblage support this age assignment. In addition to the age-diagnostic acritarchs, *Taeniatum* sp. (a cyanophytic filament) is also present. An exceptionally well-preserved assemblage dominated by *Arctacellularia tetragonala* and *Navifusa majensis* (Fig. 10) have a first appearance in the bottom part of the sequence, below 2783 m. These morphotypes may support an Early Riphean age for the terminal part of the sequence, but this latter age interpretation must wait until further investigations are undertaken on a wider geographical scale in the basin. The presence of acritarchs and leiospherid clusters indicate a depositional environment within the photic zone in a shallow-marine setting.

These biostratigraphic data support the correlation proposed in Figure 9, and confirm the existence of a pre-‘Snowball’ (Tonian–Mid Cryogenian) hydrocarbon play in the region. The presence of a significant amount of AOM (Type 1) associated with the recovered palynomorph assemblage in the drilled sequence is consistent with the fact that organic-rich shales are widely recognized in the age-equivalent outcrops of the Atar Group sequence and with anecdotal evidence of oil-prone black shales with total organic carbon (TOC)

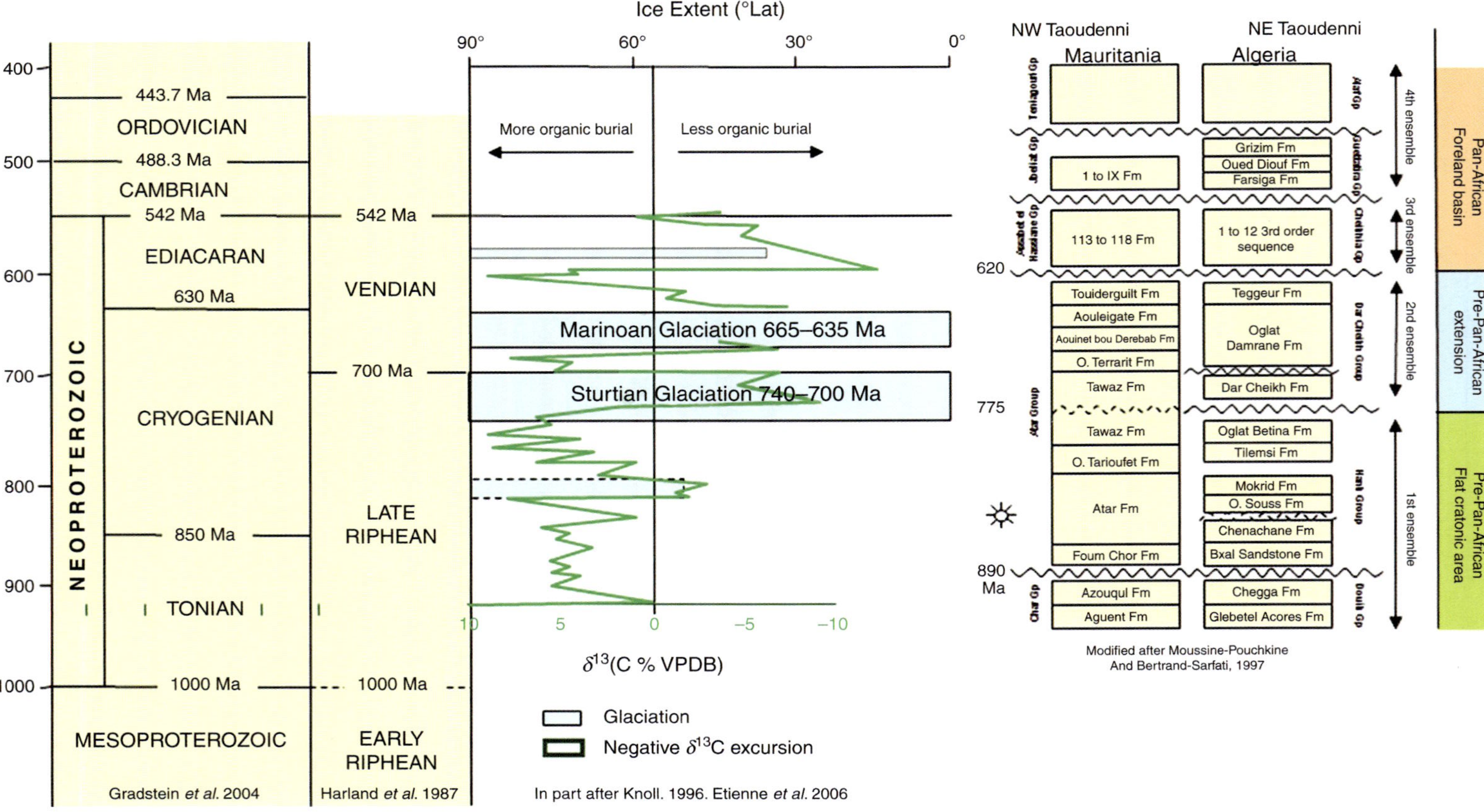

Fig. 8. Neoproterozoic–Early Palaeozoic sequences in the Taoudenni Basin (Mauritania, Mali and Algeria).

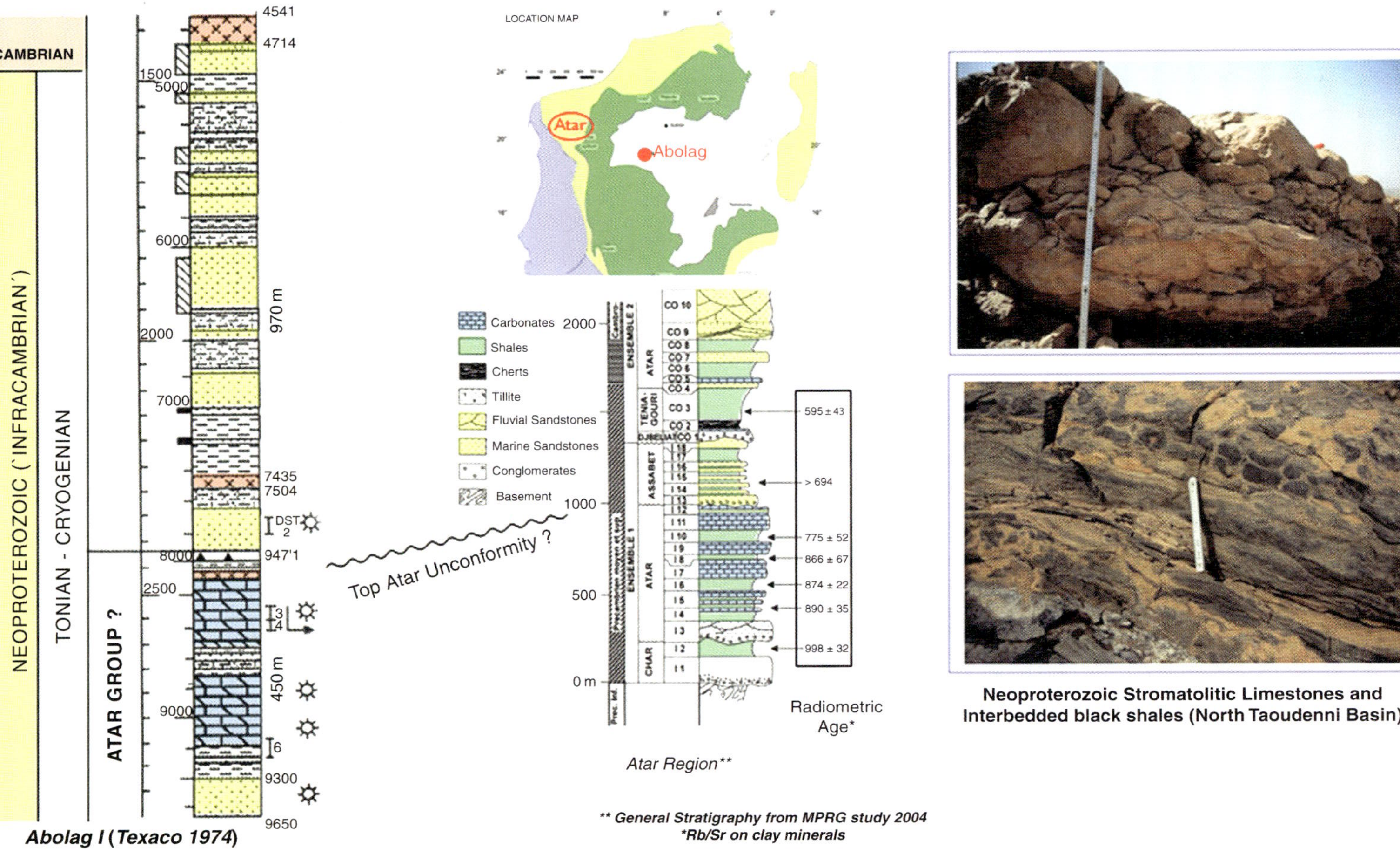

Fig. 9. Taoudenni Basin (Mauritania/Mali): tentative correlation between the Tonian–Cryogenian sequence in the Abolag-1 well and the outcrops of the Atar region.

1
2
3
4
5
6
7
8
9
Neoproterozoic Palynomorphs, Abolag 1 Well, Taoudenni Basin, Mauritania

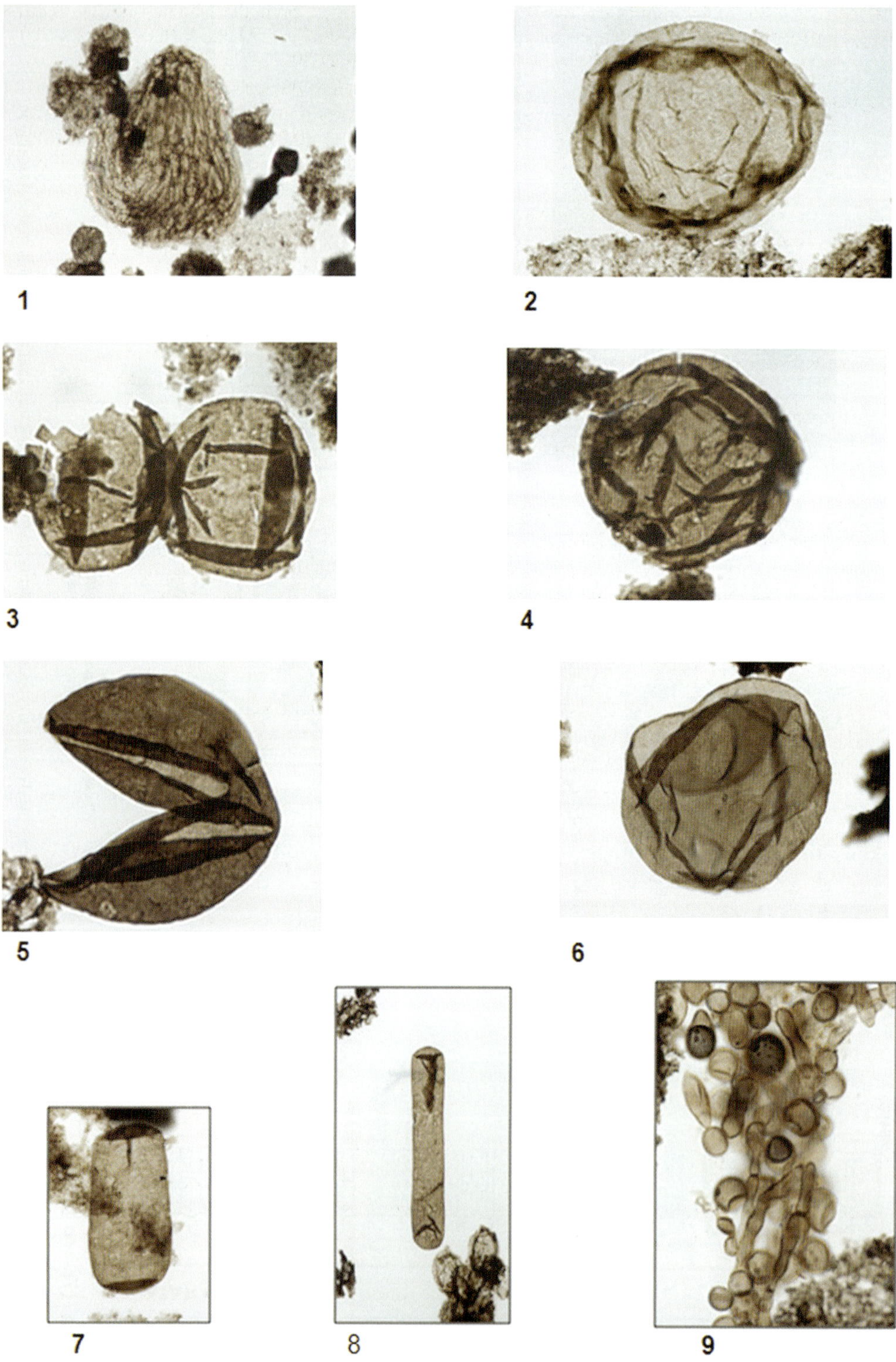

Fig. 10.

contents of 10–20% ('burning shales') within the Atar Group in the vicinity of Aguelt El Mabha.

Distribution of Infracambrian stromatolitic carbonates reservoirs in North Africa

Stromatolites were widely distributed along the entire Peri-Gondwanan Margin during the Neoproterozoic with occurrences reported from North Africa, India, Pakistan and the Middle East (e.g. Chakraborty 2004). In North Africa, in particular, stromatolites occur in different basins and at different stratigraphic levels within the Neoproterozoic sequence.

Anti-Atlas, Tindouf Basin and the Ougarta Range, Morocco and Algeria

The Neoproterozoic is the dominant stratigraphic unit cropping out in the Anti-Atlas Morocco, and extending into the Ougarta Range in Algeria. The sequence is broadly divided in two megasequences (Fig. 11). The lower unit, highly overprinted by metamorphic and tectonic events linked to the Pan-African orogenic phase, comprises the ophiolitic complex of the Bleida–Tachdamt Group and the volcanics and clastics of the Ouarzazate Group, overlying Precambrian basement (Leblanc & Moussine-Pouchkine 1994; Bouima & Mezghache 2002). The upper unit consists of a thick (3000 m) alternation of carbonate platform deposits and clastic sediments. Planar and domal stromatolites are common in the lower dolomitic units (Adoudou Formation) as well as in the upper units, and are locally associated with black shales (e.g. Igoudine–Lower Tislit Formation: Geyer & Landing 1995).

Gas shows are reported in the Neoproterozoic sequence in the subsurface on the northern margin of Tindouf basin (AZ-1 well), where the sequence cropping out in the Anti-Atlas dips into the subsurface. This confirms the existence of a proven Neoproterozoic petroleum system in this basin.

Murzuq and Al Kufrah basins, Libya

In southeastern Libya, Neoproterozoic–Early Cambrian strata have been described in outcrop from the margins of both the Murzuq and Al Kufrah basins (Bellini & Massa 1980; Hallet 2002; Aziz & Ghnia 2009). The 'Mourizidie Formation', described from the eastern margin of the Murzuq Basin (Jacqué 1962), is probably correlatable with sediments cropping out on the southern basin margin in Niger (Deynoux *et al.* 1985), and consists of interbedded conglomerates, sandstones and siltstones. Similar lithologies have been reported from the Neoproterozoic succession on the western margin of the Al Kufrah Basin (Selley 1971). Perhaps the most interesting Infracambrian sequence is that exposed locally on the eastern margin of Al Kufrah Basin. Here, along with the siliciclastic units, lenses of marble up to 500 m long and 75 m thick have been described (Saïd *et al.* 2000; El-Mehdi *et al.* 2004) and recently confirmed by fieldwork (Le Heron *et al.* 2009) (Fig. 12). These are the only potential Neoproterozoic carbonates in central eastern North Africa and, despite the strong metamorphic overprint, they locally retain a fabric reminiscent of primary stromatolitic lamination, although their precise age remains a matter of some conjecture. In the subsurface, remnants of presumed Neoproterozoic basins underlying the Palaeozoic succession in the Al Kufrah Basin, are well imaged on 2D seismic data (Benshati 2009, Fig. 11). The surface analogues contribute positively to a potential Neoproterozoic play in this area.

Potential new 'Infracambrian' plays in North Africa: an example from Cyrenaica–Sirte margin, NE Libya

Tonian–Mid Cryogenian sediments have also been recognized and documented in the subsurface of the central Cyrenaica Platform of NE Libya (Fig. 13), where they comprise a thick marine siliciclastic sequence dominated by feldspatic sandstones, locally interbedded with siltstones and shales (Arnauti & Shelmani 1988), and typically unconformably overlain by sandstones and shales of Late Ordovician age (Caradocian–Ashgillian). Late Riphean acritarchs and cyanophytic algae have been recovered in three wells (Baudet 1988; Arnauti & Shelamni 1988) on the Cyrenaica Platform (A1-33, B1-31 and C1-82), and two more wells on the adjacent Jagbub High and

Fig. 10. (*Continued*) Palynomorphs recorded in cuttings samples from Abolag-1 well, Taoudenni Basin, Mauritania. All figures are ×1000 unless otherwise stated. **1**, *Obruchevella parvissima*, Song 1984. **2**, *Leiosphaeridia crassa* (Naumova), 1949 emend. Yankauska 1989. **3**, **4**, *Leiosphaeridia asperata* (Naumova), Lindgren 1982. **5**, **6**, *Kildinosphaera chagrinata*, Vidal 1983. **7**, *Arctacellularia tetragonala* (Maithy) Hoffman & Jackson 1994. **8**, *Navifusa majensis*, Pyatiletov 1980 (×600). **9**, *Myxococcoides cantabrigiensis*, Knoll 1982.

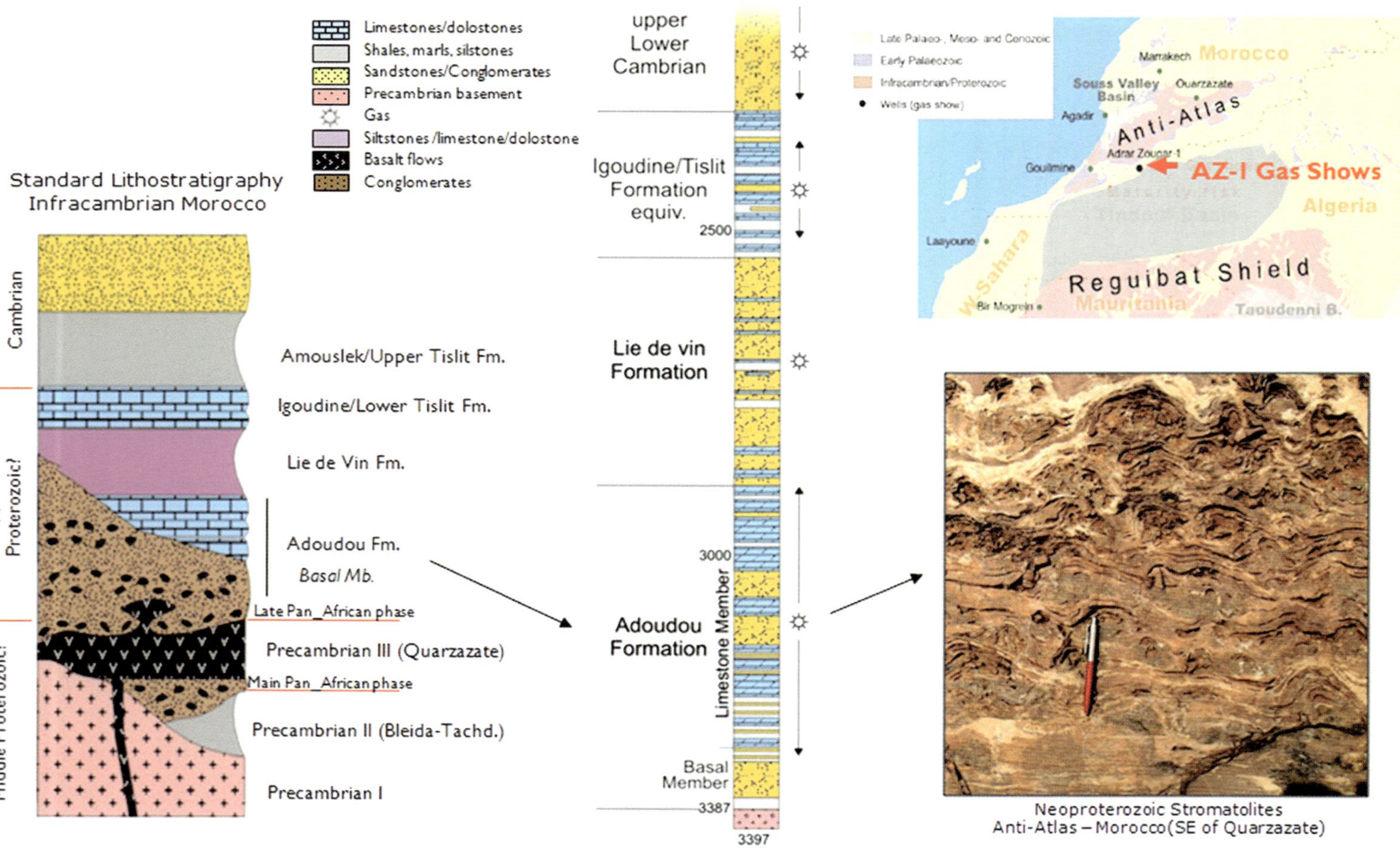

Fig. 11. Neoproterozoic–Early Cambrian sequences in the Anti-Atlas of Morocco and in the Tindouf Basin subsurface.

Fig. 12. Al Kufrah Basin (east Libya): possible stromatolitic limestones on the eastern basin margin.

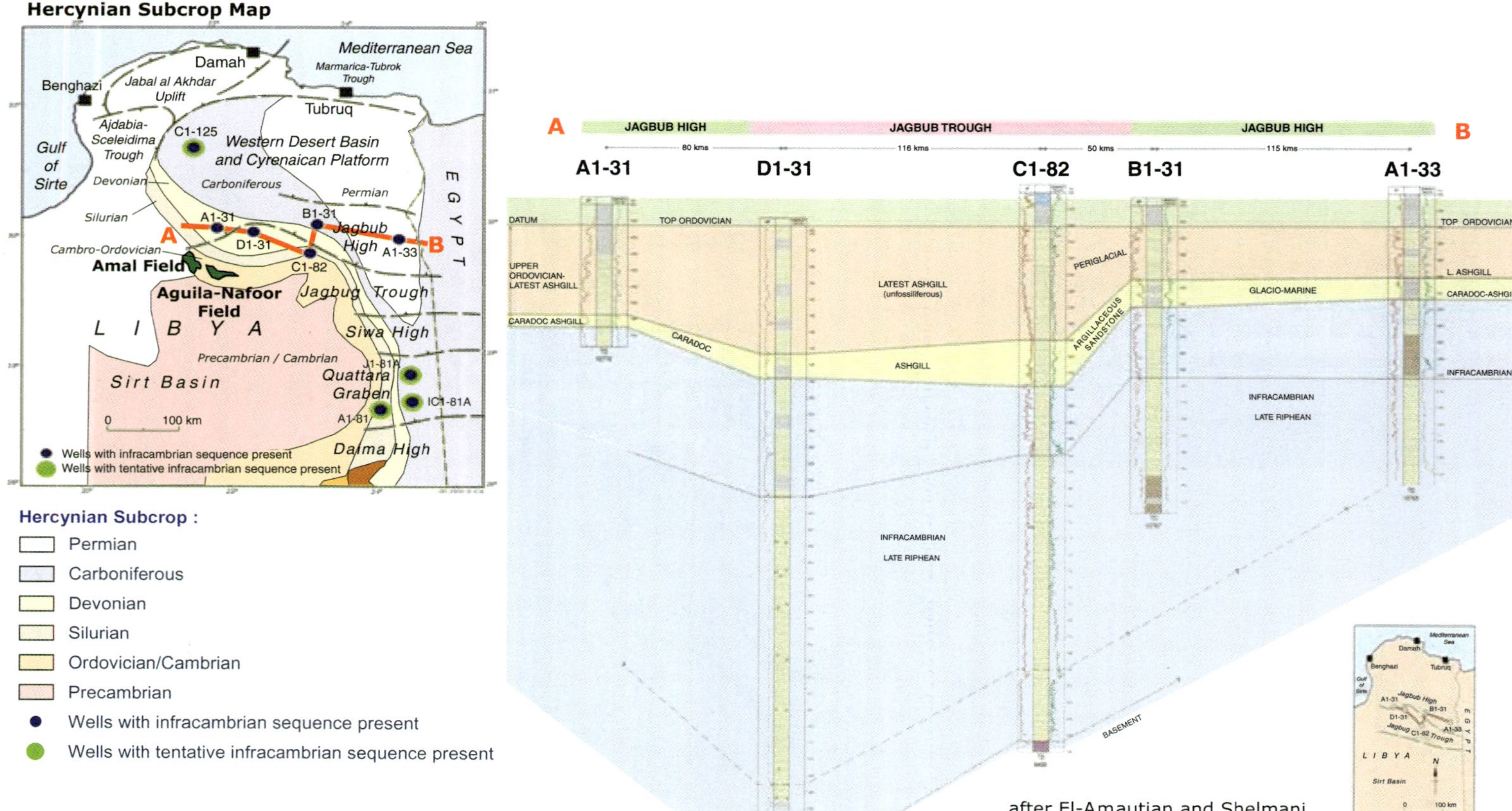

Fig. 13. Infracambrian strata in Cyrenaica, NE Libya.

Jagbub Trough have also recently yielded definitive Early–Mid Proterozoic-age microfossils. The assemblages recovered for these wells include the typical Late Riphean acritarchs *Leiosphaeridia asperta*, *Kildinosphaera chagrinata*, *K. verrucata*, *K. riphaeica* and the cynophytic filament *Taeniatum* sp. The assemblages are very similar to those recovered from the Taoudenni Basin in Mauritania. In addition, another well located further west in the NW Sirte Basin (A-10) is also reported to have penetrated an Infracambrian succession (Benshati *et al.* 2009). The Neoproterozoic succession penetrated in these wells locally exceeds 610 m in thickness.

The Augila Field contains the only proven Neoproterozoic fractured and weathered granite reservoir in the East Sirte Basin. K–Ar age dating of the granitic basement varies from 402 to 568 Ma. Considering that this could be a minimum age (Williams 1972), there is a possible broad correspondence with the Late Neoproterozoic age of the siliciclastic sequence of the Cyrenaica Platform. Hydrocarbon production in the Augila Field is not confined to the basement. Several wells also produce commercial quantities of hydrocarbons from the overlying siliciclastic and carbonate rocks. While the carbonate part of the reservoir is well dated as Late Cretaceous, the siliciclastic part remains largely undated. This basal clastic unit also contains oil in the Amal Field. The 'basal' sands are often thin or absent across the palaeo-highs, but occur in the down-thrown portions of fault blocks (Fig. 14). Subsurface data in the East Sirte Basin (Thusu 1996) indicates the presence of extensive reworked Devonian and Carboniferous palynomorphs within the Triassic and Jurassic age synrift sediments. This supports the concept that much of the Palaeozoic and possibly pre-Palaeozoic (Neoproterozoic?) sedimentary cover has been removed from the palaeo-highs and redeposited during the synrift tectonic phase. In this respect, the thick Neoproterozoic–Early Palaeozoic sequence encountered on the Cyrenaica Platform could be regarded as having been preserved from erosion and is also likely to be partially preserved in the down-faulted blocks adjacent to palaeo-highs in the East Sirte Basin.

In conclusion, the occurrence of a thick Late Riphean sequence in Cyrenaica would suggest that sediments of Infracambrian–Early Cambrian age may be preserved beneath Late Ordovician and younger successions in lows within the East Sirte Basin, especially close to the Cyrenaica Platform. The best example of this is probably the Maragh Trough where the potential existence of Infracambrian plays has been highlighted previously (FJA 1995), but is still to be fully investigated.

Neoproterozoic–Early Cambrian organic-rich strata in North Africa and their play potential

Information about the organic-richness and thermal maturity of the Neoproterozoic–Early Cambrian succession in North Africa is still patchy. Despite this, the existing data suggest that organic-rich strata are best developed in confined half-graben or over large shelfal areas such as those described for the Taoudenni Basin. Younger Neoproterozoic–Early Cambrian organic-rich deposits in Oman are considered to have been deposited in similar environments. Organic-rich strata are likely to be better developed in areas like the Taoudenni Basin, that are far from the Pan-African mountain belts. Close to the mountain belts coarse clastic sediment derived from post-Pan-African uplift and erosion is likely to dominate the depositional systems (Lüning *et al.* 2009). This applies particularly to basins such as Reggane, and Ahnet in west Algeria, where Neoproterozoic half-grabens are almost entirely filled with molasse.

Most of the Neoproterozoic organic-rich units that occur across North Africa (e.g. in the Tindouf, Reggane and Ahnet basins) were deposited between 850 and 540 Ma. There is increasing evidence that deposition of these Neoproterozoic organic-rich units was triggered by strong post-glacial sea-level rise on a global scale following major glaciations, coupled with basin development and rifting on a local scale. Post-glacial black shale deposition is also well documented in northern Gondwana during the subsequent Late Ordovician Saharan glaciation (Lüning *et al.* 2000). Detailed models of potential Neoproterozoic source rock distribution and the relationship between Neoproterozoic glacial phases and source rock deposition are still to be developed, but this concept is an attractive aspect of the Infracambrian play in North Africa.

In eastern Libya information is still too sparse to allow a proper assessment of the potential and distribution of possible Neoproterozoic source rocks. Some positive evidence has been documented for the Al Kufrah Basin, but in Cyrenaica and in the Murzuq Basin Neoproterozoic sedimentation appears to be dominated by coarse clastic deposition. However, on the Cyrenaica–Sirte basin margin, potential 'Infracambrian' reservoirs could also be charged by younger Cretaceous proven source rocks.

Hydrocarbon generation from Neoproterozoic source rocks in Oman occurred initially during Cambro-Ordovician time (Pollastro 1999) and early-generated oils have been preserved in areas of relative structural stability. In North Africa, the high maturity reached by some potential Neoproterozoic source rocks and destruction of traps during

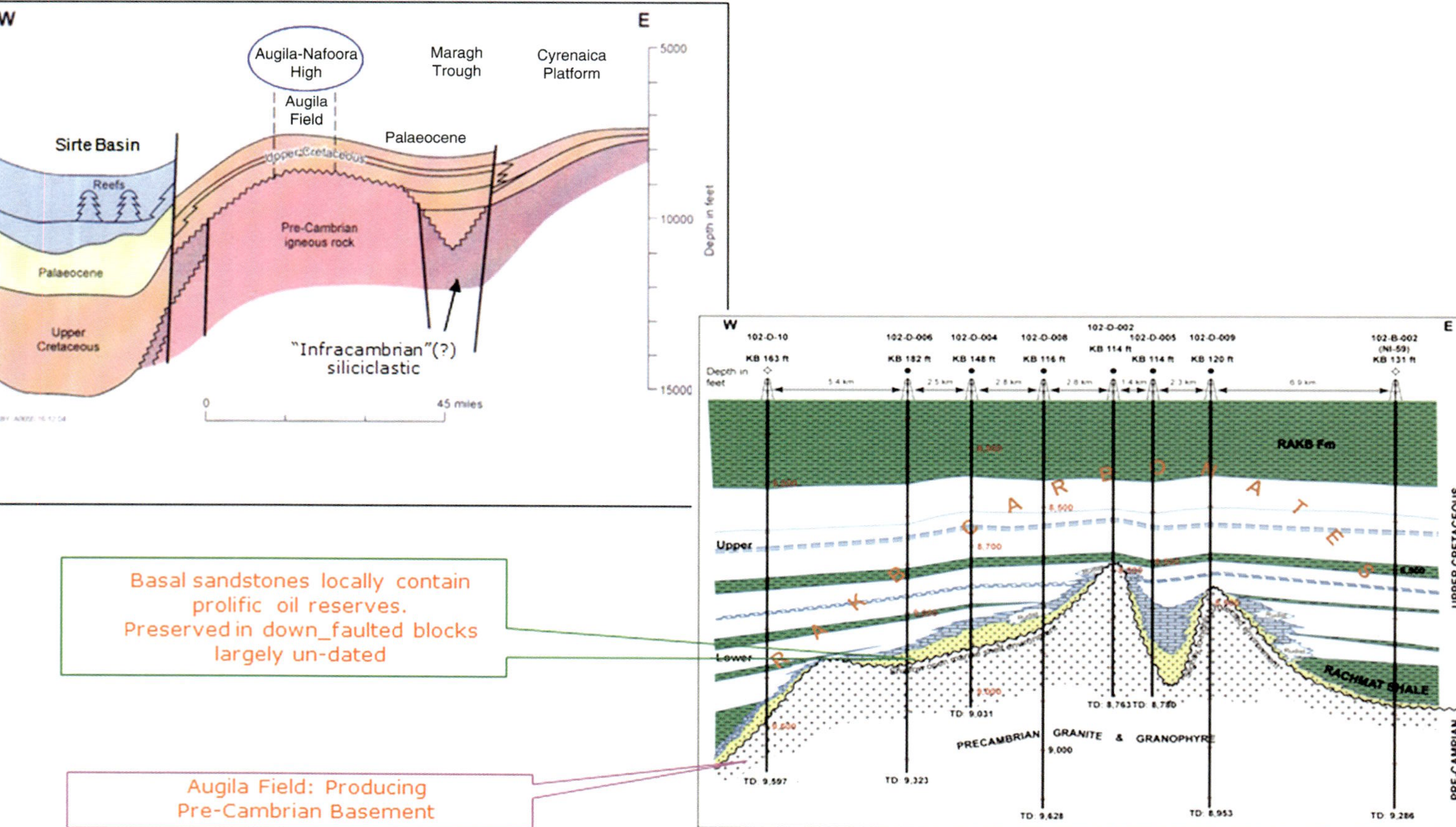

Fig. 14. Sirte–Cyrenaica Rift Margin, NE Libya. The left-hand diagram shows possible preservation of siliciclastic sediments on the Cyrenaica Platform in down-faulted blocks.

the Pan-African, Hercynian and later tectonic events are matters of concern for the preservation of early-generated hydrocarbon accumulations. However, the preservation of early-generated hydrocarbons in the Taoudenni Basin, as well as preserved pre-Hercynian hydrocarbons generated from Silurian source rocks in some North African Palaeozoic basins (Lüning *et al.* 2000), provides some reason for optimism. Areas of post-Hercynian generation (Al Kufrah Basin?) and where the source rocks have been protected from intense Pan-African, Hercynian or Alpine deformation (e.g. central Taoudenni basin?) are likely to be the most prospective regions.

Authors thank Eni for giving permission to publish this paper. A. Cozzi (Eni) and A. Whitham (CASP) are thanked for their effort in reviewing the manuscript. J. Thurow (UCL), S. Lüning (RWE), M. Geiger (Statoil), S. Kolonic (Shell), J. Bell (Hess) and S. Rasul are also thanked, along with contributors to the Maghreb Petroleum Research Group early work on North Africa 'Infracambrian'.

References

AL-SIYABI, H. A. 2005. Exploration history of the Ara intrasalt carbonate stringers in the South Oman Salt Basin. *GeoArabia*, **10**, 39–72.

AMRAD, B. 1986. Microfossiles (Acritarches) du Proterozoique superieur dans les shales de la Formation D'Atar (Mauritanie). *Precambrian Research*, **31**, 69–95.

AMTHOR, J. E., RAMSEYER, K., FAULKNER, T. & LUCAS, P. 2005. Stratigraphy and sedimentology of a chert reservoir at the Precambrian–Cambrian Boundary: the Al Shomou Silicilyte, South Oman Salt Basin. *GeoArabia*, **10**, 89–122.

ARNAUTI, A. & SHELMANI, M. 1988. A contribution to the northeast Libyan substratigraphy with emphasis on Pre-Mesozoic. *In*: ARNAUTI, A., OWENS, B. & THUSU, B. (eds) *Subsurface Palynostratigraphy of Northeast Libya*. Garyounis University, Benghazi, Libya.

AZIZ, A. & GHNIA, S. 2009. Distribution of Infracambrian rocks and the hydrocarbon potential within the Murzuq and Al Kufrah basins, NW Africa. *In*: CRAIG, J., THUROW, J., THUSU, B., WHITHAM, A. & ABUTARRUMA, Y. (eds) *Global Neoproterozoic Petroleum Systems: The Emerging Potential in North Africa*. Geological Society, London, Special Publications, **326**, 211–219.

BAUDET, D. 1988. Precambrian palynomorphs from northeast Lybia. *In*: EL-ARNAUTI, A., OWENS, B. & THUSU, B. (eds) *Subsurface Palynostratigraphy of Northeast Libya*. Garyounis University, Bengazhi, Libya, 17–25.

BELLINI, E. & MASSA, D. 1980. A stratigraphic contribution to the Palaeozoic of the southern basins of Lybia. *In*: SALEM, M. J. & BESREWIL, M. T. (eds) *The Geology of Libya*, Volume **1**, Academic Press, New York, 3–65.

BENAN, C. A. A. & DEYNOUX, M. 1998. Facies analysis and sequence stratigraphy of Neoproterozoic platform deposits in Adrar of Mauritania, Taoudeni Basin, West Africa. *Geologische Rundschau*, **87**, 283–302.

BENSHATI, H., KHOJA, A. & SOLA, M. 2009. Infracambrian sediments in Libyan sedimentary basins. *In*: CRAIG, J., THUROW, J., THUSU, B., WHITHAM, A. & ABUTARRUMA, Y. (eds) *Global Neoproterozoic Petroleum Systems: The Emerging Potential in North Africa*. Geological Society, London, Special Publications, **326**, 181–191.

BERTRAND-SARFATI, J. & MOUSSINE-POUCHKINE, A. 1988. Is cratonic sedimentation consistent with available models? An example from the Upper Proterozoic of the West African craton. *Sedimentary Geology*, **58**, 255–276.

BOUIMA, T. & MEZGHACHE, H. 2002. Les formations infracambriennes des monts de l'Ougarta (Algerie) et leur correlation avec celles de l'Anti-Atlas central (Maroc). *Memoir Service Geologique de l'Algérie*, **11**, 33–44.

CABY, R. & MONIÈ, P. 2003. Terrane assembly and geodynamic evolution of central-western Hoggar: a synthesis. *Journal of African Earth Science*, **37**, 133–159.

CHAKRABORTY, P. P. 2004. Facies architecture and sequence development in a Neoproterozoic carbonate ramp: Lakheri Limestone Member, Vindhyan Supergroup, Central India. *Precambrian Research*, **132**, 29–53.

CONDIE, K. C. 2003. Supercontinents, superplumes and continental growth: the Neoproterozoic record. *In*: YOSHIDA, M., WINDLEY, B. F. & DASGUPTA, S. (eds) *Proterozoic East Gondwana: Supercontinent Assembly and Breakup*. Geological Society, London, Special Publications, **206**, 1–21.

COZZI, A. & AL-SIYABI, H. A. 2004. Sedimentology and play potential of the late Neoproterozoic Buah Carbonates of Oman. *GeoArabia*, **9**, 11–36.

CRAIG, J. 2006. Global Neoproterozoic petroleum systems: snowballs, source rocks and the search for 'Old Oil' (Abstract). *Petroleum Exploration Society of Great Britain Newsletter*, August/September, 8–10.

CRAIG, J., RIZZI, C., ET AL. 2008. Structural styles and prospectivity in the Precambrian and Palaeozoic hydrocarbon systems of North Africa. *In*: *Geology of East Libya*, **4**, 51–122.

DEYNOUX, M., AFFATON, P., TROMPETTE, R. & VILLENEUVE, M. 2006. Pan-African tectonic evolution and glacial events registered in Neoprotozoic to Cambrian cratonic and foreland basins of West Africa. *Journal of African Earth Science*, **46**, 397–426.

DEYNOUX, M., SOUGY, J. & TROMPETTE, R. 1985. Lower Palaeozoic rocks of West Africa and the western part of Central Africa. *In*: HOLLAND, C. H. (ed.) *Lower Palaeozoic Rocks of North-western Africa and West Central Africa*. Wiley, Chichester, 337–495.

EL-ARNAUTI, A. & SHELMANI, M. 1988. A contribution to the northeast Libyan substratigraphy with emphasis on Pre-Mesozoic. *In*: ARNAUTI, A., OWENS, B. & THUSU, B. (eds) *Subsurface Palynostratigraphy of Northeast Libya*. Garyounis University, Benghazi, Libya, 1–15.

EL-MEHDI, B., TURKI, S. M., SUWESI, S. K. & OWEISS, K. 2004. Short notes and guidebook on the geology of

Al Kufrah Basin, Al'Awaynat area. *In*: *Sedimentary Basins of Libya, Third Symposium: Geology of East Libya Fieldtrip, November 2004*. Earth Science Society of Libya, Tripoli, 68.

EYLES, N. 1993. Earth's glacial record and its tectonic setting. *Earth-Science Reviews*, **35**, 1–248.

EYLES, N. & JANUSZCZAK, N. 2004. 'Zipper-rift': a tectonic model for Neoproterozoic glaciations during the break up of Rodinia after 750 Ma. *Earth-Science Reviews*, **65**, 1–73.

EVANS, D. A. D. 2000. Stratigraphic, geochronological, and palaeomagnetic constraints upon the Neoproterozoic climatic paradox. *American Journal of Science*, **300**, 347–433.

FJA CONSULTING. 1995. *The Palaeozoic Hydrocarbon Potential and Prospectivity of Eastern Algeria, Southern Tunisia and Western Egypt*. FJA Consulting, Llandudno, UK.

GEYER, G. & LANDING, E. 1995. The Cambrian of the Moroccan Atlas regions. *In*: GEYER, G. & LANDING, E. (eds) *Morocco '95; The Lower–Middle Cambrian standard of western Gondwana*. Berengeria, Wurzburg. Special Issue, **2**, 7–46.

GENNA, A., NEHILIG, P., LE GOFF, E., GUERROT, C. & SHANTI, M. 2002. Proterozoic tectonism of the Arabian Shield. *Precambrian Research*, **117**, 21–40.

GHORI, K. A. R., CRAIG, J., THUSU, B., LÜNING, S. & GEIGER, M. 2009. Global Infracambrian petroleum systems: a review. *In*: CRAIG, J., THUROW, J., THUSU, B., WHITHAM, A. & ABUTARRUMA, Y. (eds) *Global Neoproterozoic Petroleum Systems: The Emerging Potential in North Africa*. Geological Society, London, Special Publications, **326**, 109–136.

GRADSTEIN, F. M. & OGG, J. G. 2004. *Geologic Time Scale 2004*. Cambridge University Press, Cambridge.

HALLET, D. 2002. *Petroleum Geology of Libya*. Elsevier, Amsterdam.

HARLAND, W. B. & RUDWICK, M. J. S. 1964. The great Infra-Cambrian ice age. *Scientific American*, **211**, 28–36.

HOFFMAN, P. F. & SHRAG, D. P. 2002. The Snowball Earth hypothesis: testing the limits of global change. *Terra Nova*, **14**, 129–155.

HOFFMAN, P. F., KAUFMAN, A. J., HALVERSON, G. P. & SCHRAG, D. P. 1998*a*. Comings and goings of global glaciations on a Neoproterozoic tropical platform in Namibia. *GSA Today*, **8**, 1–9.

HOFFMAN, P. F., KAUFMAN, A. J., HALVERSON, G. P. & SCHRAG, D. P. 1998*b*. A Neoproterozoic Snowball Earth. *Science*, **281**, 1342–1346.

HUSSEINI, M. I. & HUSSEINI, S. I. 1990. Origin of the Infracambrian salt basins of the Middle East. *In*: BROOK, J. (ed.) *Classic Petroleum Provinces*. Geological Society, London, Special Publications, **50**, 279–292.

JACOBS, J. & THOMAS, R. J. 2004. Himalayan-type indenter-escape tectonics model for the southern part of the late Neoproterozoic–early Palaeozoic East African–Antarctic orogen. *Geology*, **32**, 721–724.

JACQUÉ, M. 1962. Reconnaissance geologique du Fezzan oriental. *Notes et Memoires Compagnie Française des Petroles*, **5**, 44.

KNOLL, A. H. & CARROLL, S. B. 1999. Early animal evolution: emerging views from comparative biology and geology. *Science*, **284**, 2129–2137.

LEBLANC, M. & MOUSSINE-POUCHKINE, A. 1994. Sedimentary and volcanic evolution of a Neoproterozoic continental margin (Bleida, Anti-Atlas, Morocco). *Precambrian Research*, **70**, 25–44.

LE HERON, D. P., HOWARD, J. P., ALHASSI, A. M., ANDERSON, L., MORTON, A. & FANNING, C. M. 2009. Field-based investigations of an 'Infracambrian' clastic succession in SE Libya and its bearing on the evolution of the Al Kufrah Basin. *In*: CRAIG, J., THUROW, J., THUSU, B., WHITHAM, A. & ABUTARRUMA, Y. (eds) *Global Neoproterozoic Petroleum Systems: The Emerging Potential in North Africa*. Geological Society, London, Special Publications, **326**, 193–210.

LÜNING, S., CRAIG, J., LOYDELL, D. K., STORCH, P. & FITCHES, B. 2000. Lower Silurian 'hot shales' in North Africa and Arabia: regional distribution and depositional model. *Earth Science Reviews*, **49**, 121–200.

LÜNING, S., KOLONIC, S., GEIGER, M., THUSU, B., BELL, J. S. & CRAIG, J. 2009. Infracambrian hydrocarbon source rock potential and petroleum prospectivity of NW Africa. *In*: CRAIG, J., THUROW, J., THUSU, B., WHITHAM, A. & ABUTARRUMA, Y. (eds) *Global Neoproterozoic Petroleum Systems: The Emerging Potential in North Africa*. Geological Society, London, Special Publications, **326**, 157–180.

MACGREGOR, D. S. 1996. The hydrocarbon systems of North Africa. *Marine and Petroleum Geology*, **13**, 329–340.

MOUSSINE-POUCHKINE, A. & BERTRAND-SARFATI, J. 1997. Tectonosedimentary subdivisions in the Neoproterozoic to early Cambrian cover of the Taoudenni Basin (Algeria–Mauritania–Mali). *Journal of African Earth Science*, **7**, 57–75.

POLLASTRO, R. M. 1999. *Ghaba Salt Basin Province and Fahud Salt Basin Province, Oman – Geological Overview and Total Petroleum Systems*. US Geological Survey, 25.

SAÏD, M. M., OWEISS, K. A. & MEHIDI, B. O. 2000. *Jabal Arkenu Sheet: Explanatory Booklet*. Joint Geological Mapping Project between the Egyptian Geological Survey and Mining authority (EGSMA) and the Industrial Research Centre (IRC) of Libya, Cairc.

SANKARAN, A. V. 2003. Neoproterozoic 'snowball earth' and the 'cap carbonate' controversy. *Current Science*, **84**, 7.

SELLEY, R. C. 1971. *Preliminary report of a reconnaissance study of western Kufra Basin, Southern Libya*. Unpublished Stratigraph Report, **95**. Oasis Oil Company of Libya.

SHANDELMEIER, H. 1988. Pre-Cretaceous intraplate basins of NE Africa. *Episodes*, **11**, 270–274.

THUSU, B. 1996. Implication of the Discovery of Reworked and in situ Late Palaeozoic and Triassic palynomorphs on the evolution of Sirt Basin, Libya. *In*: SALEM, M. J., MOUZUGHI, A. J. & HAMMUDA, O. S. (eds) *The Geology of Sirt Basin, Volume 1*. Elsevier, Amsterdam.

WILLIAMS, J. J. 1972. Augila Field Libya: depositional Environment and diagenesis of sedimentary reservoir and description of igneous reservoir. *AAPG Memoir*, **16**, 623–632.

Infracambrian hydrocarbon source rock potential and petroleum prospectivity of NW Africa

S. LÜNING[1,2]*, S. KOLONIC[1,3], M. GEIGER[1,4], B. THUSU[5], J. S. BELL[6,7] & J. CRAIG[8]

[1]*Department of Geosciences – FB 5, PO Box 330 440, University of Bremen, 28334 Bremen, Germany*

[2]*Present address: RWE Dea, Überseering 40, 22297 Hamburg, Germany*

[3]*Present address: Shell, Business Development Middle East Ltd, PO Box 926438, Ibu Arabi Street, Amman 11190, Kingdom of Jordan*

[4]*Present address: StatoilHydro ASA, 4035 Stavanger, Norway*

[5]*MPRG (Maghreb Petroleum Research Group), University College, London, Gower Street, London WC1E 6BT, UK*

[6]*Eni-LTE, Bowater House East, 68 Knightsbridge, London SW1X 7BN, UK*

[7]*Present address: Hess, London, UK*

[8]*Eni Exploration and Production Division, Via Emilia 1, 20097 San Donato Milanese, Milan, Italy*

**Corresponding author (e-mail: sebastian.luning@gmx.net)*

Abstract: Proven Infracambrian hydrocarbon plays occur in various parts of the world, including Oman, the former Soviet Union, India, Pakistan and Australia. Organic-rich strata also occur in NW Africa, and gas shows originating from Infracambrian hydrocarbon source rocks are known from well Abolag-1 in the Mauritanian part of the Taoudenni Basin. The distribution of Infracambrian source rocks in North Africa is patchy and deposition commonly occurred in half-graben and pull-apart basins. In these intra-shelf basins, marine, organic-rich shales and limestones were deposited beneath the turbulent wave zone, away from the coarse siliciclastic Pan-African molasse detritus. On the West African Craton (including the Taoudenni Basin) organic-rich horizons were also deposited earlier, in pre- and syn-Pan-African times between 0.5 and 2 Ga (Ga is 10^9 years). The long-lasting sedimentation history in this area contrasts with that of the Pan-African regions, such as Oman, which lies in the Pan-African province of the East African Orogen, where preserved sediments are rarely older than 640 Ma. Infracambrian black phyllites in the Anti-Atlas region of Morocco were deposited on a continental slope of a short-lived ocean lying to the north of the West African Craton. Hydrocarbons generated during Infracambrian times from these deposits, however, have a low preservation potential.

Infracambrian organic-rich and/or black-pyritic deposits in North Africa are proven in the Taoudenni Basin, the Anti-Atlas and the Ahnet Basin. Thick carbonate successions exist in the Taoudenni Basin, indicating deposition in areas some distance from contaminating coarse siliciclastic hinterland influx. Infracambrian strata may also occur in the Tindouf Basin. However, their deep burial and consequent early maturation history may be unfavourable for the preservation of Infracambrian-sourced hydrocarbons in this area. Local development of Infracambrian source facies may also occur in the Reggane, Ahnet, Mouydir and Iullemeden basins, as indicated by black shales in wells MKRN-1 and MKRS-1 in the Ahnet Basin. Generally, however, these basins appear to be close to the active Pan-African orogenic belt and, consequently, probably received large quantities of coarse siliciclastic sediment, largely of continental facies, which may have diluted any significant hydrocarbon source potential.

The stratigraphic term 'Infracambrian' is loosely defined. It generally refers to rocks immediately underlying the Cambrian (Wyatt 1986), but also partly extends into the early Cambrian. In North Africa and Arabia, the Infracambrian succession can be defined as comprising rocks of Neoproterozoic (1000 Ma–end Precambrian, Fig. 1) to early Cambrian age. Infracambrian strata were deposited as pre-Pan-African platform sediments and syn- to post-Pan-African molasse sediments, overlain by Cambrian and younger, cratonic strata that formed on a peneplained substrate.

From: CRAIG, J., THUROW, J., THUSU, B., WHITHAM, A. & ABUTARRUMA, Y. (eds) *Global Neoproterozoic Petroleum Systems: The Emerging Potential in North Africa.* Geological Society, London, Special Publications, **326**, 157–180.
DOI: 10.1144/SP326.8 0305-8719/09/$15.00

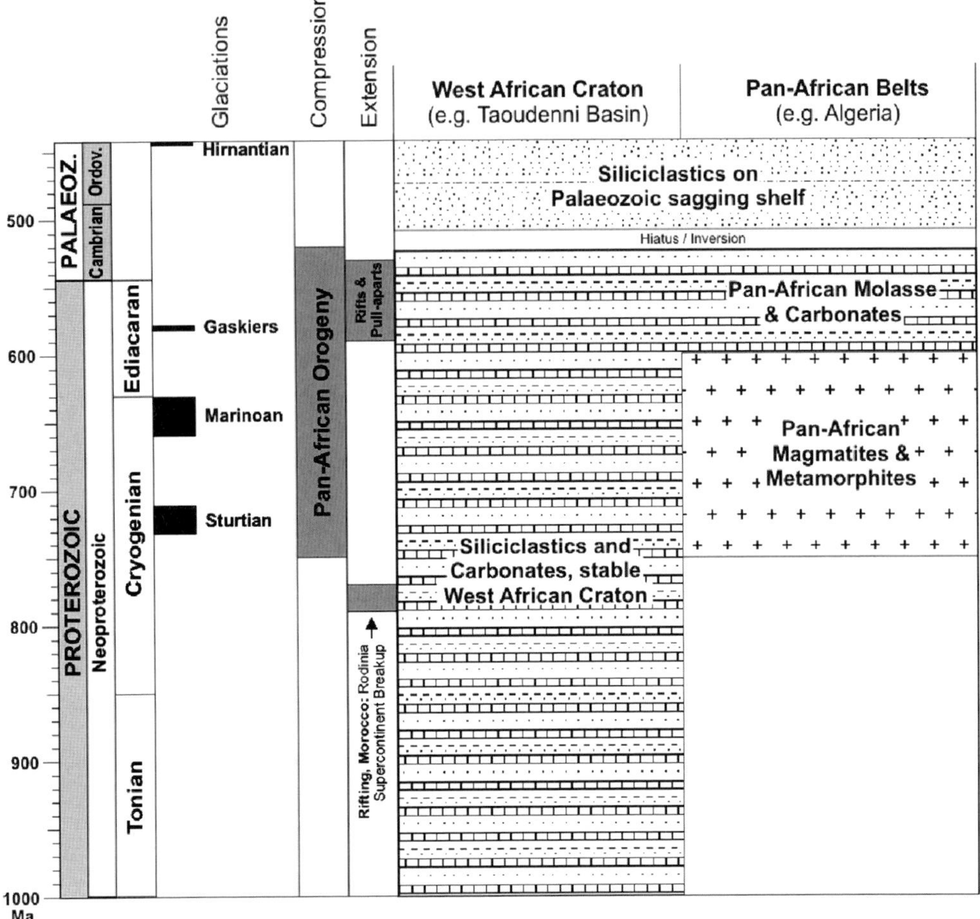

Fig. 1. Chronostratigraphic chart of the Infracambrian in NW Africa. Deposition on the West African Craton commenced significantly earlier than over the Pan-African belts. Extensional features such as half-graben and pull-apart basins were active during the late Pan-African phase and subsequently were locally inverted. The Infracambrian is characterized by abundant carbonates, while Lower Palaeozoic strata are generally dominated by siliciclastic strata. The chart is strongly simplified, sedimentation on the West African Craton may not have been continuous and it remains unclear whether glacial deposits associated with the Gaskiers and Sturtian glaciations are present in North Africa.

The Infracambrian successions across Gondwana were deposited in basins of different tectonic origin, including half-graben, pull-apart basins, foreland basins, intramontane molasse basins and continental slope basins. The timing of the onset of Infracambrian sedimentation varies greatly, with the oldest preserved sediments in NW Africa being 2000–1000 Ma in age (on the West African Craton), while in the Pan-African zone of Oman the earliest Infracambrian sediments are generally not older than 640 Ma.

In parts of Gondwana, the Infracambrian succession contains organic-rich horizons, which in several areas (e.g. Oman, Pakistan, India and Australia) have sourced significant hydrocarbon accumulations (Warren *et al.* 1998). Deposition of these marine organic-rich strata is often restricted to the deeper parts of half-graben and pull-apart basins where anoxic environments developed under low-energy conditions. Consequently, the lateral extent and thickness of these Infracambrian organic-rich strata are highly variable and patchy.

In this contribution we review the structural framework and facies of the known Infracambrian organic-rich strata in NW Africa. We also define tectono-sedimentary domains in the region based on the likelihood that: (1) depositional basins were formed here during the Infracambrian; and (2) a low-energy, marine facies with dysoxic–anoxic conditions was developed. Criteria used in this evaluation include the plate tectonic position (Pan-African belts v. cratons), proximity to large-scale wrench systems, type of sedimentary basin, stratigraphic age of sediments and dominant sediment type.

The existing subsurface database for the Infracambrian successions in NW Africa is rather poor, mostly owing to the great depth of the succession in the Palaeozoic basins with consequent rare well penetrations. Large-scale Infracambrian surface exposures occur in the Moroccan Anti-Atlas and around the Taoudenni Basin (Mauritania, Mali and Algeria). These outcrops form a valuable analogue for the subsurface. Nevertheless, the patchy primary depositional nature of the Infracambrian organic-rich strata as well as desert weathering processes complicates a reliable extrapolation of organic-richness trends into the subsurface. Robust tectono-depositional models are therefore essential for reliable regional Infracambrian source quality predictions and the development of associated Infracambrian play concepts in NW Africa.

Materials and methods

Materials

The present review is mainly based on published information and, to a minor extent, on proprietary information, including some localized field sampling and analysis.

Wireline log analysis

Hydrocarbon source rocks can often be identified in wireline logs, either by cross-plotting resistivity/sonic data (delta log R method: Passey *et al.* 1990) or from their characteristic elevated gamma-ray values. A strong correlation between gamma-ray and total organic carbon (TOC) has been reported for some Infracambrian organic-rich strata. In Oman, total gamma-ray values may be as high as 400 API with organic-richnesses of 8% and more, and the positive correlation of the two parameters forms the basis of source rock quality predictions based on wireline logs. A similarly good correlation between gamma-ray values and TOC was reported from the 1.4 Ga Proterozoic Velkerri Formation of Australia, where organic content reaches maximum values of 7% with associated gamma-ray readings of up to 300 API (Warren *et al.* 1998).

Gamma-ray values of black shales in the Infracambrian succession of the Ahnet Basin wells MKRN-1 and MKRS-1, however, do not exceed 120 and 60 API, respectively. Furthermore, resistivity values in the Mauritanian Abolag-1 well (see later) are rather low throughout the Infracambrian succession, which includes black shales, although analysis suggests these are organically lean.

Stratigraphic tools

Precise dating of Infracambrian successions is generally problematic mainly due to the scarcity of suitable stratigraphically diagnostic micro- and macrofossils. Palynomorphs provide valuable age information where sphaeromorph acritarchs are present in marine sediments. Palynomorph biostratigraphy is also largely restricted to subsurface samples because the microflora are commonly destroyed by the intense near-surface oxidization in desert environments. In outcrops, therefore, a pragmatic stromatolite 'biostratigraphic' scheme has been developed that works well on a local scale (Bertrand-Sarfati 1972; Bertrand-Sarfati & Moussine-Pouchkine 1991) and forms an important basis for lithostratigraphic subdivision of the Infracambrian successions.

Another useful age-dating technique is stable isotope chemostratigraphy, mainly based on $\delta^{13}C$ and $^{87}Sr/^{86}Sr$ isotopes (e.g. Brasier & Shields 2000; Halverson *et al.* 2005, 2007; e.g. Kaufman & Knoll 1995; Melezhik *et al.* 2001; Shields 1999, 2007). In the Anti-Atlas Mountains of Morocco, the Infracambrian (Ediacaran–Early Cambrian) interval exhibits distinctive cycles in the carbon isotope curve that match well with similar patterns from age-equivalent strata on the Siberian Platform (Magaritz *et al.* 1991; Maloof *et al.* 2005). Similar studies have been carried out for worldwide correlations. The cause of these carbon isotope oscillations, however, remains unclear. They could, for example, be the result of changes in the deposition of organic matter in ocean basins or in the biological productivity of the ocean surface.

The ages of some of the Infracambrian successions in NW Africa are also constrained by a framework of radiometric dates on interbedded volcanic and volcaniclastic rocks (Clauer 1976; Clauer & Deynoux 1987; Lahondère *et al.* 2005; Callec *et al.* 2006).

Tectonic framework

Deposition of Infracambrian strata occurred during the life cycle of the two supercontinents Rodinia and Pannotia. The supercontinent Rodinia aggregated between 1300 and 950 Ma, and broke up between 850 and 600 Ma. A new supercontinent (Pannotia–Gondwana) then formed between 680 and 550 Ma (Condie 2003). The continental crust of present-day North Africa includes the Archaean–Palaeoproterozoic West Africa and East Sahara cratons (Fig. 2), both of which belonged to the West Gondwana Province during Rodinia times and at which time they were surrounded by oceanic spreading zones (e.g. Bertrand-Sarfati *et al.* 1987).

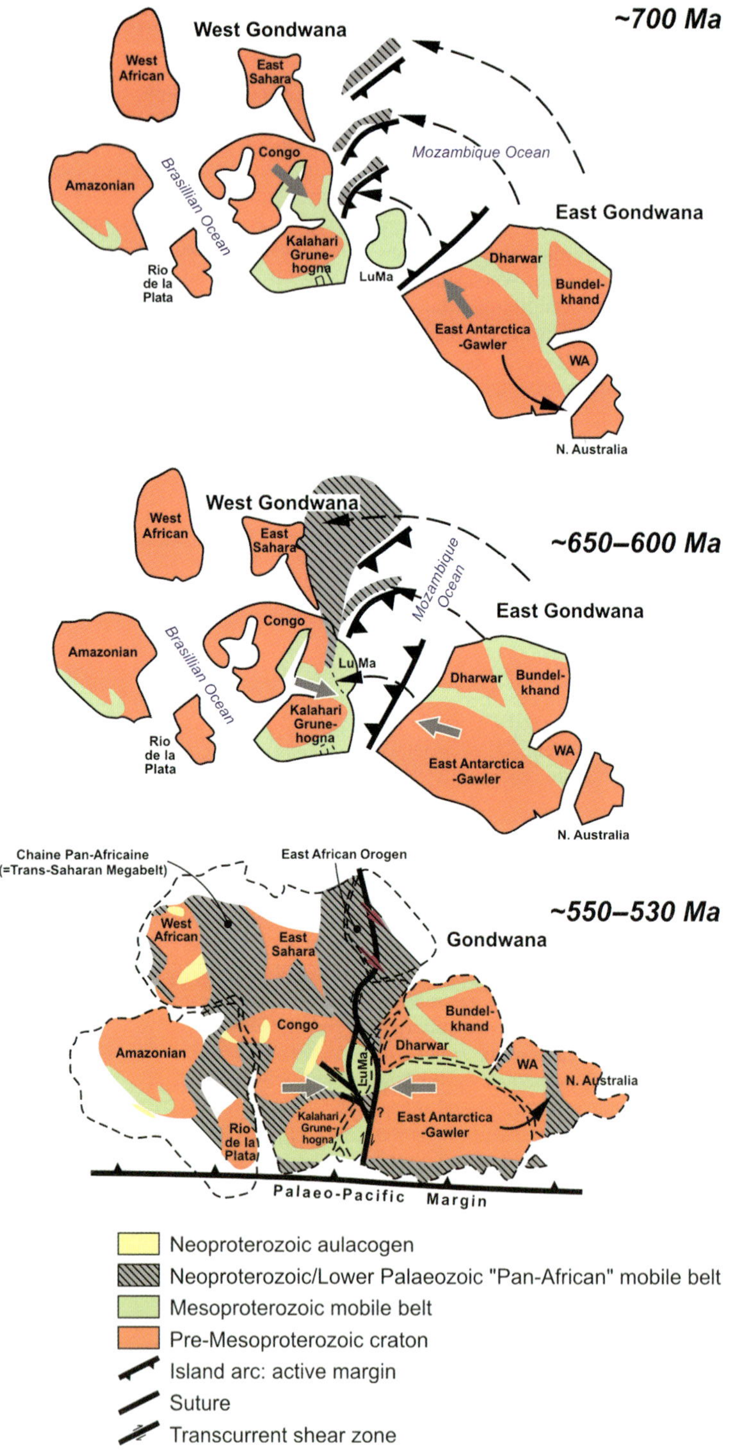

Fig. 2. Collision of West and East Gondwana during the Neoproterozoic–Cambrian, leading to the formation of the East African–Antarctic Orogen. The western and central parts of North Africa are dominated by the Archaean–Palaeoproterozoic cratons 'West Africa' and 'East Sahara' (from Jacobs & Thomas 2002).

Pan-African Orogeny

Following the split-up of Rodinia, East Gondwana separated from Laurentia and at around 600 Ma collided with West Gondwana to form Pannotia–Gondwana (Fig. 2). The East African Orogen formed along the collision zone between West and East Gondwana, including the present-day eastern margin of the East Saharan Craton (Fig. 2). Today, in western Arabia, eastern Egypt and Sudan, the northern part of the East African Orogen is represented by the Arabian–Nubian Shield. The amalgamation of different cratons and terranes at this time culminated in a major phase of deformation and uplift that is referred to as the 'Pan-African Orogeny'. In western North Africa, a second Pan-African orogenic belt developed along the collision zone between the West African and East Saharan cratons. This is termed the 'Trans-Saharan Megabelt' or, simply, the 'Chaine Pan-Africaine' (Pan-African Belt) (Fig. 2).

Chaine Pan-Africaine

The Chaine Pan-Africaine (or Trans-Saharan Megabelt) is located in parts of Algeria, Mali and Niger. It formed between 750 and 520 Ma through the collision of more than 20 terranes between the West African and East Saharan cratons (Villeneuve & Cornee 1994; Azzouni-Sekkal *et al.* 2003; Bournas *et al.* 2003; Caby 2003) (Fig. 2). The Chaine Pan-Africaine is exposed in the Anti-Atlas, Ougarta, Pharusian-Tuareg (Hoggar Massif), Gourma and Dahomeyan belts (Villeneuve & Cornee 1994; Condie 2003). The Touareg Shield consists of 'Pharusian' microplates from the so-called Pharusian Ocean, which lay between the West Africa Craton and Eastern Sahara Craton (Hallett 2002). Furthermore, Pan-African orogenic belts appear to have completely surrounded the West African Craton during the Neoproterozoic. Outcrops of Pan-African rocks also occur on the western side of West African Craton in the Mauritanian, Bassaride and Rokelide belts (Condie 2003; Deynoux *et al.* 2006).

The boundary between the Chaine Pan-Africaine and the West African Craton can also be identified in the Moroccan Anti-Atlas, which has been divided into three structural domains (Fekkak *et al.* 2004; Gasquet *et al.* 2005): the southern Anti-Atlas, which represents the northern border of the West African Craton; the central Anti-Atlas, which represents the suture zone with an accretionary complex; and the eastern Anti-Atlas where only Neoproterozoic rocks occur. In contrast, Ennih & Liégeois (2001) consider the whole Anti-Atlas system to be part of the West African Craton. According to these authors, the Pan-African units in the Anti-Atlas represent ocean slices that were thrust onto the craton approximately 685 Ma ago as a result of Pan-African accretion tectonics.

Intramontane molasse basins are also developed in the Chaine Pan-Africaine. Large amounts of Pan-African molasse accumulated in the NW Hoggar, where it fills residual basins and graben with up to 6000 m of red and green clastic sediments together with limestones/dolomites (Caby & Monié 2003). On the nearby stable West African Craton, Infracambrian deposition was also influenced by the tectonic activity in the Chaine Pan-Africaine. For example, there are considerable variations in the thickness of Neoproterozoic sediments within the Taoudenni Basin, ranging from 1000 m in Adrar to 100 m in Algeria (Moussine-Pouchkine & Bertrand-Sarfati 1997). Part of these thickness changes may also be related to a variable degree of erosion by Neoproterozoic glacial erosion, as well as regionally non-consistent differentiation of Neoproterozoic and Palaeozoic strata in general.

Infracambrian extension

Pull-apart and rift basins, half-graben and strike-slip zones formed in North Africa, Arabia and other parts of Gondwana (e.g. along its Pacific margin) during the latest phase of the Pan-African Orogeny and during earliest Cambrian times (590–530 Ma) (Sharland *et al.* 2001; Kusky & Matsah 2003; Boger & Miller 2004; Thomas *et al.* 2004) (Fig. 1). In West Gondwana, several thousand kilometres from North Africa–Arabia, the Iapetus Ocean began to open at about this time (e.g. Cawood *et al.* 2001).

The extensional movements in NE Africa and Arabia are interpreted as resulting from escape tectonics associated with the Pan-African–East African–Antarctic Orogen (Kusky & Matsah 2003; Jacobs & Thomas 2004). In Arabia, the formation of the latest Neoproterozoic and earliest Cambrian salt basins was governed by a system of transcurrent faults known as 'Najd Faults' (Husseini & Husseini 1990; Talbot & Alavi 1996; Droste 1997). The Najd Fault System strikes NW–SE across some 1100 km of the exposed Arabian Shield, and was active during various phases between 680 Ma and the early Cambrian. The exposed fault system is about 300–400 km wide in the Arabian Shield, with about 300 km of left-lateral displacement (Husseini 1988). The origin of the Najd Fault System is strongly debated, and both extensional (e.g. Husseini 1988) and compressional (Johnson & Woldehaimanot 2003) models have been proposed. According to one model, the deep basins formed as continental pull-apart basins along transform faults. Listric or gently dipping faults are interpreted as traces of gravitational sliding in the uppermost part of the crust. The geometry of these normal faults and of associated dyke swarms indicates extension in many different directions.

Some of the Infracambrian half-graben in North Africa, such as those in southern Algeria (Moussine-Pouchkine & Bertrand-Sarfati 1997), may have formed as a result of extensional collapse of the Pan-African Orogen (Greiling *et al.* 1994). The Pan-African Orogeny was followed by continental-scale uplift and erosion throughout North Africa and Arabia. This resulted in the development of a vast peneplain cutting across the exhumed Precambrian basement from Morocco in the west to Oman in the east (Avigad *et al.* 2003). In the Taoudenni Basin, two Late Neoproterozoic rifting phases have been proposed, taking place after the Marinoan glaciation and around the Neoproterozoic–Cambrian boundary.

Climate and sea-level history

Plate reconstructions suggest that North Africa was located at high latitudes at 600 Ma (Dalziel 1997) and was affected by several episodes of glaciations during the Neoproterozoic (Deynoux *et al.* 1978, 2006; Hoffman & Schrag 2002; Johnson & Woldehaimanot 2003; Miller *et al.* 2003; Jenkins *et al.* 2004). According to the 'Snowball Earth' hypothesis of Hoffman & Schrag (2002), large parts of the Earth's surface were covered by ice at about 700 Ma (Sturtian glaciation) and again at about 635 Ma (Marinoan glaciation) (Fig. 1). Both of these glacial episodes are thought to have ended abruptly and were followed by periods of very warm climate. Extreme $\delta^{13}C$ excursions coincide with these glacial 'snowball' events.

Glacial deposits associated with the Marinoan glaciation are exposed in outcrops around the Taoudenni Basin. The Bakoye Group is represented by 500 m of interbedded diamictites, glacio-aeolian, glaciofluvial and glaciomarine deposits (Deynoux & Trompette 1981; Deynoux *et al.* 1989, 2006; Proust & Deynoux 1994). In the Mauritanian Adrar region, the glaciation left behind an irregular, post-glacial topographic relief infilled by sediment showing significant lateral facies variations (Shields *et al.* 2007*a*, *b*). The eustatic sea-level rise resulting from deglaciation commenced with deposition of a thin, but widespread, limestone unit ('cap carbonate'). Isostatic rebound simultaneously lifted the highest parts of the craton above sea level (Shields *et al.* 2007*b*).

The combination of post-glacial transgression, variable palaeorelief, rise in water temperature and a possibly stratified shelfal sea potentially provides favourable conditions for the development of anoxia. The early Silurian organic-rich black shales in northern Gondwana were formed in a similar setting following the Late Ordovician glaciation (Lüning *et al.* 2000). It has yet to be determined whether the organic-rich Infracambrian sediments that occur in Oman and Mauritania are related to the Neoproterozoic (de-)glaciations.

A different climatic extreme during Infracambrian times is recorded by the deposition of salt in the Infracambrian basins of Arabia and Pakistan. Infracambrian salt deposition may have been restricted to moderate–low latitudes. In the 600 Ma plate tectonic reconstruction of Dalziel (1997), Arabia is located between 50° and 30°S. This contrasts with the high southern latitude location of West Africa at this time, which may explain the absence of evaporitic deposits, and hence a regional Infracambrian super-seal, in this region.

Significant sea-level changes occurred during the Infracambrian as a consequence of the pronounced glacial–interglacial climatic oscillations and the contemporaneous extensional tectonic movements. Sequence stratigraphic interpretations of the Infracambrian sediments of the Arabian Peninsula and Morocco were given by Sharland *et al.* (2001) and Bouougri & Saquaque (2004), respectively.

Infracambrian strata in NW Africa

Infracambrian rocks have been described from various parts of NW Africa. Stratigraphic ages, facies and geodynamic setting vary greatly across northern Gondwana, which complicates direct comparison and correlation of the successions. Palaeogeographical maps were published by Villeneuve & Cornee (1994).

Morocco

The Neoproterozoic and, especially, Cambrian are the dominant stratigraphic units in the Anti-Atlas, where they surround geological 'windows' of Precambrian crystalline basement. The 'Infracambrian' succession in the Anti-Atlas (Fig. 3) and its extension into the Algerian Ougarta Range consists of, from base to top: (1) an ophiolitic complex more than 2000 m thick; (2) the Ouarzazate Series, which is 800 m thick, and is composed of volcanics (andesites, basalts, rhyolites) (Alvaro *et al.* 2006*b*) with interbedded conglomerates and greywacke; and (3) a 3000 m-thick series that consists of an alternation of platform carbonate and detrital deposits, with the top of the 'Calcaires Inférieur' series (Boudda *et al.* 1979) or Adoudou Formation (Geyer 1990) marking the boundary between the Neoproterozoic and the Lower Cambrian (Fig. 1) (Marzela 1981; Boudda *et al.* 1987; Magaritz *et al.* 1991; Leblanc & Moussine-Pouchkine 1994; Bouima & Mezghache 2002; Bouougri & Porada

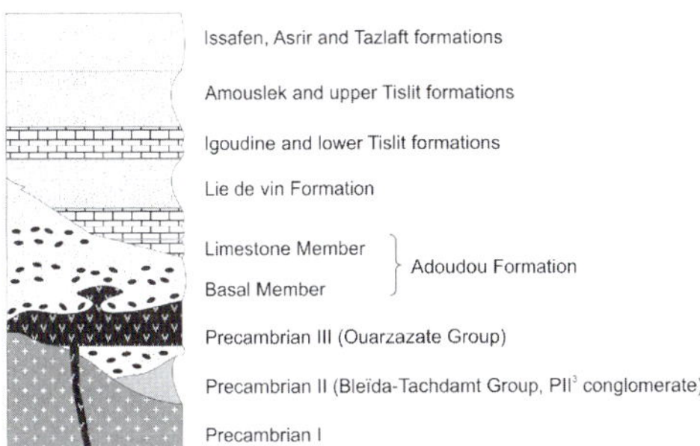

Fig. 3. Generalized Infracambrian stratigraphy, Anti-Atlas, Morocco. Not to scale. Dots, sandstones; brick pattern, carbonates; V, volcanic rocks; +, crystalline basement.

2002; Bouougri & Saquaque 2004; Thomas *et al.* 2004; Maloof *et al.* 2005). Other authors interpret the oldest sediments of the Anti-Atlas as Early Cambrian in age (Geyer & Landing 1995; Benssaou & Hamoumi 2001; Landing *et al.* 2006). Infracambrian deposits, similar to those in the Anti-Atlas, also occur in the High Atlas (El Archi *et al.* 2004*b*).

The ophiolitic complex at the base of the succession in the Anti-Atlas contains black sericitic phyllites, which belong to a monotonous succession of organic-rich shales and siltstones, with some basaltic flows and basic sills and dykes (Leblanc & Moussine-Pouchkine 1994). At outcrop scale, the 'black phyllites' consist of non-laminated to well laminated, dark grey–black schists (metamorphosed shales) that have been tightly folded and refolded. Penetrative cleavage obliterates all mesoscopic sedimentary structures.

The two main alternating facies types in these phyllites are:

- homogeneous black shales, enriched in organic material and containing numerous small pyritic framboids, but lacking clear sedimentary lamination;
- laminated dark-grey–black shales, alternating on a millimetre scale, between black, organic-rich, fine grained mudstone and slightly thicker layers of grey siltstone.

The phyllitic sequence terminates with a 20 m-thick volcano-sedimentary horizon (jasper beds, keratophyric tuffs, siltstones) that hosts the northernmost copper-rich lenses of the Bleida mining district. This unit is well exposed in the Bleida and Tachdamt areas, and thickens from 300–400 m in the east to about 1000 m to the west around Tachdamt.

The pyritic black shales and siltstones were interpreted by Leblanc & Moussine-Pouchkine (1994) as deep-water turbidite-related sediments deposited either on the continental slope or on the floor of a marginal basin. This deep-water oceanic basin formed as a result of rifting along the northern border of the West African Craton around 790 Ma (Fig. 1) associated with the break-up of the Rodinia supercontinent (Condie *et al.* 2001; Fekkak *et al.* 2004). The basinal sediments were later incorporated into the Bou Azzer ophiolite and deformed during the subsequent Pan-African Orogeny (Leblanc & Moussine-Pouchkine 1994; Gasquet *et al.* 2005).

Organic-rich strata of significantly younger, post-Pan-African age also occur in the Anti-Atlas (Buggisch *et al.* 1978; Geyer & Landing 1995). Dark stromatolitic limestones occur in several horizons in the Lower Cambrian in the central and western Anti-Atlas, namely in the Adoudou, Igoudine and Amouslek formations (Figs 3 & 4). The maximum, present-day TOC was 0.2% for 37 Lower Cambrian dark-coloured samples collected at outcrop from these formations (Fig. 1 and Table 1). The original organic-richness, however, might have been significantly higher, because the high thermal maturity of the samples indicates that much of the organic matter may have already been converted into hydrocarbons and expelled. In addition, the samples were obtained from desert surface exposures where oxidation may have destroyed any remaining organic matter. Furthermore, the extensive recrystallization of the limestones and late-diagenetic dolomitization by fluids suggest that the preservation potential of organic matter in the samples would be low, especially along the southern margin of the Anti-Atlas. In contrast, Infracambrian sediments along the northern margin of the Anti-Atlas, and especially in areas in the south of the High Atlas, show least alteration.

The Infracambrian carbonate stromatolite facies of Morocco (e.g. Raaben 1980; Alvaro *et al.* 2006*a*) is widely distributed across North Africa and Gondwana, and also occurs, for example, in India (Chakraborty 2004). The biohermal forms of stromatolites

Fig. 4. Dark grey limestones of the Adoudou Formation at a roadcut along the road 7025 Taroudant–Igherm.

Table 1. *Total organic carbon (TOC) content values of Neoproterozoic–Lower Cambrian outcrop samples from the western Anti-Atlas of Morocco. Values from the Tisrouima Basin*

Sample	Locality	Formation	TOC (%)
1	Tiout	Barre de Tata, Lie de vin	0.10
2	Tiout	Igoudine	0.07
3	Tiout	Igoudine	0.07
4	Tiout	Igoudine	0.07
5	Tiout	Igoudine	0.10
6	Tiout	Igoudine	0.10
7	Tiout	Igoudine	0.09
8	Tiout	Igoudine	0.09
9	Tiout	Amouslek	0.07
10	Lemdad	Lemdad	0.11
11	Lemdad	Lemdad	0.13
12	Lemdad	Lemdad	0.10
13	Lemdad	Lemdad	0.15
14	Lemdad	Lemdad	0.16
15	Lemdad, lower archaeocyathid limestone	Lemdad	0.23
16	Lemdad, lower archaeocyathid limestone	Lemdad	0.17
17	Taroudant-Igherm	Adoudou	0.15
18	Taroudant-Igherm	Adoudou	0.05
19	Taroudant-Igherm	Adoudou	0.07
20a	Igherm-Taliouine	Adoudou	0.13
20b	Igherm-Taliouine	Adoudou	0.09
21	Igherm-Tata	Adoudou	0.09
22a	Igherm-Tata	Lie de vin, mudstones	0.05
22b	Igherm-Tata	Lie de vin	0.05
23c	Igherm-Tata	Lie de vin	0.07
24	Igherm-Tata	Igoudine	0.15
25	Aguerd	Amouslek, mudstones	0.04
26a	Timolaye Izder	Amouslek	0.10
26b	Timolaye Izder	Amouslek	0.09
27	Timolaye Izder	Amouslek	0.10
28	5 km N Ifrane de l' Anti-Atlas	Lie de vin	0.06
29a	Ejjate plateau	Adoudou	0.09
29b	Ejjate plateau	Adoudou	0.22
30	6 km E Col de Kerdous	Anezi Group (Neoproterozoic)	0.05
31a	Road Tirhmi-Assaka	Adoudou	0.10
31b	Road Tirhmi-Assaka	Adoudou	0.09
32a	Oued Boutergui	Adoudou	0.12
32b	Oued Boutergui	Adoudou	0.07
33a	Oued Boutergui	Igoudine	0.13
33b	Oued Boutergui	Igoudine	0.07
34a	Oued Boutergui	Amouslek	0.09
34b	Oued Boutergui	Amouslek	0.08
34c	Oued Boutergui	Amouslek	0.07
35	W Tiznit	Adoudou	0.06
36a	Bleida	Bleida Group (Neoproterozoic)	0.05
36b	Bleida	Bleida Group (Neoproterozoic)	0.05
36c	Bleida	Bleida Group (Neoproterozoic)	0.18
36d	Bleida	Bleida Group (Neoproterozoic)	0.11
37a	Ait-Abdalah	Lie de vin	0.08
37b	Ait-Abdalah	Lie de vin	0.09
From Literature	Tisrouima	Pre-Amouslek (Cambrian)	1.20

are generally interpreted as indicative of relatively deeper water conditions with respect to stratiform stromatolites (Beukes & Lowe 1989).

Infracambrian strata are known from the well AZ-1 (28°59.54′N, 8°36.23′W) in the Tindouf Basin from southern Morocco, on the southern margin of the Anti-Atlas. The succession exhibits the typical Infracambrian lithostratigraphy (Fig. 5), known from outcrop. The basal section from a depth of 3387 m upwards comprises sandstones and carbonates of the Adoudou Formation, overlain by a siliciclastic sequence attributed to the Lie de vin Formation. A second overlying carbonate-dominated unit corresponds with the Igoudine and Lemdad formations at outcrop, while overlying massive sandstones may represent upper Lower–lower Middle Cambrian units. Gas shows have been reported from throughout the Infracambrian and Lower Cambrian succession in the well.

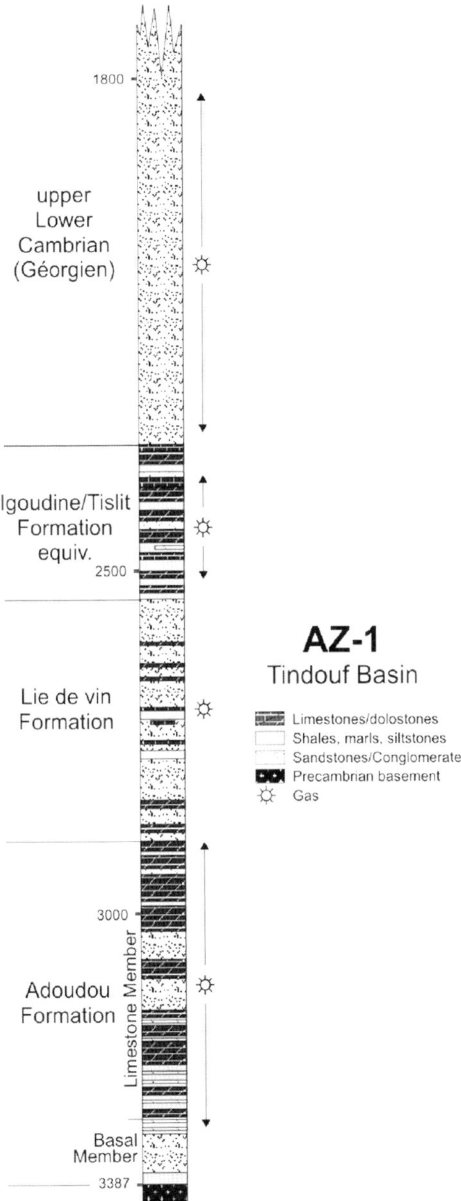

Fig. 5. Well AZ-1 from the Tindouf Basin, southern Morocco. The well contains the typical Infracambrian lithostratigraphy with two carbonate horizons. The limestones contain gas.

Late to post-Pan-African deposition in Morocco was associated with the formation of half-graben and/or pull-apart basins (Fig. 1) (Azizi Samir *et al.* 1990; Algouti *et al.* 2000; Soulaimani *et al.* 2003; El Archi *et al.* 2004*a*; Soulaimani & Pique 2004; Thomas *et al.* 2004; Gasquet *et al.* 2005; Chèvremont *et al.* 2006). Locally, the tectonics led to strong lateral facies variations between basins and ridges. The tectonic movements are also thought to have triggered large slumps, and led to the formation of ball-and-pillow structures and mud diapers (Buggisch & Heinitz 1984). Structural analysis of the Infracambrian volcanic rocks in the Moroccan Anti-Atlas by Azizi Samir *et al.* (1990), indicates that synvolcanic tectonic activity between 580 and 560 Ma included: (1) NW–SE distension; and (2) the generation of N100°–120° left-lateral strike-slip to normal faults, parallel to the Pan-African Suture. Alvaro *et al.* (2006*b*) studied volcanic rocks of the Anti-Atlas at the Neoproterozoic–Cambrian boundary and found geochemical patterns characteristic of rift-related activity.

Algeria (excluding the Taoudenni Basin)

Well and outcrop data suggest the presence of Infracambrian sediments in the Boukaï Basin, in the Ougarta Range, and in the Tindouf and Reggane basins (all in west Algeria), the Ahnet Basin (Central Algeria) and along the southern flanks of the Illizi Basin in SE Algeria (FJA 1995 unpublished report, Futyan Jawzi and Associates, *The Palaeozoic Hydrocarbon Potential and Prospectivity of Eastern Algeria, Southern Tunisia and Western Egypt*). In parts of Algeria the Infracambrian succession is termed 'Gres de Djorf Formation'.

In the Ougarta Range, the Infracambrian rocks are exposed within the Guettara and Damrane anticlines (Bouima & Mezghache 2002). In the former, the succession is more than 2000 m thick and consists of volcanic rocks (rhyolites and ignimbrites), stratiform stromatolites and fluvial–shallow-marine red conglomerates. The top of the Infracambrian succession is marked by an erosional unconformity. Organic-rich strata have not been reported from the Ougarta Range (Chikhaoui & Donzeau 1972; Bouima & Mezghache 2002). In other respects, the Infracambrian succession in the Ougarta Range is similar to the succession in the Moroccan Anti-Atlas (Bouima & Mezghache 2002).

The Boukaï Inlier is located east of the Anti-Atlas and north of the Ougarta Range. The Infracambrian volcano-sedimentary strata are generally similar in the three outcrop areas (Seddiki *et al.* 2004).

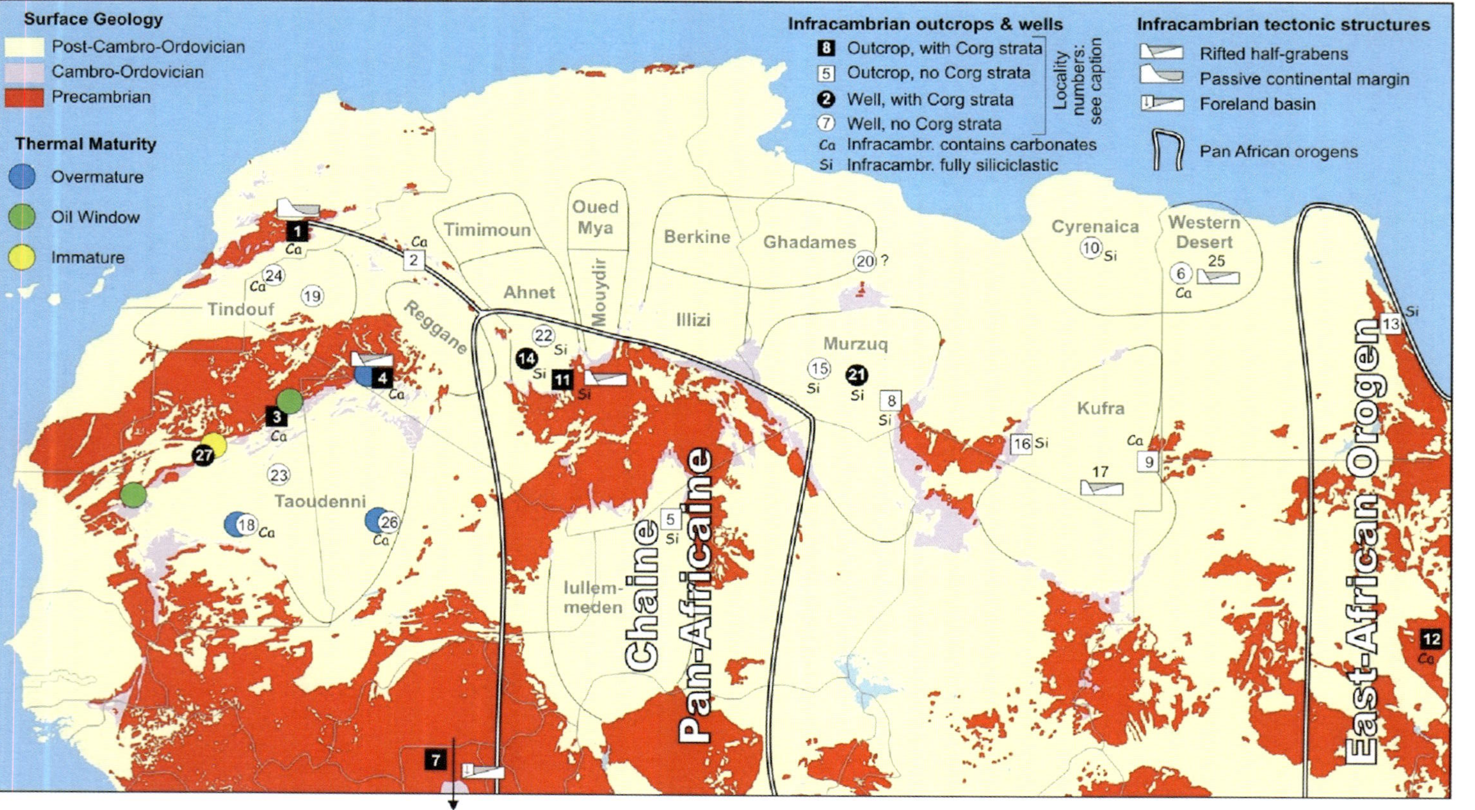

Fig. 6. Overview of lithologies and organic richnesses of the Infracambrian strata in surface exposures and wells. Also shown are Infracambrian extensional structures and thermal maturities for the Taoudenni Basin (see the text for references). Organic-rich strata and features possibly hinting at the presence of organic-rich strata are marked in bold. C_{org} strata, organic-rich strata. Infracambrian sediments are included in both Precambrian and Cambro-Ordovician. Locality ID-numbers: (1) black sericitic phyllites, turbidites on Pre-Pan-African continental slope, younger carbonates probably organically lean (Anti-Atlas, Morocco); (2) organically lean carbonates, siliciclastics and volcanics (Ougarta Range, West Algeria); (3) organic-rich strata in multiple Infracambrian horizons, organic-rich Oued Sous Formation best developed between El Mzereb and El Mreiti. 'Burning Shales' (Mauritanian Taoudenni Basin); (4) black shales in half-graben basins (Algerian Taoudenni Basin); (5) organically lean sandstones (Iullemeden Basin, Niger); (6) infracambrian strata penetrated by wells Ammonite-1, Bahrein-1, Dessouky-1 and Siwa-1; no information on lithologies (Western Desert Basin, Egypt); (7) Palaeoproterozoic black shales (Gabon); (8) organically lean shales (west margin Murzuq Basin); (9) marble and siliciclastics (east margin Al Kufrah Basin); (10) mostly continental (organically lean) siliciclatics; few thin beds with dark shale horizons (NE Libya); (11) pre-Pan-African black limestones, underlain by 1.8 Ga quartzites; post-Pan-African half-graben basins with mainly continental infill (Ahnet Basin); (12) diamictite and black limestones (Ethiopia); (13) conglomerates, greywackes and siltstones deposited as Pan-African molasse in intramontane

In the Reggane Basin the Infracambrian sandstones, shales and stromatolitic limestones have only been described from the southern margin towards the Taoudenni Basin (see below) where similar strata exist.

In the Sbaa Basin the Infracambrian strata have been penetrated by wells TK-103, TK-104, OTRA-1 and LT-1BIS. The last well was drilled into the top 33 m of the Infracambrian, which consists here of brownish-red shale and sandstone.

In the Tindouf Basin there are two wells that have been reported as having reached the Infracambrian succession: HTN-1 (Hamada de Tindouf, 27°22′28″N 8°33′52″W), drilled by SSRP; and GSL-3 (Ghassel, 27°39′20″N 6°13′43″W) drilled by Sonatrach. An Infracambrian succession consisting of more than 800 m of dolomites and siliciclastics has also been penetrated at the northern margin of the Tindouf Basin in well AZ-1 (Figs 5 & 6). It remains unclear whether this succession contains organic-rich horizons. Infracambrian sediments are absent along the southern margin of the Tindouf Basin.

Infracambrian intramontane molasse in the Ahnet Basin was deposited in more or less isolated half-graben formed over reactivated Pan-African faults (Ait Kaci Ahmed & Moussine-Pouchkine 1987; Fabre *et al.* 1988). Each of these molasse basins has a slightly different tectono-sedimentary history, with partly asynchronous developments. The molasse deposits are generally folded and consist of mainly continental (alluvial, fluvial, lacustrine) facies, with deltaic and marine slope sediments only deposited in the suture area (Fabre *et al.* 1988). The sigmoidal geometry of dyke swarms, which have intruded some of the basins, suggests that a sinistral transtensional regime may have existed during their emplacement in the Late Neoproterozoic (Caby & Monié 2003).

Black shales have not been reported from the In Semmen and Ouallen basins (Ait Kaci Ahmed & Moussine-Pouchkine 1987; Caby & Monié 2003), but may have been deposited in deeper parts of some other similar graben. Black and dark-grey pyritic shales have, for example, been reported from wells MKRS-1 and MKRN-1, where the Infracambrian succession consists of shales, siltstones and sandstones (Figs 6 & 7). Slightly older Infracambrian black shales of Upper Riphean age occur in the upper part of the Foum el Agbet Formation of the Amasine Group and are associated with stromatolites (Moussine-Pouchkine *et al.* 1988). This unit is directly underlain by 1.8 Ga-age quartzites.

Taoudenni Basin

The Taoudenni Basin lies south of the Tindouf Basin and west of the Chaine Pan-Africaine. The basin is located on the West African Craton in southern and eastern Mauritania, and extends into western Mali and southwestern Algeria (Fig. 6). Beicip (1980; *Synthèse géologique et pétrolière du bassin de Taoudeni.* Unpublished report for DMG Mauritania and DNGM, Mali) produced a synthesis of the geology of both the Mali and Mauritanian parts of the Taoudenni Basin; Villeneuve (2005) recently described the Proterozoic–Palaeozoic structural and sedimentary history of West Africa, including a detailed overview of the sub-basins and highs of the Taoudenni Basin; and Caruba & Dars (1991) have published a good overview of Mauritanian geology. To the west of the Taoudenni Basin lie the Mauritanides, a Hercynian orogen, deformed during the late Carboniferous as part of the formation of Pangaea. The intensity of the deformation decreases rapidly towards the Taoudenni Basin. Since 2000 several internationally funded projects, carried out by BGS and BRGM, facilitated the production of a series of updated Mauritania geological maps at scales 1/200 000 and 1/500 000, which helped to increase the geological and sedimentological knowledge about Neoproterozoic–Palaeozoic series of the Taoudenni Basin.

Fig. 6. (*Continued*) basins, probably organically lean; Hammamat Formation (Eastern Desert, Egypt); (14) black shales in wells MKRN-1 and MKRS-1 (South Ahnet Basin); (15) organically lean siliciclastics in wells C1-NC-58, D1-NC115 and other NC115 wells (North Murzuq Basin); (16) organically lean siliciclastics (west margin Al Kufrah Basin); (17) Infracambrian graben on seismic (South Al Kufrah Basin); (18) organically lean siliciclastics and carbonates (Abolag-1 well, Mauritanian Taoudenni Basin); (19) wells GSL-3 and HTN-1 supposedly containing Infracambrian strata (lithologies unknown). However, no infracambrian strata are included in the regional well correlations; (20) black, metamorphic schist attributed to basement (but potentially could also be Infracambrian), well A1-69 (Ghadames Basin); (21) dark grey–black siltstone, some calcite, well A1-NC101 (Murzuq Basin); (22) green shales and sandstones (but only top few metres of Infracambrian penetrated), well GF-2 (Ahnet Basin); (23) thick Infracambrian (>1760 m, TD in middle? part of Infracambrian) sandstone and shale (green, partly black) succession; 17 organically lean samples, only one sample with 0.95% TOC; Ouasa-1 well (Mauritanian Taoudenni Basin); (24) more than 800 m of Infracambrian dolomites and siliciclastics; organic richness not reported, well AZ-1 (North Tindouf Basin, not the well with the same name in the Sbaa Basin!); (25) Infracambrian graben interpreted on seismic (Western Desert Basin); (26) siliciclastics and carbonates inYarba-1 well (Taoudenni Basin, Mali): high gamma-ray values between Infracambrian markers M2 and M3, peaking at 290 API, it is unclear whether this corresponds to organic-rich strata (Mali Taoudenni Basin); (27) bituminous shales with TOC values of up to 30% were collected by J. Fabre from a fresh-water well near the Aguelt el Mabha Section (300 km SW of El Mreiti) in the Hank (Mauritanian Taoudenni Basin).

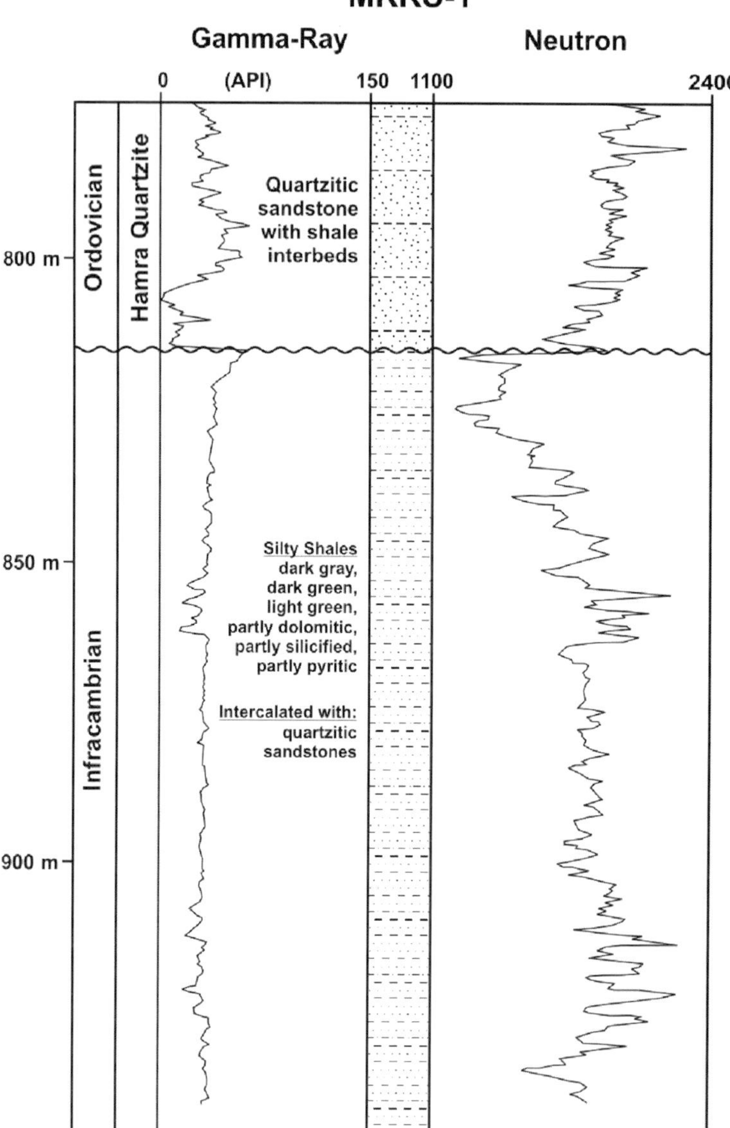

Fig. 7. Shales, siltstones and sandstones of possibly late Neoproterozoic–Early Cambrian age, penetrated in well MKRS-1 in the Ahnet Basin.

Mauritania

The Neoproterozoic–early Cambrian succession of the Taoudenni Basin is more than 2000 m thick and consists of shelfal–terrestrial deposits. It has been penetrated by the two deep exploration wells, Abolag-1 (Texaco) and Ouasa-1 (AGIP), both drilled in 1974 (Fig. 6). In both wells the Infracambrian succession is organically lean, based on the geochemical analysis of 170 samples (Figs 8 & 9). Only one sample in the Ouasa-1 well reached a TOC content of 0.95%. Nevertheless, 480 000 scfd (standard cubic feet day^{-1}: 163 m^3 day^{-1}) of gas flowed from fractured dolomites in the uppermost Middle Infracambrian in Abolag-1 during an 'open-hole' test, indicating the general viability of the Infracambrian play in the Taoudenni Basin.

Infracambrian Type I hydrocarbon source rocks with TOC values of around 17–20% and a hydrogen index (HI) of 800, indicating a good oil generating potential, were penetrated in a water well drilled in the northwestern part of the Taoudenni Basin

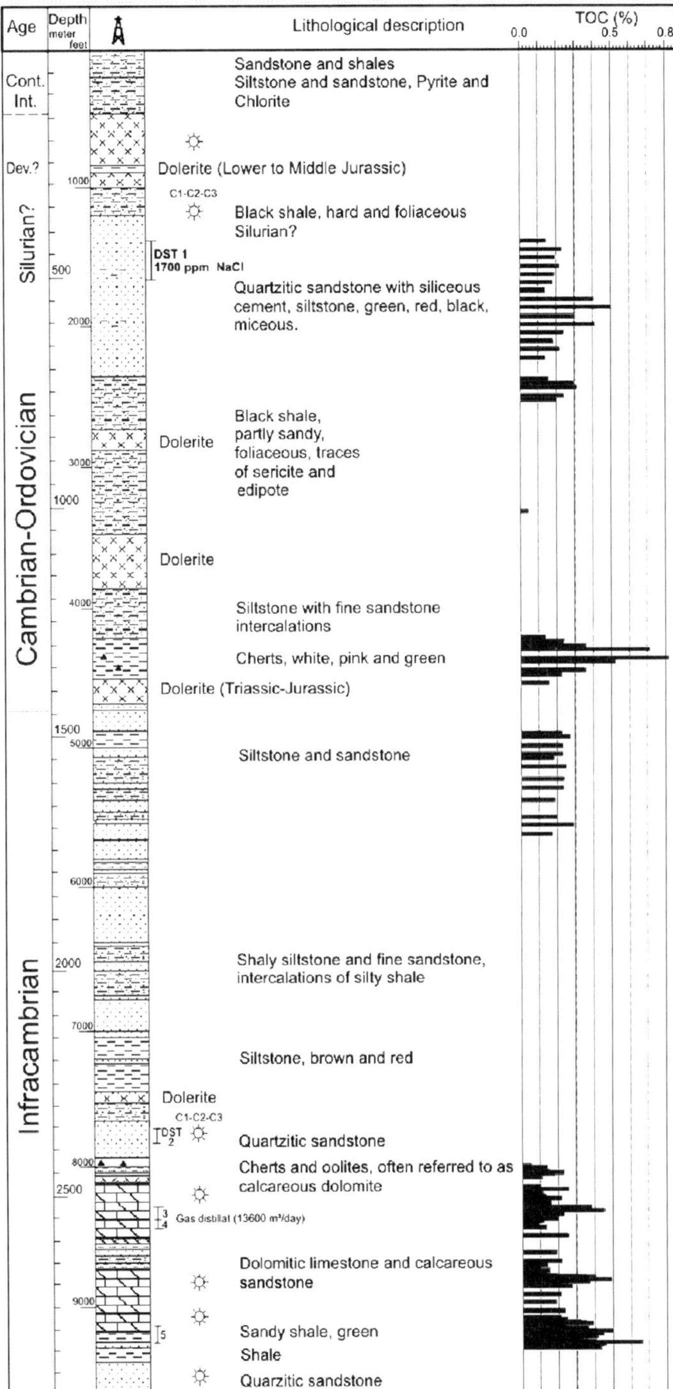

Fig. 8. Organic richness profile in the Abolag-1 well, based on the geochemical analysis of 170 samples. The TOC values are poor and vary between 0.03 and 0.66% in the Infracambrian succession, but exceed 0.81% in the lower Cambro-Ordovician. Pyrolysis shows S1 values of only up to 0.07 and S2 values of less than 0.1. T_{max} varies between 300 and 500 °C for the Infracambrian samples. Low TOC may be a result of maturation and subsequent hydrocarbon migration from the source rock. In general, Abolag-1 samples are highly mature. Most of the TOC values in the Infracambrian interval are less than 0.3% and therefore all other parameters in the Rock-Eval are questionable (Tissot & Welte 1978; Espitalié *et al.* 1985).

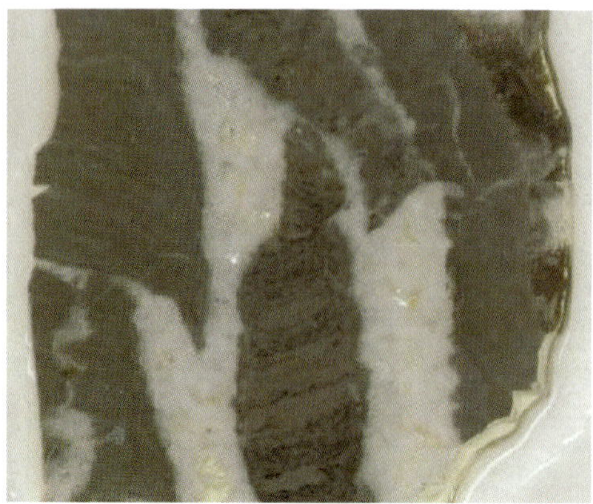

Fig. 9. Brecciated Infracambrian stromatolitic limestone from the Abolag-1 well. Pores of central block are infilled by a black substance, possibly hydrocarbons.

(Gates 2005). Assuming present-day heat flow and a simple burial model, the Infracambrian appears to be in the gas-generating window in the deeper parts of the Taoudenni Basin (Gates 2005).

Benan (1995) and Benan & Deynoux (1998) have published facies and sequence stratigraphic models for the oldest (and organically lean) Infracambrian sediments in the Taoudenni Basin, the Char Group. A facies analysis of younger Infracambrian strata of Ediacaran age was carried out by Moussine-Pouchkine & Bertrand-Sarfati (1980) who described coastal marsh, tidal and lagoonal (stromatolitic) carbonate depositional settings from the Mali portion of the Taoudenni Basin. The Algerian and Mauritanian Infracambrian lithostratigraphic units have been correlated by various authors (Bense 1964; Bertrand-Sarfati & Moussine-Pouchkine 1978; Le Page 1978; Rossi 1982; Deynoux 1985; Culver & Hunt 1991; Moussine-Pouchkine & Bertrand-Sarfati 1997; Lahondère *et al.* 2003; Callec *et al.* 2005).

The oldest Infracambrian sediments in the Taoudenni Basin have an age of about 1000 Ma (Moussine-Pouchkine & Bertrand-Sarfati 1997) and are significantly older than the oldest Infracambrian sediments in Oman (610 Ma: Sharland *et al.* 2001) where, nevertheless, broadly similiar facies are developed (Fig. 1). The age difference is due to the location of the Taoudenni Basin on the Archaean–Palaeoproterozoic West African Craton, while the Oman succession lies above the younger Pan-African orogenic belt.

Organic-rich shales and limestones in the Taoudenni Basin have been reported from the Infracambrian Atar Group (Mauritanian nomenclature; equivalent to the Hank and Dar Cheikh groups in the Algerian nomenclature, and to the El Mreiti Group in Khatt-Mauritanian Hank: Lahondère *et al.* 2003). These sediments are believed to have been deposited between 890 and 620 Ma, mostly predating the main Pan-African orogenic events in NW Africa (Bertrand-Sarfati & Moussine-Pouchkine 1988; Moussine-Pouchkine & Bertrand-Sarfati 1997). Carbon isotope stratigraphy is considered a promising tool for intra-basinal correlations of this unit (Fairchild *et al.* 1990; Teal & Kah 2005; Kah *et al.* 2009). Strontium isotope stratigraphy data for the Taoudenni Basin was published by Veiser *et al.* (1983), Shields (2002) and Shields *et al.* (2007*b*). A low relief, cratonic setting has been proposed for lower Atar (= Hank) Group times, while pre-Pan-African extension, associated with rifting around the West African Craton (Fig. 1), led to a half-graben development during deposition of the upper Atar Group (= Dar Cheikh Group) (Figs 10 & 11) (Benan & Deynoux 1998; Moussine-Pouchkine & Bertrand-Sarfati 1997).

The Atar Group in east central Mauritania in the Adrar region preserves a reef-to-basin transition. This transition is represented by a rapid facies change from a stromatolite-dominated carbonate reef to organic-rich basinal shales with TOC values of up to 20% (McInnish *et al.* 2002). In the Mauritanian Hank, between Tenoumer/ Aguelt el Mabha and El Mreiti (Lahondère *et al.* 2003), a

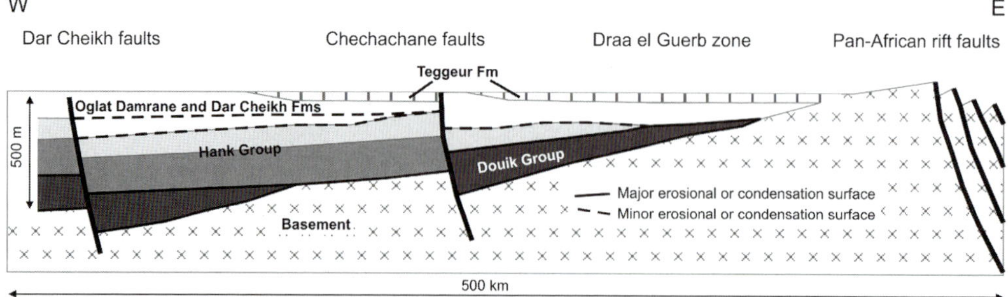

Fig. 10. Infracambrian half-graben at the eastern edge of the West African Craton during the deposition of the Dar Cheikh Group. After Moussine-Pouchkine & Bertrand-Sarfati (1997).

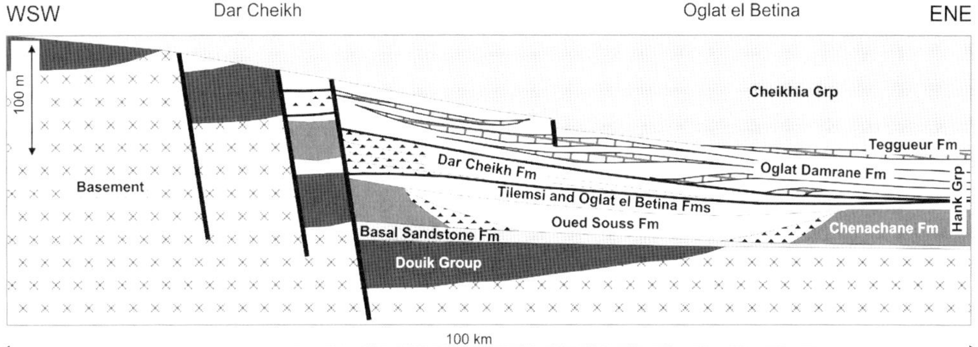

Fig. 11. Synsedimentary Infracambrian graben system in southwestern Algeria with the deposition of organic-rich strata in the Oued Souss, Dar Cheikh, Oglat Damrane and Teggeur formations (from Moussine-Pouchkine & Bertrand-Sarfati 1997).

ramp morphology explains that black shale facies constitutes a continuous band recognized at the base of the El Mreiti Group. The organic-rich clayey sedimentation is amplified in the trough centred along the axis Fdérik-Maqteir (Callec *et al.* 2005).

Bituminous shales with TOC values of up to 30% were collected by J. Fabre from a water well near the Aguelt el Mabha Section (300 km SW of El Mreiti) in the Mauritanian Hank Group succession. These shales are in the transition between thermally immature and early mature. Similar Infracambrian black shales occur in the Mauritanian Adrar area (pers. comm. Moussine-Pouchkine in co-operation with BRGM) and in the Algerian Hank Group between Ogla el Betina and Gara Guenifior, where the organic-rich horizon dips into the subsurface and is overlain by Infracambrian glacial sediments.

Depositional energy on the Taoudenni Infracambrian shelf was generally low with sedimentation in an epeiric sea (Bertrand-Sarfati & Moussine-Pouchkine 1988). Almost 75% of the carbonates are stromatolitic and at least 95% of the stromatolites are columnar, with a very high variability of morphologies related to hydrodynamic conditions (Bertrand-Sarfati & Moussine-Pouchkine 1998). Within the mixed siliciclastic–carbonate facies, the siliciclastics are mainly represented by shales and siltstones. The shales are often black and rich in pyrite and organic matter when fresh, and are dark grey–purple when weathered. The shale layers are up to 10 m thick (Bertrand-Sarfati & Moussine-Pouchkine 1988).

The organic-rich horizons occur mainly in the Oued Souss Formation (Hank Group = lower Atar Group) and the Dar Cheikh, Oglat Demrane and Teggeur formations (Dar Cheikh Group = upper Atar Group). The Oued Souss Formation in SW Algeria consists of pyritic, blue marls and black shales deposited in a quiet, relatively deep intrashelf basin. The basin was bordered by a 60–100 km-long, 20–25 m-high stromatolite biostrome (Chenachène Formation), which acted as a barrier to oceanic circulation (Bertrand-Sarfati & Moussine-Pouchkine 1992; Moussine-Pouchkine & Bertrand-Sarfati 1997). The shales are finely laminated, paper-sheet-like and organically rich. They are best developed in Mauritania between Mreiti and El Mzereb, but fresh unweathered exposures are rare (Fig. 6) (Moussine-Pouchkine & Bertrand-Sarfati 1997; Villemur 1967). The organic-rich horizon is reported to extend between 150 km NE of El Mreiti and 60 km SW of El Mzereb. In Algeria, the organic-rich unit in the Hank Group occurs between Chegga and Gara Guenifior (pers. comm. Dr S. Sacko 2005). Deposition of the upper Atar Group was associated with synsedimentary faulting and half-graben development along the Dar Cheikh Fault (Moussine-Pouchkine & Bertrand-Sarfati 1997).

Heat flow in the Taoudenni wells ranges from 15 °C km^{-1} in the Abolag-1 well in Mauritania to 19 °C km^{-1} in the Yarba-1 and Ouassa-1 wells in Mali).

Sonatrach–DNGM assume that Infracambrian black shale deposition in the Taoudenni Basin was favoured by: (1) high sea level; (2) large-scale depositional basinal lows; and (3) small-scale (tens to hundreds of kilometres) palaeogeographical depressions (pers. comm. Dr S. Sacko 2005). Notably, the black shale distribution appears to mirror the isopach of the underlying interbedded shale–sandstone unit of Douik-Char. This may reflect either continued subsidence of the Douik-Char sub-basins and/or enhanced compaction of the Douik-Char siliciclastics relative to the stromatolitic reefs on the adjacent palaeohighs.

The Sonatrach–DNGM (2003) document suggests the existence of two large WNW–ESE-trending black shale basins (pers. comm. Dr S. Sacko 2005):

- the 'Black Shale d'Aguelt Mabha' Basin (Mauritania), 800 km long, 400 km wide and containing organic-rich shale some 50 m thick, with a maximum TOC of 20–30%;
- the 'Black Shale de l'Oued Souss' Basin (Algeria, Mali), more than 150 km long, 100 km wide and containing organic-rich shale 30 m thick, and with proven TOC values of at least 5%.

It is likely that other Infracambrian 'black shale' depocentres occur in the subsurface of the Taoudenni Basin.

The thermal maturity of the Infracambrian black shales in the Hank Group in the Taoudenni Basin (Mauritania and Mali) is highly variable. They are apparently immature at Aguelt Mabha in the Mauritania, 300 km SW of El Mreiti, in the oil-generation window at El Mzereb, and overmature in many other areas including the Algerian outcrops and the Yarba-1 and Abolag-1 wells. Unweathered samples from a water well in the Atar region were analysed by Amard (1986), who recovered brown coloured acritarchs, the colour indicating maximum burial temperatures of between 75 and 150 °C within the oil-generation window. For Abolag-1 a thermal maturity consistent with gas to gas condensate generation was established for samples for the interval *c.* 2930–3093 m, based on a golden brown colour of amorphous organic matter. In the future fission-track analysis may help to unravel the complex denudation and uplift history of the Taoudenni basin and its margins.

Mali

Two deep hydrocarbon exploration wells have been drilled in the Mali part of the Taoudenni Basin, Atouila-1 (TD: Ordovician) and Yarba-1 (TD: Infracambrian) (Fig. 6).

Infracambrian organic-rich strata with TOC values of up to 5% (Pélites de la série de Nara) have been described from outcrops in northern Mali (Dars 1960, p. 65, fig. 3; Villemur 1967). Dars (1960, p. 334) reported that the foliation surfaces of the 'schistes de Drabegue' (equivalent of the 'pélites de l'Azlaf' in the Mauritanian Hank area) in the region of Bamako in southern Mali are coated in bitumen.

In 2002–2003 Sonatrach-DNGM studied the Infracambrian in outcrops in Mali and produced an unpublished report '*Evaluation du Potential Pétrolier des Bassins Sédimentaires du Mali*'. In the Gara Assaba area (25°18′28″N, 3°48′27″W) they sampled black shales interbedded with fine-grained sandstones belonging to the Formation 'Pélites de l'Azlaf', which yielded a residual (post-weathering, post-maturation) TOC of 1.4%. In the Hank area (28°45′N, 30°31′E) Sonatrach–DNGM noted probable lagoonal black shales onlapping a conophyton reef and a sedimentary breccia (pers. comm. Dr S. Sacko 2005).

The top of the Infracambrian sequence in the Taoudenni Basin is locally characterized by an angular unconformity, overlain by Cambro-Ordovician strata. In the eastern part of the Taoudenni Basin (Yarba-1 area) the upper part of the Infracambrian sequence is eroded beneath the Hercynian unconformity.

Iullemeden Basin (western Niger and eastern Mali)

The major part of the Iullemeden Basin is located in western Niger and eastern Mali. Infracambrian strata exposed in the Gourma area in NW Niger and in SE Niger consist mainly of quartzites and sandstones (Greigert & Pougnet 1967; Moody 1997), representing typical molasse sedimentation resulting from the erosion of the Pan-African mountain range (Zanguina *et al.* 1998). A carbonate–dolomite–shale series has also been described from NW Niger. Although neither Greigert & Pougnet (1967) nor Machens (1968) mention organic-rich strata in their lithostratigraphic descriptions, the presence of these, at least locally, may be assumed by analogy with the similar carbonate–shale succession in the neighbouring Taoudenni Basin.

The only well drilled to date in the Mali part of the Iullemeden Basin is Tahabanat-1 (1967–1968). This well reached a total depth (TD) of 2011 m, with sediments of Jurassic age directly overlying crystalline basement. Of the nine tests carried out over the Jurassic–Cretaceous succession in this well, none produced hydrocarbons. The Tamesna Basin may be viewed as independent or as the northern extension of the Iullemeden Basin (Fig. 6). It straddles the Mali–Algerian border and is bounded to the north by the Hoggar Massif. Only one exploration well, Tamat-1, has been drilled in the Tamesna Basin (1969–1970; TD 1170 m in crystalline basement). The Palaeozoic sequence in this well consists of a Cambro-Ordovician sandstone-dominated succession between 1141 and 800 m, overlain by nearly 170 m of Silurian shales that coarsen upwards (800–631 m), overlain by Devonian sands (631–597 m) and upper Carboniferous shales (597–544 m). The rest of the succession, from 544 m to the surface, consists of

Permo-Triassic to Lower Cretaceous alternating sandstones and shales. The Iullemeden and Tamesna basins differ primarily in that the Iullemeden is dominantly a Mesozoic basin with little to no Palaeozoic strata, while the Tamesna Basin is a dominantly Palaeozoic basin with only a thin Mesozoic succession.

West Africa and Volta Basin (Ghana and western Niger)

Black shales occur in the Palaeoproterozoic (2000 Ma) Francevillian formations (FB, FC and FD) in Gabon and are associated with cherty stromatolites (Bertrand-Sarfati & Potin 1994). These organic-rich black shales are considered to be the likely source of the bitumen occurring in the Francevillian FA sandstones. Deposition is thought to have occurred in isolated, rifted/faulted basins. Similar Palaeoproterozoic black shales have also been described from a greenstone belt in Burkina Faso (Kribek *et al.* 2004).

Organic-rich strata also occur in the Neoproterozoic (Infracambrian) Volta Foreland Basin. This basin formed as a result of crustal loading by the Chaine Pan-Africaine to the east (Flicoteaux & Trompette 1998), it is, therefore, notably different in origin to other Infracambrian basins, which are typically either pull-apart or half-graben rift basins. The Infracambrian succession of the Volta Foreland Basin in Burkina Faso, Niger and Togo contains black, pyrite-rich, silty shale and stromatolitic carbonates associated with phosphorite deposited on a distal shelf ramp during a period of eustatic transgression (Flicoteaux & Trompette 1998). It must remain speculation whether deposition of the phosphorite reflects high primary productivity conditions in the basin. On the proximal shelf ramp, in contrast, only reworked basinal phosphorite and chert were deposited.

The sea-level rise is related to the melting of ice following a period of glaciation originally dated as Late Neoproterozoic (630–595 Ma) (Culver & Hunt 1991), but is believed by Bertrand-Sarfati *et al.* (1995) to be of Early Cambrian age and by Trompette (1996) to be of latest Ediacaran age. The age of the black shales, therefore, ranges from Ediacaran to Early Cambrian, depending on the assumed age of the glaciation (Flicoteaux & Trompette 1998). New chemostratigraphic data of the Neoproterozoic in the Volta Basin has been recently published by Porter *et al.* (2004) and may help to better age-constrain the timing of organic-rich strata deposition here. Notably, other descriptions of the Infracambrian strata of the Volta Basin do not mention organic-rich sediments (Machens 1973; Deynoux 1985; Graef 2000; Shields *et al.* 2007*b*).

Organic-richness and structural control

Information about the organic-richness of the Infracambrian succession in NW Africa is patchy. Weathering has often destroyed organic matter in surface exposures and well penetrations of the Infracambrian succession are rare. Consequently, the following synthesis is far from comprehensive. Nevertheless, the existing data do allow some reassessment of areas with respect to the hydrocarbon source potential of the Infracambrian succession.

The known Infracambrian source rocks in NW Africa can be broadly divided into two main types.

- *Shelf type*: the majority of the Infracambrian organic-rich strata in North Africa were deposited in half-graben or stable basins in large shelf seas (e.g. Ahnet Basin, Taoudenni Basin). Similar shelf deposits in Oman are thought to have formed in pull-apart basins.
- *Slope type*: Infracambrian black phyllites in the Anti-Atlas (e.g. Bleida) were deposited on the continental slope of a short-lived ocean lying to the north of the West African Craton. These slope deposits differ from the cratonic–shelf deposits in terms of the preservation potential of the organic matter. Infracambrian slope sediments in North Africa were often strongly deformed and metamorphosed when they were obducted during the Pan-African Orogeny. Hydrocarbons generated during this process are unlikely to have been preserved owing to the breaching of structural traps during subsequent tectonic events.

There is both direct and indirect evidence for the presence of organic-rich Infracambrian strata in several areas of North Africa.

Taoudenni Basin

Organic-rich strata occur in parts of the Infracambrian–Palaeozoic outcrop belt, but are absent in the wells Abolag-1 and Ouasa-1 (Fig. 6). This may be explained by the fact that the Ouasa-1 well stopped in the Bir Amran–Assabet el Hassiane sandstones, before the Atar Group, and the Abolag-1 well crossed only the upper part of Atar Group, which is organically less rich in general.

The distribution of the Infracambrian potential source rocks in parts of the Taoudenni Basin is, therefore, believed to be patchy. In the northern part of the basin in the Hank, however, Infracambrian black shale distribution appears laterally more continuous. Deposition may have taken place in a tectonically rather quiet regime.

The Infracambrian organic-rich strata in the Taoudenni Basin form part of the shale–dolomite interval, reflecting a distal position relative to the influx of coarse-grained siliciclastics sediments from the hinterland. Sands and gravels are likely to dilute any existing organic matter and are often associated with high-energy conditions that are unfavourable for organic matter preservation. The identification of, and transport directions from, detrital source areas (e.g. based on sedimentary provenance studies combined with palaeocurrent measurements) may help to predict the location of distal, fine-grained facies, which may be enriched in organic material. The Taoudenni Basin is located on the stable West African Craton and a significant part of the Infracambrian succession is of pre-Pan-African age. However, even during the Pan-African times much of the Taoudenni Basin area was probably sufficiently far from the Chaine Pan-Africaine to the east that coarse erosional products from this Pan-African orogenic belt are unlikely to have dominated deposition in the more distal parts of the basin (Fig. 6).

Tindouf Basin

Infracambrian strata are known from several wells and are exposed in the Anti-Atlas region (Fig. 6). Information about their organic content, however, is generally lacking. In the Anti-Atlas, there are distinct dark-grey stromatolitic horizons. The colour may indicate an originally elevated organic-richness, but this cannot be proven because of the high thermal maturity and desert weathering at outcrop.

The presence of Infracambrian carbonates in the subsurface of the Tindouf Basin may indicate that deposition occurred in areas remote from the coarse siliciclastic sediment influx. If suitably deep half-graben existed in this region, they are likely to contain organic-rich strata by analogy with the Taoudenni Basin to the south.

Reggane, Ahnet, Mouydir and Iullemeden basins

These basins are located close to the Pan-African orogenic belt and consequently accumulated large volumes of coarse siliciclastic sediment in late–post-Pan-African times, often of continental facies. Infracambrian half-graben exist within these basins, and erosional sources and detrital transport directions during the Infracambrian are likely to have been complex and controlled by local morphology. The presence of black shales in the Ahnet Basin wells MKRN-1 and MKRS-1 indicates that some areas were, nevertheless, dominated by pelitic sedimentation, which may have been associated with, at least local, deposition of organic-rich strata. Thick carbonates sequences, however, are lacking in these wells, probably owing to the proximity to the Chaine Pan-Africaine. Black shales of pre-Pan-African age occur at outcrop in the Hoggar Massif, but it remains unclear whether these are the same age as the black shales in the MKRN-1 and MKRS-1 wells.

Thermal maturity and hydrocarbon generation

Infracambrian-sourced hydrocarbons have been recovered on test in the Abolag-1 well in the Taoudenni Basin, indicating that a proven Infracambrian petroleum system occurs in parts of NW Africa. The presence of a thick Infracambrian succession in the Taoudenni Basin suggests that hydrocarbon generation may have been started as early as Infracambrian or Early Palaeozoic times. While the preservation potential of any early formed hydrocarbon accumulation within the Infracambrian succession is uncertain, it is apparent that Hercynian and Alpine tectonic movements had little effect in the Taoudenni Basin so that early formed traps may not have been breached.

A good example for a successful Infracambrian-sourced play with Early Palaeozoic hydrocarbon generation occurs in Oman. Here, generation is thought to have commenced as early as during Cambro-Ordovician times (Pollastro 1999; Terken *et al.* 2001). The structural stability of Oman, south of the Alpine Oman Mountains, and the presence of a salt super-seal facilitated the preservation of these early generated oils.

In the Tindouf and Ahnet basins the high maturity of potential Infracambrian source rocks and the potential destruction of earlier traps during intense Hercynian deformation are a matter of concern. However, some pre-Hercynian hydrocarbons generated from the Silurian source rock appear to have survived Hercynian structural events, and there may also be areas where Infracambrian-sourced hydrocarbons could be preserved in these basins.

Early Triassic–Late Jurassic (*c.* 200 Ma) dolerite dyke and sill intrusions associated with the opening of the Atlantic are common in the Taoudenni, Reggane and Tindouf basins (Beicip 1985, p. 44; unpublished report *Taoudeni—Evaluation du Potential Petrolier du Permis Esso*). The dykes can be effectively imaged on seismic data, and their distribution and frequency varies markedly across the Taoudenni Basin. They are common along the northern Mali boundary of the basin and in the Atouila-1 area (Beicip 1980, unpublished, *Synthèse géologique et pétrolière du bassin de Taoudeni.* Unpublished report for DMG, Mauritania and DNGM, Mali). In Mauritania, the dykes appear to be rare in the northwestern part of the Taoudenni Basin (Richat, Adrar) (Beicip 1980, unpublished), but very frequent in the Mauritanian Hank (Lahondère *et al.* 2003). Source rocks within a few metres of the dykes are prone to be overcooked due to contact metamorphism. More important for hydrocarbon generation, however, is the strong increase in regional heat flow that accompanied the intrusive activity (Beicip 1985, unpublished report).

Traces of dry gas (H_2 and CH_4) have been found at a depth of 110 m in the (?) grès de Sadiola at Bourakebougou (4°N, 8°15′30″W), 10 km SE of Faladié to the west of Bamako (S. Sacko, DNGM pers. comm. 2005). Speculatively, a sill may have locally heated an Infracambrian organic-rich horizon here.

At the Yarba-1 well location, significant hydrocarbon generation and migration is anticipated to have occurred during the late Neoproterozoic (563 Ma, based on K–Ar dating) (pers. comm. Dr S. Sacko 2005). Overmaturity in the Algerian part of the Hank, however, is considered to be the result of (pre-) Hercynian burial, while the Infracambrian in the Mali part of the Taoudenni Basin is considered to have significant remaining potential for gas (pers. comm. Dr S. Sacko 2005).

Infracambrian hydrocarbon potential in NW Africa

Infracambrian-sourced hydrocarbon plays are ranked most favourable in the Taoudenni Basin based on the distribution of organic-rich strata, structural setting and maturation considerations. Organic-rich strata are known to occur in parts of the Taoudenni Basin and some hydrocarbon generation may have occurred locally in post-Hercynian times. Higher risks are attributed to the Tindouf and Ahnet basins because of deep burial and proximity to areas of intense Hercynian deformation. Detailed analysis of burial and uplift history of the Infracambrian sequences in these areas may help to quantify these risks.

There are little or no data on the Infracambrian succession in several established hydrocarbon provinces in NW Africa, including the Berkine/Ghadames, Illizi, Mouydir, Timimoun and Oued Mya basins. The Infracambrian prospectivity in these basins remains largely unknown. Petroleum systems modelling may help to identify areas that reached high thermal maturity during the Palaeozoic and may have suffered breaching of traps during the Hercynian Orogen.

Outlook

More refined depositional models are needed to improve Infracambrian facies predictions in North Africa. In particular, there is a need to develop detailed models for the deposition of the distal, fine-grained and calcareous lithologies, which are often organically enriched. Furthermore, a better understanding of depositional processes leading to lateral facies transitions into coarse molasse-type strata is required. A North African-wide study should include the mapping of lithofacies and depositional environments. Sedimentary provenance studies and systematic palaeocurrent measurements would be useful in helping to determine larger-scale sediment transport paths and to identify the main erosional sources. Currently, it remains a matter of speculation how much and what kind of sediment was derived from each of the two Pan-African orogens in North Africa, the 'Chaine Pan-Africaine' and the 'East African Orogen' (Fig. 6).

Multiple approaches should be adopted to date the Infracambrian successions. In calcareous successions isotope chemostratigraphy has proved to be valuable. Palynomorph biostratigraphic interpretation is currently hindered by the low resolution of the existing biozonal schemes. Additional work is needed to improve biostratigraphic resolution in the Infracambrian.

We are indebted to the late Dr S. Sacko (Direction Nationale de la Geologie et de Mines, DNGM, Bamako) who kindly discussed the Infracambrian and related project plans with the first author of this paper in Bamako in December 2005, only a few days before he passed away. His generosity and great hospitality will not be forgotten.

We also thank G. Lassana (DNGM) and A. Ouedraogo (Shell, Bamako) for their valuable help during this visit to Mali. We are grateful to A. Diadié Cissé and M. Simpara (Autorité pour la Promotion de la Recherche Pétrolière au Mali, AUREP, Bamako) for fruitful discussions about the petroleum geology of Mali and their friendly invitation for co-operation. We thank W. Ibrahima Lamine, I. Ould Ahmed (both Directorate of Mines and Geology, DMG, Nouakchott), M. Abdellahi O. Bouamato (Geophysics Consultant, Nouakchott) and Dr A. Mahfoud (University of Nouakchott) for generous support and co-operation during visits in Mauritania. We sincerely hope to co-operate more extensively with our colleagues in Mauritania and Mali in the near future. Professor G. Geyer (University of Würzburg) is thanked for introducing us to the Infracambrian of the Moroccan Anti-Atlas. We thank Dr N. L. Frewin (Shell, Rijswijk) for useful discussions and project support. The study was kindly funded by ENI, with a follow-up study funded by Shell. The manuscript benefited greatly from reviews of G. Shields (University of Münster) and an anonymous colleague.

References

AIT KACI AHMED, A. & MOUSSINE-POUCHKINE, A. 1987. Lithostratigraphie, sédimentologie et evolution de deux bassins molassiques intramontagneux de la chaîne Pan-Africaine; la Serie Pourpree de l'Ahnet, Nord-Ouest de Hoggar, Algérie. *Journal of African Earth Sciences*, **6**, 525–535.

ALGOUTI, A., ALGOUTI, A., BEAUCHAMP, J., CHBANI, B. & TAJ-EDDINE, K. 2000. Paléogéographie d'une plateforme infracambrienne en dislocation: série de base adoudounienne de la région Waoufengha-Igherm, Anti-Atlas occidental, Maroc. *Compte Rendu de l'Academie des Sciences, Paris, Earth and Planetary Sciences*, **330**, 155–160.

ALVARO, J. J., CLAUSEN, S., ALBANI, A. E. & CHELLAI, E. H. 2006*a*. Facies distribution of the Lower Cambrian cryptic microbial and epibenthic archaeocyathan–microbial communities, western Anti-Atlas, Morocco. *Sedimentology*, **53**, 35–53.

ALVARO, J. J., EZZOUHAIRI, H. *ET AL.* 2006*b*. The Early-Cambrian Boho volcano of the El Graara massif, Morocco: Petrology, geodynamic setting and coeval sedimentation. *Journal of African Earth Sciences*, **44**, 396–410.

AMARD, B. 1986. Microfossiles (Acritarches) du Protérozoique supérieur dans les shales de la formation d'Atar (Mauritanie). *Precambrian Research*, **31**, 69–95.

AVIGAD, D., KOLODNER, K., MCWILLIAMS, M., PERSING, H. & WEISSBROD, T. 2003. Origin of northern Gondwana Cambrian sandstone revealed by detrital zircon SHRIMP dating. *Geology*, **31**, 227–230.

AZIZI SAMIR, M. R., FERRANDINI, J. & TANE, J. L. 1990. Tectonique et volcanisme tardi-Pan Africains (580–560 Ma) dans l'Anti-Atlas Central (Maroc): interprétation géodynamique à l'échelle du NW de l'Afrique. *Journal of African Earth Sciences*, **10**, 549–563.

AZZOUNI-SEKKAL, A., LIÉGEOIS, J.-P., BECHIRI-BENMERZOUG, F., BELAIDI-ZINET, S. & BONIN, B. 2003. The 'Taourirt' magmatic province, a marker of the closing stage of the Pan-African orogeny in the Tuareg Shield: review of available data and Sr–Nd isotope evidence. *Journal of African Earth Sciences*, **37**, 331–350.

BENAN, C. A. A. 1995. *Géométrie et modélisation des corps sédimentaires d'une plate-forme cratonique d'age Protérozoique supérieur. Exemple de l'Adrar mauritanien, bassin de Taoudéni, Afrique de l'Ouest.* PhD thesis, Université Louis Pasteur, Strasbourg.

BENAN, C. A. A. & DEYNOUX, M. 1998. Facies analysis and sequence stratigraphy of Neoproterozoic platform deposits in Adrar of Mauritania, Taoudéni Basin, West Africa. *Geologische Rundschau*, **87**, 283–302.

BENSE, C. 1964. *Les formations sédimentaires de la Mauritanie méridionale et du Mali nord-occidentale.* Mémoires du Bureau de Recherches Géologique et Minières, Paris, **26**.

BENSSAOU, M. & HAMOUMI, N. 2001. L'Anti-Atlas occidental du Maroc: Ètude sedimentologique et reconstitutions paéogéographiques au Cambrien inferieur. *Journal of African Earth Sciences*, **32**, 351–372.

BERTRAND-SARFATI, J. 1972. *Stromatolites columnaires du Précambrien supérieur du Sahara nord-occidental; Inventaire, morphologie et microstructure des laminations; corrélations stratigraphiques. Géologie, 14.* Éditions du Centre National de la Récherche Scientifique, Paris.

BERTRAND-SARFATI, J. & MOUSSINE-POUCHKINE, A. 1978. Le groupe d'Achaikar à l'Ouest du Timetrine (Mali); un témoin de l'aire sédimentaire cratonique saharienne au Précambrien supérieur. *Bulletin de la Societé Géologique de France*, **20**, 62–66.

BERTRAND-SARFATI, J. & MOUSSINE-POUCHKINE, A. 1988. Is cratonic sedimentation consistent with available models? An example from the upper Proterozoic of the West African Craton. *Sedimentary Geology*, **58**, 255–276.

BERTRAND-SARFATI, J. & MOUSSINE-POUCHKINE, A. 1991. Les stromatolites de Mauritanie et leur role dans la sedimentation proterozoique du craton Ouest-Africain. *In*: CARUBA, R. & DARS, R. (eds) *Geologie de la Mauritanie [Geology of Mauritania]*, 195–215.

BERTRAND-SARFATI, J. & MOUSSINE-POUCHKINE, A. 1992. Formation et comblement d'une dépression intraplateforme engendrée par la croissance d'un biostrome stromatolitique, Protérozoique supérieur, Sahara algérien. *Comptes Rendus des l'Academie des Sciences, 2, Mecanique, Physique, Chimie, Sciences de l'Univers, Sciences de la Terre*, **315**, 837–843.

BERTRAND-SARFATI, J. & MOUSSINE-POUCHKINE, A. 1998. *Mauritanian Microbial Buildups; Meso-Neoproterozoic Stromatolites and their Environment; Six Days Field Trip on the Mauritanian Adrar.* Association des Sedimentologistes Francais, Paris, **31**.

BERTRAND-SARFATI, J., MOUSSINE-POUCHKINE, A., AMARD, B. & AIT KACI AHMED, A. 1995. First Ediacaran fauna found in Western Africa and evidence for an Early Cambrian glaciation. *Geology*, **23**, 133–136.

BERTRAND-SARFATI, J., MOUSSINE-POUCHKINE, A. & CABY, R. 1987. Les corrélations du Protérozoique au Cambrien en Afrique de l'Ouest; nouvelle interprétation géodynamique. *Bulletin de la Societé Géologique de France, Huitième Serie*, **3**, 855–865.

BERTRAND-SARFATI, J. & POTIN, B. 1994. Microfossiliferous cherty stromatolites in the 2000 Ma Franceville Group, Gabon. *Precambrian Research*, **65**, 341–356.

BEUKES, N. J. & LOWE, D. R. 1989. Environmental control on diverse stromatolite morphologies in the 3000 m. yr. Pongola Supergroup, South Africa. *Sedimentology*, **36**, 383–397.

BOGER, S. D. & MILLER, J. M. 2004. Terminal suturing of Gondwana and the onset of the Ross-Delamerian Orogeny: the cause and effect of an Early Cambrian reconfiguration of plate motions. *Earth and Planetary Science Letters*, **219**, 35–48.

BOUDDA, A., CHOUBERT, G. & FAURE-MURET, A. 1979. Essai de stratigraphie de la couverture sédimentaire de l'Anti-Atlas: Adoudounien–Cambrien inférieur. *Notes et Mémoires du Service des Mines et de la Carte Géologique du Maroc*, **271**, 1–96.

BOUDDA, A., CHOUBERT, G. & FAURE-MURET, A. 1987. Anti-Atlas. *In*: CORNELIUS, C.-D., JARNAZ, M. & LEHMANN, E. P. (eds) *Geology and Culture of Morocco.* Earth Science Society of Libya, Tripoli, 37–61.

BOUIMA, T. & MEZGHACHE, H. 2002. Les formations 'infracambriennes' des monts de l'Ougarta (Algérie) et leur corrélation avec celles de l'Anti-Atlas central (Maroc). *Mémoires du Service Géologique de l'Algerie*, **11**, 33–44.

BOUOUGRI, E. & PORADA, H. 2002. Mat-related sedimentary structures in Neoproterozoic peritidal passive margin deposits of the West African Craton (Anti-Atlas, Morocco). *Sedimentary Geology*, **153**, 85–106.

BOUOUGRI, E. H. & SAQUAQUE, A. 2004. Lithostratigraphic framework and correlation of the Neoproterozoic northern West African Craton passive margin sequence (Siroua–Zenaga–Bouazzer Elgraara Inliers, Central Anti-Atlas, Morocco): an integrated approach. *Journal of African Earth Sciences*, **39**, 227–238.

BOURNAS, N., GALDEANO, A., HAMOUDI, M. & BAKER, H. 2003. Interpretation of the aeromagnetic map of Eastern Hoggar (Algeria) using the Euler deconvolution, analytic signal and local wavenumber methods. *Journal of African Earth Sciences*, **37**, 191–205.

BRASIER, M. D. & SHIELDS, G. 2000. Neoproterozoic chemostratigraphy and correlation of the Port Askaig glaciation, Dalradian Supergroup of Scotland. *Journal of the Geological Society, London*, **157**, 909–914.

BUGGISCH, W. & HEINITZ, W. 1984. Slumpfolds and other early deformations in the early Cambrian of the Western and Central Antiatlas (Morocco). *Geologische Rundschau*, **73**, 809–818.

BUGGISCH, W., MARZELA, C. & HÜGEL, P. 1978. Die fazielle und paläogeographische Entwicklung der infrakambrischen bis ordovizischen Sedimente im Mittleren Antiatlas um Agdz (S-Marokko). *Geologische Rundschau*, **68**, 195–224.

CABY, R. 2003. Terrane assembly and geodynamic evolution of central-western Hoggar: a synthesis. *Journal of African Earth Sciences*, **37**, 133–159.

CABY, R. & MONIÉ, P. 2003. Neoproterozoic subductions and differential exhumation of western Hoggar (southwest Algeria): new structural, petrological and geochronological evidence. *Journal of African Earth Sciences*, **37**, 269–293.

CALLEC, Y., ROGER, J. *ET AL*. 2006. Notice explicative de la carte géologique de la République du Mali à 1/200 000, Feuille n° ND-29-XIV, Sandaré. Ministère des Mines, de l'Energie et de l'Eau, Bamako.

CALLEC, Y., SERRANO, O., ROGER, J., MÉTOUR, J. L., MOUSSINE-POUCHKINE, A. & MERZERAUD, G. 2005. Relations entre morphologie d'une plate-forme carbonatée et anatomie des édifices stromatolitiques: Groupes mésonéo-protérozoïques d'Atar et d'El Mreiti à la bordure nord du bassin de Taoudéni, Mauritanie [Abstract]. *In*: *10ème Congrès Français de Sédimentologie*. ASF, Paris, 51, 56.

CARUBA, R. & DARS, R. 1991. *Geologie de la Mauritanie*. Institut Supérieur Scientifique de Nouakchott.

CAWOOD, P. A., MCCAUSLAND, P. J. A. & DUNNING, G. R. 2001. Opening Iapetus: Constraints from the Laurentian margin in Newfoundland. *Geological Society of America Bulletin*, **113**, 443–453.

CHAKRABORTY, P. P. 2004. Facies architecture and sequence development in a Neoproterozoic carbonate ramp: Lakheri Limestone Member, Vindhyan Supergroup, Central India. *Precambrian Research*, **132**, 29–53.

CHÈVREMONT, P., RAZIN, P., BAUDIN, T., GABUDIANU, G., ROGER, J., THIÉBLEMONT, D., CALVES, G. ANZAR-CONSEIL. 2006. Notice explicative, carte géol. Maroc (1/50 000), feuille Awkarda. *Notes et Mémoires du Service Géologique, Maroc, MEM/BRGM.*

CHIKHAOUI, M. & DONZEAU, M. 1972. Le passage précambrien-cambrien dans les monts d'Ougarta: le conglomerat du Djebel ben Tadjine (Saoura–Sahara algérien nord-occidental). *Bulletin de la Societé d'Histoire Naturelle de l'Afrique du Nord*, **63**, 51–62.

CLAUER, N. 1976. *Géochimie isotopique du Strontium des milieux sédimentaires. Application à la géochronologie de la couverture du craton ouest africain*. Thèse University Strasbourg, France, Sciences Géologiques, **45**.

CLAUER, N. & DEYNOUX, M. 1987. New information on the probable isotopic age of the Late Proterozoic glaciation in West Africa. *Precambrian Research*, **37**, 89–94.

CONDIE, K. C. 2003. Supercontinents, superplumes and continental growth: the Neoproterozoic record. *In*: YOSHIDA, M., WINDLEY, B. F. & DASGUPTA, S. (eds) *Proterozoic East Gondwana: Supercontinent Assembly and Breakup*. Geological Society, London, Special Publications, **206**, 1–21.

CONDIE, K. C., DES MARAIS, D. J. & ABBOTT, D. 2001. Precambrian superplumes and supercontinents: a record in black shales, carbon isotopes & palaeoclimates? *Precambrian Research*, **106**, 239–260.

CULVER, S. J. & HUNT, D. 1991. Lithostratigraphy of the Precambrian–Cambrian boundary sequence in the southwestern Taoudéni Basin, West Africa. *Journal of African Earth Sciences*, **13**, 407–413.

DALZIEL, I. W. D. 1997. Neoproterozoic–Palaeozoic geography and tectonics: review, hypothesis and environmental speculation. *Geological Society of America, Bulletin*, **109**, 16–42.

DARS, R. 1960. *Les formations sédimentaires et les dolérites du Soudan occidental (Afrique de l'ouest)*. PhD thesis, University of Paris South, XI.

DEYNOUX, M. 1985. Stratigraphie, sédimentologie, paléoclimatologie et diagenèse des couvertures sédimentaires protérozoiques et paléozoiques des platesformes cratoniques africaines. *Colloque d'évaluation de l'ASP Afrique*, 47–49.

DEYNOUX, M. & TROMPETTE, R. 1981. Late Precambrian tillites of the Taoudéni Basin, West Africa. *In*: HAMBREY, M. J. & HARLAND, W. B. (eds) *Earth's Pre-Pleistocene Glacial Record*. Cambridge University Press, Cambridge, 123–131.

DEYNOUX, M., AFFATON, P., TROMPETTE, R. & VILLENEUVE, M. 2006. Pan-African tectonic evolution and glacial events registered in Neoproterozoic to Cambrian cratonic and foreland basins of West Africa. *Journal of African Earth Sciences*, **46**, 397–426.

DEYNOUX, M., KOCUREK, G. & PROUST, J. N. 1989. Late Proterozoic periglacial aeolian deposits on the West African platform, Taoudéni Basin, western Mali. *Sedimentology*, **36**, 531–549.

DEYNOUX, M., TROMPETTE, R., CLAUER, N. & SOUGY, J. 1978. Upper Precambrian and Lowermost

Palaeozoic correlations in West Africa and in the western part of Central Africa. Probable diachronism of the Late Precambrian tillite. *Geologische Rundschau*, **67**, 615–630.

DROSTE, H. H. J. 1997. Stratigraphy of the Lower Palaeozoic Haima Supergroup of Oman. *GeoArabia*, **2**, 419–472.

EL ARCHI, A., EL HOUICHA, M., EL ATTARI, A. & JOUHARI, A. 2004*a*. Le Rift syncambrien du Haut Atlas occidental et les domaines voisins, Maroc: sédimentologie, stratigraphie et magmatisme associés (Abstract). *In*: *20th Colloque Géologie Africaine, Orléans, France, 2–7 June, Abstracts Volume*, 149.

EL ARCHI, A., HOUICHA, M. E., JOUHARI, A. & BOUABDELLI, M. 2004*b*. Is the Cambrian basin of the Western High Atlas (Morocco) related either to a subduction zone or a major shear zone? *Journal of African Earth Sciences*, **39**, 311–318.

ENNIH, N. & LIÉGEOIS, J.-P. 2001. The Moroccan Anti-Atlas: the West African craton passive margin with limited Pan-African activity. Implications for the northern limit of the craton. *Precambrian Research*, **112**, 289–302.

ESPITALIÉ, J., DEROO, G. & MARQUIS, F. 1985. La pyrolyse Rock-Eval et ses applications. *Revue de l'institut Français du Petrole*, **40**, 563–579.

FABRE, J., AIT KACI AHMED, A., BOUIMA, T. & MOUSSINE-POUCHKINE, A. 1988. Le cycle molassique dans le Rameau trans-saharien de la chaîne panafricaine. *Journal of African Earth Sciences*, **7**, 41–55.

FAIRCHILD, I. J., MARSHALL, J. D. & BERTRAND-SARFATI, J. 1990. Stratigraphic shifts in Carbon isotopes from Proterozoic stromatolitic carbonates (Mauritania): influences of primary mineralogy and diagenesis. *American Journal of Science*, **290-A**, 46–79.

FEKKAK, A., BENHARREF, M. & POUCLET, A. 2004. The Neoproterozoic faulting activity in the Eastern Anti-Atlas (Morocco) (Abstract). *In*: *20th Colloque Géologie Africaine, Orléans, France, 2–7 June, Abstracts Volume*, 160.

FLICOTEAUX, R. & TROMPETTE, R. 1998. Cratonic and foreland Early Cambrian phosphorites of West Africa; palaeoceanographical and climatical contexts. *Palaeogeography, Palaeoclimatology, Palaeoecology*, **139**, 107–120.

GASQUET, D., LEVRESSE, G., CHEILLETZ, A., AZIZI-SAMIR, M. R. & MOUTTAQI, A. 2005. Contribution to a geodynamic reconstruction of the Anti-Atlas (Morocco) during Pan-African times with the emphasis on inversion tectonics and metallogenic activity at the Precambrian–Cambrian transition. *Precambrian Research*, **140**, 157–182.

GATES, T. 2005. *Hydrocarbon Projects in Mauritania and Mali, West Africa*. Technical Experts Report in Farmout Brochure by Baraka Petroleum, South Perth, Australia.

GEYER, G. 1990. Revised Lower to lower Middle Cambrian biostratigraphy of Morocco. *Newsletter on Stratigraphy*, **22**, 53–70.

GEYER, G. & LANDING, E. 1995. MOROCCO '95 – The Lower–Middle Cambrian standard of western Gondwana. *In*: GEYER, G. & LANDING, E. (eds) *Morocco '95 – The Lower–Middle Cambrian Standard of Western Gondwana*. Beringeria Special Issue 2. Freunde der Wurzburger Geowisser schaften, Wurzburg, 7–46.

GRAEF, F. 2000. Geology of West Niger. *In*: GRAEF, F., LAWRENCE, P. & VON OPPEN, M. (eds) *Adapted Farming in West Africa: Issues, Potentials and Perspectives*. Ulrich E. Grauer, Stuttgart, 35–37.

GREIGERT, J. & POUGNET, R. 1967. *Essai de description des formations géologiques de la République du Niger*. Mémoires du Bureau de Recherches Géologique et Minières, Paris, 48.

GREILING, R. O., ABDEEN, M. M. *ET AL*. 1994. A structural synthesis of the Proterozoic Arabian–Nubian Shield in Egypt. *Geologische Rundschau*, **83**, 484–501.

HALLETT, D. 2002. *Petroleum Geology of Libya*. Elsevier, New York.

HALVERSON, G. P., DUDAS, F. Ö., MALOOF, A. C. & BOWRING, S. A. 2007. Evolution of the $^{87}Sr/^{86}Sr$ composition of Neoproterozoic seawater. *Palaeogeography, Palaeoclimatology, Palaeoecology*, **256**, 103–129.

HALVERSON, G. P., HOFFMAN, P. F., SCHRAG, D. P., MALOOF, A. C. & RICE, A. H. N. 2005. Toward a Neoproterozoic composite carbon-isotope record. *GSA Bulletin*, **117**, 1181–1207.

HOFFMAN, P. F. & SCHRAG, D. P. 2002. The snowball Earth hypothesis: testing the limits of global change. *Terra Nova*, **14**, 129–155.

HUSSEINI, M. I. 1988. The Arabian Infracambrian extensional system. *Tectonophysics*, **148**, 93–103.

HUSSEINI, M. I. & HUSSEINI, S. I. 1990. Origin of the Infracambrian salt basins of the Middle East. *In*: BROOKS, J. (ed.) *Classic Petroleum Provinces*. Geological Society of America, Special Publications, **50**, 279–292.

JACOBS, J. & THOMAS, R. J. 2002. The Mozambique Belt from an East Antarctic perspective. *Royal Society of New Zealand Bulletin*, **35**, 3–18.

JACOBS, J. & THOMAS, R. J. 2004. Himalayan-type indenter-escape tectonics model for the southern part of the late Neoproterozoic–early Palaeozoic East African–Antarctic orogen. *Geology*, **32**, 721–724.

JENKINS, G. S., MCMENAMIN, M. A. S., MCKAY, C. P. & SOHL, L. 2004. *The Extreme Proterozoic: Geology, Geochemistry & Climate*. American Geophysical Union, Geophysical Monograph, **146**, 229.

JOHNSON, P. R. & WOLDEHAIMANOT, B. 2003. Development of the Arabian-Nubian Shield: perspectives on accretion and deformation in the northern East African Orogen and the assembly of Gondwana. *In*: YOSHIDA, M., WINDLEY, B. F. & DASGUPTA, S. (eds) *Proterozoic East Gondwana: Supercontinent Assembly and Breakup*. Geological Society, London, Special Publications, **206**, 289–325.

KAH, L. C., BARTLEY, J. K. & STAGNER, A. F. 2009. *Re-interpreting a Proterozoic Enigma: Conophyton–Jacutophyton Stromatolite Reefs of the Mesoproterozoic Atar Group, Mauritania*. *In*: SWART, P. K., EBERLI, G. & MCKENZIE, J. A. (eds) *Perspectives in Carbonate Geology. A Tribute to the Career of*

Robert Nathan Ginsburg, IAS Special Publication, **41**, 277–296.

Kaufman, A. J. & Knoll, A. H. 1995. Neoproterozoic variations in the C-isotopic composition of seawater: stratigraphic and biogeochemical implications. *Precambrian Research*, **73**, 27–49.

Kribek, B., Pasava, J., Sykorova, I. & Wenmenga, U. 2004. Organic matter in Palaeoproterozoic black shales of the Birmian Greenstone belts in Burkina Faso: Environment of deposition and thermal history (Abstract). *In*: *20th Colloque Géologie Africaine, Orléans, France, 2–7 June, Abstracts Volume*, 237.

Kusky, T. M. & Matsah, M. I. 2003. Neoproterozoic dextral faulting on the Najd Fault System, Saudi Arabia, preceded sinistral faulting and escape tectonics related to closure of the Mozambique Ocean. *In*: Yoshida, M., Windley, B. F. & Dasgupta, S. (eds) *Proterozoic East Gondwana: Supercontinent Assembly and Breakup*. Geological Society, London, Special Publications, **206**, 327–361.

Lahondère, D., Roger, J. *et al.* 2005. *Notice explicative des cartes géologiques à 1/20 000 et 1/500 000 de l'extrême sud de la Mauritanie*. DMG, Ministère des mines et de l'Industrie, Nouakchott, 610.

Lahondère, D., Thiéblemont, D. *et al.* 2003. *Notice explicative des cartes géologiques et gîtologiques à 1/200 000 et 1/500 000 du Nord de la Mauritanie. Vol. 1, Partie 2*. DMG, Ministère des Mines et de l'Industrie, Nouakchott, 161.

Landing, E., Geyer, G. & Heldmaier, W. 2006. Distinguishing eustatic and epeirogenic controls on Lower–Middle Cambrian boundary successions in West Gondwana (Morocco and Iberia). *Sedimentology*, **53**, 899–918.

Le Page, A. 1978. Sur la lithostratigraphie de la bordure occidentale plissée du bassin de Taoudéni dans le Guidimakha (Rep. Islamique de Mauritanie). *Bulletin de la Societé Géologique de France*, **20**, 345–354.

Leblanc, M. & Moussine-Pouchkine, A. 1994. Sedimentary and volcanic evolution of a Neoproterozoic continental margin (Bleida, Anti-Atlas, Morocco). *Precambrian Research*, **70**, 25–44.

Lüning, S., Craig, J., Loydell, D. K., Štorch, P. & Fitches, W. R. 2000. Lowermost Silurian 'hot shales' in North Africa and Arabia: regional distribution and depositional model. *Earth-Science Reviews*, **49**, 121–200.

Machens, E. 1968. Die neugefundenen Deckgebirgsreste im westlichen Niger als Zeugen einer Verbindung des Taoudéni-Beckens der westlichen Sahara mit dem Volta-Becken von Ghana. *Zeitschrift der deutschen geologischen Gesellschaft*, **117**, 469–478.

Machens, E. 1973. *Contribution à l'étude des formations du socle cristallin et de la couverture sédimentaire de l'Ouest de la République du Niger*. Mémoires du Bureau de Recherches Géologique et Minières, Paris, **82**, 145.

Magaritz, M., Kirschvink, J. L., Latham, A. J., Zhuravlev, A. Y. & Rozanov, A. Y. 1991. Precambrian/Cambrian boundary problem: Carbon isotope correlations for Vendian and Tommotian time between Siberia and Morocco. *Geology*, **19**, 847–850.

Maloof, A. C., Schrag, D. P., Crowley, J. L. & Bowring, S. A. 2005. An expanded record of Early Cambrian carbon cycling from the Anti-Atlas Margin, Morocco. *Canadian Journal of Earth Sciences*, **42**, 2195–2216.

Marzela, C. 1981. Klimazeugen aus dem jüngsten Präkambrium und Unterkambrium in S-Marokko. *Geologische Rundschau*, **70**, 473–479.

McInnish, B., Dodge, R., Bartley, J. K. & Kah, L. C. 2002. Using Landsat imagery for geological mapping in remote desert regions, Saharan West Africa (Abstract). *In*: *Geological Society of America / North-Central Section, 36th and Southeastern Section (51st) GSA Joint Annual Meeting, Lexington, Kentucky, April 3–5, 2002*, Volume 34, No. 2, GSA, Boulder.

Melezhik, A., Gorokhov, I. M., Kuznetsov, A. B. & Fallick, A. E. 2001. Chemostratigraphy of Neoproterozoic carbonates: implications for 'blind dating'. *Terra Nova*, **13**, 1–11.

Miller, N. R., Alene, M., Sacchi, R., Stern, R. J., Conti, A., Kröner, A. & Zuppi, G. 2003. Significance of the Tambien Group (Tigrai, N. Ethiopia) for Snowball Earth events in the Arabian–Nubian Shield. *Precambrian Research*, **121**, 263–283.

Moody, R. T. J. 1997. The Iullemmeden Basin. *In*: Selley, R. C. (ed.) *African Basins. Sedimentary Basins of the World, 3*. Elsevier, Amsterdam, 89–103.

Moussine-Pouchkine, A. & Bertrand-Sarfati, J. 1980. Sequences sedimentaires algo-laminaires littorales; les dolomies de Sarnyere du Proterozoique supérieur (Vendien, Gourma–Mali). *Revue de Geologie Dynamique et de Geographie Physique*, **22**, 89–99.

Moussine-Pouchkine, A. & Bertrand-Sarfati, J. 1997. Tectonosedimentary subdivisions in the Neoproterozoic to Early Cambrian cover of the Taoudenni Basin (Algeria–Mauritania–Mali). *Journal of African Earth Sciences*, **24**, 425–443.

Moussine-Pouchkine, A., Bertrand-Sarfati, J., Ball, E. & Caby, R. 1988. Proterozoic sedimentary and anorogenic volcanic series in the Pan-African range; Adnar Ahnet, northwestern Ahaggar, Algeria. *Journal of African Earth Sciences*, **7**, 57–75.

Passey, Q. R., Creabeay, S., Kulla, J. B., Moretti, F. J. & Stroud, J. D. 1990. A practical model for organic richness from porosity and resistivity logs. *AAPG Bulletin*, **74**, 1777–1794.

Pollastro, R. M. 1999. *Ghaba Salt Basin Province and Fahud Salt Basin Province, Oman – Geological Overview and Total Petroleum Systems*. US Geological Survey, Denver, CO, 25.

Porter, A. M., Knoll, A. H. & Affaton, P. 2004. Chemostratigraphy of Neoproterozoic cap carbonates from the Volta Basin, West Africa. *Precambrian Research*, **130**, 99–112.

Proust, J. & Deynoux, N. M. 1994. Marine to nonmarine sequence architecture of an intracratonic glacially related basin. Late Proterozoic deposits of the West African Taoudeni Basin in western Mali. *In*: Deynoux, M., Miller, J. M. G., Domack, E. W., Eyles, N., Fairchild, I. J. & Young, G. M. (eds) *Earth's Glacial Record*. Cambridge University Press, Cambridge, 121–145.

RAABEN, M. E. 1980. Some stromatolites of the Precambrian of Morocco. *Earth-Science Reviews*, **16**, 221–234.

ROSSI, P. 1982. *Lithostratographie et cartographie des formations sédimentaires du pourtour du massif du Kaarta, Mali occidental: Précambrian terminal-Paléozoique inférieur du sud-ouest du bassin du Taoudéni*. Saint-Jérôme à Marseille.

SEDDIKI, A., REMACI-BENAOUDA, N., COTTIN, J.-Y., MOINE, B. N., MENOT, R.-P. & PERRACHE, C. 2004. The volcano-sedimentary Boukaıs inlier (southwestern Algeria): evidence for lithospheric thinning during the Late Neoproterozoic. *Journal of African Earth Sciences*, **39**, 257–266.

SHARLAND, P. R., ARCHER, R. ET AL. 2001. *Arabian Plate Sequence Stratigraphy*. GeoArabia Special Publication, 2.

SHIELDS, G. 1999. Working towards a new stratigraphic scheme for the Neoproterozoic-Cambrian. *Eclogae Geologicae Helvetiae*, **92**, 221–233.

SHIELDS, G. A. 2002. A chemical explanation for the mid-Neoproterozoic disappearance of molar-tooth structure ~750 Ma. *Terra Nova*, **14**, 108–113.

SHIELDS, G. A. 2007. A normalised seawater strontium isotope curve and the Neoproterozoic–Cambrian weathering event. *eEarth Discussions*, **2**, 69–84.

SHIELDS, G. A., DEYNOUX, M., CULVER, S. J., BRASIER, M. D., AFFATON, P. & VANDAMME, D. 2007*a*. Neoproterozoic glaciomarine and cap dolostone facies of the Southwestern Touadeni Basin (Walidiala Valley), Senegal/Guinea, NW Africa. *Compte Rendus Geoscience*, **339**, 186–199.

SHIELDS, G. A., DEYNOUX, M., STRAUSS, H., PAQUET, H. & NAHON, D. 2007*b*. Barite-bearing cap dolostones of the Taoudeni Basin, northwest Africa: Sedimentary and isotopic evidence for methane seepage after a Neoproterozoic glaciation. *Precambrian Research*, **153**, 209–235.

SOULAIMANI, A. & PIQUE, A. 2004. The Tasrirt structure (Kerdous inlier, Western Anti-Atlas, Morocco): a late Pan-African transtensive dome. *Journal of African Earth Sciences*, **39**, 247–255.

SOULAIMANI, A., BOUABDELLI, M. & PIQUÉ, A. 2003. L'extension continentale au Néo-Protérozoique supérieur–Cambrien inférieur dans l'Anti-Atlas (Maroc). *Bulletin de la Société Géologique de France*, **174**, 83–92.

TALBOT, C. & ALAVI, J. M. 1996. The past of a future syntaxis across the Zagros. *In*: ALSOP, G. I., BLUNDELL, D. J. & DAVIDSON, I. (eds) *Salt Tectonics*. Geological Society, London, Special Publications, **100**, 89–109.

TEAL, D. J. & KAH, L. C. 2005. Using C-isotopes to constrain interbasinal stratigraphic correlations, Mesoproterozoic Atar Group, Mauritania (Abstract). Geological Society of America/Southeastern Section, 54th Annual Meeting 17–18 March 2005. *Geological Society of America, Abstracts with Program*, **37**, 45.

TERKEN, J. M. J., FREWIN, N. L. & INDRELID, S. L. 2001. Petroleum systems of Oman: Charge timing and risks. *AAPG Bulletin*, **85**, 1817–1845.

TISSOT, B. & WELTE, D. H. 1978. *Petroleum Formation and Occurrence*. Springer Verlag, Berlin.

THOMAS, R. J., FEKKAK, A. ET AL. 2004. A new lithostratigraphic framework for the Anti-Atlas Orogen, Morocco. *Journal of African Earth Sciences*, **39**, 217–226.

TROMPETTE, R. 1996. Temporal relationship between cratonization and glaciation: the Vendian-early Cambrian glaciation in western Gondwana. *Palaeogeography, Palaeoclimatology, Palaeoecology*, **123**, 373–383.

VEISER, J., COMPSTON, W., CLAUER, N. & SCHIDLOWSKI, M. 1983. $^{87}Sr/^{86}Sr$ in Late Proterozoic carbonates: evidence for a 'mantle' event at ~900 Ma ago. *Geochimica et Cosmochimica Acta*, **47**, 295–302.

VILLEMUR, J.-R. 1967. *Reconnaissance géologique et structurale du nord du bassin de Taoudéni*. Mémoires du Bureau de Recherches Géologique et Minières, Paris, 51.

VILLENEUVE, M. 2005. Palaeozoic basins in West Africa and the Mauritanide thrust belt. *Journal of African Earth Sciences*, **43**, 166–195.

VILLENEUVE, M. & CORNEE, J. J. 1994. Structure, evolution and palaeogeography of the West African Craton and bordering belts during the Neoproterozoic. *Precambrian Research*, **69**, 307–326.

WARREN, J. K., GEORGE, S. C., HAMILTON, P. J. & TINGATE, P. 1998. Proterozoic source rocks: Sedimentology and organic characteristics of the Velkerri Formation, Northern Territory, Australia. *AAPG Bulletin*, **82**, 442–463.

WYATT, A. 1986. *Challinor's Dictionary of Geology*, 6th edn. University of Wales Press, Cardiff.

ZANGUINA, M., BRUNETON, A. & GONNARD, R. 1998. An introduction to the petroleum potential of Niger. *Journal of Petroleum Geology*, **21**, 83–103.

Infracambrian sediments in Libyan sedimentary basins

H. BENSHATI[1,2]*, A. KHOJA[3] & M. SOLA[3]

[1]*National Oil Corporation, Tripoli, Libya*

[2]*Present address: BG Group, Thames Valley Park, Reading RG6 1PT, UK*

[3]*British Gas, Tripoli, Libya*

**Corresponding author (e-mail: Hesham.Benshati@bg-group.com)*

Abstract: Infracambrian sediments are widely distributed in Libya, outcropping on the eastern and western margins of Al Kufrah Basin and the eastern margin of Murzuq Basin. The sediments have been penetrated in the Central Cyrenaica Platform, Concession 10, NW Sirte Basin, and Block NC115, NW Murzuq Basin.

There are two main subdivisions. The first is metamorphosed due to local volcanism in the Murzuq and Al Kufrah basins. The second is unaltered and has been penetrated in the NW Sirte Basin. It occurs as outcropping limestone on the eastern margin of the Murzuq Basin.

These sediments generally show lateral thickness variability, with the thickest section, approximately 991–1067 m, in the Cyrenaica Platform. Individual units show thinning towards the Precambrian basement highs, and both fining-up and coarsening-up successions. Two-dimensional seismic data acquired by the operating companies AGIP and AGOCO in the southern Al Kufrah Basin image strata presumed to be Infracambrian.

The Infracambrian sediments were probably deposited as lens-shaped bodies in palaeo-lows (graben, half-graben and troughs) alongside Precambrian basement highs, trending NW–SE in the Cyrenaica platform and NE–SW in the Al Kufrah and Murzuq basins. These sediments show lateral facies changes and their nomenclature differs across the basins. Palynological and palaeontological studies suggest a Late Riphean age.

Libya, on the southern Mediterranean coast, covers an area of nearly 1.8 million km^2 (Fig. 1). A series of major tectonic events have led to the development of five sedimentary basins. A single Palaeozoic intracratonic basin was fragmented into the Ghadames, Murzuq and Al Kufrah basins in the Hercynian Orogeny. The Botnan Basin is also Palaeozoic, while the Sirte Basin is Mesozoic, as is the Cyrenaica Platform (Fig. 1).

Drilling activities in the Libyan basins and the Cyrenaica Platform – Concession 10 (NW Sirte Basin) and Block NC115 (NW Murzuq Basin) – have penetrated the widely distributed Infracambrian sediments. The sediments also crop out on the eastern and western margins of Al Kufrah Basin in the Jabal Arkenu and Jabal Nuqay areas, respectively, and on the eastern margin of Murzuq Basin in the Mourizidie area (Sola & Worsley 2000; Fig. 1).

The nomenclature of these sediments varies across the basins. They are assigned to the Infracambrian Sediments in the Cyrenaica Platform and the Sirte Basin, the Arkenu Formation in the eastern Al Kufrah Basin, the Bir Bayia Formation near Bir Bayia in the western Al Kufrah Basin and the Mourizidie Formation on the eastern margin of the Murzuq Basin.

Palynological and palaeontological studies on ditch cutting and core samples of these sediments collected from wells drilled in the Cyrenaica Platform and Concession 10, NW Sirte Basin indicate a Precambrian, Late Riphean age (El-Arnauti *et al.* 1988; Underwood 1991. *Geophysical and geological assessment of Al Kufra Basin, Southeast Libya*. Unpublished Forum Exploration Petroleum Report, NIOC Libya).

The sediments are unconformably overlain by Cambrian or younger sediments, and they unconformably overlie Precambrian basement (Mamgain 1980; El-Mehdi *et al.* 2004).

Cyrenaica Platform

The Cyrenaica Platform lies in NE Libya and extends east into Egypt. It is bounded to the north by Jabal Akhdar and to the south by Jabal Dolma, and separated from the Sirte Basin (Ajedabia Trough) to the west by a major NW–SE-trending fault.

Wells drilled in the Cyrenaica Platform have encountered Infracambrian sediments (Fig. 2). C1-82 well, central Cyrenaica Platform, reached the Precambrian basement rock beneath approximately 670 m of Infracambrian sediments; while, to the west in D1-31 well, the thickness of the Infracambrian is more than about 990 m and, to the north in C1-125 well, it is more than approximately 230 m thick, but neither of these reached the Precambrian basement (El-Arnauti *et al.* 1988).

From: CRAIG, J., THUROW, J., THUSU, B., WHITHAM, A. & ABUTARRUMA, Y. (eds) *Global Neoproterozoic Petroleum Systems: The Emerging Potential in North Africa*. Geological Society, London, Special Publications, **326**, 181–191.
DOI: 10.1144/SP326.9 0305-8719/09/$15.00

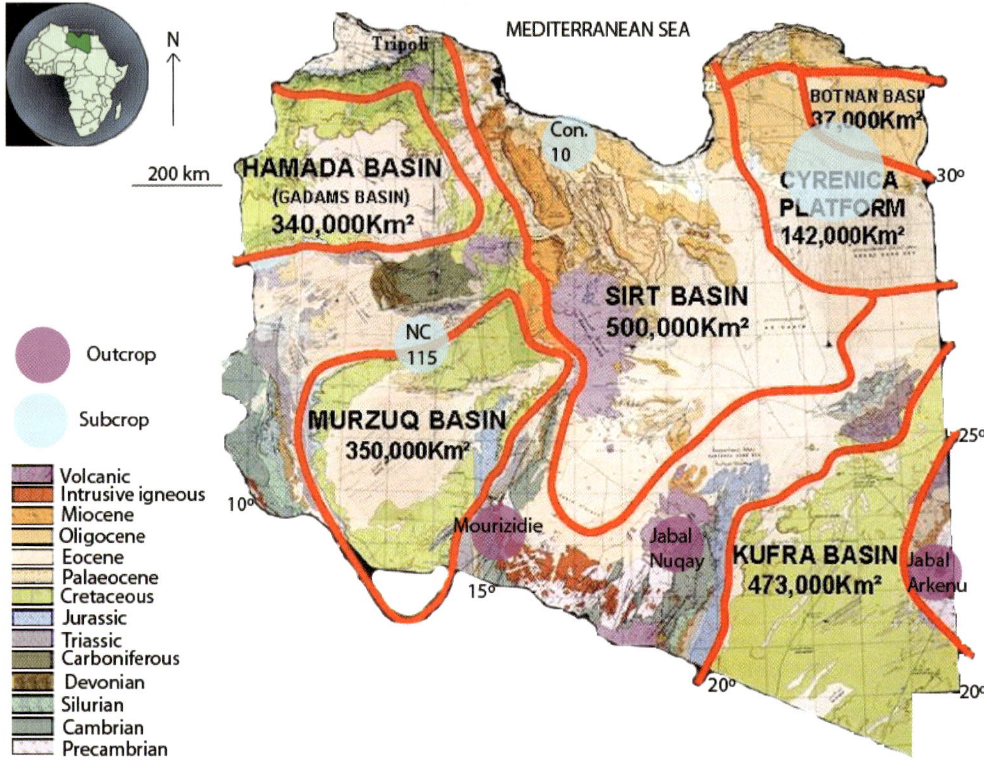

Fig. 1. The major sedimentary basins of Libya, illustrating the distribution of Infracambrian sediments, both subcrop and outcrop.

An isopach map of the Infracambrian sediments reveals the development of two possibly fault-bounded lows between the Precambrian basement highs. One low, in the Central Cyrenaica Platform, has a NW–SE trend with a maximum thickness of around 1070 m. The second low, in the south, has a NE–SW trend and is more than 1060 m thick (El-Arnauti *et al.* 1988) (Fig. 3).

An east–west section in the central Cyrenaica Platform shows that the Infracambrian sediments are unconformably overlain by Upper Ordovician formations – the Cambrian and Lower Ordovician formations are absent due to erosion or non-deposition. The Infracambrian sediments rest unconformably on Precambrian basement rock and thin toward the basement high, suggesting the synrift deposition of likely fluvial and alluvial sediments with an erosional base. These have been subdivided into three main units (El-Arnauti *et al.* 1988) (Fig. 4).

The lower unit is comprised of red–brown, coarse- to very-coarse-grained, angular–subangular, subrounded, conglomeratic sandstones. The middle unit consists of of red, medium- to coarse-grained, subangular, subrounded, micaceous sandstones. The upper unit comprises light grey–white, fine- to very-fine-grained, micaceous, glauconitic sandstones. The overall package is a fining-up succession suggesting an alluvial–fluvial type deposition.

The localities where Infracambrian sediments are deposited match the lows between the Precambrian basement highs, with the thickest section being present in the deepest part of these lows (Mamgain 1980) (Fig. 3).

Sirte Basin – Concession 10

Sirte Basin is a rift basin that evolved during the Late Cretaceous, in Cenomanian time. Concession 10 lies in the NW of the basin (Klitzsch 1971, 2000; Fig. 1).

A1-10 is a dry well drilled in Central Concession 10 by Veba Oil Operation in 1958. Drilling encountered approximately 106 m of Infracambrian sediments. These are unconformably overlain by Lower Ordovician (Tremadocian) sediments, and are composed of shales with thin quartzites. The shales are dark grey–black, hard, non-calcareous, micaceous and arenitic. The quartzite is light grey, fine–medium grained, argillaceous and fractured, with abundant feldspar grains up to 5 cm.

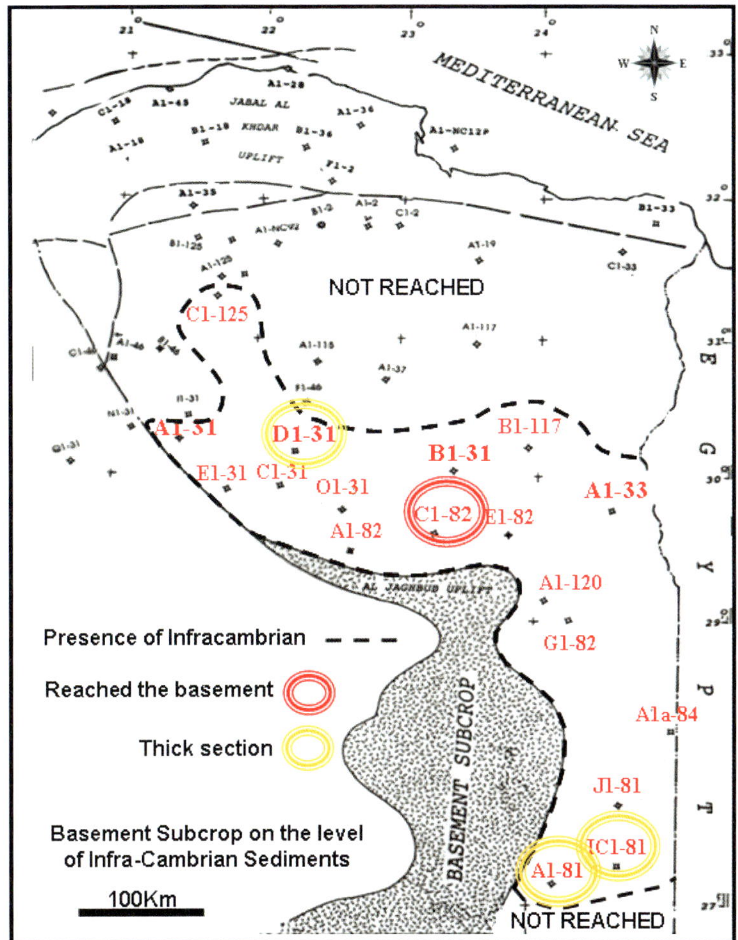

Fig. 2. Cyrenaica Platform: the locations of wells penetrating Infracambrian sediments.

Correlation between A1-10 Well in the Sirte Basin and B1-31 Well in Central Cyrenaica (*c.* 650 km apart) showed a SP (spontaneous potential) log response match (Fig. 5).

Palynological data (Underwood 1991. *Geophysical and geological assessment of Al Kufra Basin, Southeast Libya*. Unpublished Forum Exploration Petroleum Report, NIOC Libya) on Core 26, from 3195.2 to 3200.4 m, suggest a Precambrian age for these sediments. No Tremadocian microfossils have been found below 3094.6 m, and so this depth is considered to be the top of the Infracambrian.

Murzuq Basin – Block NC115

Murzuq Basin is a Palaeozoic intracratonic basin that extends south into Niger and west into Algeria. It is separated from the Hamada Basin (Ghadames) to the north by the Gargaf Arc and Atshan Saddle and from the Al Kufrah Basin to the east by the Tripoli–Tibisti Massif, and bounded by the Hoggar Massif to the SW. Block NC115 is a Repsol-operated licence located in the NW Murzuq Basin (Jacqué 1962; Sola & Worsley 2000; Fig. 1) and information related to this concession is derived from an unpublished NOC database that includes well reports.

Drilling activities have provided almost complete cover of Block NC115. Some of these wells, such as B31-NC115, S1-NC115 and H27-NC115, have penetrated the Infracambrian Mourizidie Formation (Aziz 2000; Burollet 1963; Fig. 6).

The Mourizidie Formation is unconformably overlain by the lower member of the Cambrian Hassaouna Formation. In proximal areas, the sediments are light grey conglomeratic sandstones, bedded sandstones, siltstone and claystone. The sandstones are grey, fine–medium grained, poorly

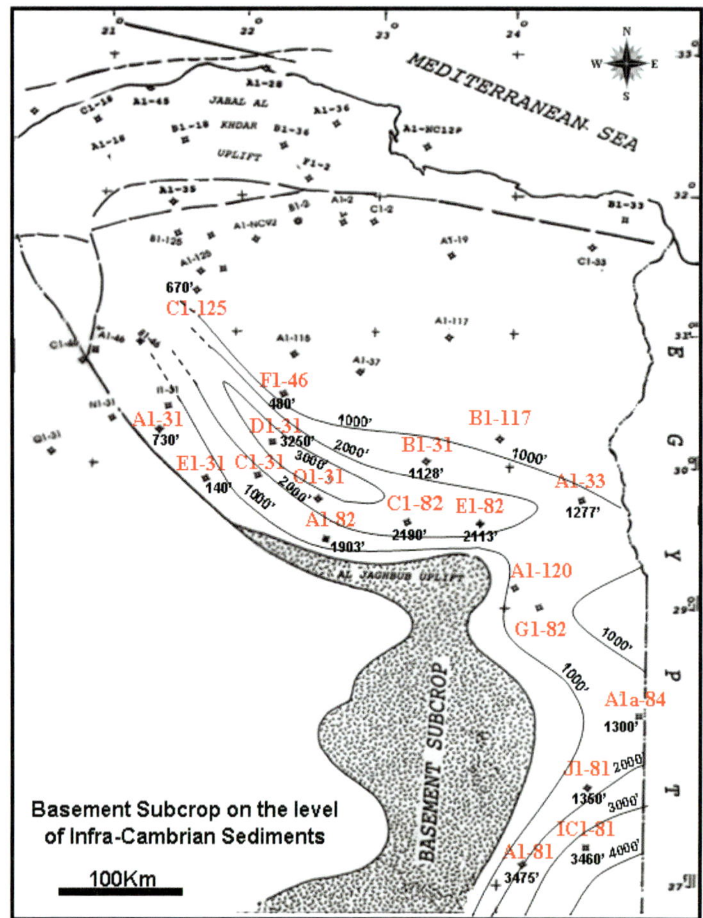

Fig. 3. Cyrenaica Platform: Infracambrian sediments isopach map.

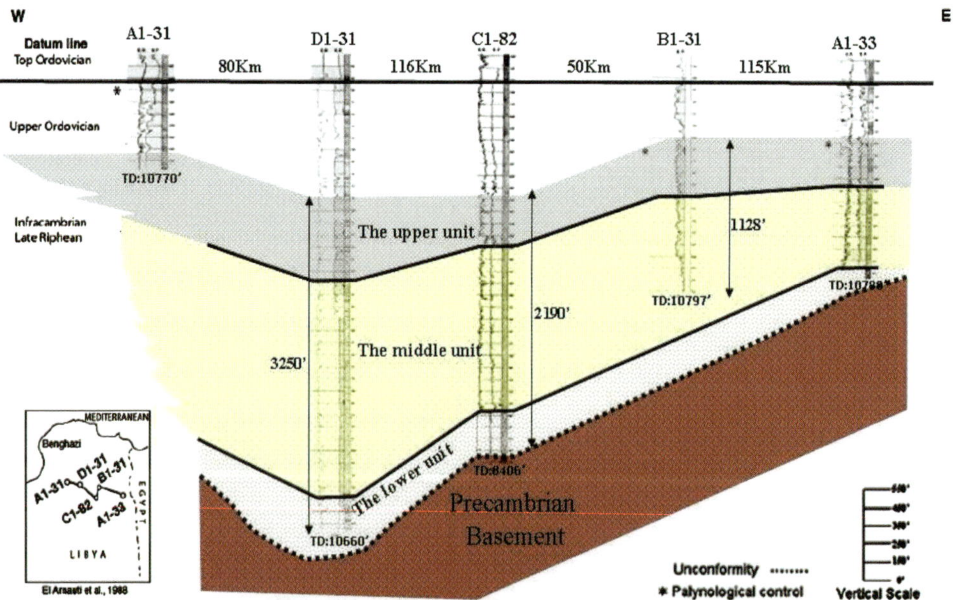

Fig. 4. East–West Stratigraphic section across the Central Cyrenaica Platform.

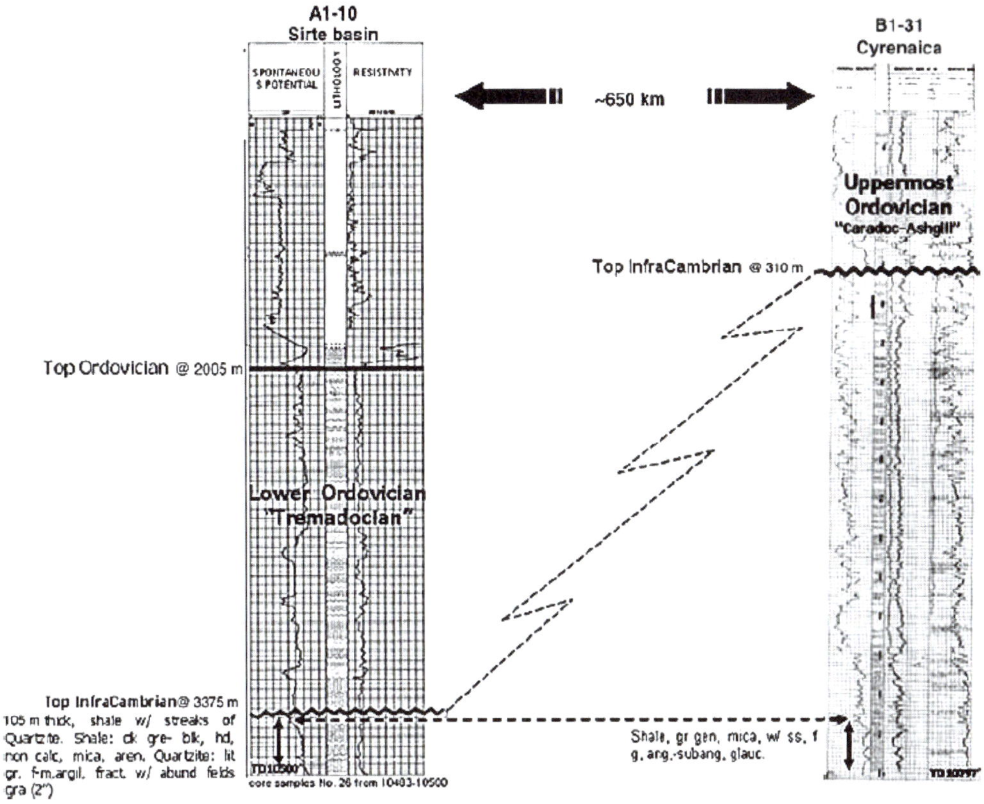

A1-10
Sirte basin
SPONTANEOUS POTENTIAL
LITHOLOGY
RESISTIVITY
~650 km
B1-31
Cyrenaica
Uppermost Ordovician
"Caradoc-Ashgill"
Top InfraCambrian @ 310 m
Top Ordovician @ 2005 m
Lower Ordovician
"Tremadocian"
Top InfraCambrian@ 3375 m
105 m thick, shale w/ streaks of Quartzite. Shale: dk gre- blk, hd, non calc, mica, aren. Quartzite: lit gr. f-m.argil. fract. w/ abund felds gra (2")
TD10500'
core samples No. 26 from 10403-10500
Shale, gr gen, mica, w/ ss, f g, ang.-subang, glauc.

Fig. 5. SP log correlation between A1-10 and B1-31.

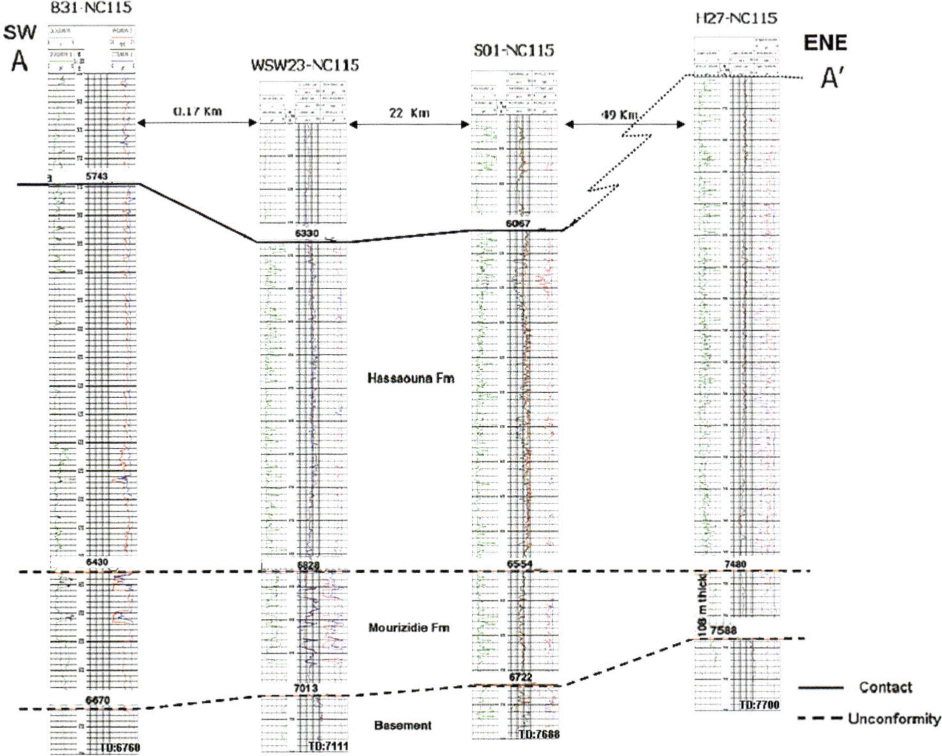

Fig. 6. NE-SW stratigraphic cross-section across NC115.

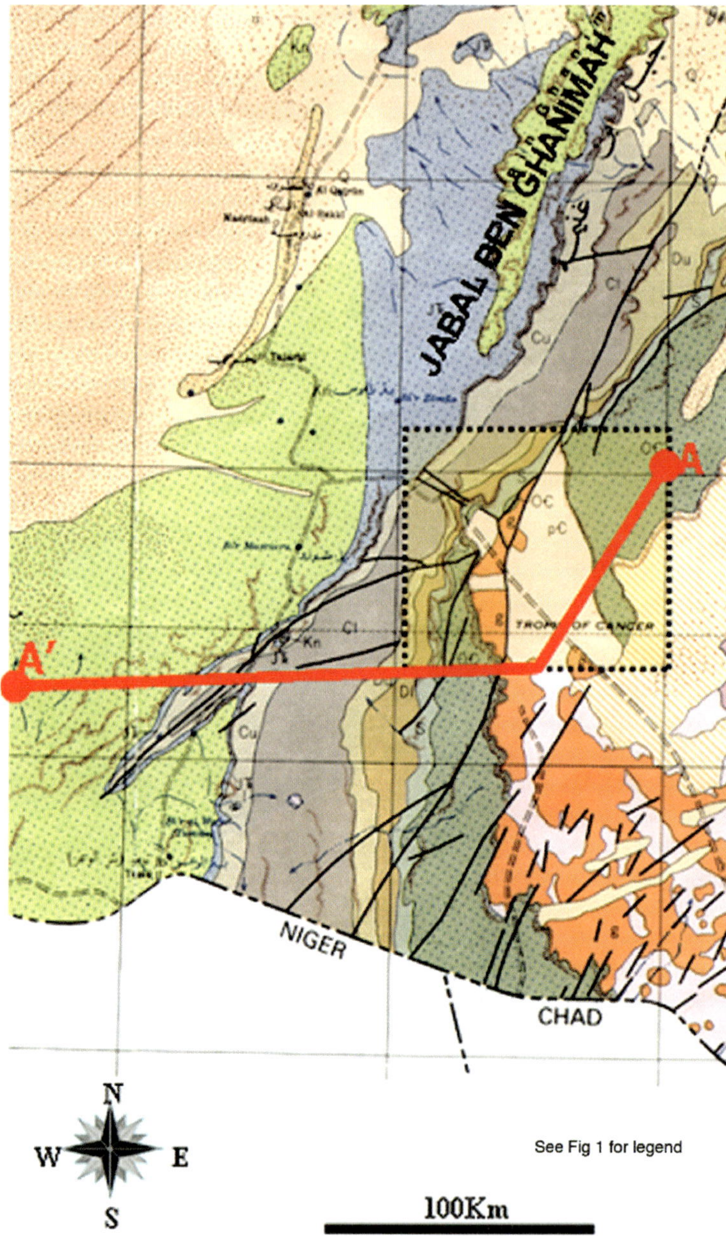

Fig. 7. Location and geological map of the Mourizidie area and the line of the cross-section in Figure 6. There is a small outcrop a few metres thick of sandstones and limestone (see Fig. 8).

to moderately sorted, well cemented, kaolinitic and micaceous, with quartz overgrowth and no porosity. The siltstones are dark grey, blocky, very fine and micaceous, with siderite concretions, and are well cemented (Davidson *et al.* 2000; Sola & Worsley 2000; Fig. 6). The geological map of the area and the line of the cross-section are shown in Figure 7.

East Murzuq Basin – the Mourizidie area

Infracambrian sediments, known here as the Mourizidie Formation (Burollet 1963), crop out in the Mourizidie area, in the east Murzuq Basin (Jacqué 1962; Aziz 2000; Fig. 7).

Fig. 8. Limestone of the Mourizidie Formation, Murzuq Basin.

These Infracambrian sediments are composed of two different sedimentary units. There is a very small exposure, 2–5 m thick, consisting of brown–white, fine–medium grained, subangular, subrounded, quartzitic sandstones, and a light grey–whitish, fractured, cryptocrystalline limestone (Aziz 2000; Fig. 8).

East Al Kufrah Basin – Jabal Arkenu area

Al Kufrah Basin, SE Libya, is a Palaeozoic intracratonic basin. It extends north to Jabal Dalma, NE into Egypt, east to Jabal Awainat, south and SE into Sudan and Chad, and west to the Sirte–Tibisti Massif (Klitzsch 1981, 2000; El-Mehdi *et al.* 2004; Fig. 1).

Jabal Arkenu is an igneous ring complex situated in the east Al Kufrah Basin, where Infracambrian sediments crop out as a NE-trending belt deposited alongside the Precambrian basement highs. The Infracambrian lies unconformably on the Precambrian basement and is unconformably overlain by the lower member of the Cambrian Hassaouna Formation (Saïd *et al.* 2000; Fig. 9).

South of Jabal Arkenu, the sediments are alluvial–fluvial sandstones. To the NE of Jabal Arkenu they change to shallow-marine sandstones and thermally metamorphosed limestones (marbles), and further NE to metamorphosed limestones underlain by marginal-marine iron-bearing quartzite. This lateral change is evidence for an interfingering relationship between these sedimentary facies (Fig. 9). A depositional model for Infracambrian sediments has been constructed to illustrate their distribution and lateral facies changes (Saïd *et al.* 2000; Fig. 10).

Two intersecting seismic lines, SL KU91-03 and SL KU89-06 in the Al Kufrah Basin demonstrate a NE-trending half-graben filled with Infracambrian sediments that can be subdivided into three main units based on seismic attributes and sequence stratigraphy. These sediments have been folded and intensively faulted by the still-active Trans-African Sahara lineament, which extends from West Africa to the Middle East. They also show thinning towards the Precambrian basement highs that might provide stratigraphic traps (Fig. 11a, b).

West Al Kufrah Basin – Jabal Nuqay area

Jabal Nugay, west Al Kufrah Basin, is a Cenozoic (post-Eocene) extrusive volcanic composed of basalt and phonolite (Fig. 12).

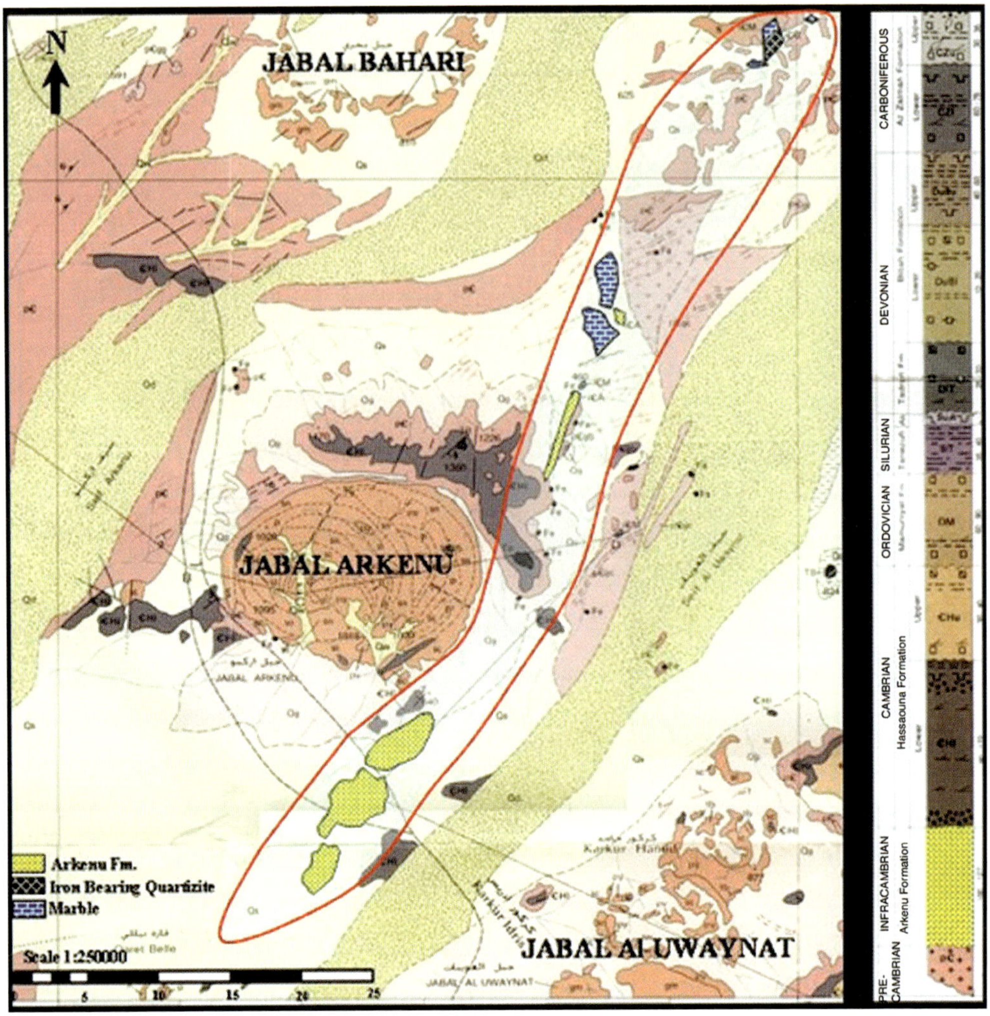

Fig. 9. NE-trending belt of Infracambrian sediments, showing lateral facies variability, alongside the Precambrian Jabal Arkenu igneous ring complex in the eastern Al Kufrah Basin (Saïd *et al.* 2000).

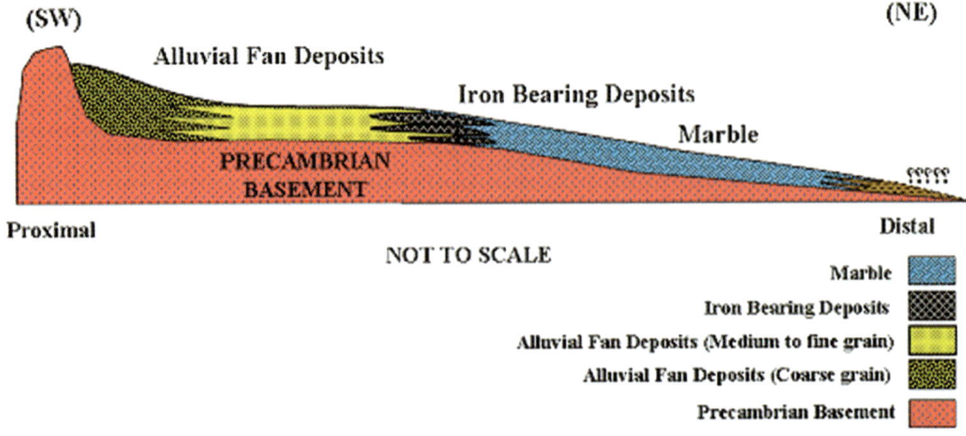

Fig. 10. Depositional model of the Infracambrian sediments in Figure 9: proximal and distal alluvial fan deposits interfinger with iron-bearing quartzites, which in turn interfinger with slightly metamorphosed limestone.

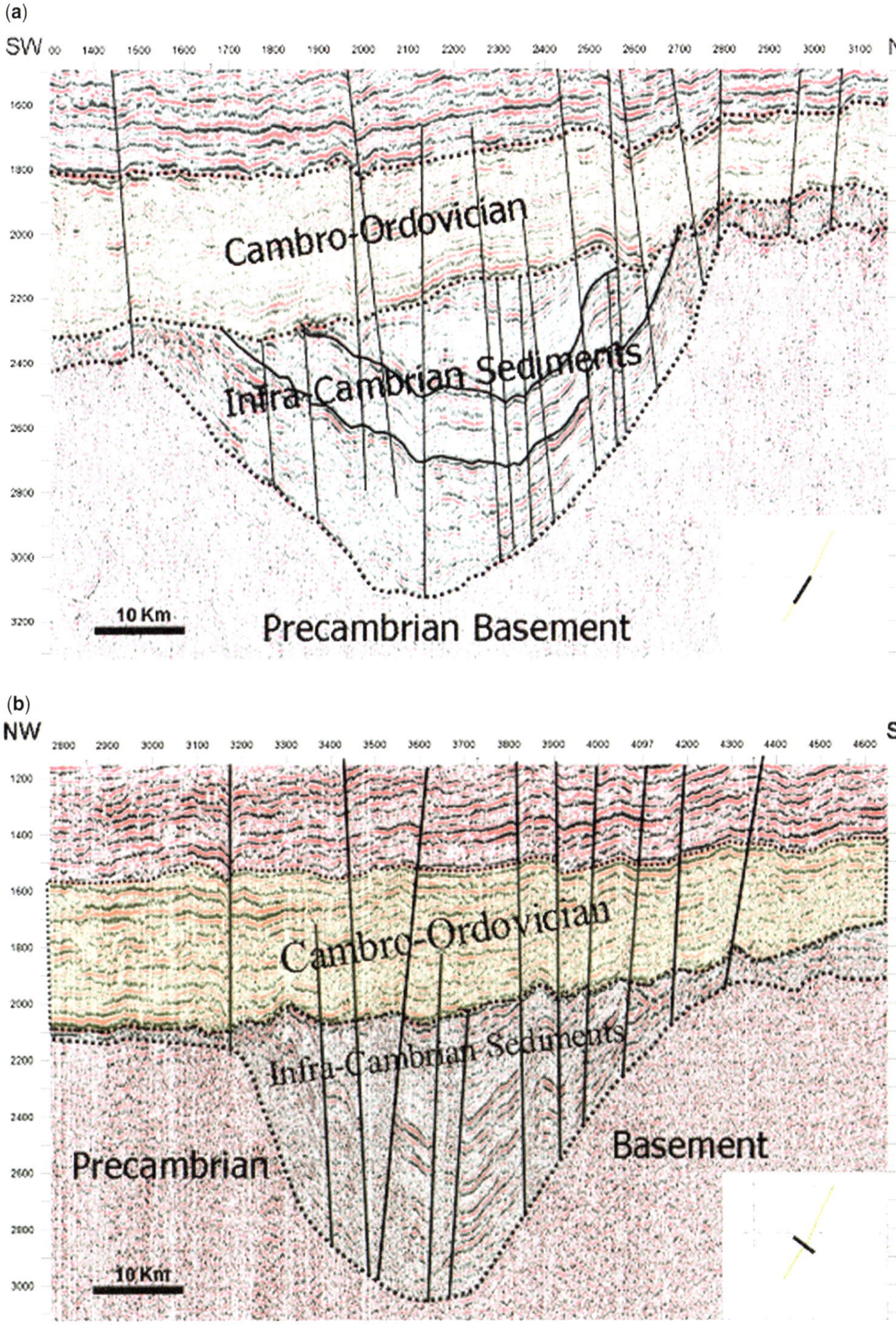

(a)

SW NE

Cambro-Ordovician

Infra-Cambrian Sediments

10 Km

Precambrian Basement

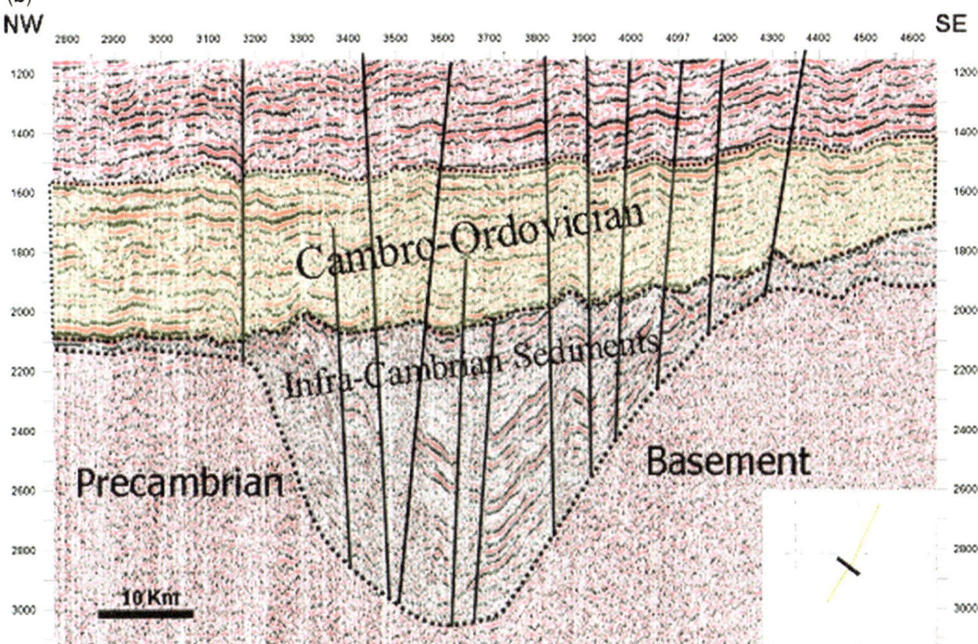

Fig. 11. (**a**) NE–SW seismic line KU91-03 across an Infracambrian graben in the north Al Kufrah Basin. Thinning towards the Precambrian highs may provide stratigraphic traps. (**b**) NW–SE seismic line KU89-06 across an Infracambrian graben in the north Al Kufrah Basin. Thinning towards the Precambrian highs may provide stratigraphic traps.

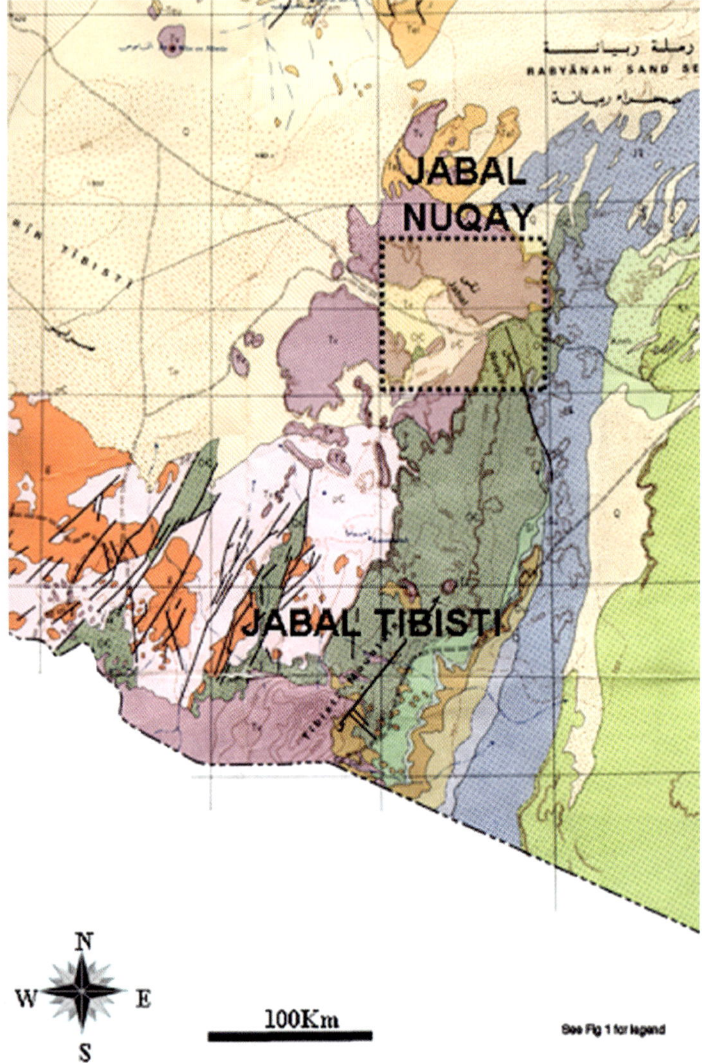

Fig. 12. Location and geological map of the Jabal Nuqay area. A few metres of Infracambrian sediments – sandstone, siltone and claystone – represent a marine transgression over an alluvial braided plain.

Infracambrian sediments, known here as the Bir Bayia Formation (Selley 1971), crop out near Bir Bayia in the Jabal Nuqay area as a small exposure a few metres thick unconformably overlain by the Cambrian Hassaouna Formation.

The sediments are red–violet feldspathic sandstones, interbedded with thinly bedded siltstone and claystone, representing a marine transgression on a braided alluvial plain (Selley 1971).

Conclusions

Infracambrian sediments are widely distributed in onshore Libyan basins and are interpreted as lens-shaped bodies deposited in palaeo-lows alongside the Precambrian basement highs, having a NW–SE trend in the Cyrenaica Platform, and a NE–SW trend in the Al Kufrah and Murzuq basins.

They are mainly clastic in the Cyrenaica Platform, and both clastic and carbonate in Murzuq and Al Kufrah basins. Limestones near the Jabal Arkenu intrusion are now considered to be marbles.

Palaeontological and palynological studies on ditch cuttings and core samples indicate a Precambrian, Late Riphean age for these sediments. They show lateral facies changes from continental alluvial sands to marginal marine sandstones, iron-bearing quartzite and limestone to marine sandstones.

Seismic data show a graben structure with Infracambrian sediment being deposited, demonstrating thickening and thinning of this sedimentary unit.

We would like to express our thanks and appreciation to the Exploration Department of the National Oil Corporation, Libya, the Earth Sciences Society of Libya and the Geological Society of London for their support and encouragement to in producing this paper.

References

AZIZ, A. 2000. Stratigraphy and hydrocarbon potential of the Lower Palaeozoic succession of licence NC115, Murzuq Basin, SW Libya. *In*: SOLA, M. A. & WORSLEY, D. (eds) *Geological Exploration in Murzuq Basin*. Elsevier, Amsterdam, 349–368.

BUROLLET, P. F. 1960. *Libye, lexique stratigraphique internationale, 4, Afrique, Pt 4a, Libye.* Commission de Stratigraphie Centre National de la Recherche Scientifique, **62**.

BUROLLET, P. F. 1963. Reconnaissance geologique dans le sud-est du bassin de Kufra. *Revue de l' Institut Français du Petroles*, **5**, 1537–1545.

DAVIDSON, L., & BESWETHERICK, S. *ET AL*. 2000. The structure, stratigraphy and petroleum geology of the Murzuq Basin, southwest Libya. *In*: SOLA, M. A. & WORSLEY, D. (eds) *Geological Exploration in Murzuq Basin*. Elsevier, Amsterdam, 295–320.

EL-ARNAUTI, A., OWENS, B. & THUSU, B. 1988. *Subsurface Palynostratigraphy of Northeast Libya*. Garyounis University Publications, Libya.

EL-MEHDI, B., TURKI, S. M., SUWESI, S. K. & OWEISS, K. 2004. Short notes and guidebook on the geology of Al Kufrah Basin, Al'Awaynat area. *In*: *Sedimentary Basins of Libya, Third Symposium, Geology of East Libya Fieldtrip, November 2004*. Earth Science Society of Libya, Tripoli.

JACQUÉ, M. 1962. *Reconnaissance geologique du Fazzan oriental*. Notes et Memoires du Compagnie Française du Petroles (TOTAL) Paris, 5–43.

KLITZSCH, E. 1971. The structural development of parts of Africa since Cambrian time. *In*: CRAY, C. (ed.) *Symposium on the Geology of Libya*. University of Libya, Tripoli, 253–252.

KLITZSCH, E. 1981. Lower Palaeozoic rock of Libya, Egypt, and Sudan. *In*: HOLLAND, C. H. (ed.) *Lower Palaeozoic of the Middle East, Eastern and Southern Africa, and Antarctica*. Wiley, New York, 131–163.

KLITZSCH, H. 2000. The structural development of Murzuq and Kufra basins – significant for oil and mineral exploration. *In*: SOLA, M. A. & WORSLEY, D. (eds) *Geological Exploration in Murzuq Basin*. Elsevier, Amsterdam, 134–150.

MAMGAIN, V. D. 1980. *The Pre-Mesozoic (Precambrian to Palaeozoic) Stratigraphy of Libya*. Industrial Research Centre, Tripoli.

SAÏD, M. M., OWENS, K. A & MEHIDI, B. O. 2000. *Jabal Arkenu Sheet: Explanatory Booklet*. Joint Geological Mapping Project between the Egyptian Geological Survey and Mining Authority (EGSMA) and the Industrial Research Centre (IRC) of Libya, Cairo.

SELLEY, R. C. 1971. Structural control of Miocene sedimentation in the Sirt Basin. *In*: *Symposium on the Geology of Libya*. University of Libya, Tripoli.

SOLA, M. A. & WORSLEY, D. (eds) 2000. *Geological Exploration in the Murzuq Basin*. Elsevier, Amsterdam.

Field-based investigations of an 'Infracambrian' clastic succession in SE Libya and its bearing on the evolution of the Al Kufrah Basin

DANIEL PAUL LE HERON[1]*, JAMES P. HOWARD[2], AIYAD MOHAMED ALHASSI[3], LESTER ANDERSON[2], ANDREW MORTON[2] & C. MARK FANNING[4]

[1]*Department of Earth Sciences, Royal Holloway University of London, Egham, Surrey TW20 0EX, UK*

[2]*CASP (Cambridge Arctic Shelf Programme), Department of Earth Sciences, 181A Huntingdon Road, Cambridge CB3 0DH, UK*

[3]*Arabian Gulf Oil Company, Exploration Division, P.O. Box 263, Benghazi, Libya*

[4]*Research School of Earth Sciences, The Australian National University, Canberra, ACT 0200, Australia*

**Corresponding author (e-mail: d.leheron@es.rhul.ac.uk)*

Abstract: Field-based investigation of 'Infracambrian' rocks cropping out on the eastern flank of Al Kufrah Basin (area 500 000 km^2) reveals a an approximately 500 m-thick clastic succession of massive and cross-bedded sandstones, separated by 60 m-thick mudrock intervals. New zircon age data indicate a maximum age of deposition of approximately 950 Ma; furthermore, the absence of zircons of Pan-African age suggests a minimum depositional age older than the Pan-African Orogeny. Previously unreported folding and spaced cleavage affects these deposits to produce a pronounced NE–SW-striking tectonic grain that is interpreted to result from NW–SE-directed orthogonal compression during the Pan-African Orogeny. These Infracambrian rocks are therefore unlikely to be suitable analogues for weakly deformed strata shown to exist beneath the Cambro-Ordovician strata of the Al Kufrah Basin. Earlier work mapped a series of Infracambrian marble outcrops along strike of the clastic deposits; thin section petrography reveals that some of these are basic igneous rocks metamorphosed to greenschist facies. Interpretation of gravity data over the Al Kufrah Basin shows NE–SW-striking faults, parallel to outcrop structures, and secondary NW–SE faults. The data do not support earlier interpretations of a rhomboidal geometry in the deep subsurface of the basin, which has previously been attributed to strike-slip (pull-apart) processes. This research impacts on earlier suggestions that the Al Kufrah Basin opened as one of a series of en echelon pull-apart basins situated along a 6000 km-long shear zone known as the Transafrican Lineament, stretching from the Nile to the Niger Delta.

This paper provides a detailed description and interpretation of an 'Infracambrian' clastic succession that crops out in the eastern Libyan Sahara (Fig. 1). A full investigation of these rocks is especially important because they are the oldest mapped (non-crystalline) stratigraphic unit in the region (Saïd *et al.* 2000) and, hence, have the potential to provide insights into the origins of the huge (500 000 km^2) Al Kufrah Basin that straddles SE Libya, northern Chad, easternmost Egypt and part of northern Sudan (Fig. 1). This sedimentary basin is filled with a predominantly clastic succession of Infracambrian–Cenozoic age (Fig. 2) reaching up to 4 km thick in the basin centre (Bellini & Massa 1980). The study of Infracambrian rocks on the flanks of the Al Kufrah Basin also has a strong economic incentive. At the time of writing (2007), the region is undergoing intense hydrocarbon exploration, and the major risk is that a viable and mature hydrocarbon source rock has not yet been shown to be present in the basin centre. Published data are extremely few and restricted to proprietary, unpublished seismic lines, two deep exploration wells in the north of the basin, relatively low-resolution aeromagnetic studies (Bellini *et al.* 1991) and scattered outcrop observations (e.g. Lüning *et al.* 1999).

Regional cross-sections across the basin (Bellini & Massa 1980) indicate that any candidate source rocks, principally Late Devonian, within the post-Silurian succession are currently buried to only about 1 km (Bellini *et al.* 1991). Therefore, these younger rocks may never have entered the oil window, in contrast to Late Devonian organic-rich shales elsewhere in North Africa (Lüning *et al.* 2003). Consequently, papers evaluating the source

From: CRAIG, J., THUROW, J., THUSU, B., WHITHAM, A. & ABUTARRUMA, Y. (eds) *Global Neoproterozoic Petroleum Systems: The Emerging Potential in North Africa.* Geological Society, London, Special Publications, **326**, 193–210.
DOI: 10.1144/SP326.10 0305-8719/09/$15.00

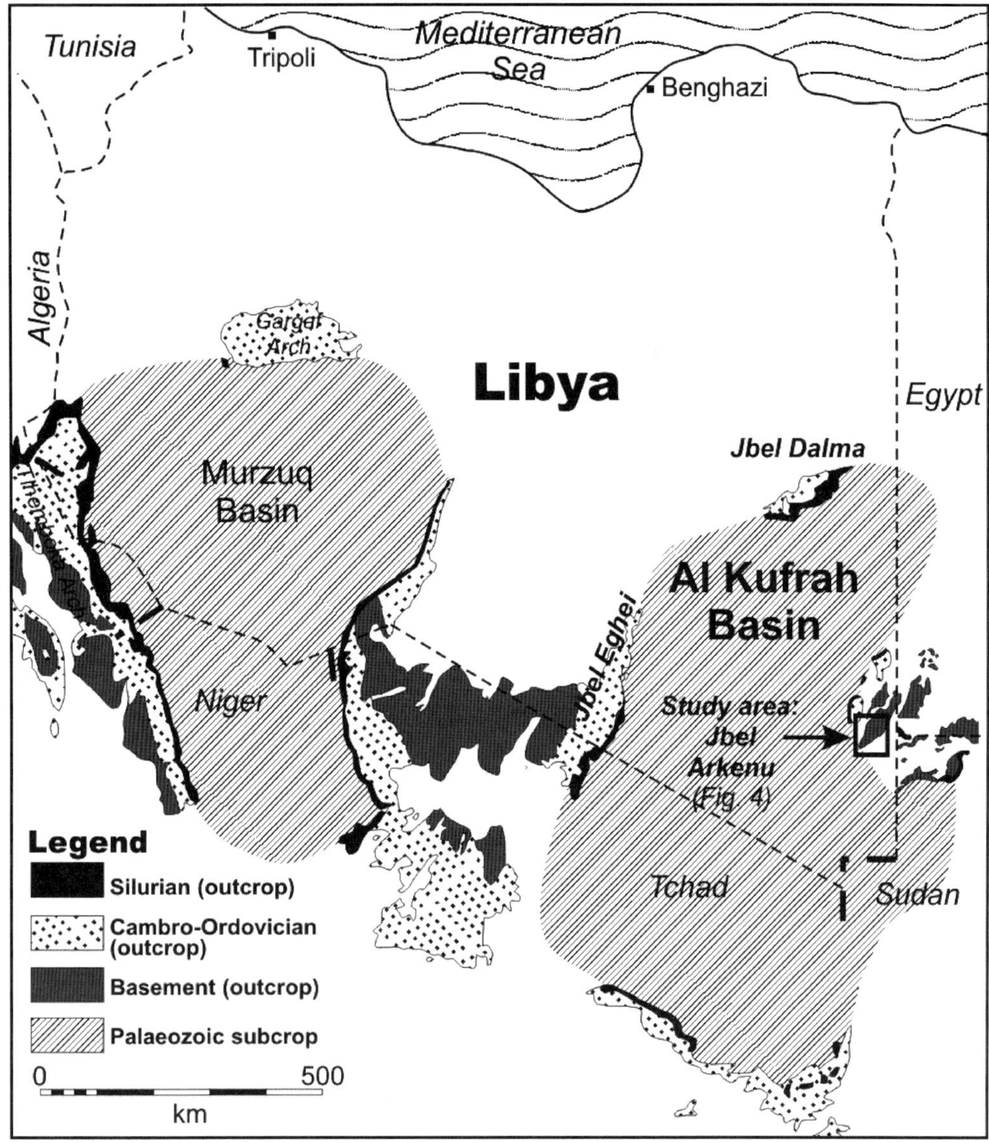

Fig. 1. Geological sketch map of Libya, showing the location of the main sedimentary basins, including the Al Kufrah Basin. The outcrop study area, in the Jabal Arkenu region, is on the border with Egypt and Sudan. See Figure 4 for a detailed interpretation of the Landsat data covering the Jabal Arkenu region.

rock potential of the Al Kufrah Basin have stressed the importance of exploring intervals of Early Silurian and 'Infracambrian' age (Lüning *et al.* 1999) (Fig. 2), because shale of this age is most likely to be sufficiently buried to have reached thermal maturity for oil generation in the basin centre. Investigation, and reinterpretation, of outcrops at Jabal Arkenu, at the eastern basin margins (Fig. 1), might be expected to provide critical new information on elements of potential 'Infracambrian' petroleum system.

This article presents new outcrop data and new interpretations for rocks of supposed 'Infracambrian' age in the Al Kufrah Basin, and considers them in several contexts, namely: (1) their sedimentology and structural history; (2) constraints on their age; (3) their validity as analogues to depocentres in the deep basin subsurface; and (4) their relationship

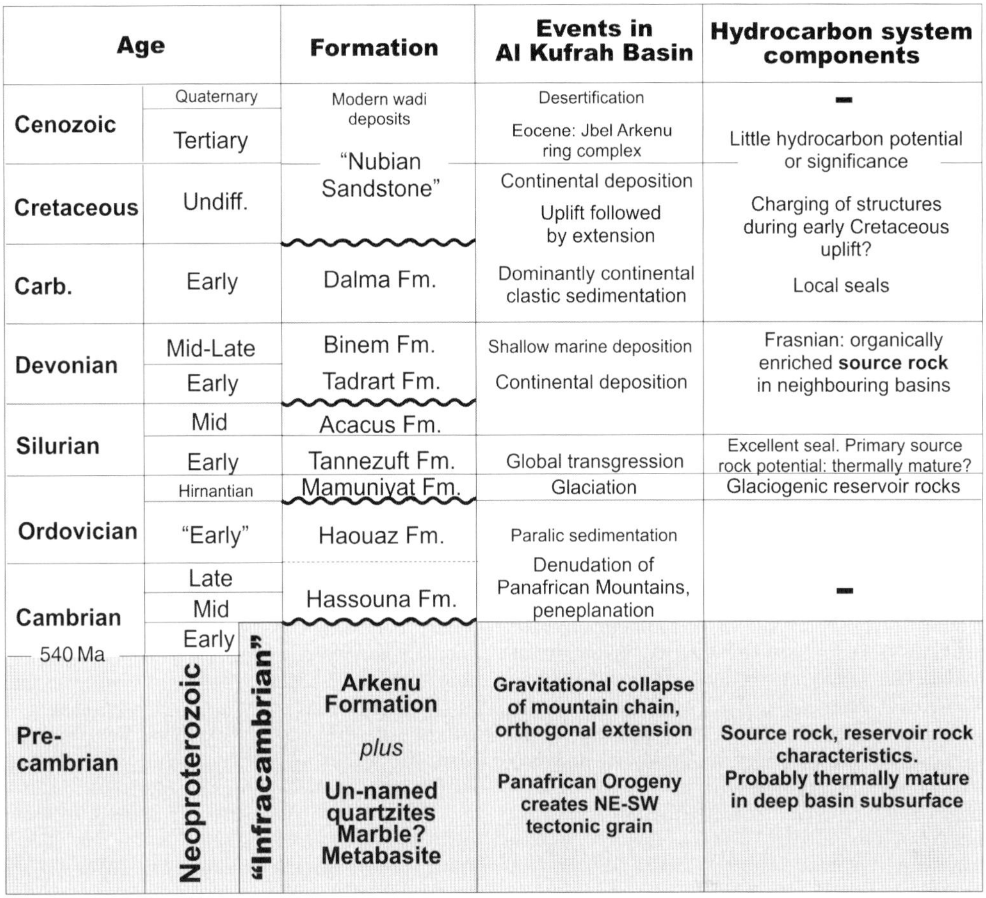

Age		Formation	Events in Al Kufrah Basin	Hydrocarbon system components
Cenozoic	Quaternary	Modern wadi deposits	Desertification	–
	Tertiary	"Nubian Sandstone"	Eocene: Jbel Arkenu ring complex	Little hydrocarbon potential or significance
Cretaceous	Undiff.		Continental deposition Uplift followed by extension	Charging of structures during early Cretaceous uplift?
Carb.	Early	Dalma Fm.	Dominantly continental clastic sedimentation	Local seals
Devonian	Mid-Late	Binem Fm.	Shallow marine deposition	Frasnian: organically enriched **source rock** in neighbouring basins
	Early	Tadrart Fm.	Continental deposition	
Silurian	Mid	Acacus Fm.		
	Early	Tannezuft Fm.	Global transgression	Excellent seal. Primary source rock potential: thermally mature?
	Hirnantian	Mamuniyat Fm.	Glaciation	Glaciogenic reservoir rocks
Ordovician	"Early"	Haouaz Fm.	Paralic sedimentation	
Cambrian	Late	Hassouna Fm.	Denudation of Panafrican Mountains, peneplanation	–
	Mid			
	Early			
540 Ma				
Pre-cambrian	**Neoproterozoic** / **"Infracambrian"**	**Arkenu Formation** *plus* **Un-named quartzites Marble? Metabasite**	**Gravitational collapse of mountain chain, orthogonal extension** **Panafrican Orogeny creates NE-SW tectonic grain**	**Source rock, reservoir rock characteristics. Probably thermally mature in deep basin subsurface**

Fig. 2. Stratigraphic context of Infracambrian metasediments and the Phanerozoic fill of the Al Kufrah Basin. The stratigraphic chart is based on our field observations (Infracambrian–Silurian) and literature data (Devonian–Cenozoic: Bellini *et al.* 1991) for the Jabal Arkenu region, at the eastern basin margin. Also shown are the key components of a potential hydrocarbon system in this basin (source rock, reservoir and seal horizons).

to continent-scale shear zones previously interpreted to traverse North Africa. The data are from the Jabal Arkenu region, on the southeastern flank of the basin. 'Infracambrian' rocks also crop out at Jabal Eghei/Jabal Duhūn at the western basin margins, but this area may be landmined and is thus inaccessible for field research. In this contribution, the term 'Infracambrian' is loosely applied to sediments of ill-defined age, younger than the latest phase of deformation to affect Proterozoic basement and older than those Cambrian strata that can be reliably dated using biostratigraphy. Therefore, the term has both lithostratigraphic significance (Infracambrian rocks) as well as chronostratigraphic implications (e.g. 'during the Infracambrian'). Consequently, sediments of 'Infracambrian' age could have been deposited at any time during the Neoproterozoic and Cambrian (Fig. 2).

Geological overview of the Al Kufrah Basin and the surrounding region

The Al Kufrah Basin is regarded as a classic sag basin with a fill of Infracambrian–Cenozoic age (Hallett 2002). Its present form is the result of multiple geo-tectonic events from Neoproterozoic times onward. During the Neoproterozoic, the present African landmass was formed during the Pan-African Orogeny, a collisional event spanning 200 Ma from *c.* 700 to 500 Ma (Unrug 1996). This mountain-building event is temporally ill-constrained. In the Taoudenni Basin of West Africa, where a succession comprising pre-, syn- and post-Pan-African age sediments is preserved (Lottaroli *et al.* 2009), Culver & Hunt (1991) recognized a Pan-African I Orogeny at around 665–655 Ma, separated by Marinoan glaciation at

630–610 Ma (Destombes *et al.* 2006), and a second phase of uplift in the Pan-African II Orogeny at about 560–550 Ma. However, from his studies of the Mauritinides fold belt, Villeneuve (2005) disputes the timing of this latter phase of uplift, preferring an envelope of approximately 550–500 Ma (i.e. into the Furongian).

Regardless of the exact timing of uplift, Neoproterozoic–Cambrian orogenic processes appear to have juxtaposed the Congo–Nile Craton, the East Saharan Craton and the Touareg Shield (Unrug 1996). During deformation, crystalline shields behaved rigidly, and softer Precambrian sediments around the craton margins were metamorphosed to greenschist facies within 'mobile belts' of deformation (Unrug 1996). The Al Kufrah Basin itself is underlain by the Congo–Nile Craton (Vail 1991). During the assembly of Africa, it has been suggested that the relative movement of the Congo–Nile Craton past the Toureg Shield and East Saharan Craton was accommodated by a large-scale, dextral shear zone (Unrug 1996). Large parts of the Pan-African Orogen are thought to have collapsed gravitationally, forming pull-apart basins as a large component of shear movement jostled the newly assembled continental blocks (Craig *et al.* 2008).

In the Al Kufrah Basin Cambrian–Ordovician times were characterized by minimal tectonic activity, and the deposition of fluvial and shallow-marine sandstones and shales during rare transgressions (Bellini *et al.* 1991) (Fig. 2). However, there is significant, but understated, evidence for regional tectonic activity in the North Africa and Arabian region during the Early Palaeozoic, including Cambrian horst and graben development over southern Morocco (Piqué 2001), mid-Ordovician tilting in the southern Arabian plate (Oterdoom *et al.* 1999) in response to rifting (Husseini 1990), and a significant (100 Ma long) hiatus in the southern Al Kufrah Basin, between the Cambrian and Late Ordovician (Bellini *et al.* 1991). This regional evidence for Early Palaeozoic tectonic activity across North Africa was recognized early on by Klitzsch (1971) who identified long (>1000 km) NW–SE-striking structures forming horst and graben, such as the Kalanshiyu–Awaynat Uplift skirting the field area described in this paper. These structures, Klitzsch (2000) postulated, were long-lived palaeohighs and depocentres throughout the Early Palaeozoic.

By the Late Ordovician continent-scale ice sheets grew across North Africa, depositing glaciogenic sandstones, which record evidence for soft-sediment striation, folding and thrusting of the substrate by subglacial processes (Le Heron *et al.* 2005) (Fig. 2). These glaciogenic sandstones form the principal reservoir unit in the Murzuq Basin to the west (Hallett 2002) and are expected to be of equal economic importance in the Al Kufrah Basin. During a major marine transgression, associated with melting Late Ordovician ice masses, Early Silurian marine shale of the Tanezzuft Formation was deposited (Fig. 2). The lowermost shales of this formation have been suggested to account for some 90% of North Africa's Palaeozoic derived hydrocarbons (Lüning *et al.* 2000). Regional progradation resulted in deposition of stacked shallow-marine parasequences of the Acacus Formation during the Mid–Late Silurian. Regionally, these parasequences are organized into a major third-order regressive system, which culminated in predominantly braided fluvial deposition by the earliest Devonian. Although marine deposits do occur within the Devonian–Carboniferous succession of the Al Kufrah Basin (Turner 1991), it is unclear whether any organic-rich intervals have been buried to sufficient depths to become viable source rock intervals.

The southeastern flank of the Al Kufrah Basin has been magmatically active at several times during the Phanerozoic, implying a long-lived source of heat in this region. Large acid (broadly granitic) intrusions were formed during the Cambrian (Jabal Babein), Devonian and Eocene (*c.* 44–39 Ma: Bellini *et al.* 1991 and references therein). In this paper, field data are presented from the flanks of Jabal Arkenu (Fig. 1), an Eocene ring complex comprising phonolitic, trachytic, syenitic and granitiferous materials of both intrusive and extrusive affinity (Saïd *et al.* 2000). In the eastern Sahara, the absence of obvious plate boundaries rules out an island arc-type origin for these acid igneous materials, and thus the Jabal Arkenu region can be interpreted as an intraplate hotspot (Saïd *et al.* 2000).

Sedimentology of Infracambrian rocks

Observations

At the eastern flank of the Al Kufrah Basin, a sedimentary succession of presumed Infracambrian age (Saïd *et al.* 2000) crops out discontinuously for approximately 45 km along the eastern side of Jabal Arkenu igneous complex. It is best exposed immediately south of this intrusion (Fig. 1). These rocks comprise the steeply dipping Arkenu Formation (Fig. 3a), iron-bearing quartzite (Fig. 3b) and 'Infracambrian marbles' of previous authors (Lüning *et al.* 1999; Saïd *et al.* 2000) in the northernmost part of the outcrop belt. In general, rocks of the Arkenu Formation are moderately silicified and recrystallized to a low metamorphic grade. Saïd *et al.* (2000) estimated that the Arkenu Formation was 127.5 m thick and largely undeformed, and, although the true thickness of this succession cannot be properly established due to deformation (see later), we estimate its thickness to be greater at about 500 m. For these reasons, we do not present a sedimentary log of these rocks in this paper.

The Arkenu Formation comprises slightly recrystallized, interbedded sandstones, rare conglomerates and mudrocks. Sandstones are buff yellow, fine grained and medium-bedded, often structureless and stacked in uninterrupted units at least 10–15 m thick (Fig. 3c). Within these sandstones, sedimentary structures include both tabular cross-beds (Fig. 3d) and possible ripple cross-lamination. Way up cannot be determined from these sedimentary structures. The mudrocks form thick accumulations (at least 60 m), and comprise maroon, lilac and dark grey silty shale. Subtle changes in grain size between beds include variations in silt content, reflected in the weathering pattern of the mudrocks where silty laminae stand proud of more clay-rich horizons (Fig. 3e). A single conglomerate horizon was observed on the eastern side of the exposure. The grain size of this conglomerate is variable with both pebbly (Fig. 3f) and granular clasts set within a sandstone matrix. These examples weathered out *in situ* and are not related to any obvious channel topography. Pebbles are well rounded with a maroon coloration. In thin section, the pebbles are shown to be metamorphosed quartz arenitic sandstones, comprising quartz grains with undulose extinction and sutured contacts (Fig. 3g). By comparison, the sandy matrix of the conglomerate shows some later quartz overgrowth, but sutured contacts are few.

Interpretation

A full reconstruction of the depositional environment of the Arkenu Formation is restricted by: (1) our inability to determine the correct way up, and, hence, document both the context and nature of facies changes; and (2) to accurately document the thickness of the formation owing to possible stratigraphic repetition (see the description of deformation structures below). The thick, uninterrupted units of medium-bedded, fine-grained sandstone indicate repeated deposition and reworking of sandstone in a moderate- to high-energy sedimentary environment. The general paucity of sedimentary structures suggests that the sediment surface was largely flat, although tabular cross-bedded sandstones are testament to the action of traction currents and the migration of straight crested bars across the sediment surface. As mudrocks are organized into 60 m-thick units, rather than punctuating the sandstone at intervals, a significant switch in depositional process is implied to explain the loss of the sand fraction and a transition to a mud-fallout-dominated setting. The alternation of clay-prone and silty horizons within these mudrocks appeals to a possible distal turbidity current mechanism to account for grains in the >63 μm fraction. The local preservation of conglomerates within the Arkenu Formation suggests a dramatic, temporary increase in energy within the depositional system. The distinct difference in grain-on-grain contacts between the matrix and the pebbles (namely the evidence for metamorphism in the latter: Fig. 3g) clearly indicates an extra-basinal source for some of the clasts.

Assuming that an Infracambrian age for the Arkenu Formation is correct (Saïd *et al.* 2000), the absence of bioturbation does not automatically discount a marine origin, because prolific burrowing is not generally recorded in North African sandstones until the Mid-Cambrian (Seilacher *et al.* 2002). Deposition of the Arkenu Formation is equally plausible within a marine or terrestrial environment. Thick and uninterrupted sandstone successions are recorded from braidplains and shoreface environments alike, as well as interchannel portions of subaqueous fan complexes. Moreover, 60 m-thick mudstone with subtle variations in silt–clay content could be deposited in a lacustrine setting fed by dilute underflows, in an outer-shelf setting fed ultimately by offshore-directed currents or in a barrier lagoonal environment. The facies recorded here, coupled with uncertainties in their succession, do not provide us with sufficiently diagnostic criteria to distinguish between these possibilities.

Deformation affecting Infracambrian and Cambrian sedimentary rocks

Observations

Southeast of Jabal Arkenu, the Arkenu Formation outcrop belt is about 3.5 km wide and organized into five distinct ridges, each 50 m high (Fig. 4). These stand proud of the surrounding desert peneplain. The crest lines to these ridges are generally subparallel, converging toward the southernmost part of the outcrop (Fig. 4). The metasediments discussed above are steeply dipping to vertical and strike in the range 025°–035°. Owing to the subvertical strata and irregular terrain, reliable dip values are difficult to record.

A number of structural measurements were made from Infracambrian and Cambrian strata: each has been plotted on the satellite image (Fig. 4). The westernmost two ridges are tightly folded into an antiform structure, identified by steep opposing dips on their respective crest lines (point ‘a’, Fig. 4). Locally, cleavage is developed on the east limb of the antiform (point ‘b’, Figs 4 & 5a, b). The cleavage dips 024°/40°W: strike is parallel to

Fig. 3. The Arkenu Formation of Infracambrian age, south of Jabal Arkenu (see Fig. 4 for location). (**a**) General view of the formation, comprising steeply dipping clastic sedimentary rocks, with the Jabal Arkenu ring complex in the background. Looking NE (along strike). The geologist is circled for scale. (**b**) Heavily deformed and quartzified sandstone, looking north. (**c**) Thick, uninterrupted units of medium-bedded, fine-grained sandstone. (**d**) Tabular

bedding, in the same trend as both the ridge crestlines and the overall trend of the outcrop belt. In another exposure, an approximately 10 m amplitude antiform was identified, with limbs trending 012°/60°W and 014°/67°E (point 'c', Fig. 4). Smaller, metre-scale parasitic folds are also observed, and these verge eastward (point 'd', Fig. 4), plunging 40°/020° (Fig. 5c, d). Although the poor exposure prevents fault contacts being readily identified, we infer one thrust fault, striking parallel to both the ridge crestlines, bedding and cleavage (Fig. 4). This thrust fault is interpreted on the basis of abrupt dip changes. In addition, vertical joints striking 090°–110° were identified.

The geological map sheet of Saïd *et al.* (2000) shows a slight inflection/curvature in the NE–SW outcrop pattern of Infracambrian rocks around Jabal Arkenu and a concentration of iron deposits along the length of the outcrop belt. This is supported by our discovery of highly ferruginous quartzites described earlier (Fig. 3b). Furthermore, pervasive quartz veining affects all Infracambrian rocks in the region, cross-cutting the cleavage. On a ridge NE of Jabal Arkenu, Infracambrian rocks comprise extensively silicified/recrystallized sandstones (bedding dips 020°/40°W) cut by felsic intrusions.

The contact between the Arkenu Formation and the overlying Lower Member of the Hassaouna Formation (Mid Cambrian: Saïd *et al.* 2000) is not exposed, and a gap of more than 500 m separates the most closely spaced outcrops of the two formations. To the west of the Arkenu Formation outcrop belt, the Hassaouna Formation is composed entirely of sandstone and lacks the extensive silicification/recrystallization apparent in the Arkenu Formation. Furthermore, it dips gently (25°–35°) to the SW, striking between 095° and 140° (Fig. 4). However, compared to the remainder of the exposure, sandstones cropping out in the east (i.e. closest to the Arkenu Formation) are anomalous and dip gently toward the east (Fig. 4).

Interpretation

The only detailed geological map of the Jabal Arkenu region (Saïd *et al.* 2000) indicates that the Arkenu Formation dips <5°W; only slightly greater than the regional geological dip of Phanerozoic rocks to the north. It is unclear to us how this large error arose on an otherwise reliable map. Furthermore, the existing map implies that the Arkenu Formation remained subhorizontal until it was metamorphosed in the Eocene during the intrusion of the Jabal Arkenu ring complex (Saïd *et al.* 2000).

Given the presence of cleavage, all deformation within the Arkenu Formation is interpreted to be post-depositional. As the strike directions of bedding, cleavage and fold axial planes are parallel, the deformation is interpreted to record an intense phase of NW–SE compression that produced eastward-verging tight folds, possibly underlain by thrust faults. The absence of metamorphism within the Hassaouna Formation, coupled with the lack of significant deformation, strongly suggests that the Arkenu Formation was deformed prior to the Mid Cambrian. Deformation associated with emplacement of Jabal Arkenu itself during the Eocene (Saïd *et al.* 2000) is discounted for two reasons. First, if recrystallization of the Arkenu Formation into meta-sandstones occurred within the thermal aureole of this intrusion, then similar recrystallization of the Hassaouna Formation should have occurred as both are equally proximal to the southern margins of this intrusion (Fig. 4). Second, the intensity of deformation within the Arkenu Formation, and its strikingly linear character, is incompatible with uplift caused by intrusion, which would have tilted the rocks to the SE. The tectonic driving force behind this previously unrecognized deformation event is unclear. However, we suggest that a period of compressional deformation may be compatible with the poorly constrained cratonic movements associated with the assembly of Gondwana and the final phase of the Pan-African Orogeny during the Early Cambrian (Hallett 2002). We discuss the significance of this event to the wider evolution of Al Kufrah Basin later in this paper.

The gentle eastward dip of the Hassaouna Formation closest to the Arkenu Formation outcrop (Fig. 4) hints at further structural complexity. This could be explained in terms of: (1) post-depositional, differential compaction of the Hassaouna Formation; (2) post-Cambrian reactivation of the pre-existing NE–SW-trending structures documented above; or (3) a fault contact between the two formations. We discount option (1) because the amount of compaction expected from a sand-dominated succession seems too small to account for the deformation observed. Both of the remaining options

Fig. 3. (*Continued*) cross-bedding, a common feature of sandstones in the Arkenu Formation. Note that determining way up from these structures is not possible. (**e**) Part of a 60 m-thick succession of mudrocks (dipping vertically). Note how some beds are more elevated than others, reflecting variations in silt content in the succession. (**f**) Quartz-rich conglomerates within the Arkenu Formation. (**g**) Thin section (cross-polarized light) of the conglomerate in F. Topmost part of the field of view shows a clast, made up of a meta-sandstone with irregular and sutured contacts between quartz grains. The lower part of the field of view shows the matrix, comprising subangular–subrounded sand grains, with an interstitial clay matrix. Photographs and samples from vicinity of 22°06.92′N 24°41.62′E.

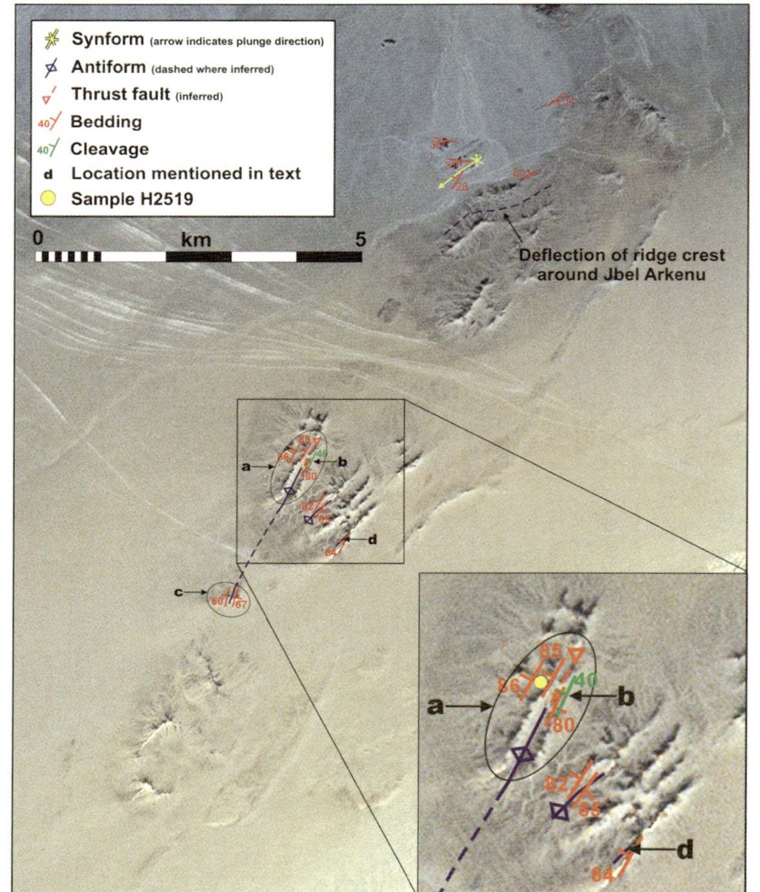

Fig. 4. Landsat data covering the exposures of Infracambrian rocks in the Jabal Arkenu region, with structural data. Locations a–d are referred to in the structural section of this paper. A yellow spot marks the location of sample H2519 on which SHRIMP U–Pb zircon age dating was carried out.

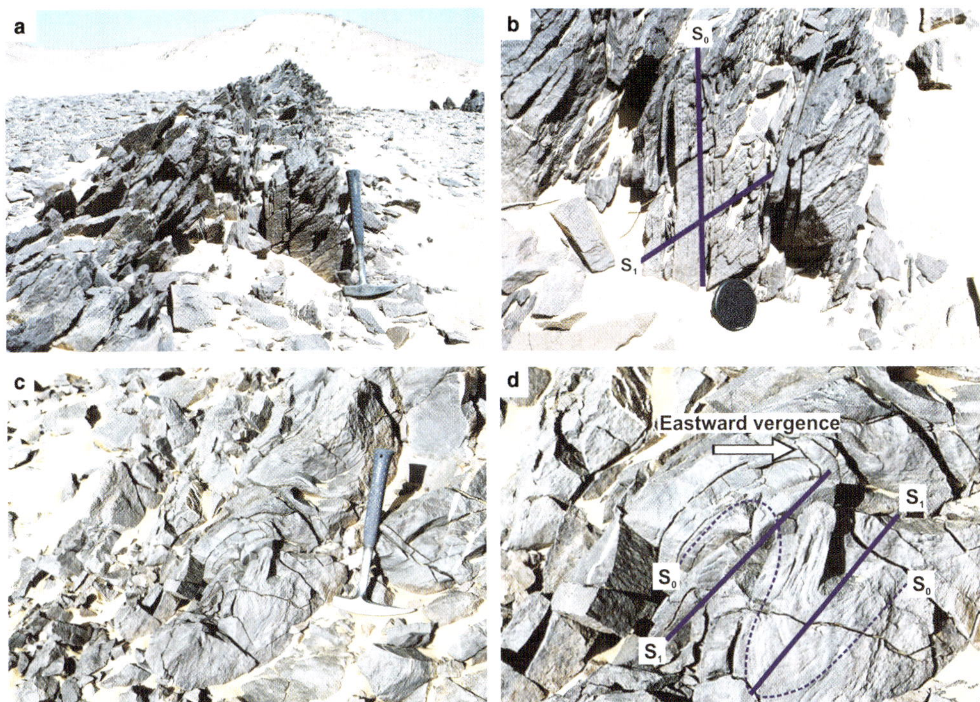

Fig. 5. Deformation affecting the Arkenu Formation. (**a**) and (**b**) are, respectively, distant and close-up shots of bedding–cleavage relationships in these Infracambrian sandstones. The bedding (S_0) is subvertical, cut by a shallower (45° dipping) cleavage (S_1). The strike of both bedding and cleavage are parallel, generally NE–SW (see Fig. 4 for detail). (**c**) and (**d**) are, respectively, distant and close-up shots of minor antiforms, interpreted as parasitic on the main antiformal structures affecting the outcrop. The strike and dip of the fold-axial surfaces is comparable to that of the cleavage planes shown in (B). Note the eastward vergence of these structures in (D). (A) and (B) taken from 22°07.22′N 24°41.70′E; (C) and (D) taken from 22°06.54′N 24°42.06′E.

require a second, post-Cambrian, deformation event and we currently prefer option (2) to option (3) as it does not require the presence of an additional, speculative fault. A geophysical survey is required to provide conclusive data on the relationship between the Arkenu and Hassaouna formations.

The intrusion of the Jabal Arkenu ring complex is interpreted to have produced the gentle SE dip of the Hassaouna Formation (Fig. 4), although contact metamorphism appears to have been minimal. The age of quartz veining is unclear: in the Arkenu Formation quartz veins cross-cut cleavage, but are not found in the Hassaouna Formation. However, we suggest that veining is associated with the Eocene intrusion based on the evidence for felsic dykes throughout the region. Vein emplacement may have focused on pre-existing lines of weakness such as the deformed Infracambrian strata and this may also explain the localization of iron mineralization in these rocks. It is possible that the second phase of deformation in the NE–SW-trending ridge discussed above occurred during the emplacement of these intrusions.

The modern relief is interesting in that, aside from Jabal Arkenu itself, the Infracambrian strata alone have substantially resisted regional peneplanation. The outcrops of the Hassaouna Formation are very small by comparison and we consider it probable that the early metamorphism affecting the Arkenu Formation has contributed to its resilience. The result of this topography is that the Infracambrian outcrop, although interesting, is isolated from its true geological context by lack of local subsurface data.

Petrology of crystalline rocks in the Infracambrian outcrop area

In the Jabal Arkenu region, the northeastern outcrop extremity of Infracambrian rocks (Fig. 4) is mapped as 'Infracambrian marble' (Saïd *et al.* 2000). In the field, outcrop quality of the largest exposure of these rocks is extremely poor, with a relief of only 5 cm exposing a green-blue weathering, crystalline lithology. This rock fails to react with HCl, and was thus studied in thin section for petrographic analysis (Fig. 6a–d). Minerals include twinned plagioclase feldspar megacrysts (3–4 mm long), sericite pseudomorphs after feldspar, quartz, extensively developed matrix sericite, pale green, fibrous–acicular amphibole, high-relief lozenges of epidote, local chlorite and a high proportion of opaques (*c.* 5%). No carbonate is present.

The assemblage of minerals described above automatically discounts a sedimentary protolith. Instead, the paragenesis of feldspar megacrysts, amphibole and epidote (Fig. 6a–d) diagnose the specimen as a porphyritic basalt. In basic igneous rocks, a fibrous–acicular amphibole (actinolite) commonly replaces pyroxene during low-grade hydrous metamorphism, or alternatively during late-stage hydrothermal modification of pyroxene in the cooling igneous body (uralitization: Allaby & Allaby 1990). Coupled with the occurrence of chlorite (Fig. 6a, b), the mineral assemblage is indicative of at least chlorite-grade greenschist metamorphism (Yardley 1998).

The age of this metabasite is unknown. However, metamorphism of the protolith to greenschist facies by the emplacement of the Jabal Arkenu ring complex is considered extremely unlikely, given that Cambrian sandstones of the Hassaouna Formation, which are much closer to it, show no signs of thermal alternation/recrystallization. Furthermore, as the only known mafic rock in the region, we discount its origin as parent magma for the felsic Jabal Arkenu intrusion. The metabasite lies approximately 45 km along strike from the Arkenu Formation. Therefore, an ancient (pre-Eocene) age is thus inferred, and noting that most Neoproterozoic 'mobile belts' in North Africa (such as those of the Tibesti, western Libya) are metamorphosed to greenschist facies (Hallett 2002), a Neoproterozoic

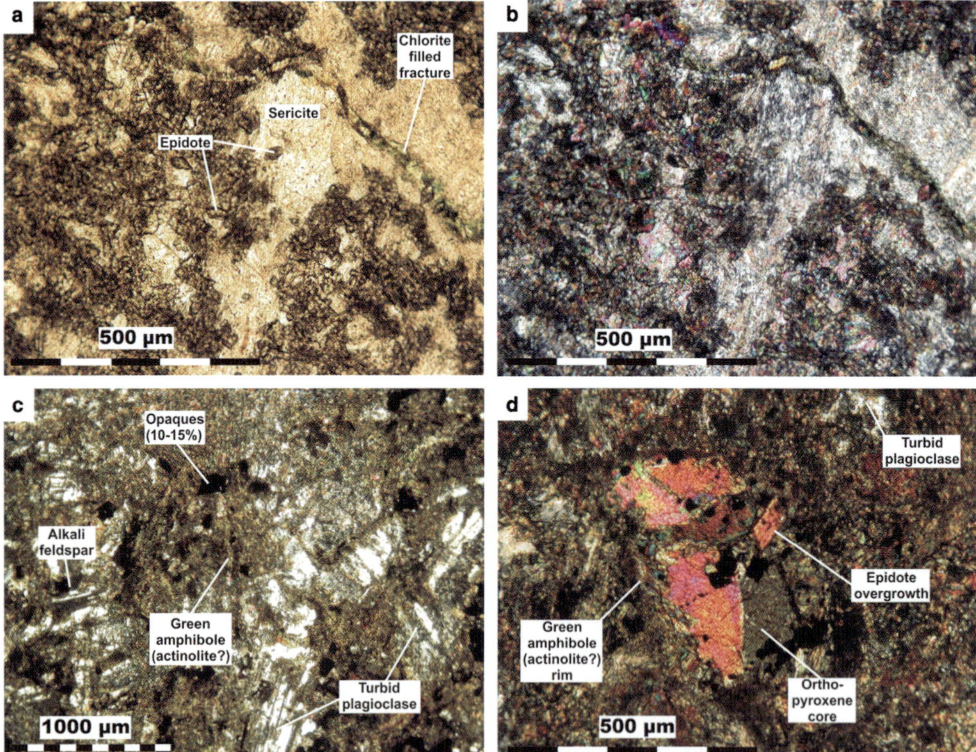

Fig. 6. Photomicrographs of a metabasite collected NE of Jabal Arkenu, from a low-lying outcrop area previously mapped as 'Infracambrian marble' (Saïd *et al.* 2000) (see Fig. 4 for location). (**a**) and (**b**), in plane polarized light and under cross-polars, respectively, illustrate a paragenesis of epidote and sericite. In addition, chlorite occurs as fracture fills. (**c**) Paragenesis of green amphibole (possibly actinolite), turbid plagioclase megacrysts and alkali feldspar. Note that opaque minerals are also important; accounting for up to 5% of the surface are of this slide. (**d**) Composite crystal comprising an orthopyroxene core and a reaction rim of green amphibole (possibly actinolite), with evidence for additional overgrowth of epidote. The relationship between this metabasite and metasediments of the Arkenu Formation cannot be confirmed in the field. Sample collected from 22°24′22″N 24°53′01″E.

age for the metabasite is also plausible. Given this information, and the along-strike relationship of the metabasite to the folded and thrust Arkenu Formation (Fig. 4), we speculate that deep crustal thrusting may have brought this rock to the surface during the same compressional phase to have deformed and slightly metamorphosed rocks to the south of the outcrop belt.

Constraining the age of the Arkenu Formation

The depositional age of the Arkenu Formation is uncertain owing to the scarcity of biostratigraphic controls. Zircon age data have been used on many occasions to constrain the maximum depositional age of sandstones (e.g. Fedo *et al.* 2003) using the principle that the sediment cannot have been deposited earlier than the youngest reliable zircon date. The Arkenu Formation zircon age spectrum contains three distinct peaks within the Archaean, at *c.* 2500, *c.* 2640 and *c.* 2720 Ma, but it lacks the peak at 500–900 Ma seen in spectra from younger formations (Fig. 7a). The youngest reliable zircon age acquired during the first analytical phase was 951 ± 11 Ma, the three grains with apparently younger ages being discarded due to discordance (Fig. 7b). In order to provide further constraints on the depositional age of the Arkenu Formation, a second SHRIMP (sensitive high-resolution ion microprobe) analytical phase was conducted, targeting colourless zircons with euhedral or angular habits, these being the most likely candidates for the youngest components of the zircon spectrum. This analytical phase failed to identify any grains younger than 951 ± 11 Ma. It therefore appears that the Arkenu Formation at Jabal Arkenu has a maximum depositional age of about 950 Ma. Although this does not prove that deposition did not take place much later, the absence of zircons in the 500–900 Ma range, which corresponds to the Pan-African Orogeny, arguably the most important crust-forming event in North Africa and the Middle East (Klusky *et al.* 2003), strongly suggests that deposition took place prior to the Pan-African series of orogenic events.

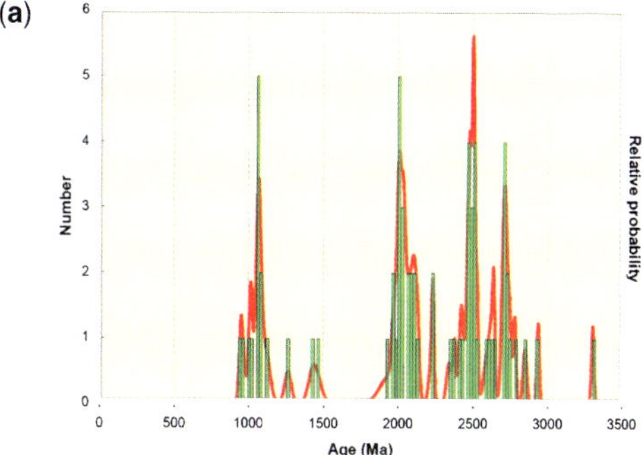

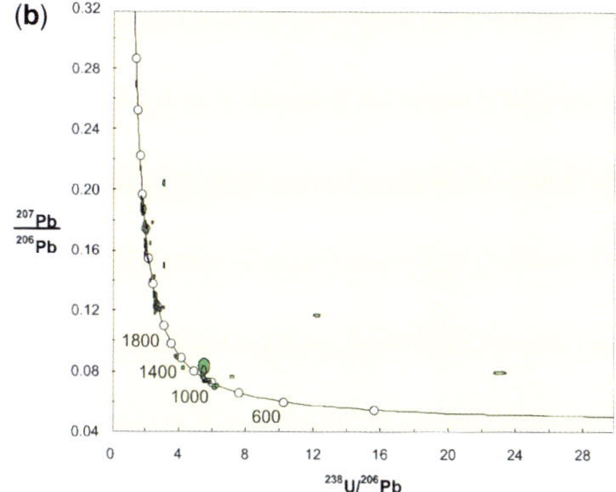

Fig. 7. (**a**) Combined histogram–relative probability plot for sample H2519 (Arkenu Formation, Jabal Arkenu). Grains with unreliable U–Pb ages are not plotted. (**b**) Tera–Wasserburg concordia diagram with U–Pb isotopic data from sample H2519 (Arkenu Formation, Jabal Arkenu). All data are plotted and were analysed in a targetted manner to find the youngest zircons in the sample. Data point ellipses are 1σ. Common Pb correction has not been applied. Data in pale green yield unreliable U–Pb ages and have not been used in the histogram–relative probability plot.

Discussion

Is the Al Kufrah Basin really a pull-apart basin?

Bellini *et al.* (1991) proposed a model of the Al Kufrah Basin as a sinistral transtensional basin based on the identification of a rhombochasm, some 150 km wide and 400 km long in magnetic datasets (Fig. 8a). By that time, however, a pull-apart origin for the basin had already been accepted (Schandelmeier 1988). Lüning *et al.* (1999) report a rift graben with identifiable syn- and post-rift fill, within the deep subsurface of the basin, based on seismic lines acquired and interpreted by AGOCO. However, these workers do not report whether faults were clearly imaged on these lines, and neither do they investigate the pull-apart basin model of Bellini *et al.* (1991). Proprietary seismic lines have subsequently shown that the Al Kufrah Basin contains two deep depocentres sealed beneath/onlapped by 'layer cake' Palaeozoic stratigraphy (Craig *et al.* 2008) but no obvious faults are imaged. Across central and eastern North Africa, the youngest widespread post-Neoproterozoic strata are generally Cambrian (Selley 1997; Tawadros 2001; Hallett 2002), and hence the deep core of the Al Kufrah Basin has been assumed to be

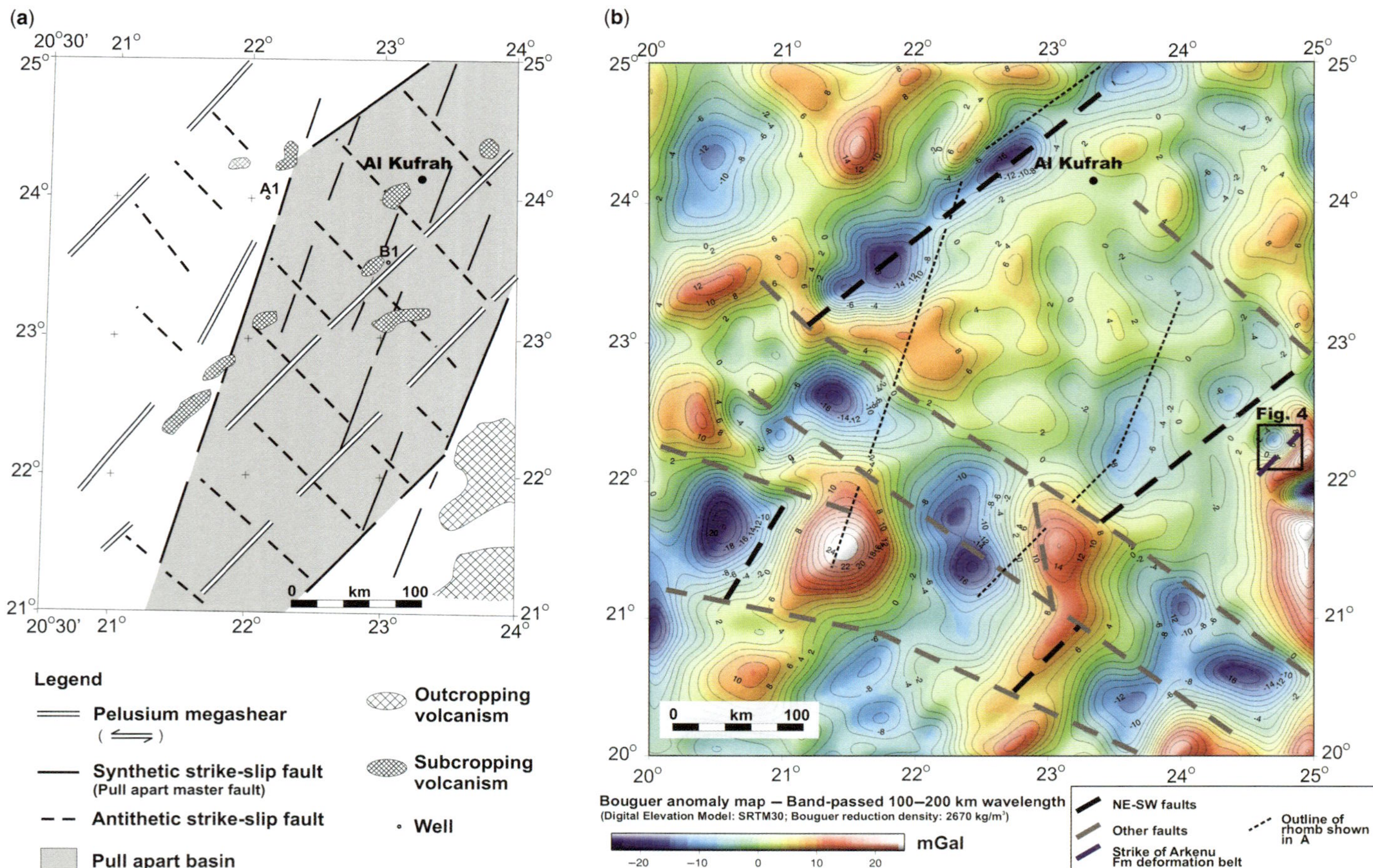

Fig. 8. (**a**) Interpretation of regional-scale fault systems in The Al Kufrah Basin from magnetic data (after Bellini *et al.* 1991). In this figure a sinistral strike-slip system is dominant (the 'Pelusium megashear'), with a rhombochasm presently in the deep subsurface of the Al Kufrah Basin defined by synthetic strike-slip master faults. Note how under this model large transcurrent faults associated with the 'Pelusium megashear' cross-cut the rhomb. This cross-cutting relationship requires a pull-apart basin to form before a parallel set

'Infracambrian'. Therefore, the pull-apart basin of Bellini *et al.* (1991) or the rift basin of Lüning *et al.* (1999) is implicitly of Infracambrian age.

Public domain gravity data processed using an EIGEN-GL04C gravity model (Förste *et al.* 2006) was acquired to generate a Bouguer anomaly map over the Al Kufrah Basin as far east as Jabal Arkenu (Fig. 8b). This map was used to investigate whether the structural interpretations of Bellini *et al.* (1991) are supported by the most up-to-date gravity models. It has a resolution of 1–200 km, and hence emphasizes large-scale crustal structures and the expanse of smaller faults. The data show the presence of four NE–SW-striking faults within the Al Kufrah Basin, approximately striking parallel to the outcrop belt of the Arkenu Formation and the structures within it (c.f. Figs 4, & 8b). These NE–SW faults are cut and offset by a later generation of NW–SE fault structures. Superimposing the rhombochasm drawn by Bellini *et al.* (1991) highlights the dramatic difference between these interpretations (Fig. 8b).

Bellini *et al.* (1991) indicate a sinistral pull-apart basin trending NE–SW, which Schandelmeier (1988) suggests was one of three en echelon NE African basins formed under a regional Late Precambrian strike-slip regime (Fig. 9). However, modern gravity data used in this study do not show a rhomb, and hence they do not support a pull-apart basin geometry (Fig. 8b).

Integrating all of the sedimentological, structural, petrographic and geophysical data available to us, we tentatively propose the following alternative model for the early evolution of the Al Kufrah Basin. Mountain building during the Pan-African Orogeny resulted in intense deformation in the Jabal Arkenu area, isoclinally folding, thrusting and cleaving the Arkenu Formation. The tectonic context of this formation is uncertain, but deformation clearly imparted a pronounced NE–SW structural grain to Infracambrian sediments. The same deformation event was probably responsible for the formation of NE–SW faults, up to 300 km long, in the subsurface of the Al Kufrah Basin (Fig. 8b). Given the context of compressional

A. Late Precambrian:
Pull-apart basin development
along a major dextral shear zone

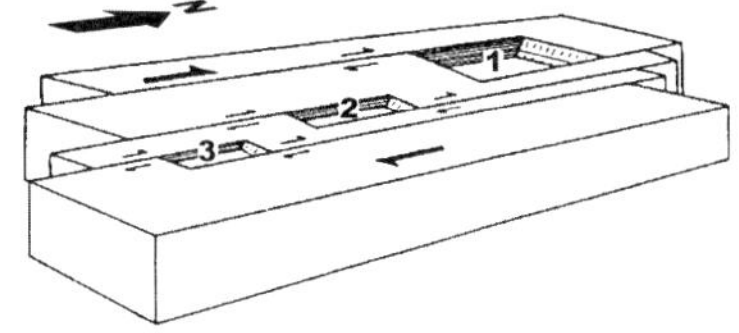

B. Mid-Late Cambrian:
ENE-WSW tension forms NNW-SSE
horsts and graben

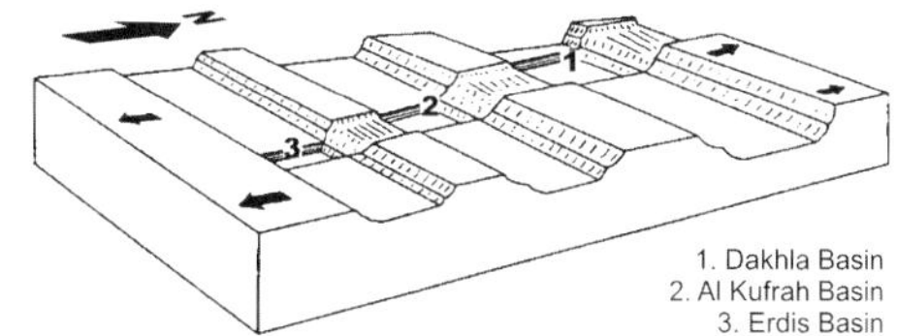

1. Dakhla Basin
2. Al Kufrah Basin
3. Erdis Basin

Fig. 9. Development of the Al Kufrah, Erdis and Dakhla basins in NE Africa, as dextral pull-apart basins (rhombochasms). Model reproduced from Schandelmeier (1988).

deformation affecting the Arkenu Formation at the basin margins, it is possible that these originated as thrusts.

Following the main mountain-building event, collapse and, possibly, rifting gave rise to what is now the deep subsurface of the basin, with extension perpendicular to the same NE–SW structural grain. Collapse along NW–SE-striking faults may have occurred during this process (second generation of faults: Fig. 8b), although a later origin cannot be discounted for these structures. Therefore, in our model, we propose: (1) predominantly orthogonal compression during the Pan-African Orogeny to account for the style of deformation in the Arkenu Formation; and (2) orthogonal extension along the same regional tectonic grain, accompanied by possible oblique transcurrent faulting, to form the main part of Al Kufrah Basin (e.g. Fig. 8b). Under this

Fig. 8. (*Continued*) of faults is superimposed under continuing strike-slip deformation. However, under this model it is unclear why ongoing extension would not continue pull-apart basin development, and produce the Pelusium megashear faults instead. (**b**) Bouguer gravity anomaly map, calculated by subtracting the Bouguer correction from the free air anomaly map of the Al Kufrah Basin area. The map was computed from publicly available data of the International Centre for Global Earth Models (ICGEM) (http://icgem.gfz-potsdam.de/ICGEM/ICGEM.html), using the EIGEN-GL04C gravity model (Förste *et al.* 2006). This gravity model, referred to as EIGEN GL04C, combines data from GRACE and LAGEOS satellite missions with terrestrial data at 0.5° × 0.5° resolution. Data for the Al Kufrah Basin (5° × 5°) were resampled to 1 arc minute resolution for processing. The resulting gravity/geoid anomaly wavelength resolution achieved is about 110 km. This process emphasizes long-wavelength crustal structure at the expense of smaller-scale faults. Interpretation shows two major NE–SW faults: one north of Al Kufrah settlement, and another immediately north of Jabal Arkenu. Both NE–SW faults are remarkably parallel to the belt of deformation shown in Figure 4. No rhomb is apparent in these data.

model it is apparent that sediments in the deep subsurface of the basin are highly unlikely to be younger and, hence, are correlatives to those at the surface.

It should be stressed that the intensity of deformation in the Arkenu Formation (Figs 4 & 5) is relatively low. In the core of modern orogens it is common to find more intense products of deformation, such as in the vicinity of the Karakoram fault zone in SW Tibet, where shear zones and mylonitic fabrics are typical (Phillips *et al.* 2004). Consequently, it is suggested that the deformation in the Arkenu Formation is much more representative of far-field stresses at the fringes of a collision zone.

Regional analogues

In view of the above discussion, our discussion of regional Infracambrian analogues is two facetted and includes, first, the search for correlative rocks to the Arkenu Formation and, secondly, analogues for the subsurface of the Al Kufrah Basin (Fig. 8a, b). Remarkably little is published on potentially correlative rocks to the Arkenu Formation in North Africa. The Mourizidie Formation, which crops out north of the Mourizidie Pass at the eastern flanks of the Murzuq Basin, may provide some comparison. This formation is sandwiched between metamorphic 'Pharusian' or Pan-African Basement and undifferentiated rocks, mapped as Cambro-Ordovician in age by Jacqué (1962) at about the same time as Klitszch (1963) in two high-quality papers that stand the test of time, almost 50 years on, as the only comprehensive geological studies in this area. Jacqué (1962) justifies the stratigraphic position of these sediments as follows: the Infracambrian rests 'in angular unconformity on folded partially granitised, Pharusian basement, and reworked into Cambro-Ordovician rocks which rest in gentle discordance above'. He places great emphasis on extrapolating stratigraphic observations from several hundred kilometres further west in Algeria, and states that 'certain petrographic characteristics closely resemble the Purple Series of Ahnet and not Cambro-Ordovician there'. In this reference to the Purple Series of Ahnet, Jacqué (1962) was alluding to Infracambrian rocks at the southern flanks of the Ahnet Basin, which have subsequently been interpreted to record the infill of rift graben formed in response to the collapse of mountains formed during the Pan-African Orogeny (Rahmani & Haouchine 1999). According to Jacqué (1962), it should be noted that 'devitirified rhyolite pebbles are found at the base of the Cambro-Ordovician sequence above', providing an upper age limit for supposed Infracambrian rocks of this region. Although he did not observe rhyolites *in situ* in Mourizidie, Jacqué (1962) notes that in other parts of the Sahara acid volcanic rocks are known beneath the Cambrian of the Anti-Atlas (Morocco), Adrar des Iforas (Mauritania) and the eastern Hoggar (Algeria). It is an unfortunate consequence of both this early and later work on the Infracambrian of North Africa that lower and upper age constraints are frequently borrowed from neighbouring basins, making their absolute ages highly uncertain.

Further afield, Lüning *et al.* (1999) suggested that pull-apart basin fills of supposed Infracambrian age in Oman could provide potential analogues for the deep Al Kufrah subsurface. Given our preferred model for orthogonal extension outlined above, we would question the validity of a comparison based on strike-slip deformation. However, the use of such analogues is attractive to the Libyan petroleum industry because Neoproterozoic ramp silicilytes (known as pthanites in North Africa) and deep-water carbonates within the Huqf Supergroup are extremely important source rocks in Oman, where oil reserves in this succession may exceed 300 million barrels (Amthor 2000). In neighbouring Saudi Arabia, similar dark grey limestones of the Jibilah Group were deposited in a similar ramp setting and are good hydrocarbon source rocks, with typical TOC (total organic carbon) values of 3% (Nicholson & Janjou 2003). Such analogue comparisons are necessarily both optimistic and speculative because the deep subsurface Al Kufrah Basin has been penetrated by only two wells. Some of the most organically enriched, and hence oil-prone, carbonates of the Huqf Supergroup in Oman and the Jibilah Group of neighbouring Saudi Arabia appear to have been deposited in a ramp setting (Nicholson & Janjou 2003; Amthor *et al.* 2005) that was structurally active, and not an interior rift basin.

Infracambrian in the Al Kufrah Basin: linchpin of a 6000 km-long Transafrican Lineament?

Schandelmeier (1988) suggested that the Al Kufrah Basin, the Erdis Basin (Chad) and the Dahkla Basin (Egypt) formed during the Neoproterozoic along a pronounced NE–SW crustal weakness in North Africa (Fig. 9). However, the concept of an even larger scale NE–SW-trending trans-continental belt of deformation extending from the Niger Delta to the Nile Delta (Fig. 10) is not new. During the mid-1970s photogeological mapping, coupled with the availability of Landsat data for the first time, resulted in the publication of ambitious, large-scale models of trans-African deformation. The belts of deformation described in these models were assumed to be long-lived and repeatedly reactivated from the Mesoproterozoic

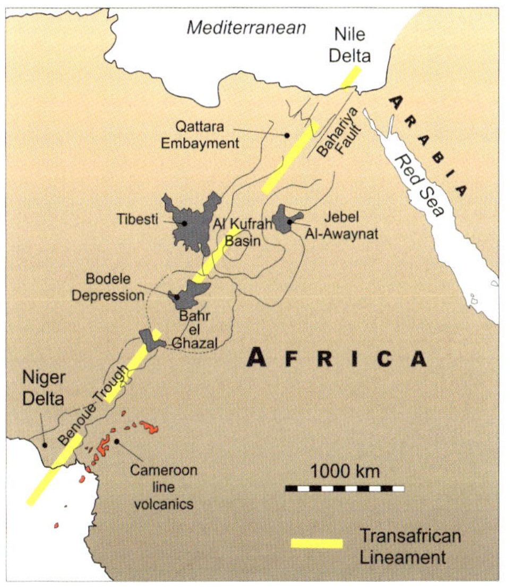

Fig. 10. Major structural features cited as evidence for a trans-continental tectonic–magmatic lineament between the Nile Delta, Egypt and the Niger Delta in West Africa. Reproduced from Nagy *et al.* (1976).

onwards, and were termed the Transafrican Lineament (Nagy *et al.* 1976) (Fig. 10), the Pelusium Line (Neev 1977; Neev *et al.* 1982) or, more recently, the Guinean–Nubian lineament (Guiraud *et al.* 1985; Rogers *et al.* 1995). In many of these models outcrops in NE Africa, on the flanks of the Al Kufrah Basin in particular, are cited as key constraints. Subsequently, numerous references to a Transafrican Lineament cutting through part of the Al Kufrah Basin have been made in the literature (e.g. Keeley 1994; Ghebreab 1998; Klitzsch 2000; Hallett 2002). In this section, we explore how these large-scale belts of deformation were defined, and whether they are fact or fiction. If real, were they active or initiated in the Infracambrian as is widely supposed (e.g. Schandelmeier 1988) and how do they relate to deformation described in the Arkenu Formation above?

Nagy *et al.* (1976) described six key geological features that together formed a Transafrican Lineament (Fig. 10), which they interpreted as a continent-scale fault or fracture zone. From NE to SW these features are: (1) the Qattara embayment, with NE–SW-trending basins and swells in rocks of Cretaceous and Cenozoic age; (2) the Bahariya Fault, presently a topographically elevated structure with a NE–SW strike and east of the Qattara embayment; (3) the Al Kufrah Basin; (4) the Bahr el Ghazal region, northern Chad, including the Bodele topographic depression; (5) Lake Chad depression; and (6) the Benoue Trough, an elongate NE–SW-striking outcrop of Cretaceous–Cenozoic rocks bounded on either side by Proterozoic deposits and interpreted as a failed rift arm of the Niger Delta triple junction (Burke & Dewey 1973).

The alignment of the above structural features was explained by Nagy *et al.* (1976) as an ancient crustal suture. To support this interpretation, they cited as key evidence systematic petrographic and geochemical changes in the composition of the Ben Ghenema batholith on the NW flank of the Tibesti Massif (central Libya) as evidence for a suture dividing ancient continental crust to the west of this intrusion from ancient oceanic crust to the east. The intrusion is dated at around 550 Ma (Pegram *et al.* 1976), and was thus emplaced in the latest Precambrian in the latter stages of the Pan-African Orogeny (Unrug 1996). If correct, the interpretation requires the closure of a former ocean to account for the presence of ocean crust. It is surprising, therefore, that the Transafrican Lineament of Nagy *et al.* (1976) is actually shown to bypass the Tibesti Massif (Fig. 10), the structure of which a suture model was intended to explain, instead of running down the 'axis' of the Al Kufrah Basin.

In his reply to Nagy *et al.* (1976), Chukwu-ike (1977) rightly cautioned that joining disparate geological features of broadly the same orientation is a highly subjective geological exercise. Moreover, he cited the lack of ophiolites and deep-sea sediments, regarded as classical diagnostic features of ocean suture zones, as a reason to question whether this zone formed during a collision process between two formerly separated crustal units. Nagy *et al.* (1976) concede that the genetic relationship between the Transafrican Lineament and igneous activity is unclear, yet they argue that Phanerozoic reactivation occurred frequently and that the most recent possible volcanic activity was in the Cenozoic.

At about the same time, Neev (1975) published evidence for a large-scale 'swarm' of tectonic lineaments in the Levant, identifying a NE–SW-oriented zone of sinistral strike-slip faults. In the eastern Nile delta he found evidence for a NE–SW-trending escarpment, which was postulated to belong to the same family of lineaments, joined them together in one shear system and coined the term 'Pelusium Line'. Later, Neev *et al.* (1982) attempted an interpretation of tectonic features observed on Landsat data, which they mapped out across North Africa, focusing on NE–SW-striking geological features. In that paper they redefined the Pelusium Line as the 'Pelusium megashear system and its associated lineament swarms' to encompass a broad and diffuse belt of deformation that extended from the Levant, through the eastern Nile Delta via the Bahariya graben (NE Egypt) and into a rhombochasm in Al Kufrah Basin. This rhomb, which was in a similar location to that later interpreted in Bellini *et al.* (1991) (Fig. 8a), was interpreted to lie in the middle of a more than 500 km-wide belt of deformation that included the Tibesti Massif (Libya) and the Ennedi–Bourkou range (Chad) within one continuous shear zone. One continuous, NE–SW-striking master fault defined the northern limit of the shear zone passing through Pic Touside in the Tibesti Mountains, with less well defined NE–SW master faults in the south. Smaller (*c.* 100 km long) fault systems were interpreted to terminate against the master faults to define en echelon, left-lateral megashears that have functioned since Precambrian times (Neev *et al.* 1982).

Like Nagy *et al.* (1976), Neev *et al.* (1982) draw attention to NE–SW-striking faults in the northern Tibesti Massif, three of which they interpret as showing 3, 4 and 6 km of left-lateral offset. Furthermore, Neev *et al.* (1982) claim that more than 500 m of vertical movements have occurred along some of the faults in the Tibesti area. More recently, Guiraud *et al.* (2000) published evidence for a 6000 km-long NW–SE lineament in northern Africa, termed the Tibesti Lineament, purported to cross Africa from the Ougarta range (Algeria), via the Tibesti Massif, into Darfur and East Africa (Fig. 11). In that paper a single line (hence lineament) was interpreted, and clearly visible on both unprocessed satellite imagery and a digital elevation model, as a result of the scaling-up process of a 200–300 km-wide belt of faults into a single line (Fig. 11). On the same digital elevation model, there is no evidence of a linear swarm of faults attributable to a NE–SW-striking Transafrican Lineament (or Pelusium Line), which we conclude to be unsubstantiated because its age, geometry, origin and, even, position over several hundred kilometres are in doubt from the previous discussion.

As a final comment, we might expect regional-scale, long-lived zones of crustal weakness such as the Transafrican Lineament to remain tectonically active to the present day, particularly if claims of a multiple Neoproterozoic–Cenozoic reactivation history (Nagy *et al.* 1976) are correct. Maps showing magnitude >4.5 earthquakes across the North African and Arabian regions illustrate a clustering of epicentres along the Red Sea spreading centre and in the eastern Mediterranean (Adams & Barazangi 1984). As a major zone of crustal weakness oblique to the Red Sea spreading centre, the Transafrican Lineament would be an excellent means of accommodating transcurrent neotectonic stresses. However, for detectable earthquakes in NE Africa, none are recorded, and hence the region appears to be currently tectonically quiescent. From the limited available data, it does not appear that significant transcurrent strain is currently occurring along the zone of the supposed Transafrican Lineament.

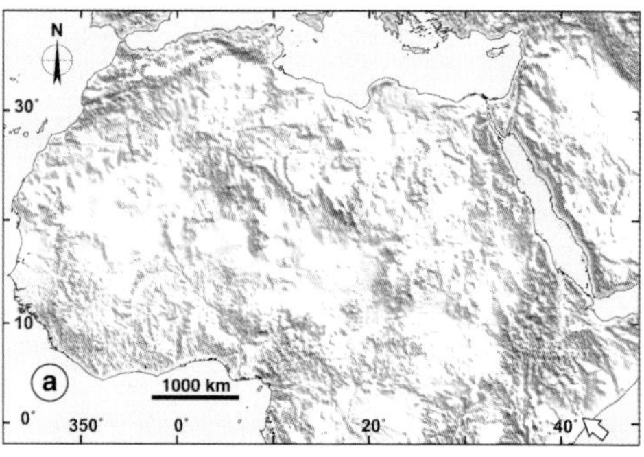

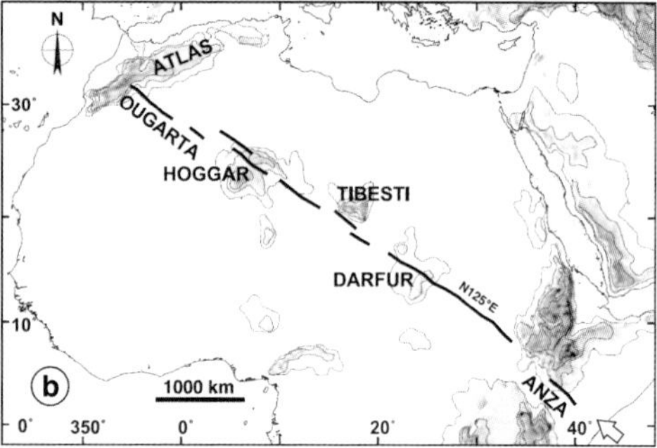

Fig. 11. (**a**) Raw (uninterpreted) digital elevation model of North Africa and (**b**) the interpretation of a NW–SE-striking 'Tibesti Lineament' from that dataset. Note that the purported NE–SW-striking Transafrican Lineament long-supposed to traverse NE Africa and the Al Kufrah Basin is not detected on this image. Both the DEM, which is based on EDC-USGS GTOPO30 data, and the interpretation are taken directly from Guiraud *et al.* (2000).

This study has shown that there is evidence for predominantly orthogonal compression in the Infracambrian succession of Jabal Arkenu during the Pan-African Orogeny, and that orthogonal extension due to gravity collapse of the Pan-African Mountains may also account for the deep basin in the subsurface of Al Kufrah. A re-evaluation of the available data suggests that previously published interpretations of the strike-slip component of deformation along local, but important, fault zones in the eastern Sahara, at least, may have been previously overstated.

Conclusions

- At the SE flank of the Al Kufrah Basin, Infracambrian rocks comprise metasediments (sandstones and mudrocks). The sandstones were deposited in a generally moderate- to high-energy setting, possibly in a shallow-marine or fluvial depositional environment. The hydrocarbon source rock potential of the interbedded mudstones at outcrop is nil due to deformation;

however, their potential at depth remains. This topic needs significant further research in the context of hydrocarbon exploration.

- New SHRIMP zircon age data for the Arkenu Formation at Jabal Arkenu indicate a maximum depositional age of approximately 950 Ma. Although this does not prove that deposition did not take place much later, the absence of zircons in the 500–900 Ma range, which corresponds to the Pan-African Orogeny, strongly suggests deposition took place prior to the Pan-African series of orogenic events.
- The Infracambrian Arkenu Formation was cleaved and deformed to produce large-scale (*c.* 10 m amplitude) antiforms and inferred thrusts. These structures are interpreted to have formed during an episode of orthogonal compression, probably in the Late Precambrian–Cambrian Pan-African Orogeny. The extent of this deformation is hard to constrain owing to limited outcrop, but the along-strike presence of greenschist-grade metabasites suggests significant uplift occurred that lends support to a regional-scale deformation event.
- The deformation process imparted a pronounced NE–SW structural grain on Infracambrian sediments. Collapse of the crust following peak orogenesis may provide an alternative model for the origins of the deep depocentre in the subsurface Al Kufrah Basin, without need to invoke formation of a series of huge en echelon pull-apart basins in NE Africa. However, the possibility remains that some of the regional faults traversing the basin were initiated or reactivated at a later stage.
- To the NE of Jabal Arkenu, an outcrop of 'Infracambrian marble' described by previous authors is shown by thin section petrography to be a metabasite. Infracambrian marbles are correctly mapped further to the north.
- In the light of our results, it seems likely that previous models for the origins of NE–SW-striking faults active in NE Africa during the Infracambrian overstate the importance of (sinistral) strike-slip tectonics during the early evolution of the Al Kufrah Basin area.

Muktar, Hussein and Enwhere of El Omerahil Tours, Sebha, co-ordinated our fieldwork. Mr Y. Abutarruma and Mr A. Asbili of the Earth Science Society of Libya facilitated permissions and dealt with logistics. The authors profited from lengthy and engaging debates with Dr S. Rice and Dr D. Strogen (CASP) about the Transafrican Lineament, and from the review comments of Prof. E. Klitzsch, Dr M. Klitzsch, Dr R. Phillips, Dr S. Lüning and Dr J. Redfern. Mr B. Vautravers is thanked for obtaining maximum resolution images for Figure 4. Professor J. Craig of Eni and Dr B. Thusu (MPRG) are acknowledged for their continued commitment to the project and determination to see it succeed. Finally, we wish to thank all oil industry sponsors of the CASP Southern Libyan Basins Project for financial support.

References

ADAMS, R. D. & BARAZANGI, M. 1984. Seismotectonics and seismology in the Arab region: A brief summary and future plans. *Bulletin of the Seismological Society of America*, **74**, 1011–1030.

ALLABY, A. & ALLABY, M. 1990. *Dictionary of Earth Sciences*, 2nd edn. Oxford University Press, Oxford.

AMTHOR, J. 2000. Precambrian carbonates of Oman; a regional perspective. *GeoArabia*, **5**, 47.

AMTHOR, J. E., RAMSEYER, K., FAULKNER, T. & LUCAS, P. 2005. Stratigraphy and sedimentology of a chert reservoir at the Precambrian–Cambrian boundary; the Al Shomou Silicilyte, South Oman salt basin. *GeoArabia*, **10**, 89–122.

BELLINI, E. & MASSA, D. 1980. A stratigraphic contribution to the Palaeozoic of the southern basins of Libya. *In*: SALEM, M. J. & BUSREWIL, M. T. (eds) *The Geology of Libya, 1*. Academic Press, London, 3–56.

BELLINI, E., GIORI, I., ASHURI, O. & BENELLI, F. 1991. Geology of Al Kufrah Basin, Libya. *In*: SALEM, M. J., SBETA, A. M. & BAKBAK, M. R. (eds) *The Geology of Libya, 6*. Elsevier, Amsterdam, 2155–2184.

BURKE, K. & DEWEY, J. F. 1973. Plume-generated triple junctions: key indicators in applying plate tectonics to old rocks. *Journal of Geology*, **81**, 406–433.

CHUKWU-IKE, I. M. 1977. A crustal suture and lineament in North Africa – discussion. *Tectonophysics*, **40**, 375–382.

CRAIG, J., RIZZI, C. *ET AL*. 2008. Structural styles and prospectivity in the Precambrian and Palaeozoic hydrocarbon systems of North Africa. *In*: SALEM, M. J., OUN, K. M. & ESSED, A. S. (eds) *The Geology of East Libya*, **4**, 51–122.

CULVER, S. J. & HUNT, D. 1991. Lithostratigraphy of the Precambrian–Cambrian boundary sequence in the southwestern Taoudeni Basin, West Africa. *Journal of African Earth Sciences*, **13**, 407–413.

FEDO, C. M., SIRCOMBE, K. N. & RAINBIRD, R. H. 2003. Detrital zircon analysis of the sedimentary record. *Reviews in Mineralogy and Geochemistry*, **53**, 277–303.

FÖRSTE, C., FLECHTNER, F. *ET AL*. 2006. A mean global gravity field model from the combination of satellite mission and altimetry/ gravimetry subsurface gravity data. *Geophysical Research Abstracts*, **8**, 03462.

GHEBREAB, W. 1998. Tectonics of the Red Sea region reassessed. *Earth-Science Reviews*, **45**, 1–44.

GUIRAUD, R., DOUMNANG MBAIGANE, J.-C., CARRETIER, S. & DOMINGUEZ, S. 2000. Evidence for a 6000 km length NW–SE-striking lineament in northern Africa: the Tibesti Lineament. *Journal of the Geological Society, London*, **157**, 897–900.

GUIRAUD, R., ISSAWI, B. & BELLION, Y. 1985. Les lineaments guineo-nubiens: un trait structural majeur a l'echelle de la plaque africaine. *Comptes Rendus des Seances de l'Academie des Sciences, Serie 2: Mecanique-Physique, Chimie, Sciences de l'Univers, Sciences de la Terre*, **300**, 17–20.

HALLETT, D. 2002. *Petroleum Geology of Libya*. Elsevier, Amsterdam.

HUSSEINI, M. I. 1990. The Cambro-Ordovician Arabian and adjoining plates: a glacioeustatic model. *Journal of Petroleum Geology*, **13**, 267–288.

JACQUÉ, M. 1962. Reconnaissance geologique du Fezzan oriental. *Notes et Memoires, Compagnie Française du Petrole*, **5**, 44.

KEELEY, M. L. 1994. Phanerozoic evolution of the basins of northern Egypt and adjacent areas. *Geologische Rundschau*, **83**, 728–742.

KLITZSCH, E. 1963. Geology of the north-east flank of the Murzuk Basin (Djebel ben Ghnema-Dor el Gussa area). *In*: *First Saharan Symposium. Revue de l'Institut Français du Petrôle*, Special Volume, 97–113.

KLITZSCH, E. 1971. The structural development of parts of North Africa since Cambrian time. *In*: *Symposium on the Geology of Libya, Proceedings Volume*. El Fateh University, Libya, Tripoli, 253–262.

KLITZSCH, E. 2000. The structural development of the Murzuq and Kufra basins – significance for oil and mineral exploration. *In*: SOLA, M. A. & WORSLEY, D. (eds) *Geological Exploration of the Murzuq Basin*. Elsevier, Amsterdam, 143–150.

KLUSKY, T. M., ABDELSALAM, M., STERN, R. J. & TUCKER, R. D. 2003. Evolution of the East African and related orogens, and the assembly of Gondwana. *Precambrian Research*, **123**, 81–85.

LE HERON, D. P., SUTCLIFFE, O. E., WHITTINGTON, R. J. & CRAIG, J. 2005. The origins of glacially related soft-sediment deformation structures in Upper Ordovician glaciogenic rocks: implication for ice sheet dynamics. *Palaeogeography, Palaeoclimatology, Palaeoecology*, **218**, 75–103.

LOTTAROLI, F., CRAIG, J. & THUSU, B. 2009. Neoproterozoic–Early Cambrian (Infracambrian) hydrocarbon prospectivity of North Africa: a synthesis. *In*: CRAIG, J., THUROW, J., THUSU, B., WHITHAM, A. & ABUTARRUMA, Y. (eds) *Global Neoproterozoic Petroleum Systems: The Emerging Potential in North Africa*. Geological Society, London, Special Publications, **326**, 137–156.

LÜNING, S., CRAIG, J. *ET AL.* 1999. Re-evaluation of the petroleum potential of the Kufra Basin (SE Libya NE Chad) does the source rock barrier fall? *Marine and Petroleum Geology*, **16**, 693–718.

LÜNING, S., CRAIG, J., LOYDELL, D. K., STORCH, P. & FITCHES, B. 2000. Lower Silurian 'hot shales' in Northern Africa and Arabia: regional distribution and depositional model. *Earth-Science Reviews*, **49**, 121–200.

LÜNING, S., KOLONIC, S., LOYDELL, D. & CRAIG, J. 2003. Reconstruction of the original organic richness in weathered Silurian shale outcrops (Murzuq and Kufra basins, southern Libya). *GeoArabia*, **8**, 299–308.

NAGY, R. M., GHUMA, M. & ROGERS, J. J. W. 1976. A crustal suture and lineament in North Africa. *Tectonophysics*, **31**, 67–72.

NEEV, D. 1975. Tectonic evolution of the Middle East and the Levantine Basin (easternmost Mediterranean). *Geology*, **3**, 683–686.

NEEV, D. 1977. The Pelusium Line; a major transcontinental shear. *Tectonophysics*, **38**, 1–8.

NEEV, D., HALL, J. K. & SALL, J. M. 1982. The Pelusium megashear system across Africa and associated lineament swarms. *Journal of Geophysical Research*, **87**(B2), 1015–1030.

NICHOLSON, P. G. & JANJOU, D. 2003. The Neoproterozoic play in Saudi Arabia. *In*: *AAPG Annual Convention with SEPM, Salt Lake City, Utah, Extended Abstracts*, **12**, 128.

OTERDOOM, H. W., WORTHING, M. A. & PARTINGTON, M. 1999. Petrological and tectonostratigraphic evidence for a Mid Ordovician rift pulse on the Arabian Peninsula. *GeoArabia*, **4**, 467–500.

PEGRAM, W. J., REGISTER, J. K., FULLAGAR, P. D., GHUMA, M. A. & ROGERS, J. J. W. 1976. Pan-African ages from a Tibesti Massif batholith, southern Libya. *Earth and Planetary Science Letters*, **30**, 123–128.

PHILLIPS, R. J., PARRISH, R. R. & SEARLE, M. P. 2004. Age constraints on ductile deformation and long term slip rates along the Karakoram fault zone, Ladakh. *Earth and Planetary Science Letters*, **226**, 305–319.

PIQUÉ, A. 2001. *Geology of Northwest Africa*. Gebrueder Borntraeger Verlagsbuchhandlung, Berlin.

RAHMANI, A. & HAOUCHINE, S. 1999. Structural evolution and implication of fracture distributions: Ahnet Basin, Algeria. *In*: *AAPG International Conference and Exhibition; Abstracts. AAPG Bulletin*, **83**, 1334.

ROGERS, J. J. W., UNRUG, E. & SULTAN, M. 1995. Tectonic assembly of Gondwana. *Journal of Geodynamics*, **19**, 1–34.

SAÏD, M. M., OWEISS, K. A. & MEHIDI, B. O. 2000. *Jabal Arkenu Sheet: Explanatory Booklet*. Joint Geological Mapping Project between the Egyptian Geological Survey and Mining Authority (EGSMA) and the Industrial Research Centre (IRC) of Libya. Cairo.

SCHANDELMEIER, H. 1988. Pre-Cretaceous Intraplate Basins of NE Africa. *Episodes*, **11**, 270–273.

SEILACHER, A., LÜNING, S., MARTIN, M. A., KLITZSCH, E., KHOJA, A. & CRAIG, J. 2002. Ichnostratigraphic correlation of lower Palaeozoic clastics in the Kufra Basin (SE Libya). *Lethaia*, **35**, 257–262.

SELLEY, R. C. 1997. The basins of Northwest Africa; structural evolution. *In*: SELLEY, R. C. (ed.) *Sedimentary Basins of the World: 3, African Basins*. Elsevier, Amsterdam, 17–26.

TAWADROS, E. 2001. *Geology of Egypt and Libya*. Balkema, Rotterdam.

TURNER, B. R. 1991. Palaeozoic deltaic sedimentation in the southeastern part of Al Kufrah Basin, Libya. *In*: SALEM, M. J. & BELAID, M. N. (eds) *The Geology of Libya, 5*. Elsevier, Amsterdam, 1713–1726.

UNRUG, R. 1996. The assembly of Gondwanaland; scientific results of IGCP Project 288; Gondwanaland sutures and mobile belts. *Episodes*, **19**, 11–20.

VAIL, J. R. 1991. The Precambrian tectonic structure of North Africa. *In*: SALEM, M. J., SBETA, A. M. & BAKBAK, M. R. (eds) *Third Symposium on the Geology of Libya*, Volume 6. Tripoli, Libya, Elsevier, Amsterdam, 2259–2268.

VILLENEUVE, M. 2005. Palaeozoic basins in West Africa and the Mauritanide thrust belt. *Journal of African Earth Sciences*, **43**, 166–195.

YARDLEY, B. W. D. 1998. *An Introduction to Metamorphic Petrology*. Longman, Singapore.

Distribution of Infracambrian rocks and the hydrocarbon potential within the Murzuq and Al Kufrah basins, NW Africa

ABDUSSALAM AZIZ[1] & SADEG GHNIA[2]*

[1]*Exploration Department, Exxon Mobil Libya, Tripoli, Libya*

[2]*Baker Hughes Geoscience, Tripoli, Libya*

**Corresponding author (e-mail: sghnia@yahoo.com)*

Abstract: This paper presents an overview of the distribution of the Infracambrian rocks within the Murzuq and Al Kufrah basins obtained by integrating existing data and new outcrop and subsurface data, and discusses the implications for the regional depositional setting and hydrocarbon potential.

In the Murzuq Basin the Infracambrian units comprise volcaniclastics, metasediments and fine- to medium-grained sandstones with an average thickness of 61 m, reflecting a proximal, clastic-dominated depositional system.

In the Al Kufrah Basin carbonate facies in Jabal Arkanu (eastern margin) indicate that the distal facies may become organic-rich. Interpreted seismic lines reveal Infracambrian graben systems similar to those found elsewhere that provide good hydrocarbon source rocks in petroliferous basins.

The Murzuq and Al Kufrah basins in southern Libya are intracratonic basins located on the Saharan Platform of North Africa (Fig. 1). The Murzuq Basin, occupying southwestern Libya, has an area of 350 000 km^2. The Al Kufrah Basin occupies southeastern Libya and northeastern Chad, with extensions into NW Sudan and SW Egypt.

Geological fieldtrips in the Murzuq Basin started in the eighteenth century. Although the first well, A1-1 drilled in 1957, has oil shows and well B1-1 was the first discovery of appreciable amounts of oil, none of the wells drilled showed oil in commercial quantities. In the 1980s a new exploration and production agreement (EPSA) was signed and, during the last two decades deep exploration wells have provided important information about the Infracambrian rocks. By comparison, the Al Kufrah Basin is relatively underexplored; however, geological field studies have been conducted recently and seismic data acquired.

The main reservoirs in the Murzuq Basin are Middle and Upper Ordovician sandstones sourced from the Lower Silurian Hot Shale. While in the Al Kufrah Basin the main candidate reservoirs are the Cambro-Ordovician sandstones.

Stratigraphy and geological setting

The North African Platform has been subjected to several phases of tectonic movement, ranging in time from Precambrian to Cenozoic. Uplifts and erosional events have removed several stratigraphic units, some partially, some totally.

The Murzuq and Al Kufrah basins have very similar stratigraphy, consisting of shallow-marine–fluvial deposits ranging in age from Infracambrian to Cretaceous (Figs 2 & 3).

While the Cambro-Ordovician deposits in the Al Kufrah Basin are as yet undifferentiated, those in the Murzuq Basin are well known. The Murzuq Basin stratigraphic column comprises Infracambrian strata unconformably overlain by the Upper Cambrian Hassaouna Formation of fluvial–shallow-marine sandstone units, followed by Achebayt and Hawaz shallow-marine, tidally influenced deposits of Middle Ordovician age. The Upper Ordovician Melaz Shuqran and Mamuniyat distal–shallow-marine deposits unconformably overlie the Hawaz Formation.

In the Murzuq and Al Kufrah basins the major Silurian transgression is represented by Tanezzuft Shale, overlain by Acacus deltaic sandstone units, locally eroded from the NC115 concession (for the location see Fig. 4) during Silurian–Devonian erosional events (Caledonian unconformity). The Devonian–Permian strata are dominated by shale with sandstone intercalations. Several erosional events occur within this section.

The Mesozoic strata, represented by the Zarzaitine, Taouratine and Messak groups, consist of shale and sandstones deposits. The Messak units form local escarpments on the southern part of the NC115 concession.

The Murzuq and Al Kufrah basins are, for the most part, covered by recent Holocene sands, in many areas forming sand dunes with considerable relief.

From: CRAIG, J., THUROW, J., THUSU, B., WHITHAM, A. & ABUTARRUMA, Y. (eds) *Global Neoproterozoic Petroleum Systems: The Emerging Potential in North Africa*. Geological Society, London, Special Publications, **326**, 211–219.
DOI: 10.1144/SP326.11 0305-8719/09/$15.00

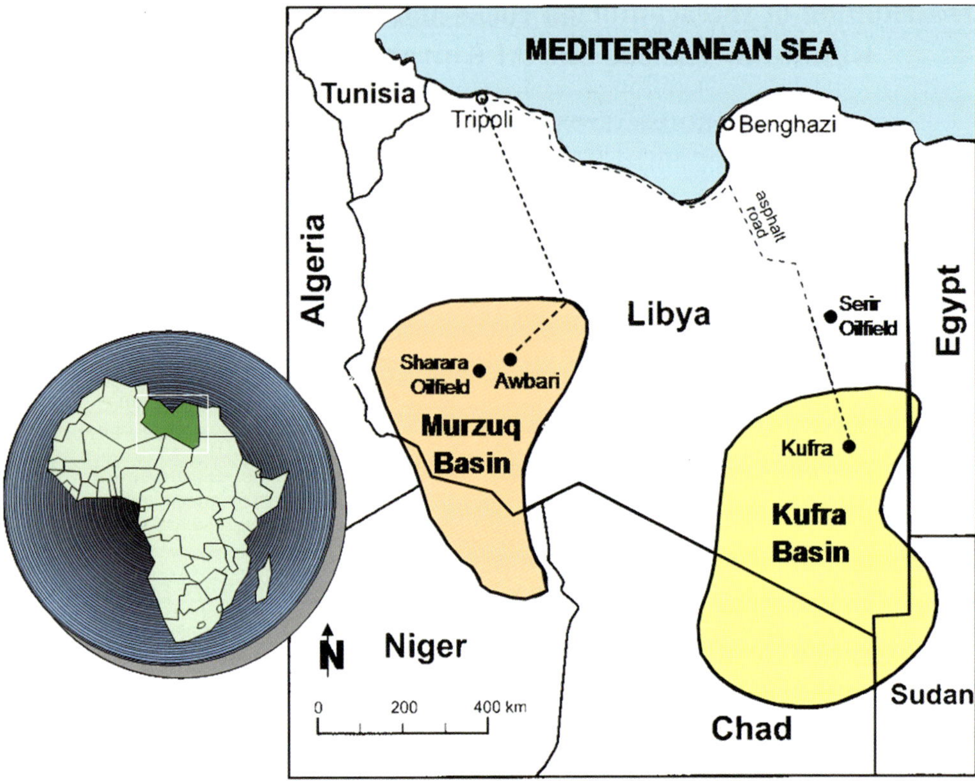

Fig. 1. Location map of the Murzuq and Al Kufrah basins within Libya.

Evolution of Infracambrian rocks

Seismic and well data, obtained from the Murzuq Basin, provide a clear picture of the subsurface Infracambrian and Palaeozoic units that constrain the geological evolution of these stratigraphic units. A great thickness of Infracambrian rocks were deposited in pre-existing graben and palaeolows, showing that tectonic activities played a role in both the deposition and the preservation of these units.

A number of tectonic events affected the area during Infracambrian and Palaeozoic times. Figure 5 illustrates the impact of the major events in the NC115 block, in the NW Murzuq Basin. The most significant of these were the Pan-African Orogeny (Late Precambrian), the Caledonian Orogeny (Late Ordovician–Early Devonian) and the Hercynian Orogeny (Carboniferous). The Gondwana supercontinent was formed as a consequence of the earliest event (Boote *et al.* 1998), and is typified by a north–south-trending system of horst and graben.

Infracambrian deposits

The Precambrian basement is characterized by an erosional peneplained surface, representing a prolonged erosional phase. This is seen in seismic sections as a high-amplitude reflector and is a clear event in well logs.

The Murzuq Basin

The Murzuq Basin Precambrian basement rocks are exposed in the Tibesti–Hoggar Uplift region and in some places around Jabal al Qarqaf (Fig. 4).

The basement rocks are mainly high-grade metamorphic rocks. In the Tibesti area these are divided into the Lower and Upper Tibesti series, and age-dating reveals 1250–820 and 820–600 Ma, respectively (El-Makhrouf 1988). The Mourizidie Formation is correlatable on the western part of NC115 concession, while to the east it has a limited lateral distribution (Fig. 6), the upper part of the Mourizidie Formation has a lightly

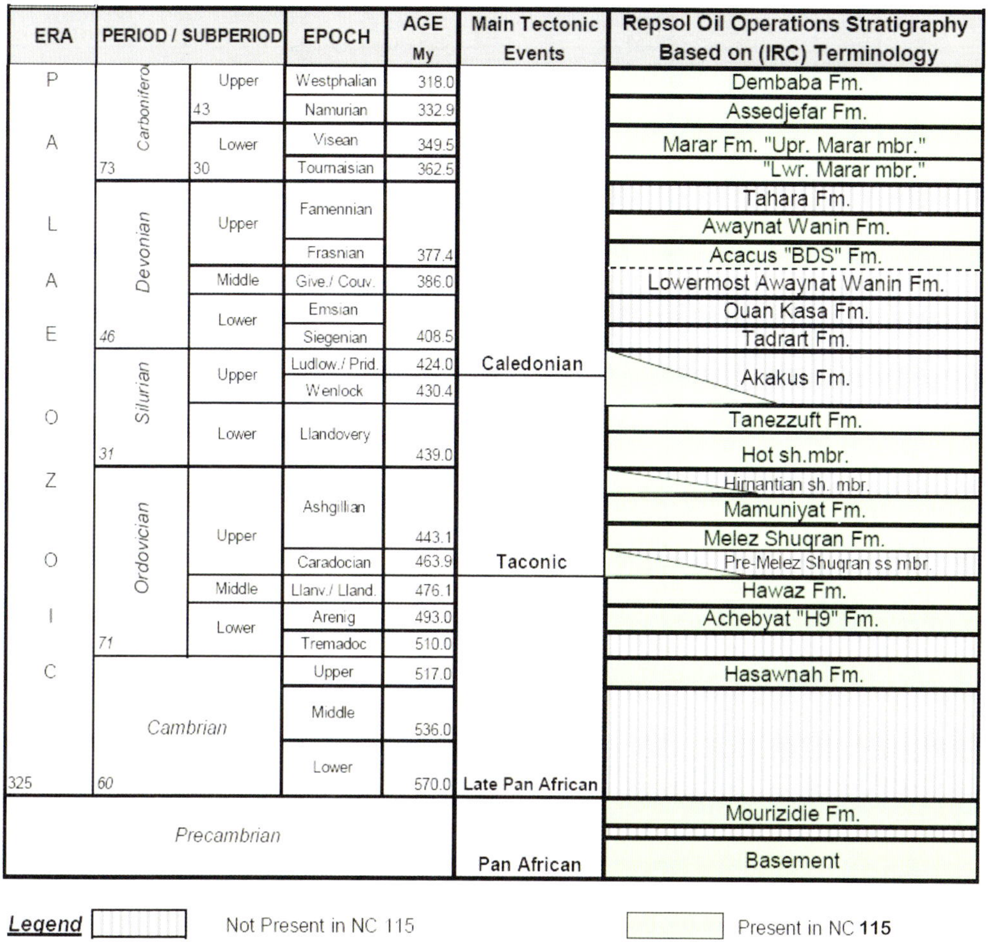

ERA	PERIOD / SUBPERIOD		EPOCH	AGE My	Main Tectonic Events	Repsol Oil Operations Stratigraphy Based on (IRC) Terminology
P A L A E O Z O I C	Carboniferous	Upper	Westphalian	318.0		Dembaba Fm.
		43	Namurian	332.9		Assedjefar Fm.
		Lower	Visean	349.5		Marar Fm. "Upr. Marar mbr."
	73	30	Tournaisian	362.5		"Lwr. Marar mbr."
	Devonian	Upper	Famennian			Tahara Fm.
						Awaynat Wanin Fm.
			Frasnian	377.4		Acacus "BDS" Fm.
		Middle	Give./Couv.	386.0		Lowermost Awaynat Wanin Fm.
		Lower	Emsian			Ouan Kasa Fm.
	46		Siegenian	408.5		Tadrart Fm.
	Silurian	Upper	Ludlow./Prid.	424.0	Caledonian	Akakus Fm.
			Wenlock	430.4		
		Lower	Llandovery			Tanezzuft Fm.
	31			439.0		Hot sh.mbr.
	Ordovician	Upper	Ashgillian			Hirnantian sh. mbr.
						Mamuniyat Fm.
				443.1		Melez Shuqran Fm.
			Caradocian	463.9	Taconic	Pre-Melez Shuqran ss mbr.
		Middle	Llanv./Lland.	476.1		Hawaz Fm.
		Lower	Arenig	493.0		Achebyat "H9" Fm.
	71		Tremadoc	510.0		
	Cambrian		Upper	517.0		Hasawnah Fm.
			Middle	536.0		
325	60		Lower	570.0	Late Pan African	
Precambrian					Pan African	Mourizidie Fm.
						Basement

Legend: Not Present in NC 115; Present in NC 115

Fig. 2. Stratigraphic column of the Murzuq Basin.

metamorphic series of lithologies and is disconformably overlain by the Hassaouna Formation (Fig. 7).

In Concession NC115 well data, the Mourizidie Formation consists of dark grey slate showing abundant fractures, most of which are filled by silica. Several wells have penetrated this sequence, showing intervals of deformed tuff fragments and less common andesitic–rhyolitic volcanic bombs in a very hard, siliceous, finely crystallized matrix. Detailed petrographic description (Fig. 8) shows volcaniclastic sandstone, which overlies the basement slates. It is moderately sorted and fine grained, and characterized by a pervasive abundant chloritic clay matrix (G4) – certainly related to the breakdown and alteration of unstable volcanic minerals. They also show dolomites (H7) and partially kaolinized muscovite (D12) (Fig. 9).

The Al Kufrah Basin

The Precambrian basement and Infracambrian strata outcrop on the eastern and western margins of the basin (Fig. 10). The Infracambrian strata are similar in composition to the clastic Mourizidie Formation of the Murzuq Basin.

The type section was measured SE of Jabal Arkenu (Saïd *et al.* 2000). It consists of several layers of conglomerate and coarse-grained sandstone with kaolinitic cement. For the western margin, Selley (1971) reported approximately 70 m of fine-grained sandstones and basal conglomerates. These were interpreted as braided alluvial plain deposits and are referred to as the 'Bir Bayai Formation'. On the eastern margin, at Jabal Arkenu, Infracambrian and Palaeozoic sedimentary rocks surround an igneous ring structure, but only the Infracambrian

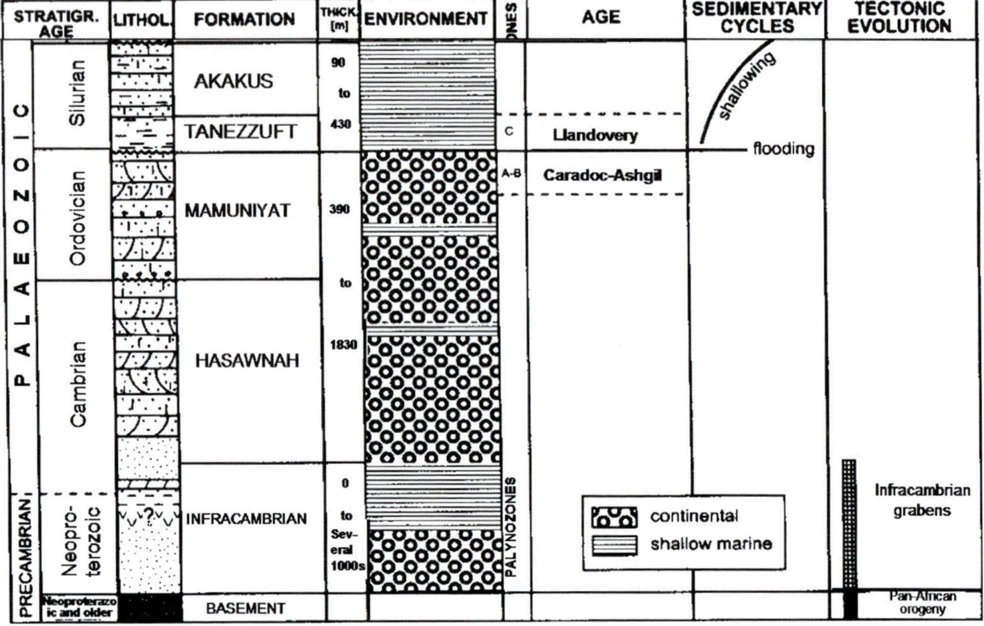

Fig. 3. Stratigraphic column of the Al Kufrah Basin.

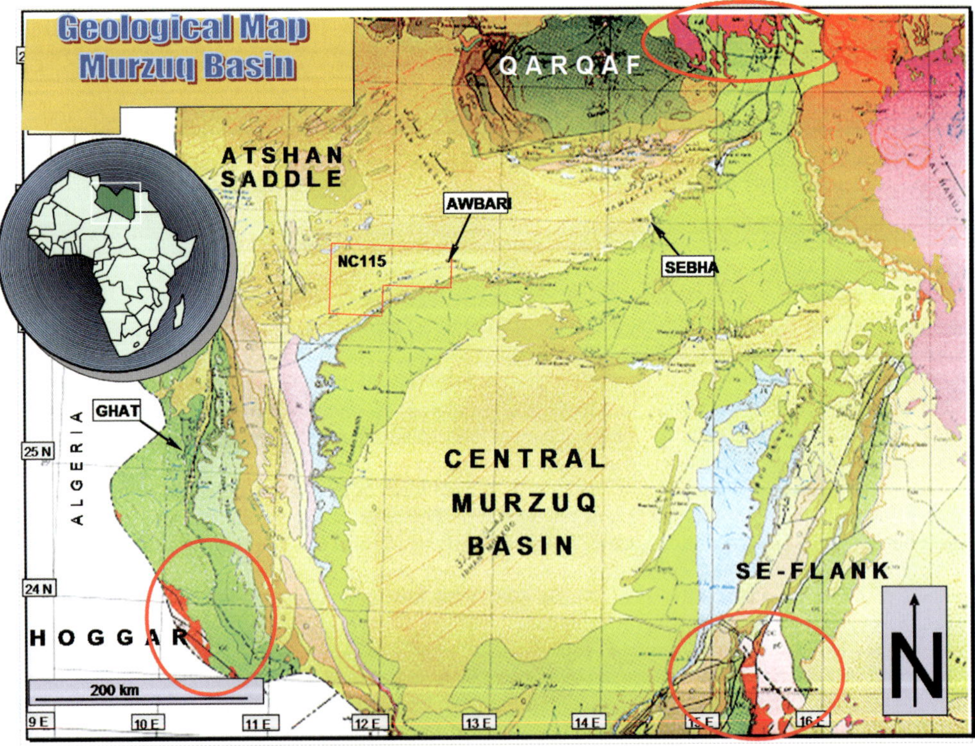

Fig. 4. Map showing Infracambrian outcrops on the flanks of Murzuq Basin (IRC).

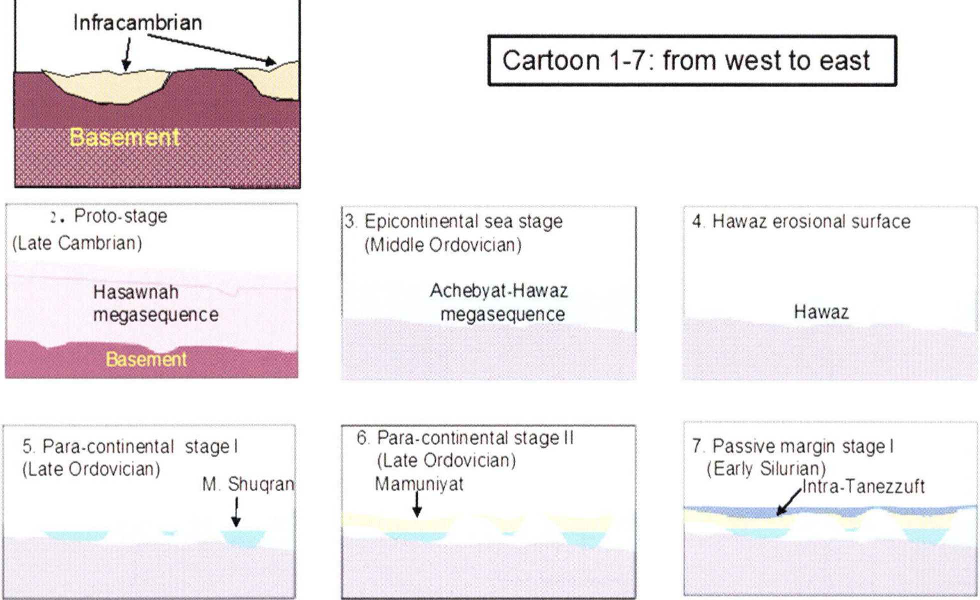

Fig. 5. Cartoon showing the evolution of Infracambrian rocks (after Ghnia & Alshaibi 2003).

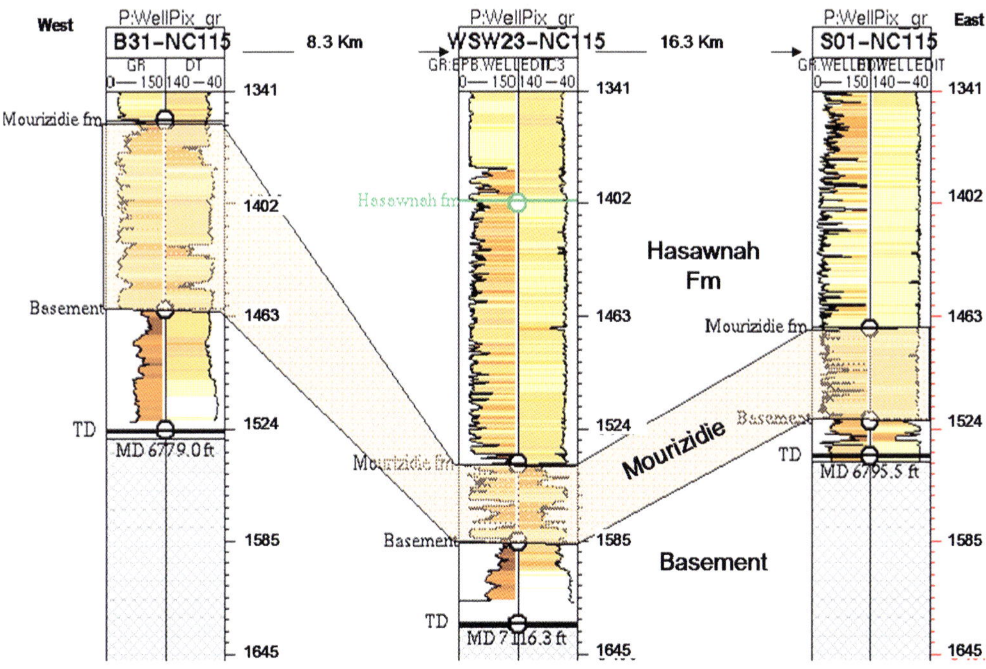

Fig. 6. The distribution of the Mourizidie Formation within the NC115 Concession.

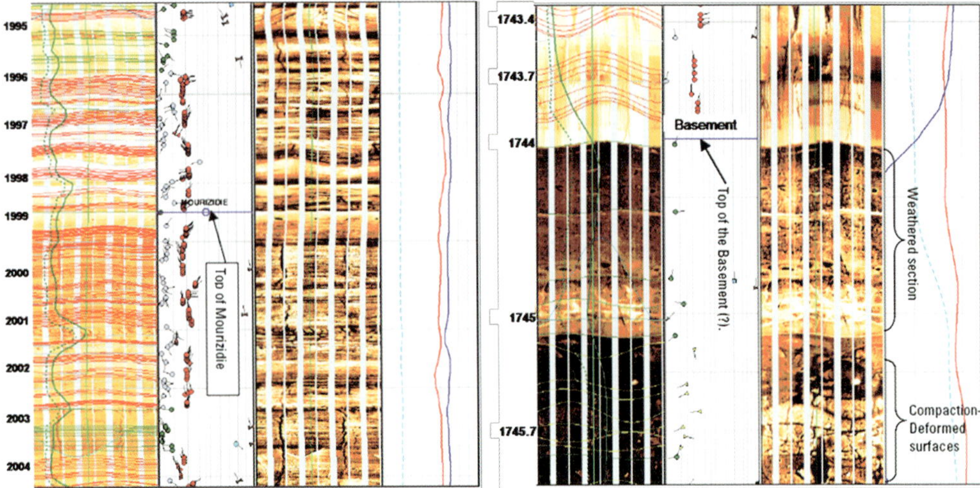

Fig. 7. The upper and lower contacts of the Mourizidie Formation.

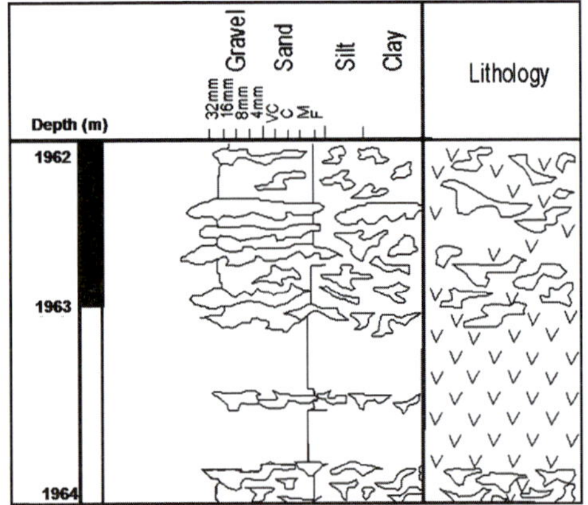

Fig. 8. Core data of the Mourizidie Formation comprises deformed tuff fragments with less common andesitic–rhyolitic volcanic bombs in a very hard siliceous finely crystallized matrix.

Fig. 9. Microphotograph of the Mourizidie Formation.

strata have been affected by contact metamorphism, indicating the timing of the intrusion. Lithologies include iron-bearing quartzites, marble and the siliciclastic Arkenu Formation (conglomerates, sandstone and feldspathic sandstone, with quartz veins). The marbles occur as isolated outcrops NNE of Jabal Arkenu and become widespread in the southern part of Jabal Barbain where lenses up to 500 m long and 75 m thick have been found (Saïd *et al.* 2000). The rocks are typified by a distinct white–bluish colouring and are highly fractured. They are composed of recrystallized limestone, and exhibit a medium-grained granular mosaic of anhedral calcite crystals enclosing tremolite, chlorite and mica.

Discussion

Worldwide, many proven hydrocarbon plays have been sourced from Infracambrian organic-rich rocks. For example, the Infracambrian black limestones and shales of West Africa and Oman have been sourced in this way and are associated with stromatolitic carbonates. The source rock potential of Infracambrian rocks has been confirmed by the discoveries of the major oil and gas fields in central Saudi Arabia (1989) and in western Oman (1978) within the Rub Al Khali Basin (Fig. 11). The Ara Formation of southern Oman contains rich hydrocarbon source rocks and consists of a cyclic succession of carbonate and evaporates deposited in a restricted basin (Mattes & Conway Morris 1990). Some geochemical analysis has been made on surface samples of the Infracambrian sequence in the Eglab range, which is located half in Mauritania (Taoudenni Basin) and half in Algeria

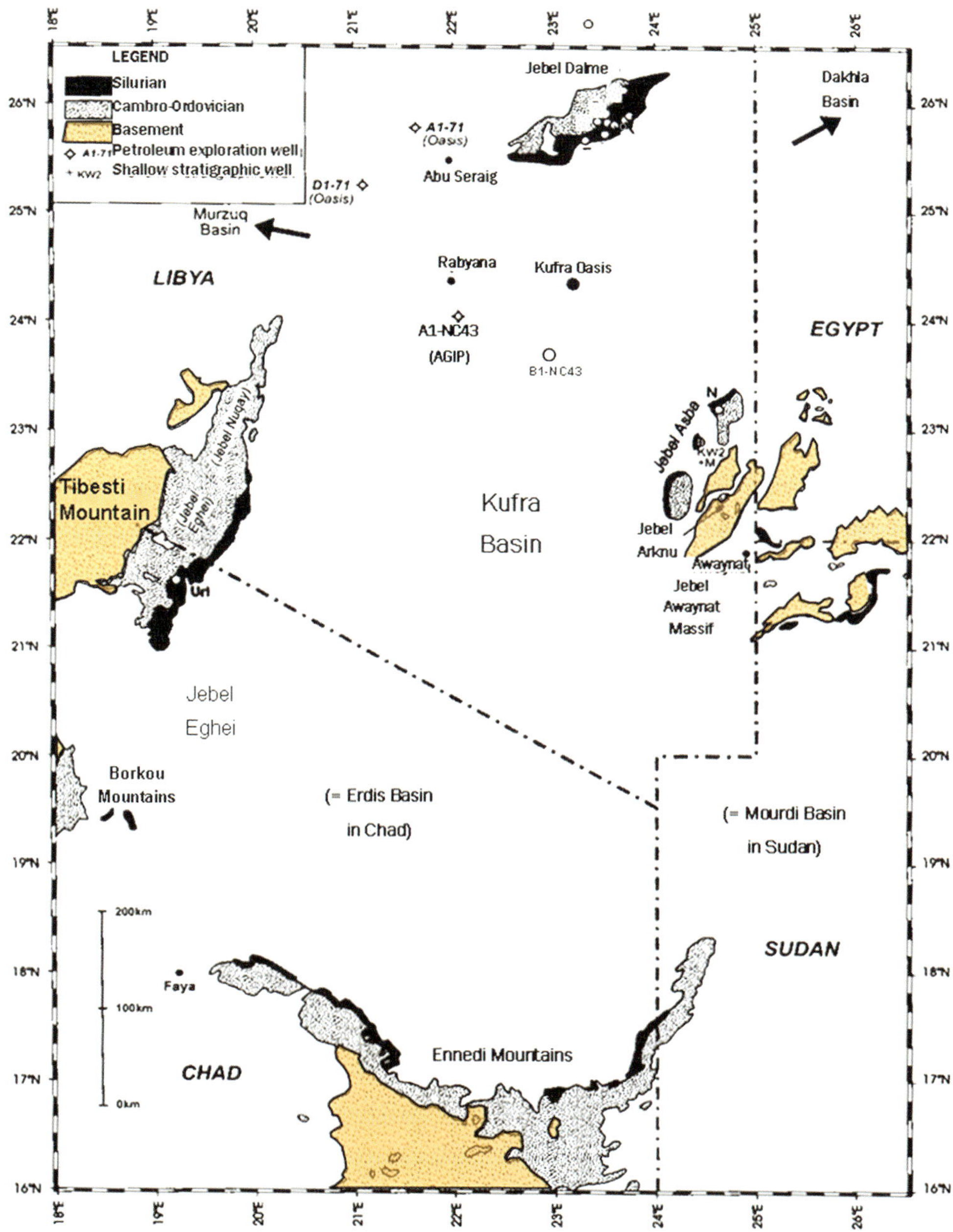

Fig. 10. Map showing Infracambrian outcrops on the flanks of the Al Kufrah Basin (Industrial Research Centre, Tripoli, Libya).

(Yetti, Iguidi). The results showed excellent source rock quality [total organic carbon (TOC) >5%] (Daoudi Mohamed 2006). This has highlighted interest in the Murzuq and Al Kufrah basins, especially the latter because it contains Infracambrian carbonates.

The Al Kufrah Basin limestones may indicate a facies-shift away from the proximal sandstone-dominated deposits of the Murzuq Basin to distal carbonate deposits. These carbonates may also accommodate localized organic-rich horizons in the deepest parts of the shelf. An Infracambrian

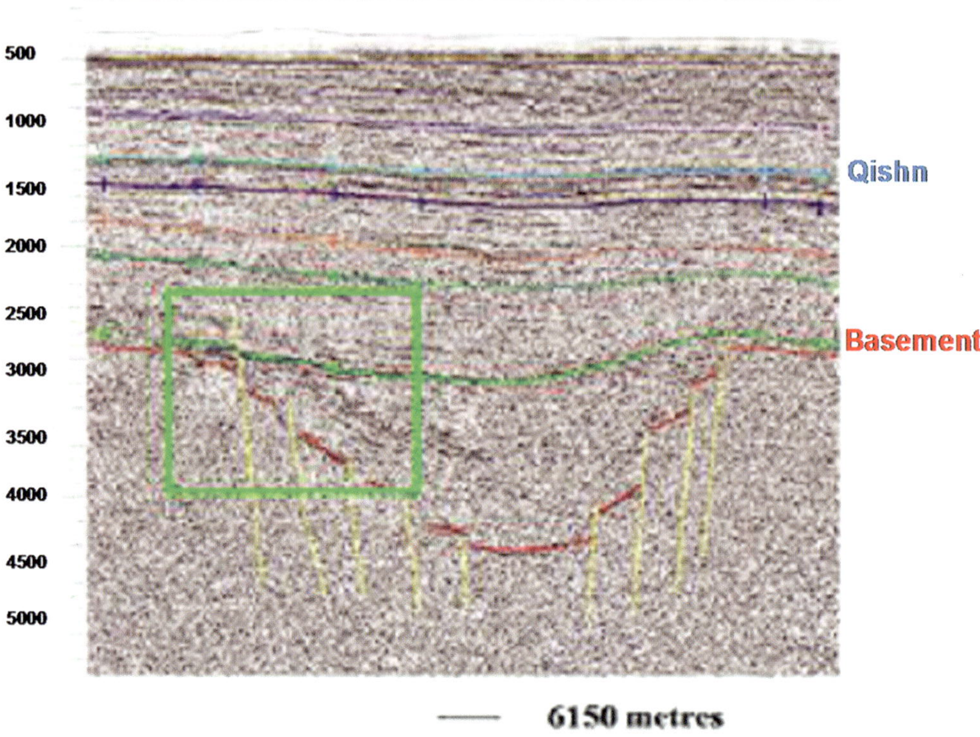

Fig. 11. Seismic section from the South Rub'al Khali Basin.

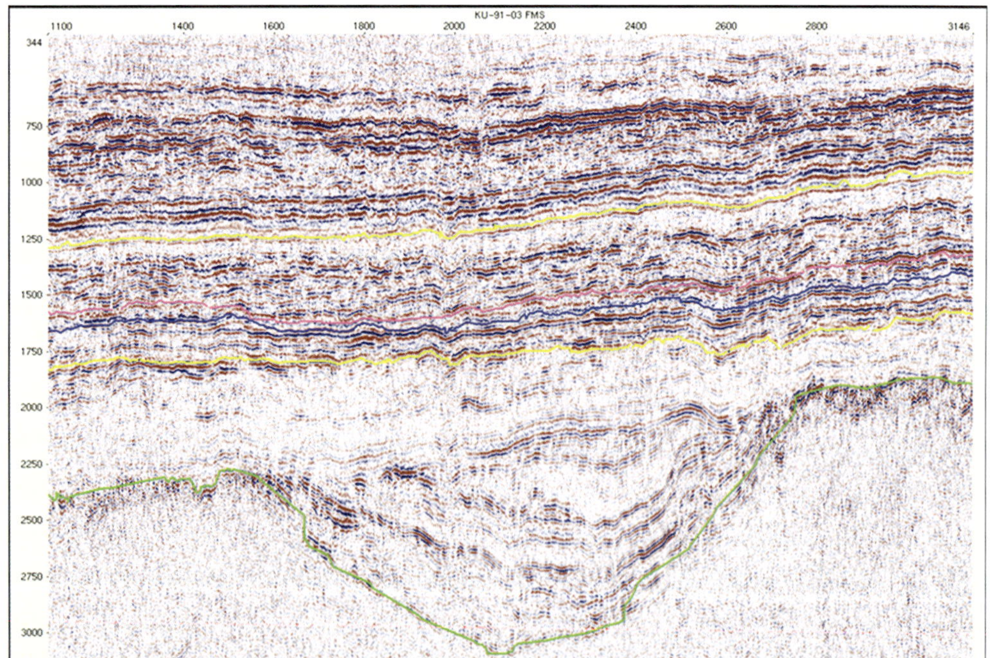

Fig. 12. Seismic section from the Al Kufrah Basin: graben with possible Infracambrian infill of deep-marine organic-rich carbonate and shale facies.

rift graben with syn- and post-rift infill of 1.5 km of undetermined lithologies has been interpreted on seismic lines from the southern Al Kufrah Basin (Fig. 12). To date, it is debatable as to whether the depositional environment within the graben was conducive to the deposition of organic-rich strata that could be a potential source rock.

Summary

Important information is obtained from deep wells drilled in the Murzuq Basin and outcrop studies in the Al Kufrah Basin. The Infracambrian rocks change in facies from proximal sandstone-dominated deposits in the Murzuq Basin to deep-marine facies in the Al Kufrah Basin.

The outcrop data from Jabal Arkenu, Al Kufrah Basin, reveal the presence of Infracambrian carbonate rocks. Seismic data reveal the presence of an Infracambrian graben in the Al Kufrah Basin, which could be infilled by deep-marine organic-rich limestone and shale that could serve as a source rock.

The Infracambrian petroleum system occurs in many parts of the world and could also be an attractive exploration target in Libya.

References

BOOTE, D. R. D., CLARK-LOWES, D. D. & TRAUT, M. W. 1998. Palaeozoic petroleum systems of North Africa. *In*: MACGREGOR, D. S., MOODY, R. T. J. & CLARK-LOWES, D. D. (eds) *Petroleum Geology of North Africa*. Geological Society, London, Special Publications, **132**, 7–68.

DAOUDI, M. 2006. Infra-Cambrian sequence: new petroleum play in the South-Western Sahara, Algeria. *In*: *GEO 2006. 7th Middle East Conference and Exhibition, 27–29 March, Manama, Bahrain.*

EL-MAKHROUF, A. A. 1988. Tectonic interpretation of Jebel Eghei area and its regional application to Tebisti orogenic belt, south central Libya (S.P.L.A.J). *Journal of African Earth Sciences*, **7**, 945–967.

GHNIA, S. & ALSHAIBI, A. 2003. Contribution to the stratigraphy of Murzuq Basin, SW Libya, from the view of NC-115. *In*: *AAPG Hedberg Conference on Palaeozoic and Triassic Petroleum Systems in North Africa, Algiers, Algeria, Abstracts.*

MATTES, B. W. & CONWAY MORRIS, S. 1990. Carbonate/evaporite deposition in the Late Precambrian–Early Cambrian Ara Formation of southern Oman. *In*: ROBERTSON, A. H. F., SEARLE, M. P. & RIES, A. C. (eds) *The Geology and Tectonics of the Oman Region*. Geological Society, London, Special Publications, **49**, 617–636.

SAÏD, M. M., OWEISS, K. A. & MEHIDI, B. O. 2000. *Jabal Arkenu Sheet: Explanatory Booklet*. Joint Geological Mapping Project between the Egyptian Geological Survey and Mining Authority (EGSMA) and the Industrial Research Centre (IRC) of Libya, Cairo.

SELLEY, R. C. 1971. *Preliminary report of a reconnaissance study of the Western Kufra Basin, southern Libya*. Oasis Oil Company of Libya, Tripoli.

Infracambrian petroleum play elements of the NE Taoudenni Basin (Algeria)

A. RAHMANI, A. GOUCEM, S. BOUKHALLAT & N. SAADALLAH

Sonatrach, Activité Amont Division Exploration, Avenue du 1ère Novembre, 35000 Boumerdes, Algeria

**Corresponding author (e-mail: jonathan.craig@eni.it)*

Abstract: The Taoudenni Basin, North Africa's largest sedimentary basin, is located in western Mauritania, northern Mali and southwestern Algeria. Of the four petroleum wildcat wells drilled to date, the Abolag-1 well, Mauritania, yielded gas shows in Infracambrian (Neoproterozoic) stromatolitic carbonates. We present details of the different plays of the basin from the Chenachène region in Algeria. The Infracambrian is generally composed of three sedimentary packages: a basal sandstone (a unit of the Douik Group), overlain by carbonates (the Hank Group), sandstones and shales (the Dar Echeikh Group). The play is sourced by Infracambrian organic-rich black shales. In neighbouring Mauritania these were penetrated by water wells and shallow boreholes, containing in places >20% TOC. In the Hank Group the best reservoirs are associated with fractured intervals. The Dar Echeikh Group includes several potential reservoir units with porosities of up to 26%. Potential petroleum trap types in the Algerian part of the Taoudenni Basin are associated with folds, the basal Palaeozoic unconformity, and Infracambrian and Triassic–Jurassic half-graben.

With an area of approximately 1 500 000 km^2, the Taoudenni Basin, which takes the form of an extensive syneclise, is the largest basin in West and North Africa. Located in southwestern Algeria, the basin also covers part of northern Mali and western Mauritania. It is situated south of the Tindouf and Reggane basins, from which it is separated by the Reguibat Shield, and to the west of the Hoggar Massif. It is topped by a very slightly deformed sedimentary cover of the Neoproterozoic and Palaeozoic series (Villemur 1967). The first hydrocarbon exploration work took place during the 1970s, when four wells were drilled: Abolag-1 and Ouassa-1 in Mauritania; and Yarba-1 and Atouila-1 in Mali. Only Abolag-1 provided a few gas shows in the stromatolite limestones of Infracambrian age.

Lithology and structure

The Neoproterozoic sedimentary cover of the northeastern part of the Taoudenni Basin, in the Chenachène region in Algeria (Fig. 1), can be subdivided into three main lithological assemblages (e.g. Caby 1965; Bertrand-Sarfati 1970; Bertrand-Sarfati & Fabre 1972; Bertrand-Sarfati *et al.* 1980, 1994, 1997) (Figs 2 & 3).

- A lower sandstone assemblage (Douik Group), which consists of coarse sandstones and conglomerates, and lies unconformably on a substratum consisting of pink granites of the Reguibat basement. Thicknesses vary from 20 to 150 m.
- A limestone assemblage (Hank Group), which consists of a superimposition of three units. One is siliciclastic at the base, and two are limestones with a generally regressive trend at the top. Thicknesses vary from 400 to 700 m.
- An upper sandstone–clay assemblage (Dar Echeikh Group), which consists of exclusively siliciclastic deposits. The overall trend in this assemblage is regressive. Thickness is on the order of 500 m.

Exposures of the Neoproterozoic formations are distributed over two major areas separated by a broad outcrop band of the crystallophyllite Reguibat basement of NW–SW orientation (Fig. 3). The first is an eastern zone, which could correspond to a small and relatively narrow basin in its northern part (at Chenachène), and which opens into a trough toward the south (at Tilemsi and Grizim). The deposits there are organized into a large syncline with an axis plunging to the SE. The western flank has been affected by a north–south fault known as the Oued Chenachène, along which there are en echelon folds. The second area is a western zone, whose limits are in the shape of a broad arc of a circle that extends beyond the boundaries of Algeria into Mauritania and Mali. Regional dips range from 50° to 80°, which gives this Neoproterozoic assemblage the appearance of a geosyncline. In addition, the Oued Souss fault is well marked in the basement, and cuts both Proterozoic and Cambrian–Ordovician formations in this western zone.

From: CRAIG, J., THUROW, J., THUSU, B., WHITHAM, A. & ABUTARRUMA, Y. (eds) *Global Neoproterozoic Petroleum Systems: The Emerging Potential in North Africa.* Geological Society, London, Special Publications, **326**, 221–229.
DOI: 10.1144/SP326.12 0305-8719/09/$15.00

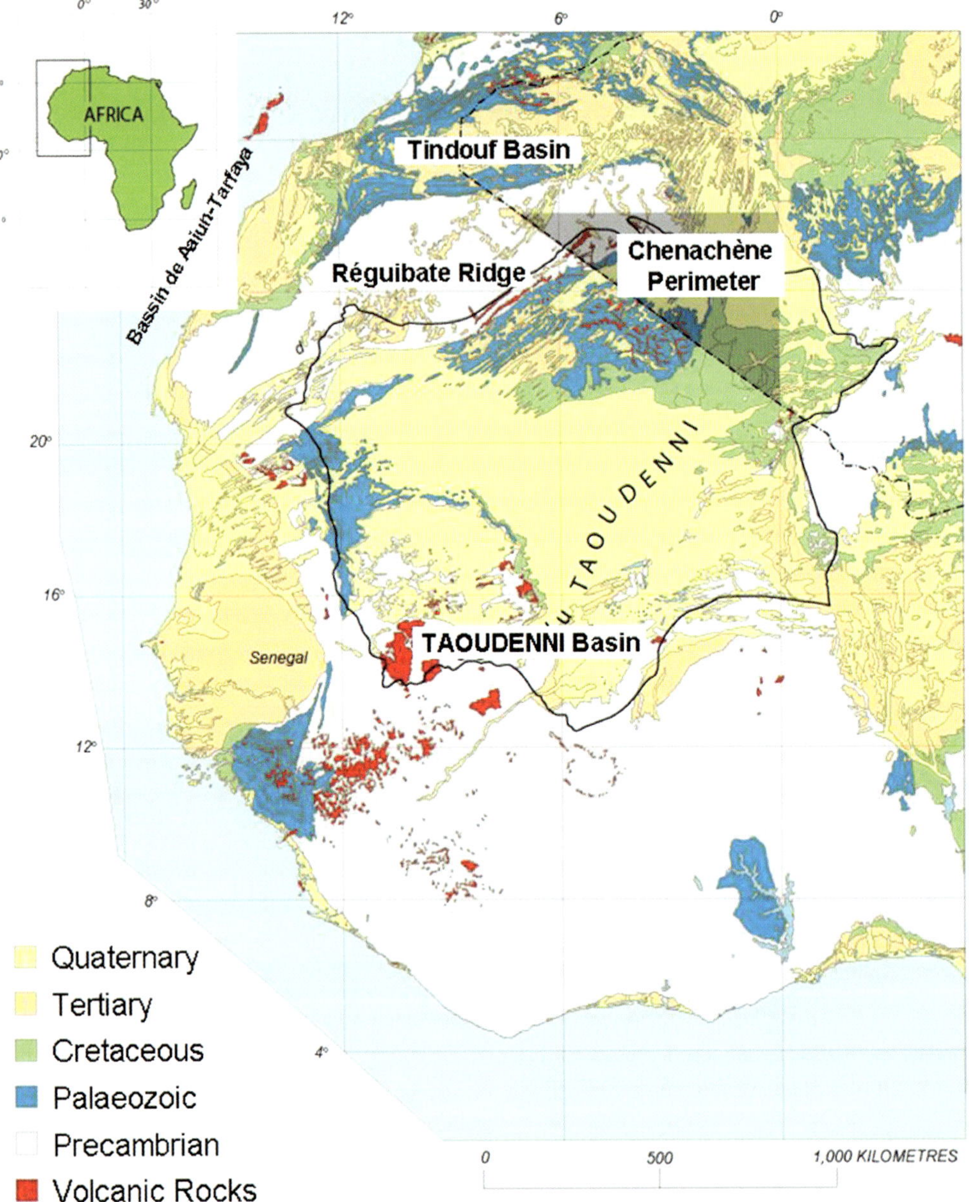

Fig. 1. Geological overview map of the Taoudenni Basin and location of the Chechachène study area in Algeria in the northeastern part of this basin.

Model and structural evolution

Preliminary analysis of these two structural elements (Fig. 4a) reveals a number of important events in the tectonic and sedimentary history of the Infracambrian sedimentary assemblages, and has made it possible to develop a chronology of their occurrence (e.g. Bertrand-Sarfati *et al.* 1978; Boullier *et al.* 1979; Bronner *et al.* 1980; Caby 2003; Caby & Monié 2003). In fact, the evolution of the northwestern zone of the Chenachène perimeter took place in four major stages (Fig. 4b).

- At the end of the cratonization of the Reguibat High (2000 Ma), submeridian basement faults were reactivated through extension or

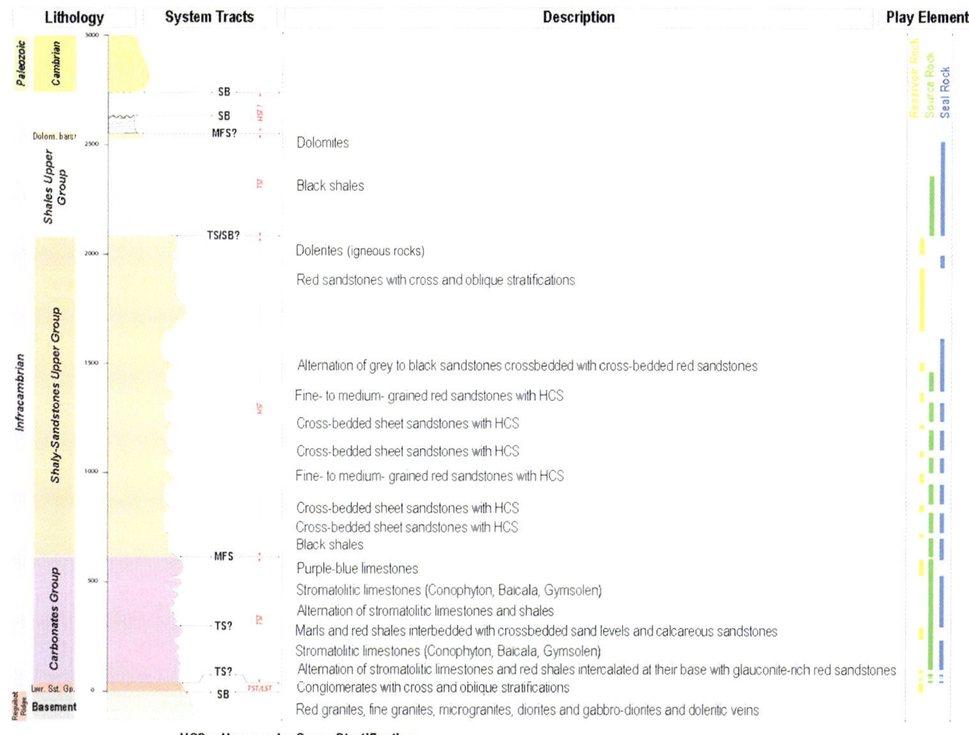

Fig. 2. Stratigraphic column of the Chenachène study area. HCS, hydrocarbon shows.

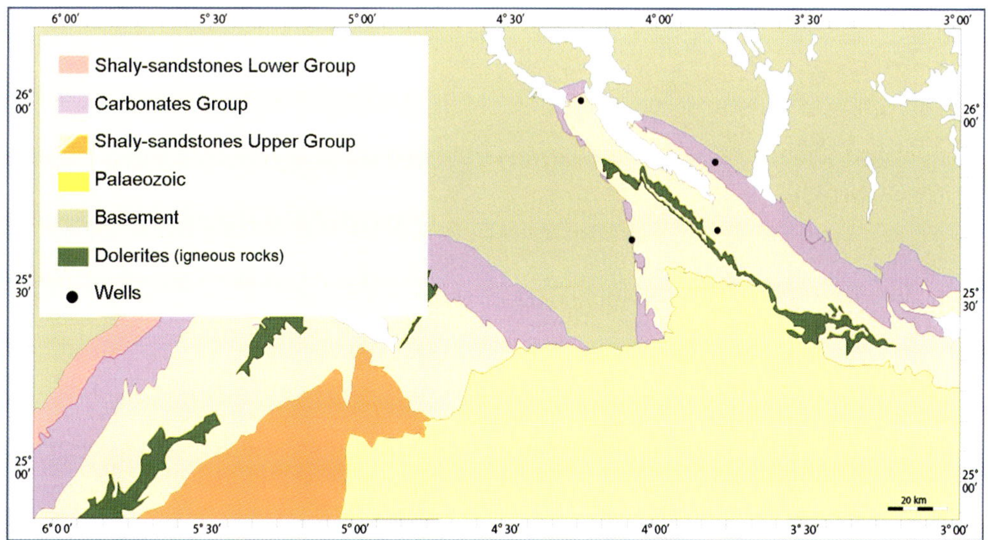

Fig. 3. Map of the Chenachène study area.

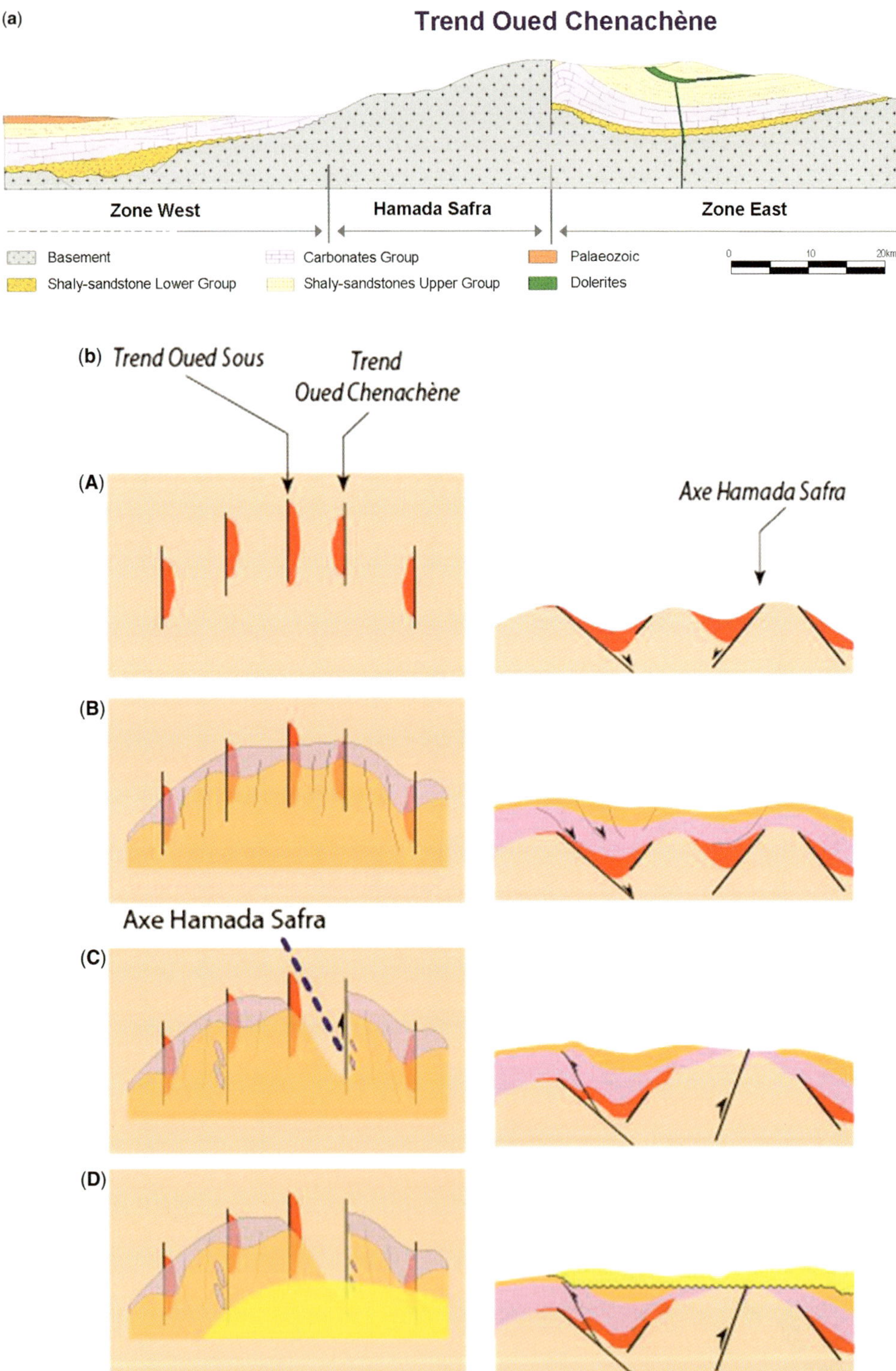

Fig. 4. (*Continued*).

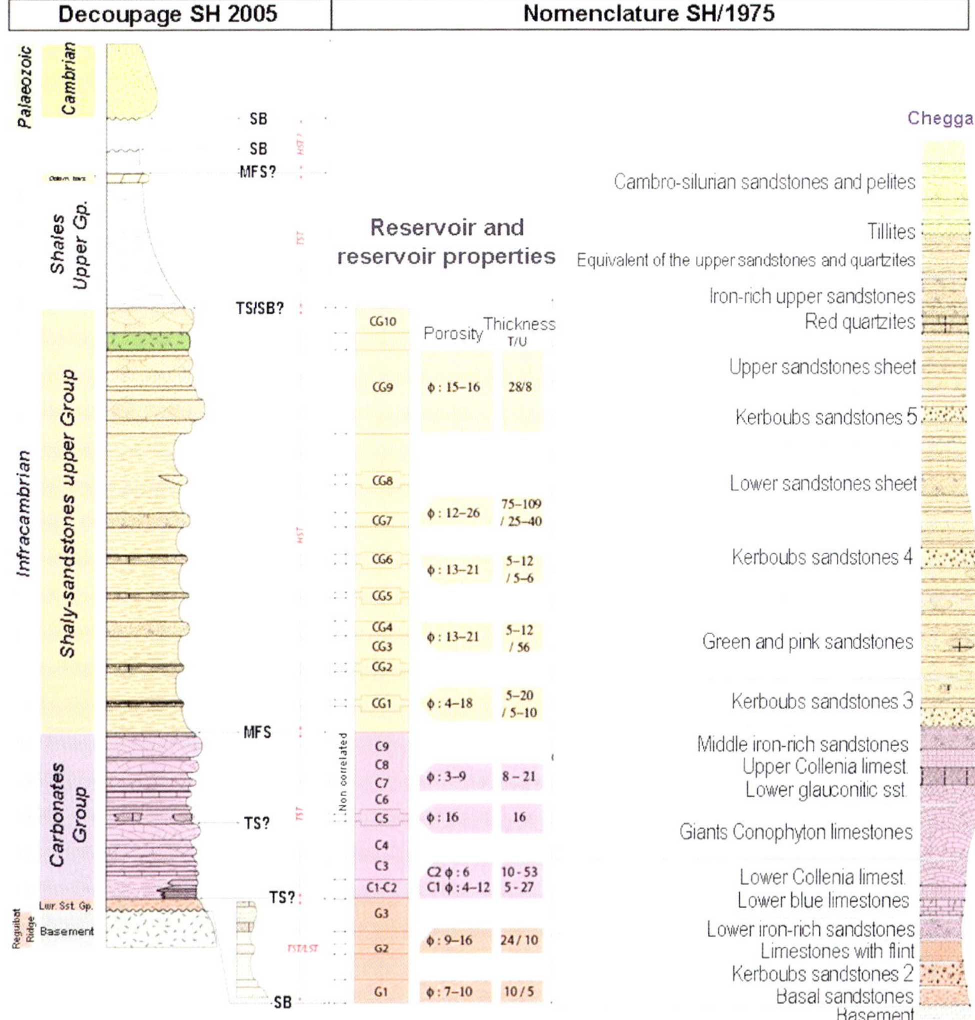

Fig. 5. Reservoir properties of the Hank Series (see the text for details).

transtension, and an unconformable basal sandstone series was deposited along these in what were probably pull-apart basins.

- A transgression with deposition of the Stromatolite Limestone Series and a regression with deposition of the Upper Clay–Sandstone Series occurred. The two assemblages were deposited with variations in thickness indicative of transpressive and synsedimentary activity of faults in the Infracambrian sedimentary cover unit (probably associated with faults in the basement).
- The Hamada Safra axis formed through transpressive (perhaps dextral) reactivation of the basement faults and formation of the folded structures of Oued Chenachène and Oued

Fig. 4. (**a**) Preliminary analysis of the structural elements of the Neoproterozoic formations. (**b**) The evolution of the northwestern zone of the Chenachène perimeter. (A) Deposition of the sandstone series in relation to the reactivation of basement faults in extensional or transtensional regimes. (B) General transpression; deposition of the carbonates and sandstones series with the development of a north–south-trending normal faults array. (C) reactivation of basement faults under a transpressional regime; formation of en echelon folds in the infracambrian series. (D) Discordant of the Cambrian over the infracambrian structures.

Souss. This axis could also represent pre-existing relief from the start of Infracambrian sedimentation, and would, therefore, have constituted a topographical boundary between the eastern and western zones.
- Termination of the transpressive tectonics and deposition of the first Palaeozoic members (Cambrian) unconformably on the Infracambrian purple series.

Petroleum systems

Source rocks

In parts of the Taoudenni Basin, an excellent, organic-rich hydrocarbon source rock, described as black shale, exists in the Infracambrian formations. Data that would allow a reliable assessment of the lateral and vertical extent of these clays are currently unavailable. Regional modelling studies have demonstrated, however, that this source rock was deposited in a synrift context, which could give rise to great thickening in the vicinity of major faults in the region.

Reservoirs

The Proterozoic series, also known as the Hank Series, is represented by detritic and limestone formations that developed between the two major unconformities on the Precambrian basement and beneath the Eocambrian tillite. This series is subdivided into three groups (Fig. 5).

- Douik Group – known as G1, G2 and G3 – or 'lower detritic assemblage'. This sandstone assemblage has a lateral extent that is impossible to predict in our Algerian study sector. It appears to disappear around the 5°W meridian at Chenachéne. Reservoir qualities are medium.
- Hank Group – known as C1–C9. This stromatolite limestone assemblage has a broad extent from the Mauritanian Adrar to Grizim, a distance of 1200 km. The thicknesses visible at

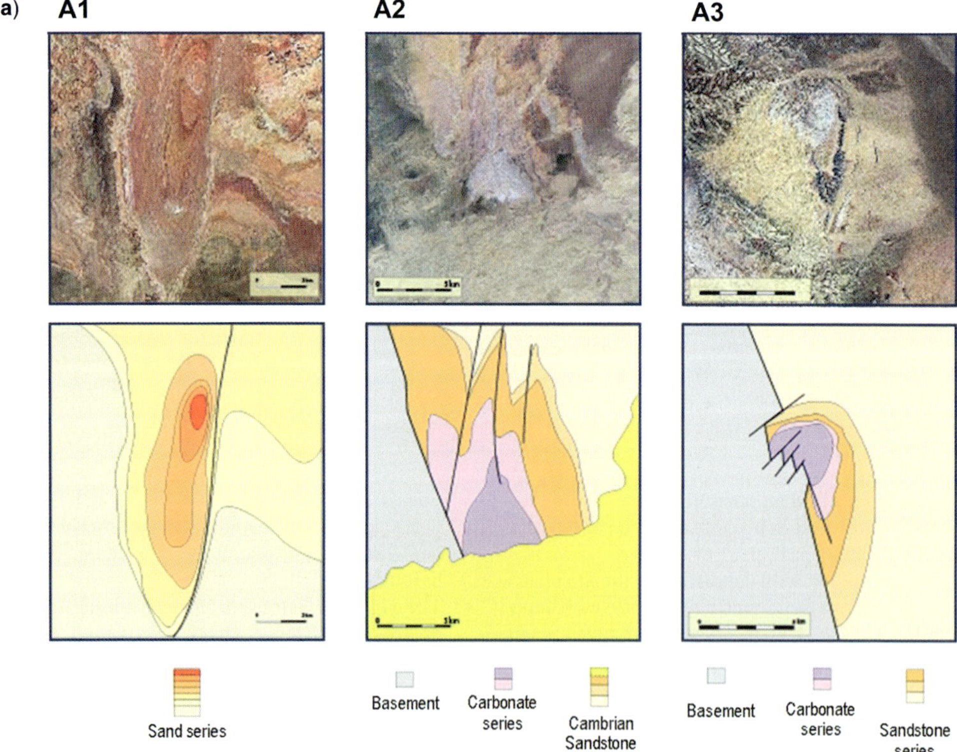

Fig. 6. (**a**) The different trapping styles present in the Chenachène area, including traps formed under the Palaeozoic deposits to the south of Chenachène. Traps can be structural, stratigraphic or combination traps. Good reservoirs with sufficient thickness potentially allow large accumulations of hydrocarbons when trapping conditions are similar to styles B and D. Trapping style A: closed or along-fault anticlinal structures, frequently associated with the north–south-trending strike-slip faults of Oued Chenachène and Oued Souss.

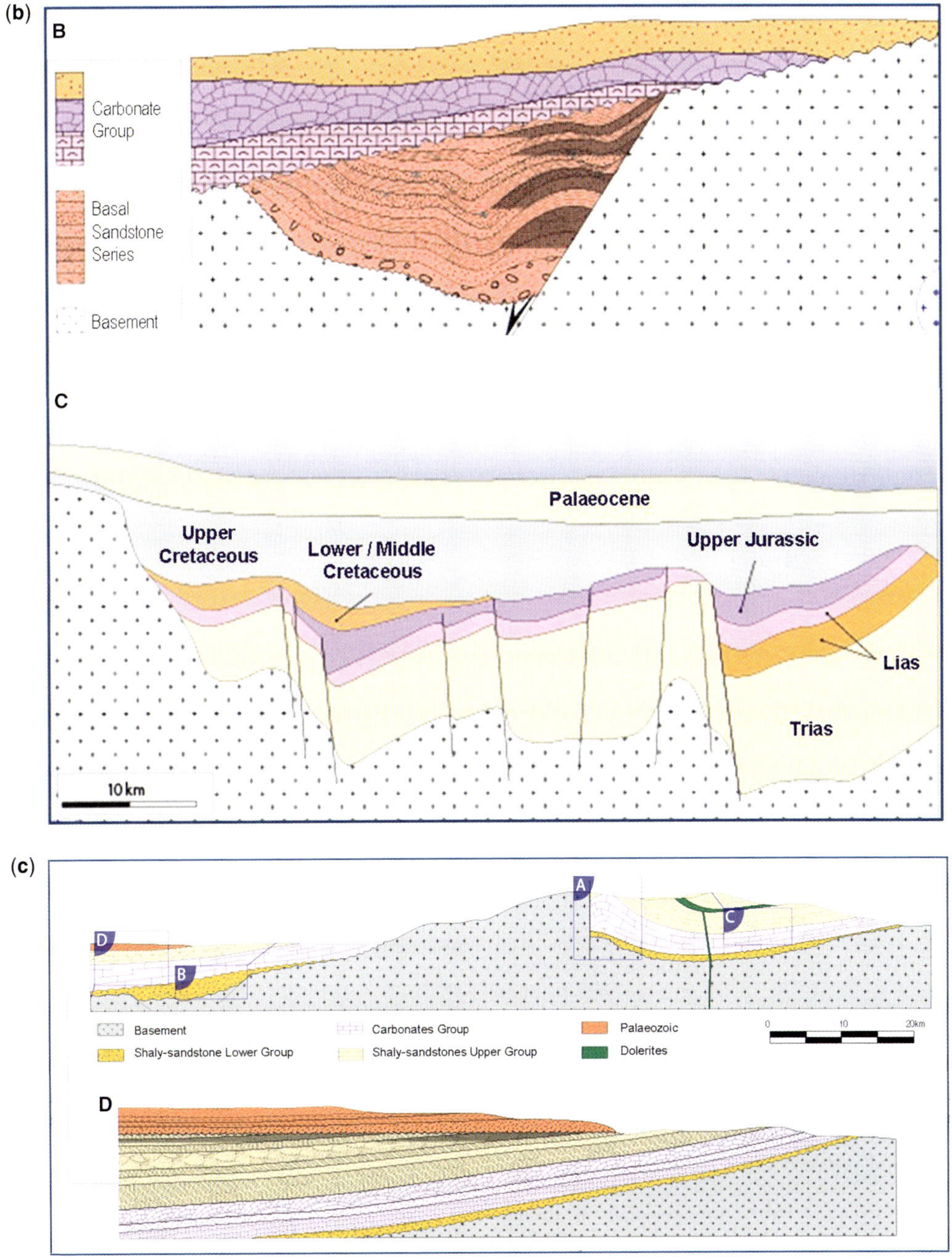

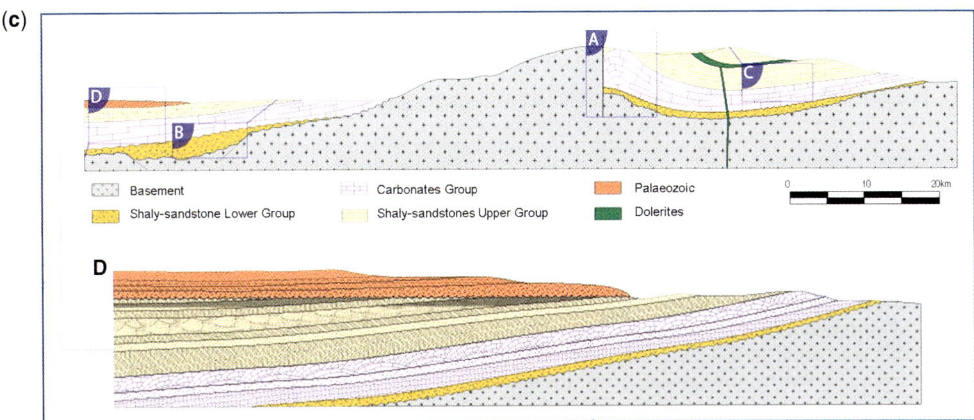

Fig. 6. (*Continued*) (**b**) Structural cross-section of the eroded and rotated fault-blocks in the Ninian region, East shetland Basin (from Albright *et al.* in Allen & Allen 1990). Trapping style B: only present in the Basal Sandstone series when in contact with source rocks (here, the Carbonates Group). Trapping style C: all traps developed in extensional or transtensional syntectonic depositional context; (**c**) Different trapping styles present in the Chenachène area, including traps formed under Palaeozoic deposits to the south of Chenachène. Traps can be structural, stratigraphic or a combination. Good reservoirs with sufficient thickness potentially allow large accumulations of hydrocarbons when trapping conditions are similar to styles B and D. Trapping style D: stratigraphic traps formed by the transpression of the Cambrian on to truncated pinched-out Infracambrian sandstones (mainly within the Uppen Sandstone series).

outcrop to the south of Eglab vary between 19 and 35 m. These limestones are often compact and hard, with poor petrophysical characteristics. When fractured they can form hydrocarbon reservoirs, as in the case of the Abolag-1 well in Mauritania, which release small amounts of gas. Above the stromatolite limestones, ferruginous sandstones appear to be continuous with fine–medium, generally friable, saccharoidal sandstones known as the Kerboub facies. Their thicknesses vary from 20 m at Chegga to 16 m at Tilemsi.

- Dar Cheikh Group, which is divided into four subgroups:
  - CG 1 and CG 3 subgroup, which consists of light–greenish, sometimes saccharoidal, medium–coarse sandstones with porosities of 7–18% and thicknesses of 5–20 m;
  - CG 4 and CG 5, a subgroup of sandstones that are generally clean, sometimes friable and quite porous, quite well developed in the region, and could form a good reservoir. Thicknesses range from 8 to 22 m. Porosities are of the order of 6–19%;
  - CG 6–CG 7 subgroup, which has two sandstone reservoir levels: a basal level with locally good reservoir characteristics (e.g. at Chegga), where mean porosity is in the order of 21%; or medium (e.g. at Mokrid), where porosity varies between 2.5 and 13%. The thickness is only 5–6 m. A friable summit reservoir sometimes occurs in the Kerboub sandstones. These form a reservoir 25–40 m thick. Porosity is 12–26% at Chegga and 13–25% at M'dennah;
  - CG 8–CG 9 and CG 10 subgroup, which is not very well developed in the east. It is 47 m thick at Chegga, and porosities vary from 8 to 15%. The sandstones are fine–medium with cross-bedding, and rapidly pass into compact limestone-containing sandstones.

Traps

Structural and sedimentological investigations have identified different types of traps that may exist within the perimeter of Chenachène. *Type A*, a purely structural trap corresponding to folding induced by the transpressive tectonics at the end of the Proterozoic (Fig. 6a). *Type B*, which corresponds to a mixed trap and affects only the basal sandstone series when it is covered by the limestone series. The extent of this type of trap is interesting because this structural feature can be found along all the north–south faults, and particularly towards the south beneath the Palaeozoic series (Fig. 6b). *Type C*, which includes all of the traps found in sedimentation contemporary with extensional tectonics (of the North Sea type) (Fig. 6b). *Type D*, which is a conventional stratigraphic trap given the arrangement of the Infracambrian series that dips southwards beneath the Palaeozoic series and bioherms to embedded stromatolites (Fig. 6b).

Conclusions

- The Taoudenni Basin, located in western Mauritania, northern Mali and southwestern Algeria, represents North Africa's largest sedimentary basin.
- The Infracambrian of the Chenachène region in Algeria is generally composed of three sedimentary packages: (1) a basal sandstone unit of the Douik Group overlain by (2) carbonates of the Hank Group and (3) sandstones and shales of the Dar Echeikh Group.
- The play is sourced by Infracambrian organic-rich black shales which in neighbouring Mauritania were penetrated by water wells and shallow boreholes, containing in places >20% TOC.
- In the carbonate-dominated Hank Group, the best reservoirs are associated with fractured intervals. The sandstones of the Dar Echeikh Group contain several potential reservoir units with porosities of up to 26%.
- Potential petroleum trap types in the Algerian Taoudenni Basin are associated with folds, the basal Palaeozoic unconformity and Infracambrian and Triassic–Jurassic half-grabens.

Special thanks go to Sebastian Lüning, without whose assistance this paper would never have materialised.

References

ALLEN, P. R. & ALLEN, J. R. 1990. *Basin Analysis: Principles and Applications*. Blackwell Scientific, Oxford.

BERTRAND-SARFATI, J. 1970. Les édifices stromatolitiques de la-série calcaire du Hank (Précambrien supérieur): déscription, variations latérales, paléoécologie; Sahara occidental, Algérie. *Bulletin de la Societé d'Histoire Naturelle de l'Afrique du Nord*, **61**, 13–38.

BERTRAND-SARFATI, J. & FABRE, J. 1972. Les stromatolites des formations lacustres post-muscoviennes du Sahara septentrional (Algérie). Proceedings of the 24th International Geological Congress, Montreal, Canada, 458–470.

BERTRAND-SARFATI, J., BERTRAND, J. M., CABY, R. & MOUSSINE-POUCHKINE, A. 1980. Evolution comparée du craton Ouest-African et de la zone mobile Pan-Africain au Proterozoique supérieur en Algérie et au Mali. *In*: MITROFANOV, F. P. (ed.) *Printsipy i kriterii raschleneniya dokembriya v mobil'nykh zonakh* [*Principles and Criteria for Subdivisions of the Precambrian in Mobile Zones*]. Nauka, Leningrad, 236–255.

Bertrand-Sarfati, J., Caby, R., Ducrot, J., Lancelot, J., Moussine-Pouchkine, A. & Saadallah, A. 1978. The late Pan-African intracontinental linear fold belt of the eastern Hoggar (central Sahara, Algeria): geology, structural development, U/Pb geochronology, tectonic implications for the Hoggar Shield. *Precambrian Research*, **7**, 349–376.

Bertrand-Sarfati, J., Flicoteaux, R., Moussine-Pouchkine, A. & Ait Kaci Ahmed, A. 1997. Lower Cambrian apatitic stromatolites and phospharenites related to the glacio-eustatic cratonic rebound (Sahara, Algeria). *Journal of Sedimentary Research*, **67**, 957–974.

Bertrand-Sarfati, J., Moussine-Pouchkine, A., Amard, B. & Ait Kaci Ahmed, A. 1994. Une faune d'Ediacara découverte pour la première fois dans le groupe de Cheikhia (Grès supérieur), Néoprotérozoique Saharien, Algérie. *Bulletin Office National de la Géologie*, **5**, 115–128.

Boullier, A., Fabre, J. & Lesquer, A. 1979. Evidence for late Precambrian plate tectonics in West Africa. *Nature*, **278**, 223–227.

Bronner, G., Roussel, J., Trompette, R. & Clauer, N. 1980. Genesis and geodynamic evolution of the Taoudenni cratonic basin (Upper Precambrian and Palaeozoic Western Africa). *In*: Bally, A. W., Bender, P. L., McGetchin, T. R. & Walcott, R. I. (eds) *Dynamic of Plate Interiors*. American Geophysical Union, Geodynamics Series, **1**, 81–90.

Caby, R. 1965. Les formations précambriennes de l'extrémité orientales de l'axe cristallin Yetti-Eglab (Sahara algérien occidental). *Bulletin de la Société Géologique de France*, **7**, 341–352.

Caby, R. 2003. Terrane assembly and geodynamic evolution of central-western Hoggar: a synthesis. *Journal of African Earth Sciences*, **37**, 133–159.

Caby, R. & Monié, P. 2003. Neoproterozoic subductions and differential exhumation of western Hoggar (southwest Algeria): new structural, petrological and geochronological evidence. *Journal of African Earth Sciences*, **37**, 269–293.

Villemur, J.-R. 1967. *Reconnaissance géologique et structurale du nord du bassin de Taoudéni*. Mémoires du Bureau de Recherches Géologique et Minières, Paris, **51**.

Upper Vendian–lowest Ordovician sequences of the western Gondwana margin, NE Spain

JOSÉ A. GÁMEZ VINTANED[1]*, ULRICH SCHMITZ[2] & ELADIO LIÑÁN[1]

[1]*Área y Museo de Paleontología, Universidad de Zaragoza, E-50009 Zaragoza, Spain*

[2]*LO&G Consultants, Baderweg 149, D-45259 Essen, Germany*

**Corresponding author (e-mail: gamez@unizar.es)*

Abstract: The intra-Vendian (Ediacaran)–intra-Tremadocian succession of the Cadenas Ibéricas in NE Spain is divided into nine sequences. Overall, these are interpreted as second-order sequences. Those that encompass Lower and lower Middle Cambrian carbonates, with identified transgressive systems tract (TST) and highstand systems tract (HST) phases, may constitute elements of composite sequences. The lowermost sequence is of Late Vendian age. In Lower and lower Middle Cambrian units, sequence tops indicate drowning, reflecting extensional tectonics. Rifting effects are traceable up to mid-Mid Cambrian times. The remaining sequences probably represent a sag phase, either accentuating the preceding extensional local basin regime or heralding the Gondwana passive margin stage. Of the southerly Gondwana deposits those of other areas of the Iberian Peninsula and of the Moroccan Anti-Atlas fold belt show similar conditions during the Early Cambrian, that is, an Early Cambrian extensional regime, and, as for the Moroccan fold belt, four sequences imaging TST and HST phases.

In Cambrian times, the northwestern margin of Gondwana comprised tectono-sedimentary units representing a ribbon-like terrane more than 10 000 km long (von Raumer *et al.* 2006). In this setting, the northern tectono-sedimentary units of Iberia adjoined the Gondwana mainland, whilst today's southern Iberian tectono-sedimentary units represented external regimes, being closest to the suture determining the margin of the Gondwana continent (Figs 1 & 2). For a more comprehensive view of the history of Iberia, see Martínez-García (2006).

The Cadenas Ibéricas, the intra-Vendian–intra-Tremadocian section of which is being dealt with herein, are located within today's northern zones of the Iberian Peninsula (Lotze 1961) in the Cantabrian and West Asturian–Leonese zones (Fig. 2). Both their palaeogeographic position and the resulting basin development with some 5500 m of sediments – divided into sequences and subdivided by stratigraphic events – are obviously favourable prerequisites for stratigraphic comparisons and basin-to-basin correlations. Previous attempts to study the sequence stratigraphy of the Cambrian of the Cadenas Ibéricas were made by Gabaldón (1990; for the whole succession) and by Álvaro & Clausen (2005; for the Lower–Middle Cambrian boundary interval).

Targets of the addressed comparisons and correlations are the Gondwana northwestern platform and the fringing fold belts.

Basin evolution

The Cadenas Ibéricas are part of the Cantabrian and West Asturian–Leonese tectono-sedimentary units. Essentially, these owe their present position to Hercynian migration and counter-clockwise rotation. During Cambrian times they adjoined the northwestern margin of the Gondwana continent and, as for the West Asturian–Leonese unit, had accumulated a succession of some 5000 m. Accommodation space during Lower Cambrian depositional history was provided through extensional tectonics, as deduced from analyses of the larger area (von Raumer *et al.* 2006). According to Liñán & Gámez Vintaned (1993), a rifting episode spreading from the Ossa–Morena Zone governed the general traits of basin evolution, indicated by volcanic activity in the latter and by generalized regression events. The Cambrian basin fill of Cadenas Ibéricas comprised shallow-water clastics until mid-Early Cambrian times, a depositional pattern dominated by carbonates until earliest Mid Cambrian time, and again clastics during the remainder of the Cambrian Period. The substantial thicknesses possibly reflect fault-rejuvenation.

The Upper Vendian (Ediacaran) sequence (S1)

Classically, authors working with rocks of the terminal Neoproterozoic system in Spain have referred

From: CRAIG, J., THUROW, J., THUSU, B., WHITHAM, A. & ABUTARRUMA, Y. (eds) *Global Neoproterozoic Petroleum Systems: The Emerging Potential in North Africa.* Geological Society, London, Special Publications, **326**, 231–244.
DOI: 10.1144/SP326.13 0305-8719/09/$15.00

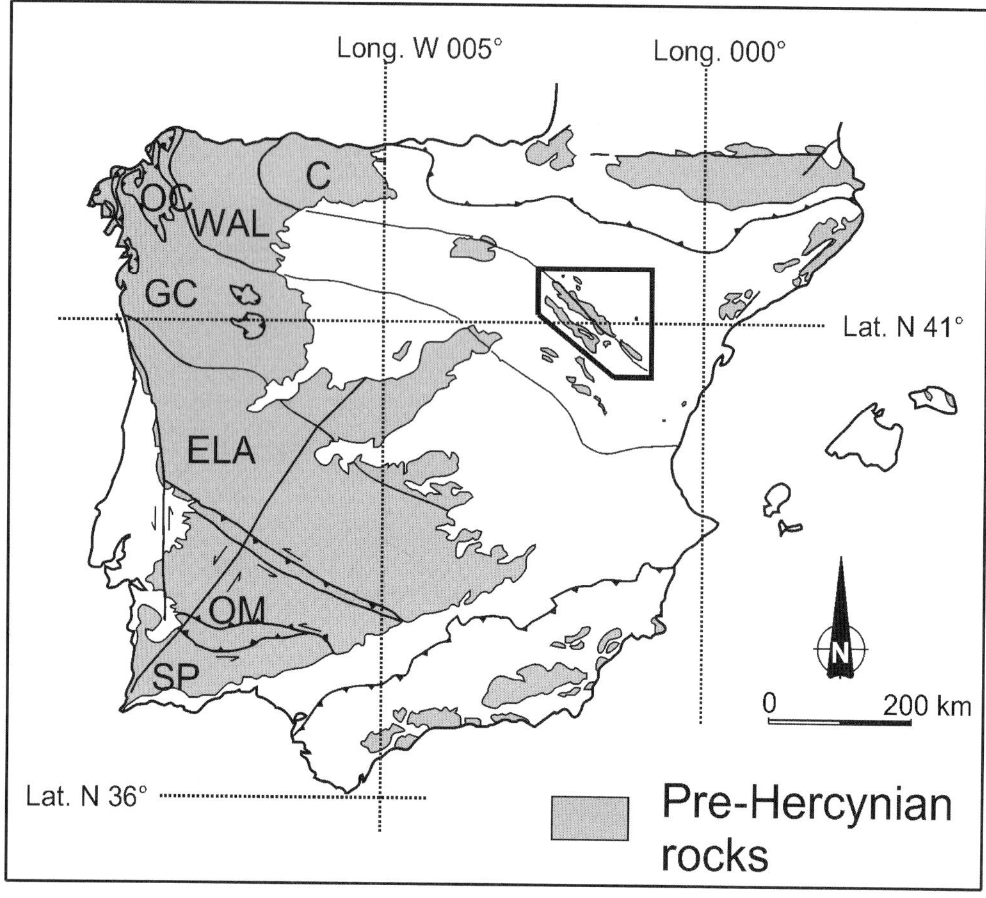

Fig. 1. Location map showing the Cadenas Ibéricas (see frame) and the pre-Hercynian outcrops and tectono-sedimentary units of Iberia. SP, South Portuguese Zone. For other abbreviations see Figure 2.

to this as Vendian, as the litho- and biostratigraphic definition of the latter allows its usage and correlation, which is not the case with the recently proposed Ediacaran System (which was defined with serious deficiencies). For this reason, the term Vendian will be used in the present text.

Exposure of Vendian sediments in the Cadenas Ibéricas is very limited compared with that of the Lower Palaeozoic, and no Neoproterozoic basement rocks – as are recorded in the neighbouring Central Iberian Zone (which includes the Galician–Castilian and East Lusitanian–Alcudian zones: for an overview see Valladares *et al.* 1998, 2002) – are at the surface. Comparison and correlation with other areas is complicated by this, and by the fact that the transition from Vendian to Cambrian sediments is obscured.

There are two Vendian outcrops. The largest one is located near the Jalón River and occupies the core of the Paracuellos de la Ribera Antiform, which is sectioned axially by the Jarque fault into two asymmetrical flanks showing differences in stratigraphy (e.g. the El Frasno Formation does not appear on the SW flank). The succession exposed is of Late Vendian age, indicated by the ichnofossil *Torrowangea rosei* found by Liñán & Tejero (1988) and by the absence of Cambrian-type trace fossils. It has been suggested that its top is affected by the latest event of the Cadomian Orogeny, expressed as the Córdoba Regression (Liñán *et al.* 2002). Litho-units concerned are those of the Paracuellos Group (Fig. 3). The lowermost unit of the exposed succession, the Sestrica Formation, is mainly composed of claystones alternating with sandstones. The following Saviñán Formation is an alternating succession of claystones and sandstones, with sandstones predominating, particularly towards the top. The descriptions of Liñán & Tejero (1988) indicate that this unit includes turbidites. The formation is succeeded by a distinct marker horizon (Liñán

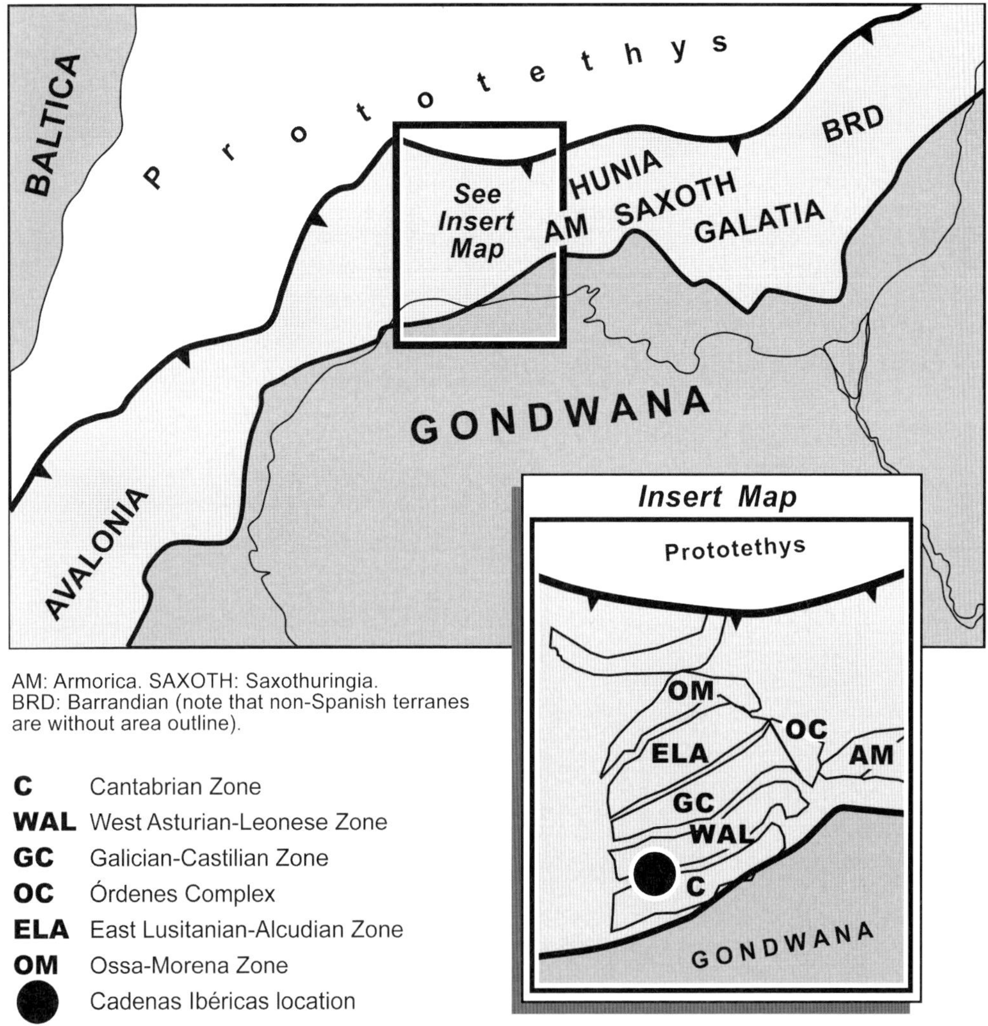

Fig. 2. Tentative reconstruction of the northwestern Gondwana margin during the Furongian, redrawn from von Raumer *et al.* 2006 (generalized).

& Tejero 1988: El Frasno Formation) composed of black chert layers. The uppermost litho-unit (Aluenda Formation), the top of which is not exposed, again consists of an alternation of claystones and sandstones.

The second Vendian outcrop is located in Codos, beside the Grío River. The succession to the NE of the Datos fault starts with phosphatic dolostones (shelly-oolitic packstone with an admixture of clay minerals) followed by sandstones and siltstones. Given the lithological dissimilarities and that its underlying part is not exposed, a straightforward correlation of the Upper Vendian of Codos with the Paracuellos Group is not possible (see Fig. 3). Nevertheless, the presence of a rich association of small shelly fossils of Late Vendian age in the Codos dolostones allows the interpretation the Codos succession as the top of the Vendian in the region, having been deposited near the Vendian–Cambrian boundary (Gámez Vintaned *et al.* 2007).

Few diagnostic sedimentary features have been described from the Paracuellos Group, that is, only the main litho-unit characteristics are available for sedimentary environment and systems tracts identifications. They suggest the two lower formations reflect lowstand systems tract (LST) conditions – the section with turbidites pertains to the slope fan phase, and the overlying coarsening-upwards

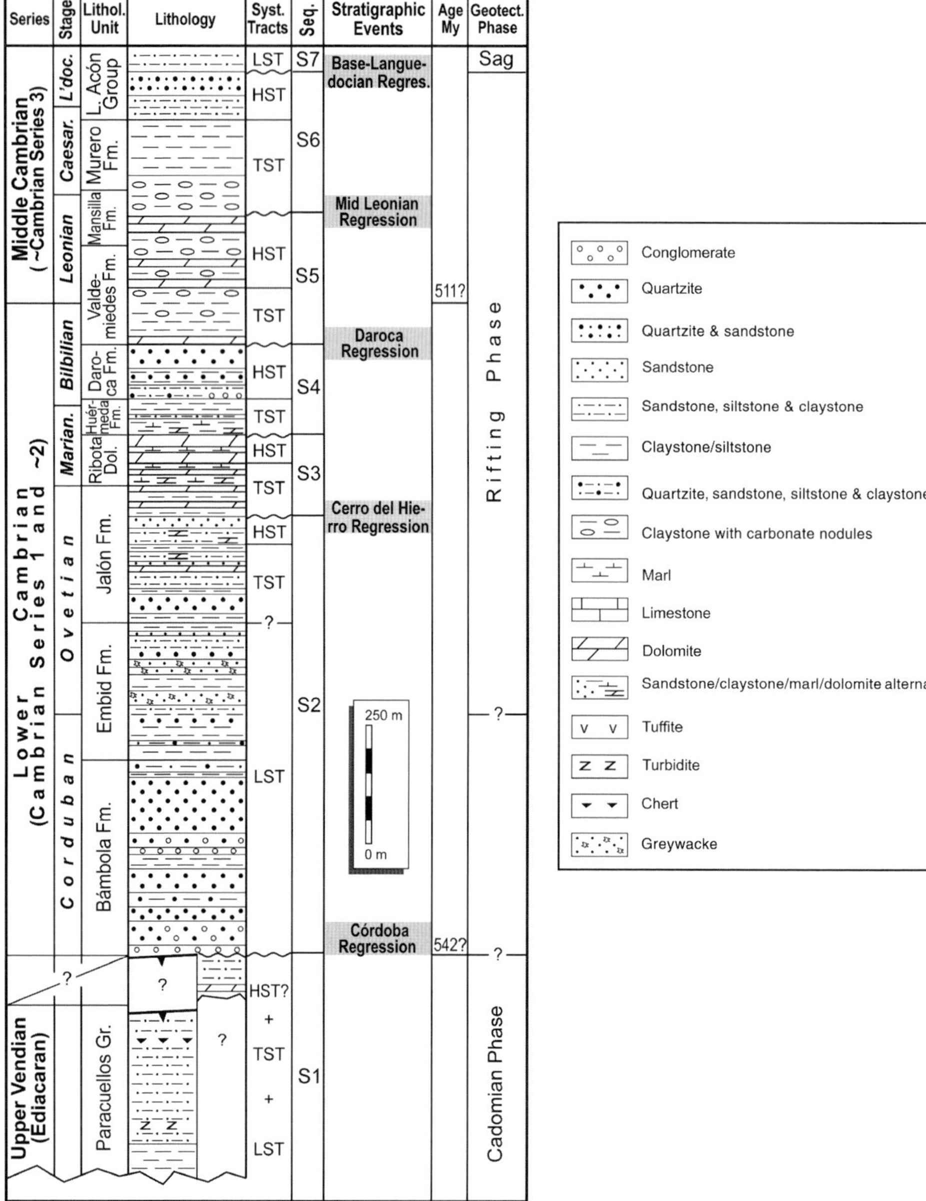

Fig. 3. Upper Vendian–middle Middle Cambrian stratigraphy of the Cadenas Ibéricas, with subdivision into sequences and systems tracts, correlated with published stratigraphic events (Liñán *et al.* 2002). For the Upper Vendian sequence, the left column applies to the area near Paracuellos de la Ribera, while the right-hand one represents the Codos area. (Sequence boundaries are just confronted with the middle of correlating stratigraphic event boxes, the vertical dimension of the latter does not represent either time or thickness.)

succession to the prograding wedge phase. The El Frasno Formation marks, accordingly, the transition from LST to transgressive systems tract (TST) conditions; the latter are represented by the Aluenda Formation. The uppermost parts of the Aluenda Formation and the succession preceding the Cambrian Bámbola Formation (Fig. 3) in Codos could represent highstand conditions. This interpretation parallels that of Valladares *et al.* (2002, fig. 2.4) covering the coeval section of the Galician–Castilian

Zone. Also, it is noted that the correlation chart of Liñán *et al.* (2002, fig. 3.2) and Valladares *et al.* (2002, fig. 2.3), which suggests direct comparison with litho-units from both areas, may be interpreted accordingly.

The lower Lower Cambrian sequence (S2)

Because of poor outcrop conditions, the basal boundary of this sequence is poorly defined, that is, either post-Cambrian thrust faulting obscures the transition from the Upper Vendian to the Lower Cambrian (Lotze 1961), or, as is the case in outcrops near the village of Codos, the basal parts of the Bámbola Formation either follow the underlying strata with an angular unconformity (Gámez Vintaned 2007) or follow conformably on a 'Precambrian' section (Teyssen 1980; Bartsch 2008; Schmidt-Thomé unpublished). The transition from the Vendian to the Cambrian System has long been debated (Teyssen 1980; Liñán & Tejero 1988) – indeed, the tectonic processes underlying the mode of transition are as yet under discussion: some authors argue that near-to-general uplifting followed the last phase of the Cadomian Orogeny and that the Córdoba Regression seemed to be related to that uplift during Early Corduban times (Liñán & Gámez Vintaned 1993; Liñán *et al.* 2002; Sánchez-García *et al.* 2003), whilst others propose that extensional tectonics prevailed in Iberia, accompanied by a possibly coeval sea-level drop (Valladares *et al.* 2002). It is probable that the models are not mutually exclusive (see Vidal *et al.* 1994, pp. 755–758; Santamaria i Casanovas 1995, p. 155).

This earliest sequence of Cambrian age covers the lower–upper part of the Corduban Stage, and the lower and middle parts of the Ovetian Stage. Its basal boundary is correlated with the Córdoba Regression. The top boundary obviously coincides with the Cerro del Hierro Regression (Liñán & Gámez Vintaned 1993; Liñán *et al.* 2002). The litho-units encompassed are, from bottom to top, the Bámbola Formation, the Embid Formation and parts of the Jalón Formation (see Fig. 3). The lowermost outcropping succession of the sequence (Bámbola Formation) mainly comprises quartzites, with conglomerate intercalations. The conglomerate intervals are marine deposits (Schmidt-Thomé 1973). It is notable that the transition from the quartzites to the conglomerates is gradual in the lower part and abrupt higher up. According to Gámez Vintaned (2007), the following ichnotaxa are present: *Arenicolites*, *Diplocraterion*, *Gordia*, *Monomorphichnus lineatus*, *Phycodes*, *Planolites*, *Rusophycus cantabricus* (= *R. bonnarensis*), *R. didymus* and the *Scolicia* group. The *Cruziana* ichnofacies accounts for most of the observed ichnocoenoses, whilst the *Skolithos* ichnofacies is present towards the top. Trace fossils are more abundant and diverse in the Embid Formation: *Arenicolites*, *Astropolichnus hispanicus* (an Ovetian index fossil), *Bergaueria*, *Cochlichnus*, *Cruziana cantabrica*, *Dictyodora*, *Didymaulichnus*, *Dimorphichnus*, *Diplichnites*, *Diplocraterion helmerseni*, *D. parallelum*, *Gordia*, *Helminthopsis*, *Monocraterion*, *Monomorphichnus*, *Palaeophycus*, *Phycodes*, *Pilichnia*, *Planolites*, *Psammichnites*, *R. cantabricus*, *R. fasciculatus* (= *R. avalonensis*), *Sericichnus*, *Skolithos* and *Teichichnus* (Liñán *et al.* 1996; Gámez Vintaned 2007).

Possibly, the succession represents a set of LST basin-floor fans, deposited on a substratum tilted at very low angle, as may be interpreted from Liñán & Tejero (1988), who suggest a cartographically identifiable unconformity. The coarse clastics are overlain by alternating sandstones, siltstones and claystones. Claystones and siltstones eventually dominate the lower part of the Embid Formation. It is suggested that these, together with the succeeding sequence of claystones, siltstones, sandstones, quartzites and greywackes, represent the LST slope fan unit and prograding wedge sedimentation. Whether, in this context, the observed greywacke levels (Lotze 1961) signal the occurrence of turbidites, cannot be judged. The *Cruziana* ichnofacies is predominant, whereas the *Skolithos* ichnofacies, recorded at the top of the Embid Formation, indicates a shallowing episode. In this, desiccation cracks, evidence of subaerial exposure, are also found.

It is worthy of note that the basal part (Bámbola Formation) has also been proposed to represent a deltaic sedimentary environment (Gabaldón 1990; cf. von Raumer *et al.* 2006; Gámez Vintaned 2007). More evidence, however, is needed to critically review the impact on the systems tracts identification.

The lower part of the Jalón Formation shows a retrogradational succession of quartzites, sandstones, siltstones, claystones and dolostones, and, following a distinct claystone interval, a progradational succession of claystones, siltstones and sandstones. Most probably, the successions represent the transgressive and highstand phases of the sequence (Fig. 3). Trace fossils are abundant. In addition to the list given by Liñán *et al.* (1996), *Conostichus* aff. *Sagittichnus*, *Skolithos* and escape structures are also present (Gámez Vintaned 2007). Trace fossil records indicate that, locally, the *Glossifungites* ichnofacies (Schmitz 1971), and, particularly close to the top of the upper part (Aliaga 2008) and in the lower member, the *Skolithos* ichnofacies prevail (associated with raindrop imprints and desiccation cracks, karstification of dolostone top beds and halite pseudomorphs). It is also worthy of note that the oldest (Upper Ovetian) Cambrian trilobites of the Cadenas Ibéricas appear near the base of the Jalón Formation in the Jalón Valley (Sdzuy 1987): *Dolerolenus*? sp. indet., *Anadoxides*? sp. indet. and *Thoralaspis* n. sp. A).

Facies and thickness changes in the Jalón Formation are particularly noticeable, and are interpreted as a reflection of the ongoing rifting.

The middle Lower Cambrian–lower Middle Cambrian sequences

The succession, which is dominated by carbonates, was subdivided by Gámez *et al.* (1991) into several succeeding transgressive–regressive cycles. Sectioning the succession into systems tracts also results in the identification of TSTs and highstand systems tracts (HSTs) (Myers & Milton 1996). It is noted, however, that the suggested subdivision into TSTs and HSTs is not identical with that of the suggested cycle phases. The basal TSTs generally follow with a distinct boundary on the preceding HSTs. In terms of sequence stratigraphic models for carbonate platforms (Emery 1996), the terminal phase, succeeding the HST, would be that of lowstand or drowning. In the case of the upper Lower Cambrian–upper Middle Cambrian succession, drowning is suggested as the dominating mechanism in terminating the sequences. It is interpreted as a reflection of accelerated subsidence or basin deepening, that is, at least a temporary increase in water column depth.

The middle Lower Cambrian sequence (S3)

This sequence covers the upper part of the Ovetian Stage and the lower part of the Marianian Stage. No stratigraphic event, known from elsewhere, has been identified as correlating with the top sequence boundary (Fig. 3). The litho-units involved are the upper part of the Jalón Formation and the Ribota Dolomite. The lower part of the sequence consists of an alternating succession of dolomites, marls, claystones, siltstones and sandstones. It reflects an environment of mixed tidal flats, with diagnostic sedimentary structures such as desiccation cracks (Schmidt-Thomé 1973), and gypsum and halite pseudomorphs. This succession, that is, the upper part of the Jalón Formation, together with the lower part of the Ribota Dolomite (see also Palacios & Moczydłowska 1998), which comprises dolomites, limestones and marls, represents the transgressive phase of the sequence. It is capped by a regionally traceable marl layer, representing the maximum flooding surface, with preserved trilobite fauna and acritarchs. The remaining dolomites of the Ribota Dolomite are interpreted as platform carbonates, representing the HST. The carbonate deposition ends abruptly with the formation of a phosphate layer, and is followed by claystones of the Huérmeda Formation. We propose that the sudden disappearance of carbonates (which coincides in time with that in central Ossa–Morena) reflects platform drowning induced by extensional tectonics.

The upper Lower Cambrian sequence (S4)

This sequence comprises the upper part of the Marianian Stage and the lower part of the Bilbilian Stage. The respective litho-units are the Huérmeda and Daroca formations (Fig. 3). The top of the sequence correlates with the Daroca Regression. The lower formation consists of claystones, which may host thin layers of dolomites and sandstones. Marls occasionally replace the claystones. Locally, the claystones are dark, kerogen-rich and laminated (Palacios & Moczydłowska 1998). Trilobites adapted to open marine conditions are present. The top layer is characterized by the occurrence of conoidal shells (Lotze 1961; Schmidt-Thomé 1973). Sedimentary structures are scarce. Ichnocoenoses indicate sublittoral conditions typical of the *Cruziana* ichnofacies (Liñan *et al.* 1996), with some episodes of opportunistic colonizers belonging to the *Skolithos* ichnoguild. Obviously, the succession represents the transgressive phase of the sequence and its top layer, the maximum flooding surface. Gámez *et al.* (1991) and Palacios & Moczydłowska (1998) suggest that the (entire) formation represents the culmination of the first Lower Cambrian transgressive episode. The succeeding Daroca Formation comprises quartzites, sandstones, siltstones and claystones. Towards the top, it shows a progradational, shallowing trend. The sedimentary structures (Schmitz 1971; Palacios & Moczydłowska 1998: cross-bedding, ripple marks, ball-and-pillow structures, load casts, reworking) support the attribution to the *Cruziana* ichnofacies, derived from analysis of trace fossil assemblages (Schmitz 1971). The sedimentary profile of the formation suggests HST conditions. It ends with a distinct boundary, but yields no evidence of emergence (Palacios & Moczydłowska 1998).

In terms of basin evolution, the top of this sequence – or, in the definition of Gámez *et al.* (1991), the upper subunit, that is, the Daroca Formation itself – is proposed as playing a significant role. It indicates, according to Gámez *et al.* (1991), a regressive event (Daroca Regression), which the authors correlate with the Hawke Bay Regression. It can also be correlated with similar coarse clastic units elsewhere in Iberia (e.g. the El Castellar Formation in the Ossa–Morena Zone), and with a number of late Early Cambrian events at global scale, from neighbouring Morocco (Asrir Regression) to the antipodal Australia (upper Beetle Creek Formation) (Liñán & Gámez Vintaned 1993, p. 838). This seems to be a global event (of eustatic and/or climatic nature), which may have been enhanced by regional factors, such as tectonic movements in Baltica. The latter, as emphasized by Nielsen & Schovsbo (2005), owes its existence to a major uplift, documented by a stratigraphic gap, whilst deposition of the Daroca Formation clastics does not reflect evidence of an associated stratigraphic gap. It is likely that the common denominator is retreating (highstand) sea level.

The uppermost Lower Cambrian–lowermost Middle Cambrian sequence (S5)

This sequence comprises the upper part of the Bilbilian Stage and two thirds of the Leonian Stage. The Valdemiedes and Mansilla formations span this time interval; both are rich in trilobites and record the *Cruziana* ichnofacies. The lower half of the Mansilla Formation coincides with the Mid Leonian Regression (Sdzuy *et al.* 1999). Claystones form most of the lower part of the Valdemiedes Formation. Towards the top, these pass into an alternating succession of claystones, bearing carbonate nodules and flaser and wavy bedded carbonate layers embedded in claystones, and carbonate layers and massive beds. The Mansilla Formation is composed of carbonate nodule-bearing reddish mudstones and nodular–massive reddish carbonates (Liñán *et al.* 1992). Tops of massive carbonate beds are characterized by corrosion surfaces (Schmitz 1971). The alternating claystone–carbonate–dolomite succession shows a cyclic pattern, with a basal claystone phase and a terminal carbonate phase (Schmitz 2008). The cyclic regressive sedimentary pattern, together with the sedimentary structures of nodular, wavy and flaser bedding that characterize the Wilson (1975) facies belt 7, suggest that deposition of that interval largely took place on the open marine platform.

In its lower part, the Valdemiedes Formation represents TST conditions (recording the Valdemiedes Event faunal turnover at the Bilbilian–Leonian boundary, that is, Lower–Middle Cambrian boundary; Fig. 3: Liñán *et al.* 2008), whereas the transitional upper part, dominated by the regressive cycles, reflects frequent sea-level fluctuations. These initiate the HST phase of the sequence, which culminates in the massive carbonates of the Mansilla Formation. It is noteworthy that the carbonate deposition ends rather abruptly, and is followed by the nodule-bearing claystones of the top of the Mansilla Formation. This suggests that the sudden disappearance of carbonates, as in the case of the top of the Ribota Dolomite, reflects drowning induced by extensional tectonics. This interpretation agrees with that of Quesada (2006) when discussing the Cambrian evolution of the Ossa–Morena Zone; he proposes that the Middle Cambrian 'upper pelite/sandstone sequence' (upper part of the Lower–Upper Cambrian Upper Detrital Formation) begins with a sudden deepening.

The lower Middle Cambrian sequence (S6)

The time interval covered by this sequence extends from the upper part of the Leonian Stage through the Caesaraugustan Stage to the lowermost part of the Languedocian Stage (Fig. 3). The Base–Languedocian Regression (or 'late Caesaraugustan regressive episode', Liñán *et al.* 2002) defines the top of the sequence. The litho-units involved are the Murero Formation and the basal part of the lower Acón Group. The succession is dominated by clastics. The lower part, representing the Murero Formation, comprises claystones and minor siltstones with embedded carbonate nodules, overlain by a sand-free claystone section; trilobites are very abundant, and ichnofossils point to environmental conditions of the *Cruziana* ichnofacies. It is suggested that, where there are interbeds of carbonate nodules, the succession is equivalent to the Wilson (1975) facies belt 2, that is, an open shelf environment. The upper part, representing the lowermost section of the lower Acón Group, is an alternating claystone–siltstone–sandstone section, which, towards the top, grades into a massive sandstone–quartzite unit. Ichnofossils listed (Liñán *et al.* 1996) point to depositional conditions characterized by the *Cruziana* ichnofacies, irregularly interspersed with beds showing *Monocraterion*, short *Skolithos* and escape structures indicating short high-energy episodes.

The lower part of the succession, that is, the Murero Formation, was deposited under TST conditions, whereas the upper part, that is, the lowermost part of the lower Acón Group, reflects HST progradation (Schmitz 2006). It ends with a distinct boundary.

The upper Middle Cambrian–intra-Tremadocian sequences

The lithostratigraphic succession, comprised by this interval, was subdivided by Schmitz (2006) into

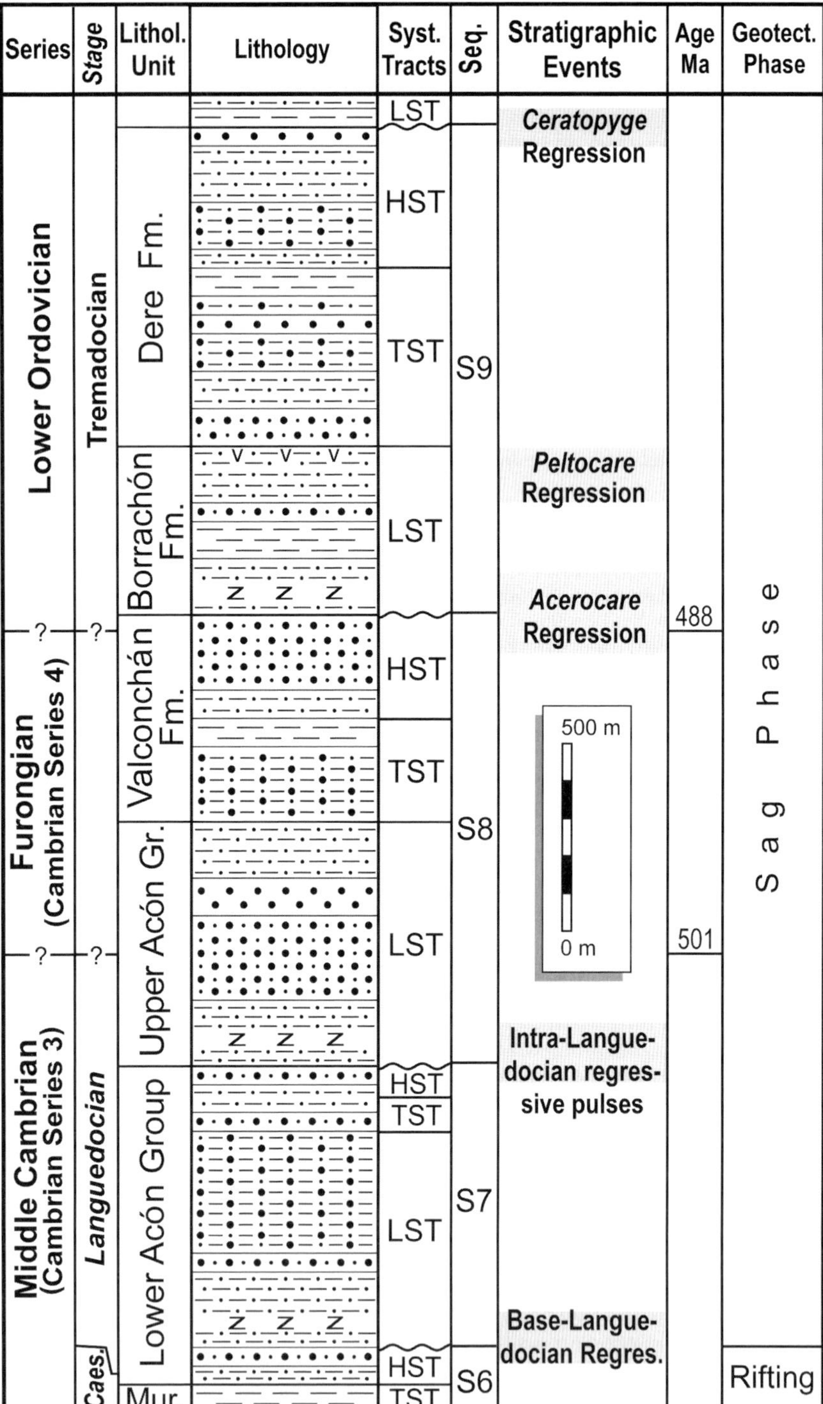

Fig. 4. Middle Middle Cambrian–Lower Tremadocian stratigraphy of the Cadenas Ibéricas, with subdivision into second-order sequences and systems tracts, correlated with published stratigraphic events (from Schmitz 2006). (Sequence boundaries are simply placed at the middle of correlating stratigraphic event boxes, the vertical dimension of the latter does not represent either time or thickness.) Legend as in Figure 3.

three second-order sequences. The characteristics common to the sequences are:

- their completeness in terms of systems tracts;
- the lithologies (which mainly comprise claystones, siltstones, sandstones and quartzites);
- the considerable thicknesses (the sequences measure many hundreds of metres each);
- their lack of terrestrial sedimentation;
- their distinct lithological breaks related to the sequence boundaries;
- their conspicuous LST turbiditic intervals;
- shallow-water quartzites with *Skolithos* burrows at TST and/or HST levels.

The upper Middle Cambrian sequence (S7)

The lower of the three sequences covers the lower and middle parts of the Languedocian Stage. Its top is correlated with the intra-Languedocian regressive pulses, which were introduced by Liñán *et al.* (2002, see also Álvaro *et al.* 2003). The formation involved is essentially the lower section of the Acón Group (Fig. 4). In the lower part it is dominated by claystones and siltstones alternating with sandstone layers. With its basal, local basin-floor fan, its turbidite interval included in the overlying claystone-dominated section and a frequently occurring prograding sandstone package at its top, this part probably represents basin, slope and platform conditions. The upper part is divided into a lower and an upper quartzite–sandstone unit and a middle claystone-dominated unit. Identification of the sedimentary environment relies on the interpretation of sedimentary structures (Schmitz 2006) and on the listing of trace fossils (Liñán *et al.* 1996). They evidence a spectrum of environments bracketing the *Cruziana* and opportunistic *Skolithos* ichnofacies.

The sedimentary profile suggests that LST conditions control the lower part, and TST and HST conditions the upper part (see Fig. 4).

The uppermost Middle Cambrian–intra-Upper Cambrian (Furongian) sequence (S8)

The sequence is of Languedocian age in its lower half and of Furongian age in its upper half. Its top is correlated with the *Acerocare* Regression (Schmitz 2006). The formations that make up this sequence are the upper part of the Acón Group and the succeeding Valconchán Formation (Fig. 4). The former is predominantly composed of claystones, within which, in particular intervals, sandstones and quartzites are intercalated. As for the lower part of the preceding sequence, locally developed basin-floor fans, a turbidite interval and a terminal, prograding sandstone–quartzite succession point to a sedimentary profile from basin through to slope and to platform sedimentation. Essentially, the Valconchán Formation consists of an upper and a lower quartzite–sandstone unit separated by a claystone-dominated unit. The lower unit typically hosts layers of oligomict conglomerates. The two quartzite–sandstone units contain thin layers of lingulid shell debris. Both the ichnofossil assemblages and the sedimentary structures suggest overall shallow-water conditions, whereby the spread of environments is represented by the *Skolithos*, *Cruziana* and *Glossifungites* ichnofacies (Schmitz 2006).

In terms of sea-level stand, it is suggested that the upper part of the Acón Group reflects LST conditions, and the Valconchán Formation TST and HST conditions (see Fig. 4).

The intra-Upper Cambrian (Furongian)–intra-Tremadocian sequence (S9)

The time covered by this sequence is the intra-Furongian–intra-Tremadocian interval. The sequence top corresponds to the *Ceratopyge* Regression (Schmitz 2006). Two formations make up the sequence: the Borrachón Formation and the Dere Formation (Fig. 4). The lower formation, that is, the Borrachón, is a claystone succession with subordinate intercalations of sandstones, quartzitic sandstones and quartzites. The local occurrence of a basin-floor fan, the occurrence of a turbidite interval within the overlying claystone section and the frequent occurrence of coarsening-upwards sandstone packages in the top part of the succession suggest an environmental development from basin through to slope and on to shallow-water conditions, the latter with local tidal sedimentation. The succeeding Dere Formation generally, yet not consistently, displays a tripartite subdivision into two quartzite–sandstone units separated by a claystone-dominated unit. Amongst environmentally diagnostic sedimentary features are transgressive surfaces of erosion, oligomict conglomerates – the sorting of which indicates shore deposits – colonies of *Skolithos*, particularly where the succession is less differentiated, and ichnofossil assemblages pertaining to the *Cruziana* ichnofacies. Noteworthy are vestiges of volcanic activity elsewhere, that is, a tuffite layer (Josopait 1972, table 1) and a locally occurring deposit of reworked tuffitic material (Scheuplein unpublished).

In the described sequence, we propose that the Borrachón Formation represents LST conditions, and the Dere Formation TST and HST conditions (see Fig. 4).

The intra-Tremadocian–Arenigian interval

The Lower Ordovician section ends with the Santed and Armorican Quartzite formations. They seem to be separated by a transgressive surface, whilst the top of the latter obviously represents a sequence boundary, at approximately the top Arenigian level. With some uncertainty, Schmitz (2006) proposes that both formations were joined in one sequence. The uncertainty reflects poor outcrop conditions, which are best explained as a 'bottom-section-up-to/top-section-down-to' situation, with an information gap in-between. Neither the lower nor the upper formation is being dealt with in this context.

Discussion

The subdivision of the Upper Vendian–Lower Tremadocian succession into sequences and systems tracts leads us to moderate previous interpretations of stratigraphic events. This, consequently, results in a somewhat modified view of the basin evolution. It holds particularly true for: (i) previous suggestions of sea-level changes; and (ii) conclusions drawn from the subsidence history of the Cadenas Ibéricas. In the following, these modified interpretations are discussed and – for verification purposes – related to the Gondwana geology.

With reference to proposed global sea-level trends showing a Cambrian transgression as part of a major first Phanerozoic cycle (Vail *et al.* 1991), it is noted that Liñán & Gámez Vintaned (1993) identified sea-level falls related to three Lower Cambrian regression events in Iberia. The herein suggested non-emergence and drowning at the sequence tops, within the interval in question, supports the classification of the three Lower Cambrian regressions, whereas a uniform transgressive deepening trend dominating the Cambrian basin evolution (see Vail *et al.* 1977; Carr 2002) is not confirmed.

In terms of subsidence history, the analysis of systems tracts suggests the subdivision of the succession into three evolutionary phases: (1) the outgoing Cadomian phase; (2) the rifting phase; and (3) the sag phase. Of these, the outgoing Cadomian phase may represent a consolidation period, honouring the concept of an Early Corduban, Cadomian-induced uplifting (Liñán *et al.* 2002; Sánchez-García *et al.* 2003), or, less likely, perhaps the link between periods of extensional tectonics, that is, between those of the Late Vendian (Valladares *et al.* 2002) and the succeeding Cambrian Period. In any case, the sedimentary pattern of Sequence 1, that is, that during Late Vendian times, resembles the one of Von Raumer *et al.* (2006), which suggests the filling of a subduction trough for the Central Iberian succession. Although uplifting and non-deposition prevailed at the time of the Vendian–Cambrian boundary – probably concurring with a sea-level fall and resulting in a lowstand phase – steep topographies were partially maintained.

Obviously, that interpretation is to be viewed against the background of newly gained insight into the events related to the change from the Precambrian to the Cambrian, which may be summarized in the following paragraph.

The main event linked to the Cadomian Orogeny in Iberia was the growth of a magmatic arc (Ossa–Morena Zone) and its accretion to the margin of Gondwana (represented by the Iberian autochthonous block) during Neoproterozoic times (Quesada 2006). According to several authors (Sánchez-García *et al.* 2003; Rodríguez Alonso *et al.* 2004; Pereira *et al.* 2006), the Cadomian pulses extended into earliest Cambrian (Early Corduban) times in the Ossa–Morena Zone. The collisional event against the Iberian autochthonous block also affected the northern (in modern co-ordinates) areas of it (i.e. the Cantabrian–Iberian Basin). According to Sánchez-García *et al.* (2003), the general uplift in Ossa–Morena and other parts of the Iberian Massif was the result of lithospheric thermal expansion in connection with massive underplating of mantle-derived magmas. Later, gravitational collapse of the expanded lithosphere led to extensional deformation and thinning, and eventually to progressive flooding of the thinned crust (Quesada 2006). Rifting propagated diachronously across the Ossa–Morena Zone from the Early Cambrian (Late Corduban) to the Late Ordovician (Liñán & Gámez Vintaned 1993; Sánchez-García *et al.* 2003), reaching most domains of the Iberian Massif in Early Cambrian times. Yet, rift-related bimodal magmatism developed only in Ossa–Morena.

The rifting phase, postulated as dominating the Cambrian evolution by most authors, seems to be confirmed by this assessment. It is suggested that it triggered the process of drowning of particular sections. The time interval mainly affected, however, may differ from those proposed by others. Several authors propose the extensional tectonic regime was effective in Iberia from the base of the Cambrian onwards (e.g. Ribeiro *et al.* 1990; Ábalos *et al.* 2002), or even earlier, and to have been effective during the entire Cambrian Period (e.g. von Raumer *et al.* 2006). As shown in this paper, the effects of rifting on stratigraphic successions in Cadenas Ibéricas are traceable at least from Early Ovetian times up to Early Languedocian. According to Liñán & Gámez Vintaned (1993), rifting started earlier in the Ossa–Morena Zone (that is, closest to the Cadomian suture), by Late Corduban times.

Evidence of the rifting phase is inferred from thickness and facies variations, and interpreted drowning events, which were most probably caused by a sudden increase in water depth. Extension is not evidenced by a respective fault pattern observed in the field. However, thickness and facies variations, particularly of the Embid, Jalón and Daroca formations (Aliaga & Schmidt-Thomé unpublished), may point to the existence of larger fault compartments, although the main faults of the rifting system are expected to line up outside the outcrop area.

The sag phase possibly controlled the basin architecture from as early as late Mid Cambrian (Early Languedocian) times onwards. It is characterized by considerable thickness, which reflects the sequential repetition of deposition, ranging from lowstand to highstand. It is likely that rejuvenation of extensional faults had amplifying effects. On the basis of local data, it is difficult to assess whether the sag phase is a local phenomenon that accentuates the preceding pull-apart/strike-slip basin setting of the West Asturian–Leonese Zone and the subsiding continental block setting of the Cantabrian Zone (von Raumer *et al.* 2006) or whether it heralds the Gondwana passive margin stage, which itself is suggested to be evidenced at (earliest) Ordovician times (Quesada 2006) or at Mid–Late Ordovician times (von Raumer *et al.* 2006). For the same reasons of ambiguity, the termination of the phase may be drawn differently, that is, either including or excluding the ubiquitous deposits of the Arenigian Armorican Quartzite Formation (Gutiérrez-Marco *et al.* 2002).

The discussion of correlation possibilities with the Gondwana geology largely leans on Deynoux *et al.* (2006), who cover, in this context, all related aspects. These are: (i) correlation of the Upper Vendian sections; (ii) the role of the Cadomian (Pan-African) Orogeny relative to the transition from Vendian to Cambrian strata; and (iii) the subdivision of supposed Cambrian successions both on the platform and in marginal fold belts.

In addition to (i): Deynoux *et al.* (2006) subdivide the Neoproterozoic–Upper Ordovician section into two megacycles, of which the second is the one with which to correlate. It covers the interval between the intra-Upper Vendian and Upper Ordovician glaciations. Consequently, its middle part is of interest, relative to the aimed-at comparison: at about 560–550 Ma the last Pan-African event (Pa II) is recorded, followed by (partly reddish) molasse sediments (e.g. Youkounkoun Basin), or volcano-clastics (Hoggar-Iforas Belt) or molasse sediments with succeeding platform carbonates (Anti-Atlas Belt). Occasionally, diamictites are encountered. As summarized by Deynoux *et al.* (2006), these sediments are generally considered as foreland or intramontane, post-collisional molasses, but may also mark – for example, in the Anti-Atlas – the beginning of an extensional regime (a similar situation is depicted in the Ossa–More Zone of Iberia where basal Cambrian, coarse siliciclastics of the Torreárboles Formation are followed by Ovetian–Marianian carbonates).

Obviously, the Paracuellos Group of the Cadenas Ibéricas has little in common with the referred African successions, yet appears to reflect a synchronous Upper Vendian basin fill in a marine environment.

In addition to (ii): the end of the Pan-African Orogeny is, according to Deynoux *et al.* (2006), highly diachronous, that is, the events of the closure and uplift period seem to cover the time span from 640–580 Ma (B2 phase of the Anti-Atlas area) to 560–550 Ma (Pa II event of the southwestern margin of the craton). Thereafter, extensional tectonic regimes continued or took over. As to the Anti-Atlas fold belt, two differing models were proposed to explain the extensional regime (for their authors see Deynoux *et al.* 2006): either extension resulted soon after collision and suturing from extensive post-orogenic uplift and subsequent collapse; or the development of extensional basins was unrelated to the Pan-African Orogeny, but, rather, was due to the dislocation of a Pan-African peneplain and the initiation of a rift system creating space for the subsequent succession. The latter suggestion is partly based on the observed presence of a normal fault system.

The terminal Pan-African events are obviously of Precambrian age, which suggests that the Early Cambrian (Early Corduban) uplift and the corresponding Córdoba Regression in Cadenas Ibéricas do not necessarily correlate, timewise, with sediments succeeding the terminal Pan-African events (which are of Vendian–Cambrian age in the above cases).

In addition to (iii): the Gondwana Cambrian formations are poorly controlled by biostratigraphy. Those successions suggested to be of Cambrian age are lacking the definition of the base, and very few individual levels within the succession rendered age-controlling information (for details see Deynoux *et al.* 2006). Also, identifications of the top appear to be of limited value. The age-dating employing the appearance of *Lingula* specimens as signifying the beginning of the Ordovician system in the northern Taoudenni Basin contradicts the evidence used by Josopait (1972) from the Cadenas Ibéricas. Likewise, the age-dating of the succession transgressively following the Série Pourprée in the Hoggar–Iforas Belt, where

510 Ma is taken as being close to the Cambrian–Ordovician boundary, contradicts timescales currently in use (Davidek *et al.* 1998).

Overall, in the absence of an accepted Cambrian stratigraphy, correlation with both stratigraphic units and sequences of the Cadenas Ibéricas will be highly speculative. The Lower Cambrian succession of the Anti-Atlas area, however, differs from that. Its Upper Limestone Formation (= Igoudine Formation: Geyer 1989), with trilobite finds (Sdzuy 1978; Geyer 1996), is described as reflecting epicontinental sedimentation at the edge of a fault-controlled subsiding basin. The rifting process controlling subsidence aborted at the end of the Early Cambrian, as deduced from thicknesses of the overlying section (for authors see Deynoux *et al.* 2006). Nevertheless, Geyer & Landing (2006) prefer a transtensional tectonic setting. Benssaou (2006) suggests a subdivision into four transgressive–regressive cycles, formed by TST and HST phases, that is, he identifies them as composite sequences controlled by third-order sea-level changes, and relates those to extensional tectonic events. There are strong similarities between these cycles and the sequences of the Cadenas Ibéricas.

Conclusions

Sufficiently reliable data allow the division of the intra-Vendian–intra-Tremadocian succession of the Cadenas Ibéricas into a set of sequences. Levels of confidence, however, vary. The sequences established by this review and adopted from a previous study cover the time intervals: (1) Late Vendian (Ediacaran); (2) early Early Cambrian; (3) mid Early Cambrian; (4) late Early Cambrian; (5) latest Early Cambrian–earliest Mid Cambrian; (6) early Mid Cambrian; (7) late Mid Cambrian; (8) latest Mid Cambrian–intra-Late Cambrian (Furongian); and (9) intra-Late Cambrian (Furongian)–intra-Tremadocian.

Overall, the sequences are considered to be second-order sequences. Those sequences comprising carbonate successions, that is, (3)–(6), may partly or entirely be interpreted as elements of composite sequences. Such interpretation accounts for the suggested role of the extensional tectonics in the modification of the sequence structures.

The subdivision of the succession into sequences improves interpretations relative to stratigraphic events and basin evolution. Key elements are:

(i) the discussion of the assumed tectonically induced break at the Vendian–Cambrian transition, suggested to reflect last Cadomian pulses occurring in Ossa–Morena;

(ii) the confirmation of a rift phase, at Early–mid Mid Cambrian level, deduced from facies and thickness variations, and the interpretation of drowning caused by extensional tectonics;

(iii) the suggestion of a sag phase at mainly late Mid and Late Cambrian levels, comprising thick clastic second-order sequences, which accentuates a preceding rifting basin setting, possibly heralding the Gondwana passive margin stage.

As to (i), it is noted that the question of the Vendian–Cambrian transition is possibly, owing to outcrop conditions, not solvable in the studied area. The lack of evidence affects the interpretation of basin evolution at that time interval. From circumstantial evidence, it appears that a break in the sedimentary development took place from before to after that transition, controlled by effects of the Cadomian phase and sea-level changes.

Correlation possibilities with Gondwana are limited as yet, owing to imprecise age determinations relative to Gondwana Cambrian successions. This, however, does not apply to the Lower Cambrian succession of the Anti-Atlas area, the interpretation of which suggests similar evolutionary conditions, that is, an Early Cambrian extensional tectonic regime, coupled with four transgressive–regressive cycles that image TST and HST phases.

The authors are much indebted to A. T. Nielsen (Copenhagen) who critically reviewed Early Ordovician correlation schemes and suggested improvements; and would like to thank E. Landing (Albany) for his critical revision, and thanks to J. Thurow and B. Thusu (both London) for their editorial assistance and patience in accompanying the preparation of this contribution. J. A. Gámez Vintaned and E. Liñán thank the Ministerio de Educación y Ciencia of Spain and FEDER-EU (project CGL2006-12975/BTE, 'MURERO') and Gobierno de Aragón (Grupo Consolidado E-17, 'Patrimonio and Museo Paleontológico'). This is a contribution to IGCP Project 493 ('The Rise and Fall of the Vendian Biota. Origin of the Modern Biosphere').

References

Ábalos, B., Carreras, J. *et al.* 2002. Variscan and Pre-Variscan Tectonics. *In*: Gibbons, W. & Moreno, T. (eds) *The Geology of Spain*. Geological Society, London, 55–83.

Aliaga, A. 2008. Geological investigations in the Eastern Iberian Chains north of the River Jalón (Zaragoza, Spain) – a summary. *In*: von Raumer, J. F. (ed.) *Franz Lotze (1903–1971). The doctoral theses of his students in Spain and Portugal*. Serie Nova Terra, **36**, Laboratorio Xeolóxico de Laxe, A Coruña, 223–228.

Álvaro, J. J. & Clausen, S. 2005. Major geodynamic and sedimentary constraints on the chronostratigraphic correlation of the lower–middle Cambrian transition in the western Mediterranean region. *Geosciences Journal*, **9**, 145–160.

Álvaro, J. J., Elicki, O., Geyer, G., Rushton, A. W. A. & Shergold, J. H. 2003. Palaeogeographical controls on the Cambrian trilobite immigration and evolutionary patterns reported in the western Gondwana margin. *Palaeogeography, Palaeoclimatology, Palaeoecology*, **195**, 5–35.

Bartsch, G. 2008. Geological investigations in the Eastern Iberian Chains between Rio Jalón and the road Miedes–Codos–Cariñena (Spain) – a summary. *In*: von Raumer, J. F. (ed.) *Franz Lotze (1903–1971). The doctoral theses of his students in Spain and Portugal*. Serie Nova Terra, **36**, Laboratorio Xeolóxico de Laxe, A Coruña, 229–234.

Benssaou, M. 2006. Geodynamic significance of the stacking of Early Cambrian sequences in the western Anti-Atlas. *In*: *Global Infracambrian Hydrocarbon Systems and the Emerging Potential in North Africa. Conference Proceedings*. Geological Society, London, Petroleum Group, 60–61.

Carr, I. D. 2002. Second-order sequence stratigraphy of the Palaeozoic of North Africa. *Journal of Petroleum Geology*, **25**, 259–280.

Davidek, K., Landing, E., Bowring, S. A., Westrop, S. R., Rushton, A. W. A., Fortey, R. A. & Adrain, J. M. 1998. New uppermost Cambrian U–Pb date from Avalonian Wales and age of the Cambrian–Ordovician boundary. *Geological Magazine*, **135**, 303–309.

Deynoux, M., Affaton, P., Trompette, R. & Villeneuve, M. 2006. Pan-African tectonic evolution and glacial events registered in Neoproterozoic to Cambrian cratonic and foreland basins of West Africa. *Journal of African Earth Sciences*, **46**, 397–426.

Emery, D. 1996. Carbonate systems. *In*: Emery, D. & Myers, K. J. (eds) *Sequence Stratigraphy*. Blackwell Science, Oxford, 211–237.

Gabaldón, V. 1990. Plataformas siliciclásticas externas: Facies y su distribución areal (Plataformas dominadas por tormentas). Parte II: Análisis de cuencas. *Boletín Geológico y Minero*, **101**, 3–33.

Gámez, J. A., Fernández-Nieto, C., Gozalo, R., Liñán, E., Mandado, J. & Palacios, T. 1991. *Bioestratigrafía y evolución ambiental del Cámbrico de Borobia (Provincia de Soria. Cadena Ibérica Oriental)*. Cuaderno Lab Xeolóxico de Laxe, Coruña, **16**, 251–271.

Gámez Vintaned, J. A. 2007. *Paleoicnología del límite Precámbrico/Cámbrico y del Cámbrico Inferior en el norte de la Cadena Ibérica Oriental*. PhD thesis, Universidad de Zaragoza.

Gámez Vintaned, J. A., Liñán, E. & Zhuravlev, A. Y. 2007. Neoproterozoic shelly fossils of Spain. *In*: Semikhatov, M. A. (ed.) *The Rise and Fall of the Vendian (Ediacaran) Biota. Origin of the Modern Biosphere*. Transactions of the International Conference on the IGCP Project 493, Moscow, August 20–31 2007. GEOS, Moscow, 10–11.

Geyer, G. 1989. Late Precambrian to early Middle Cambrian lithostratigraphy of southern Morocco. *Beringeria*, **1**, 115–143.

Geyer, G. 1996. The Moroccan fallotaspidid trilobites revisited. *Beringeria*, **18**, 89–199.

Geyer, G. & Landing, E. 2006. Latest Ediacaran and Cambrian of the Moroccan Atlas regions. *In*: Geyer, G. & Landing, E. (eds) *Morocco 2006*. Ediacaran–Cambrian Depositional Environments and Stratigraphy of the Western Atlas Regions. Explanatory Description and Field Excursion Guide. *Beringeria* Special Issue **6**, 7–46.

Gutiérrez-Marco, J. C., Robardet, M., Rábano, I., Sarmiento, G. N., San José Lancha, M. A., Herranz Araújo, P. & Pieren Pidal, A. P. 2002. Ordovician. *In*: Gibbons, W. & Moreno, T. (eds) *The Geology of Spain*. Geological Society, London, 31–49.

Josopait, V. 1972. *Das Kambrium und das Tremadoc von Ateca (Westliche Iberische Ketten, NE-Spanien)*. Muenstersche Forschungen zur Geologie und Palaeontologie, **23**, 1–121.

Liñán, E. & Gámez Vintaned, J. A. 1993. Lower Cambrian palaeogeography of the Iberian peninsula and its relation with some neighbouring European areas. *Bulletin de la Société géologique de France*, **164**, 831–842.

Liñán, E. & Tejero, R. 1988. Las formaciones precámbricas del antiforme de Paracuellos (Cadenas ibéricas). *Boletín de la Real Sociedad Española de Historia Natural (Geología)*, **84**, 39–49.

Liñán, E., Gozalo, R., Gámez, J. A. & Alvaro, J. 1992. Las formaciones del Grupo Mesones (Cámbrico Inferior-Medio) en las Cadenas Ibéricas. *In*: *Actas de las sesiones cientificas. III Congreso Géologico de España*. Tome **1**. III Congreso Geológico de España y VIII Congreso Latinoamericano de Geología, Salamanca 1992, 517–523.

Liñán, E., Villas, E., Gámez Vintaned, J. A., Álvaro, J. J., Gozalo, R., Palacios, T. & Sdzuy, K. 1996. Síntesis paleontológica del Cámbrico y Ordovícico del Sistema Ibérico (Cadenas Ibéricas y Cadenas Hespéricas). *Revista Española de Paleontología*, **3**, 21–32.

Liñán, E., Gozalo, R., Palacios, T., Gámez Vintaned, J. A., Ugidos, J. M. & Mayoral, E. 2002. Cambrian. *In*: Gibbons, W. & Moreno, T. (eds) *The Geology of Spain*. Geological Society, London, 17–29.

Liñán, E., Dies Álvarez, M. E. *et al.* 2008. Proposed Global Stratotype Section and Boundary Point (GSSP) for Cambrian System Stage 5 and Series 3 in Murero (Spain). *In*: *The 13th International Field Conference of the Cambrian Stage Subdivision Working Group. The Siberian Platform, Western Yakutia. Yakutsk, July 20th–August 1st 2008*. SNIIGGIMS, Novosibirsk, 43–48.

Lotze, F. 1961. Das Kambrium Spaniens. Teil I: Stratigraphie. *Akademie der Wissenschaften und der Literatur, Abhandlungen der mathematisch-naturwissenschaftlichen* Klasse, **1961/6**, 283–498.

Martínez-García, E. 2006. Proterozoic–Lower Paleozoic terrane accretion and Variscan domains in the Iberian Massif (Spain and Portugal). *In*: Schroeder, R. & Perejón, A. (eds) Contributions to the Geology of Spain. In Memory of Prof. Franz Lotze. *Zeitschrift der Deutschen Gesellschaft für Geowissenschaften*, **157**, 559–574.

MYERS, K. J. & MILTON, N. J. 1996. Concepts and principles of sequence stratigraphy. *In*: EMERY, D. & MYERS, K. J. (eds) *Sequence Stratigraphy*. Blackwell Science, Oxford, 11–41.

NIELSEN, A. & SCHOVSBO, N. 2005. Lower Cambrian sequence stratigraphy in Scandinavia. *Geologiska Föreningens i Stockholm Förhandlingar*, **127**, 1 (Abstract).

PALACIOS, T. & MOCZYDŁOWSKA, M. 1998. Acritarch biostratigraphy of the Lower–Middle Cambrian boundary in the Iberian Chains, province of Soria, northeastern Spain. *In*: *Revista Española de Paleontología,* N° extr. Homenaje al Prof. Gonzalo Vidal, 65–82.

PEREIRA, M. F., CHICHORRO, M., LINNEMANN, U., EGUILUZ, L. & BRANDÃO SILVA, J. 2006. Inherited arc signature in Ediacaran and Early Cambrian basins of the Ossa–Morena Zone (Iberian Massif, Portugal): paleogeographic link with European and North African Cadomian correlatives. *Precambrian Research*, **144**, 297–315.

QUESADA, C. 2006. The Ossa–Morena Zone of the Iberian Massif: a tectonostratigraphic approach to its evolution. *In*: SCHROEDER, R. & PEREJÓN, A. (eds) Contributions to the Geology of Spain. In Memory of Prof. Franz Lotze. *Zeitschrift der Deutschen Gesellschaft für Geowissenschaften*, **157**, 585–595.

RIBEIRO, A., QUESADA, C. & DALLMEYER, R. D. 1990. Geodynamic evolution of the Iberian Massif. *In*: DALLMEYER, R. D. & MARTÍNEZ GARCÍA, E. (eds) *Pre-Mesozoic Geology of Iberia*. Springer, Berlin, 399–409.

RODRÍGUEZ ALONSO, M. D., PEINADO, M., LÓPEZ-PLAZA, M., FRANCO, P., CARNICERO, A. & GONZALO, J. C. 2004. Neoproterozoic–Cambrian synsedimentary magmatism in the Central Iberian Zone (Spain): geology, petrology and geodynamic significance. *In*: DÖRR, W., FINGER, F., LINNEMANN, U. & ZULAUF, G. (eds) The Avalonian–Cadomian Belt and Related Peri-Gondwana Terranes. *International Journal of Earth Sciences (Geologische Rundschau)*, **93**, 897–920.

SÁNCHEZ-GARCÍA, T., BELLIDO, F. & QUESADA, C. 2003. Geodynamic setting and geochemical signatures of Cambrian–Ordovician rift-related igneous rocks (Ossa–Morena Zone, SW Iberia). *In*: MURPHY, J. B. & KEPPIE, J. D. (eds) *Collisional Orogenesis in the Geological Record and Modern Analogues. Tectonophysics*, **365**, 233–255.

SANTAMARIA I CASANOVAS, J. 1995. *Los yacimientos de fosfato sedimentario en el límite Precámbrico-Cámbrico del anticlinal de Valdelacasa (Zona Centro-Ibérica)*. Doctoral thesis, Facultat de Ciencies, Universitat Autònoma de Barcelona, Bellaterra.

SCHMIDT-THOMÉ, M. 1973. Beitraege zur Feinstratigraphie des Unterkambriums in den Iberischen Ketten (Nordost-Spanien). *Geologisches Jahrbuch, Reihe B*, **7**, 3–43.

SCHMITZ, U. 1971. Stratigraphie und Sedimentologie im Kambrium und Tremadoc der Westlichen Iberischen Ketten noerdlich Ateca (Zaragoza), NE-Spanien. *Muensterscher Forschungen zur Geologie und Palaeontologie*, **22**, 1–123.

SCHMITZ, U. 2006. Sequence stratigraphy of the NE Spanish Middle Cambrian to Early Ordovician section. *In*: SCHROEDER, R. & PEREJÓN, A. (eds) Contributions to the Geology of Spain. In Memory of Prof. Franz Lotze. *Zeitschrift der Deutschen Gesellschaft für Geowissenschaften*, **157**, 629–646.

SCHMITZ, U. 2008. Geological investigations in the Cambrian and Tremadocian successions north of Ateca (Western Iberian Mountain Chain, NE-Spain) – a summary. *In*: VON RAUMER, J. F. (ed.) *Franz Lotze (1903–1971). The doctoral theses of his students in Spain and Portugal*. Serie Nova Terra, **36**, Laboratorio Xeolóxico de Laxe, A Coruña, 243–248.

SDZUY, K. 1978. The Precambrian–Cambrian boundary beds in Morocco (preliminary report). *Geological Magazine*, **115**, 83–94.

SDZUY, K. 1987. Trilobites de la base de la formación del Jalón (Cámbrico inferior) de Aragón. *Revista Española de Paleontología*, **2**, 3–8.

SDZUY, K., LIÑÁN, E. & GOZALO, R. 1999. The Leonian Stage (early Middle Cambrian): a unit for Cambrian correlation in the Mediterranean subprovince. *Geological Magazine*, **136**, 39–48.

TEYSSEN, T. 1980. Acerca del problema de una discordancia assíntica en las Cadenas Ibéricas (NE-España). *Estudios geológicos*, **36**, 403–407.

VAIL, P. R., AUDEMARD, F., BOWMAN, S. A., EISNER, P. N. & PEREZ-CRUZ, C. 1991. The stratigraphic signatures of tectonics, eustasy and sedimentology – an overview. *In*: EINSELE, G., RICKEN, W. & SEILACHER, A. (eds) *Cycles and Events in Stratigraphy*. Springer, Berlin, 619–659.

VAIL, P. R., MITCHUM, R. M. & THOMPSON, S., III. 1977. *Seismic Stratigraphy and Global Changes of Sea Level. Part 4. Seismic Stratigraphy – Applications to Hydrocarbon Exploration*. AAPG Memoir, **26**, 83–97.

VALLADARES, M. I., BARBA, P., COLMENERO, J. R., ARMENTEROS, I. & UGIDOS, J. M. 1998. La sucesión sedimentaria del Precámbrico Superior-Cámbrico Inferior en el sector central de la Zona Centro Ibérica: Litoestratigrafía, geoquímica y facies sedimentarias. *Revista de la Sociedad Geologica de España*, **11**, 271–283.

VALLADARES, M. I., BARBA, P. & UGIDOS, J. M. 2002. Precambrian. *In*: GIBBONS, W. & MORENO, T. (eds) *The Geology of Spain*. Geological Society, London, 7–16.

VIDAL, G., PALACIOS, T., GÁMEZ VINTANED, J. A., DÍEZ BALDA, M. A. & GRANT, S. W. F. 1994. Neoproterozoic–early Cambrian geology and palaeontology of Iberia. *Geological Magazine*, **131**, 729–765.

VON RAUMER, J. F., STAMPFLI, G. M., HOCHARD, C. & GUTIÉRREZ-MARCO, J. C. 2006. The Early Palaeozoic in Iberia – a plate tectonic interpretation. *In*: SCHROEDER, R. & PEREJÓN, A. (eds) Contributions to the Geology of Spain. In Memory of Prof. Franz Lotze. *Zeitschrift der Deutschen Gesellschaft für Geowissenschaften*, **157**, 575–584.

WILSON, J. L. 1975. *Carbonate Facies in Geologic History*. Springer, Berlin.

Potential for oil and gas in the Proterozoic carbonates (Sirban Limestone) of Jammu, northern India

G. M. BHAT*, GHARA RAM & SUMITA KOUL

Postgraduate Department of Geology, University of Jammu, Jammu, India

**Corresponding author (e-mail: bhatgm1@rediffmail.com)*

Abstract: The Proterozoic Sirban Limestone of Jammu in northern India contains an assemblage of Neoproterozoic microflora comparable to other such assemblages from different Proterozoic oil- and gas-bearing carbonate successions in India, Morocco and Siberia. The stromatolitic Sirban Limestone succession is composed of lamina-scale organic-rich source rocks, limestone units with good-quality reservoir characteristics and seal horizons that together constitute the basic physical elements of a petroleum system. A reworked microfloral assemblage of Neoproterozoic age is also recovered from the unconformably overlying Subathu Formation (Eocene) from the area. Some *in situ* elements of this assemblage, including a few genera, were also recovered from the Sirban Limestone. This reworked assemblage may be related either to upwards migration of hydrocarbons from deeper sediments or, more likely, to the existence of the Sirban Limestone and younger Neoproterozoic formations as a positive area that sourced the microflora to the Eocene Subathu Basin.

The Proterozoic Sirban Limestone (also known as Great Limestone, Vaishnov Devi Limestone, Jammu Limestone and Trikuta Limestone) is an autochthonous unit in the Jammu Sub-Himalaya that crops out to the south of the Main Boundary Thrust belt (Fig. 1). This autochthonous limestone unit is unconformably overlain by early Cenozoic sediments (Subathu and Murree formations, and the Siwalik Group), but the base of the unit is not exposed anywhere in the region. On the basis of lithological similarity, the Sirban Limestone has been correlated with: outcrops to the NW in the Salt Range and the Muzzafarabad–Punch sector (Sirban Limestone) in Pakistan; to the NE in Dharamshala (Dharamkot Limestone); and in Shimla (Tundapather Limestone) in Himachal Pradesh, northern India. The general stratigraphy of the Jammu region is given in Table 1.

The Sirban Limestone is widely distributed in the Jammu region. It consists of thickly bedded, highly jointed, hard, dark-grey dolomite and limestone units, interbedded with thin chert, ash and shale beds. The carbonate units are characterized by microbial mats and various types of stromatolites. These rocks are quite resistant to erosion and weathering. The unconformably overlying upper part of the Subathu Formation is composed of dark-grey limestone and interbedded shale units capped by olive-green orthoquartzite beds.

Sharma (2006) processed a few carbonate samples collected from Eocene rocks (Subathu Formation) at Muthal for recovery of microflora using standard palynological techniques (Vidal 1976, 1988). In the process he recovered reworked Neoproterozoic microfloral assemblage from some of these samples (Fig. 2), which prompted a detailed investigation of the adjacent exposed sections of the Sirban Limestone. Subsequently we measured, described and sampled three sections for the microfloral investigation and characterization of source and reservoir potential, including the one located at Bidda that yielded *in situ* Neoproterozoic microfloral elements, some of which are similar to the reworked ones recovered by Sharma (2006) from Muthal. We processed the Sirban Limestone samples using the methods of Vidal (1976, 1988). All of the *in situ* Neoproterozoic microfloral elements have undergone thermal alteration. However, the thermal maturity of the microfloral elements is variable, with the majority being significantly altered. Conversely, the reworked Neoproterozoic microflora recovered is light coloured, thermally less altered and better preserved.

Field observations

The measured section at Bidda occurs in the Riasi Inlier of the Sirban Limestone. Raha (1984) reported the occurrence of three distinct stromatolite assemblage zones in the Sirban Limestone of Jammu namely the *Kussiella–Colonnella* Assemblage Zone, the *Conophyton–Colonnella* Assemblage Zone and the *Baicalia* Assemblage Zone and assigned a late Early–early Middle Riphean age to these zones. At about the same time Kumar (1984)

From: Craig, J., Thurow, J., Thusu, B., Whitham, A. & Abutarruma, Y. (eds) *Global Neoproterozoic Petroleum Systems: The Emerging Potential in North Africa.* Geological Society, London, Special Publications, **326**, 245–254.
DOI: 10.1144/SP326.14 0305-8719/09/$15.00

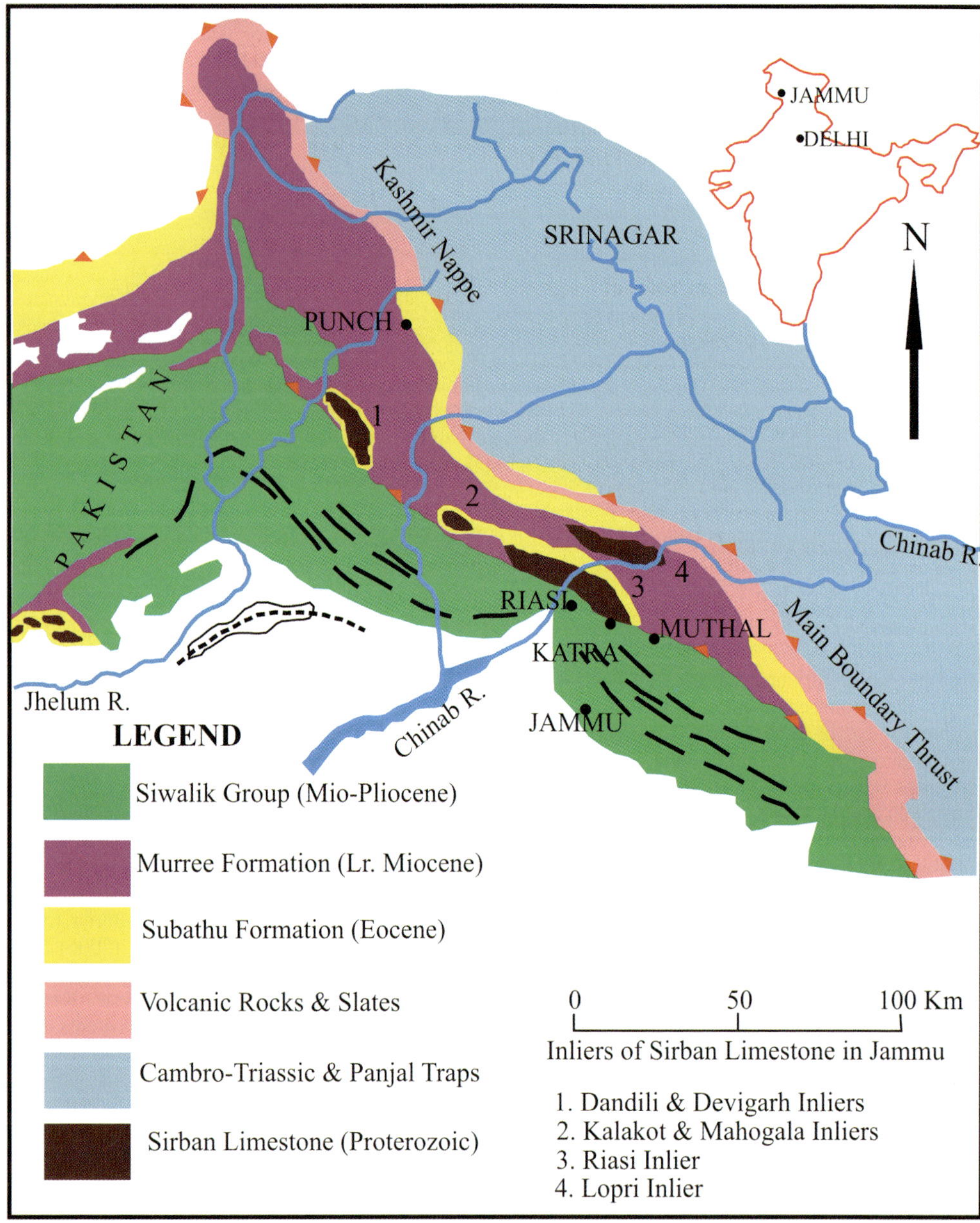

Fig. 1. Regional geological and location map, showing the location of outcrop sections discussed in text. Generalized chronostratigraphic column of the Jammu area showing the location of the Sirban Limestone and Subathu Formation.

reported two stromatolite assemblage zones from the Semri Group (Lower Vindhyan) coeval to the first two assemblage zones of Raha (1984) and assigned an Early–Middle Riphean age to these zones. He also reported another stromatolite assemblage zone in the Bhander Group (Upper Vindhyan) coeval to the third assemblage zone of Raha (1984) and assigned a Late Riphean age to this zone. However, recent studies in Australia have revealed that *Baicalia* is common in rocks younger than 750 Ma and that the other taxa reported from Sirban Limestone are long ranging. Also, the age of galena occurring in pockets and lenses within the topmost orthoquartzite beds of the Sirban

Table 1. *General stratigraphic succession of the Jammu region*

Alluvium and terrace deposits	Recent and sub-recent deposits
Siwalik Group	Miocene–Plio-Pleistocene
Murree Formation	Upper Eocene–Lower Miocene
Subathu Formation	Eocene
----------------------Unconformity------------------------ (Marked by breccia and bauxite)	
Sirban Limestone	Proterozoic
----------------------Base not exposed--------------------	

Limestone lying above the *Baicalia* Assemblage Zone at different localities (Sarsendu and Khairi) in the area has been dated at approximately 967 Ma (with minimum and maximum model ages of 925 and 1000 Ma) using Pb-isotope age determination (Raha 1984). If the Pb-isotope age is taken as correct, then the Sirban Limestone is older than Neoproterozoic. But the occurrence of the *Baicalia* Assemblage Zone below the orthoquartzite unit and recovery of some elements of Neoproterozoic microflora from the beds below the *Conophyton–Colonnella* Assemblage Zone (this study) and earlier by Venkatachala & Kumar (1996, 1997, 1998) from the chert and shale beds of the Sirban Limestone in Bidda and Muthal localities suggest a Neoproterozoic age (Cryogenian) for this part of the Sirban Limestone (Fig. 3). Therefore, the three stromatolite assemblage zones reported in the Sirban Limestone may range in age from Mesoproterozoic to Neoproterozoic and possibly younger. The measured section lies just below the *Conophyton–Colonnella* Assemblage Zone, and is composed of thickly bedded, highly jointed, hard, dark-grey and brownish dolomite units. Some beds are composed of pebble- and stromatolite-bearing dolomite (Fig. 3). The *Conophyton–Colonnella* Assemblage Zone shows lateral east–west continuity from, respectively, Muthal to north of Barakh. It is best exposed on the Jotipuram–Salal road east of Bidda village (Fig. 4). Large cylindrical and conical columnar forms of *Conophyton* and *Colonnella* dominate this stromatolite assemblage.

At Muthal, the Subathu Formation of Eocene age is composed of black carbonaceous shale, sandstone

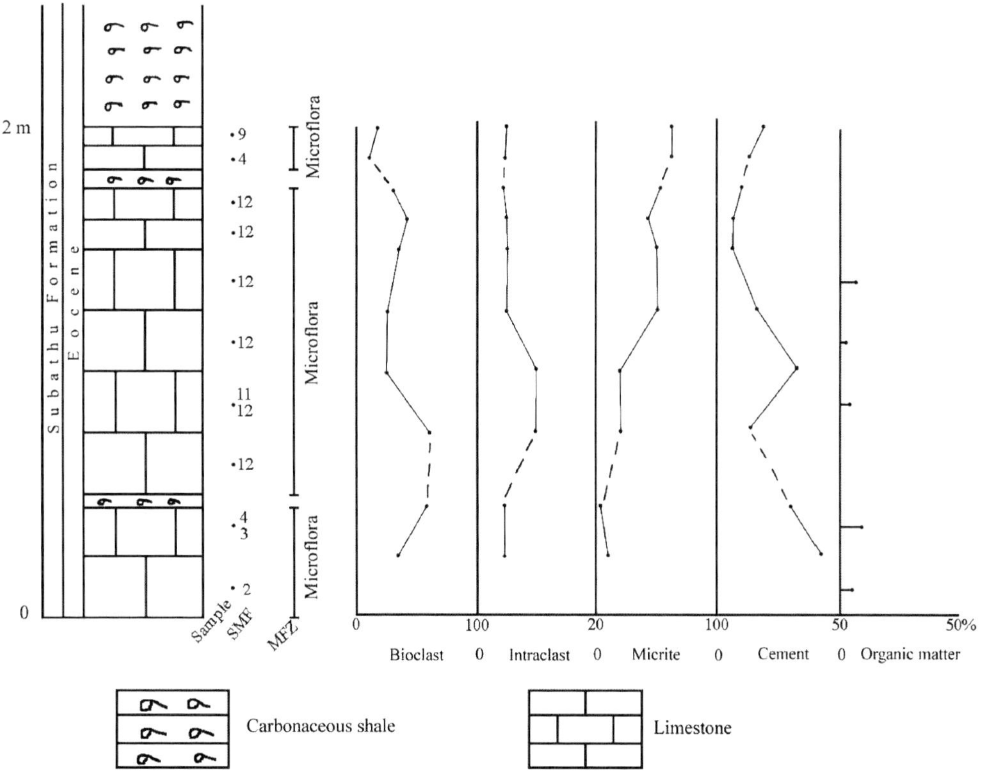

Fig. 2. Composite Eocene stratigraphic section measured at Muthal.

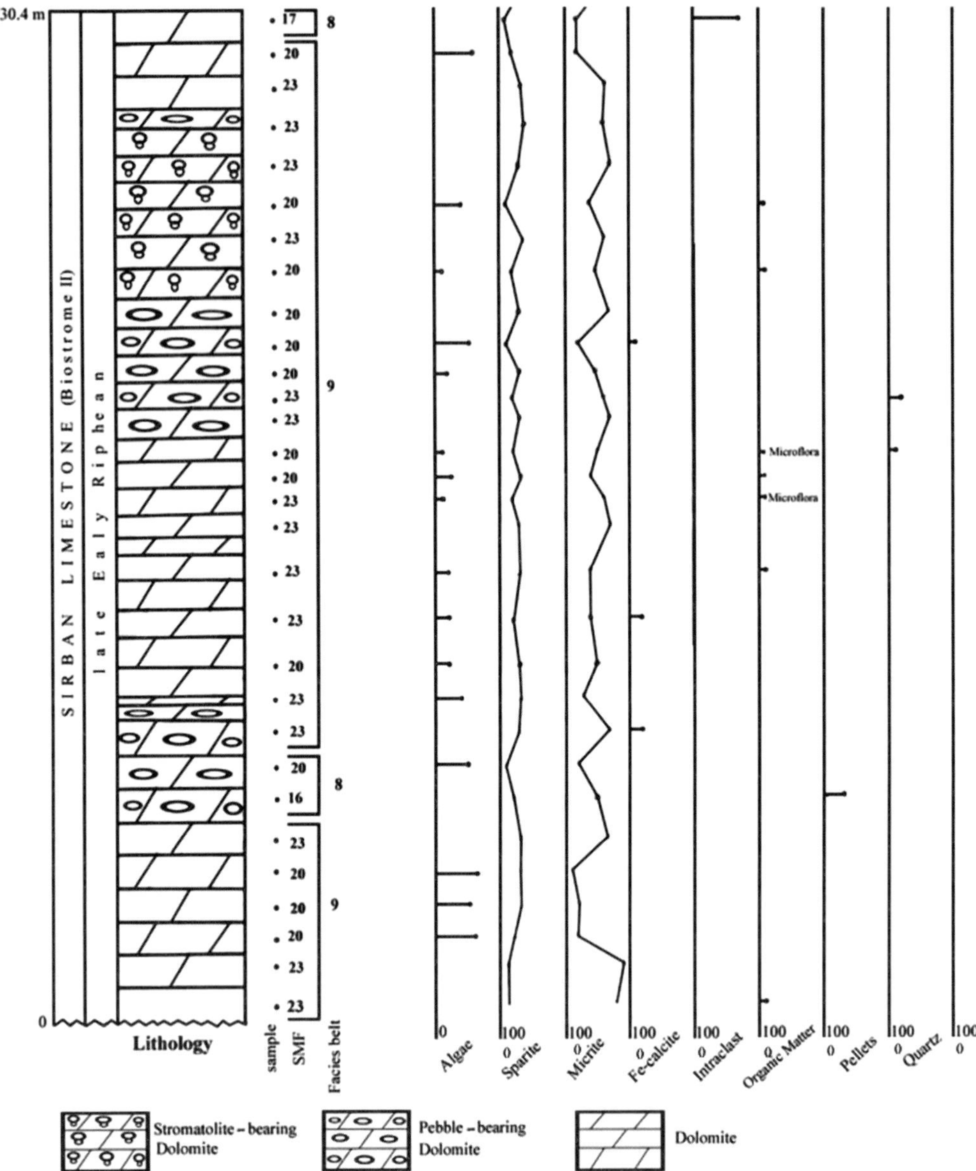

Fig. 3. Composite Proterozoic stratigraphic section measured at Bidda.

and topmost olive-green orthoquartzite units. A thin unit of bauxite occurs at its base and is capped by sandstone beds of the Murree Formation at the top. The basal contact with the Sirban Limestone represents a regional unconformity, whilst its upper contact with the Murree Formation is gradational. A 2 m-thick band of Nummulitic limestone and intervening thin beds of carbonaceous shale of the Subathu Formation was sampled for palynological analysis. The contacts between the limestone and carbonaceous shale beds are sharp. The limestone beds represent sooty black, laminated and massive limestones of Eocene age corresponding to the *Alveolina oblonga–A. dainelli* zones.

Depositional environment

The Sirban Limestone at Bidda is represented by well-sorted packstones and grainstones, belonging

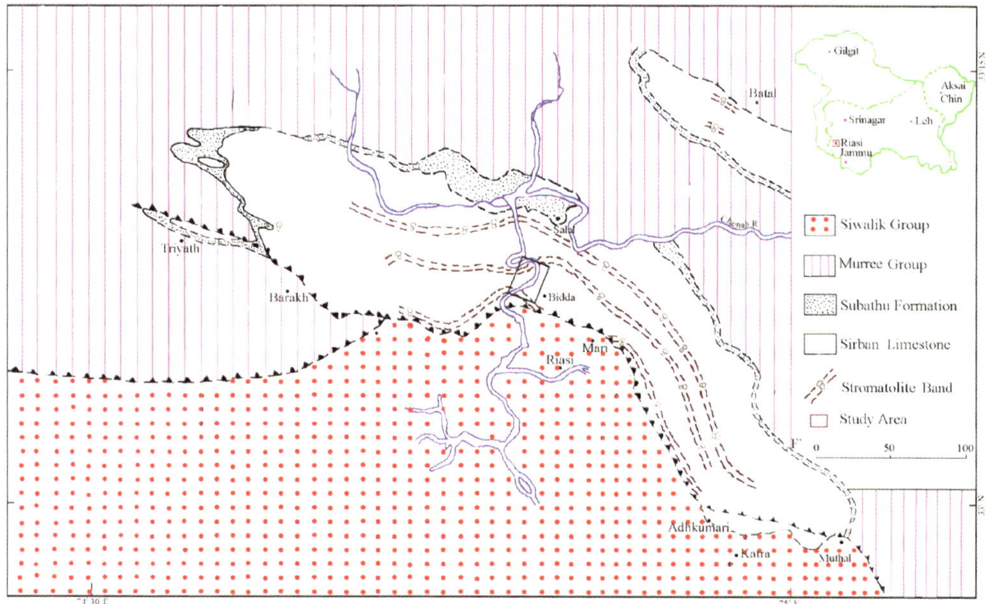

Fig. 4. Geological map of the Riasi Inlier and the adjoining area showing the continuity of biostromes.

to two standard microfacies (SMF) associations (SMF types of Wilson 1975). These microfacies associations include SMF types 16, 17 and 20 and SMF types 20 and 23 representing facies belts 8 (restricted platform environment) and 9 (platform evaporite environment), respectively. The restricted platform environments comprise cut-off lagoons and coastal ponds with restricted circulation. The platform evaporite environments consist of evaporites formed by intermittent flooding of the supratidal areas in an arid climate regime. The important facies of the platform evaporite environment are evaporitic dolomite and associated anhydrite. The biota of this environment are mostly microbial mats and stromatolites.

The Muthal section of the Eocene Subathu Formation is composed of SMF types 2, 3, 4, 9, 11 and 12. The microfacies are composed of dark-grey wackestone–packstone units with shaly intercalations. These microfacies occur in two associations: SMF types 2, 3, 4 and 9 – belonging to the facies belt 3 (deep shelf margin) – and SMF types 9, 11 and 12 – belonging to the facies belt 6 (winnowed platform edge sands). The facies belt 3 association is characterized by carbonate detritus and pelagic materials, whereas facies belt 6 association is characterized by calcarenite. The faunal elements recovered from the Muthal section include reworked shallow-marine bioclasts, some autochthonous benthic and planktonic fauna, and reworked Neoproterozoic microflora.

Microflora

A diverse assemblage of microflora, including some species of Neoproterozoic age, was recovered from the Sirban Limestone in the measured section at Bidda. This includes seven genera, which are *Leiosphaeridia* (Eisenack 1958); *Melanocyrillium* (Bloeser 1979 ex Bloeser 1985); *Vandalosphaeridium* (Vidal 1981); *Trachysphaeridium* (Timofeev 1959 ex Timofeev 1966); *Gorgonisphaeridium* (Staplin *et al.* 1965); *Arctacellularia* (German in Timofeev *et al.* 1976) and *Pterospermopsimorpha* (Timofeev 1966) (Fig. 5). Three species of the genus *Leiosphaeridia* were identified in the assemblage, which are *L. crassa* (Naumova 1949); *L. minutissima* (Naumova 1949); *L. kulgunica* (Jankauskas 1989). Other species recovered from the Sirban Limestone include *Melanocyrillium* sp., *T. laminaritum* (Timofeev 1966); *V. reticulatum* (Vidal 1981); *G. pindyium* (Zang 1995) and *P. insolita* (Timofeev 1969) (Fig. 5a–f). In addition, microstromatolites and cyanobacteria [*Rivularia haematites* (Lamarck & De Candolle 1806)] (Figs 6 & 7), *Chuaria circularia*, minor thread-like filaments with and without branching, were observed in thin sections of these carbonates. The systematic description of the genera and species is in progress and will be published separately. The reworked Neoproterozoic microfloral assemblage recovered from the samples of Eocene age from Muthal section include the same three species of the genus

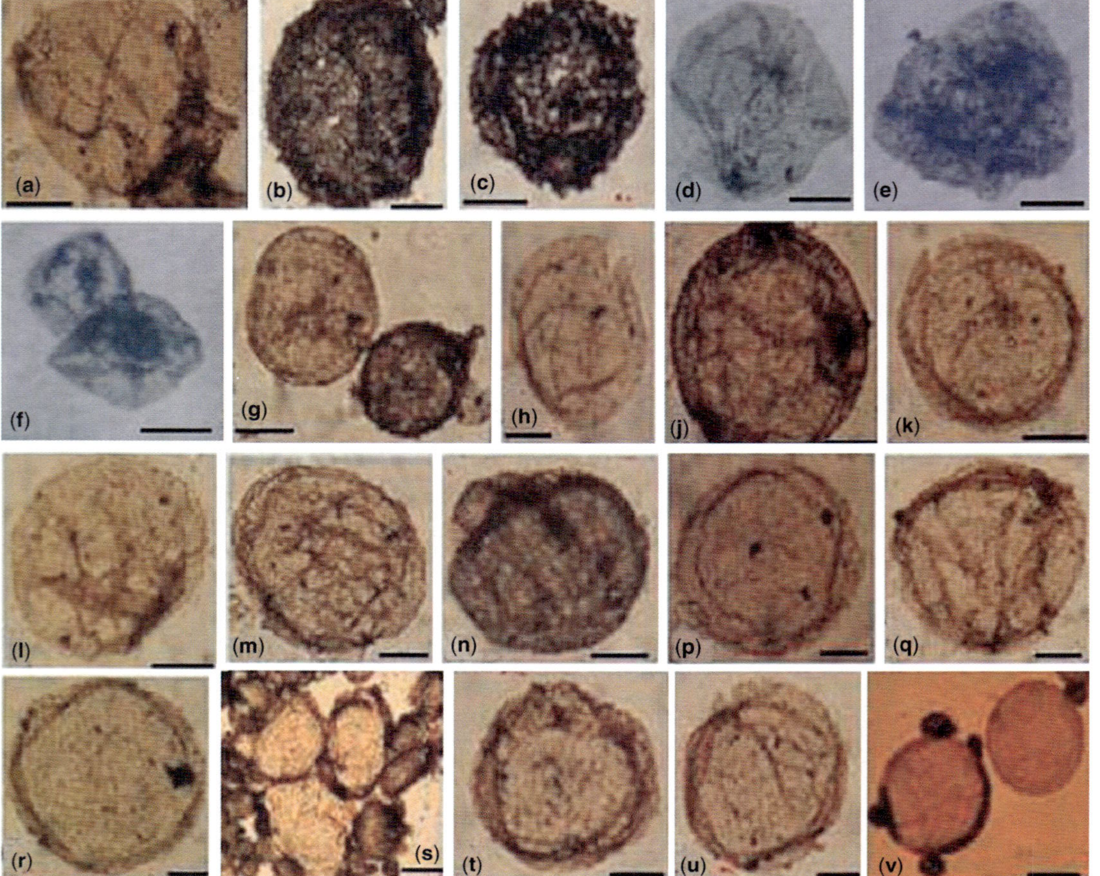

Fig. 5. Microflora recovered from Sirban Limestone (**a–f**) and from Subathu limestone (**g–v**). (a) *Leiosphaeridia crassa* (Naumova 1949); (b) *Trachysphaeridium laminaritum* (Timofeev 1966); (c) *Vandalosphaeridium reticulatum* (Vidal 1981); (d) *Leiosphaeridia Minutissima* (Naumova 1949 *emend.* Yankauskas in Yankauskas *et al.* 1989); (e) *Leiosphaeridia Kulgunica* (Yankauskas 1989); (f) *Pterospermopsimorpha* (Timofeev 1966, *emend*. Mikhaylova and Yankauskas in Yankauskas *et al.* 1989); (g) *Kildinella* sp. (Timofeev 1966); (h, k) *Leiosphaeridia anquiensis* (Yin 1991); (j) *Maculosphaera* sp.; (l, m) *Leiosphaeridia minutissima* (Naumova 1949); (n, p) *Leiosphaeridia crassa* (Pykhova 1973, *cf.* Fensome *et al.* 1990); (q) *Leiosphaeridia tenuissima* (Eisenack 1958); (r, t, u) *Navifusa majensis* (Pyatiletov 1980); (s) *Chuaria circularis* (Walcott 1899); (v) *Leptoteichos golubici* sp. Scale bar is 10 μm.

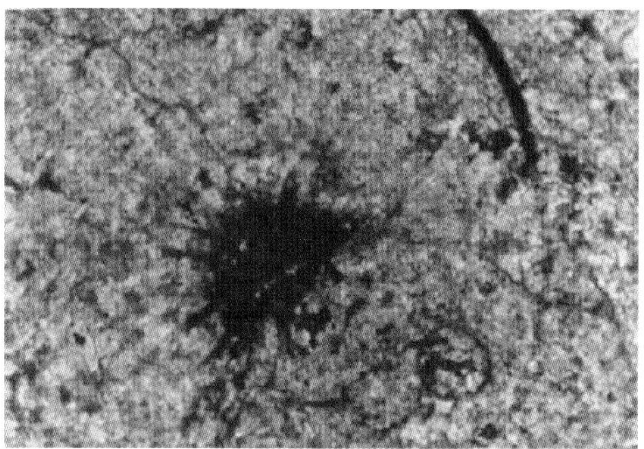

Fig. 6. *Rivularia haematites* (Lamarck & de Candolle 1806).

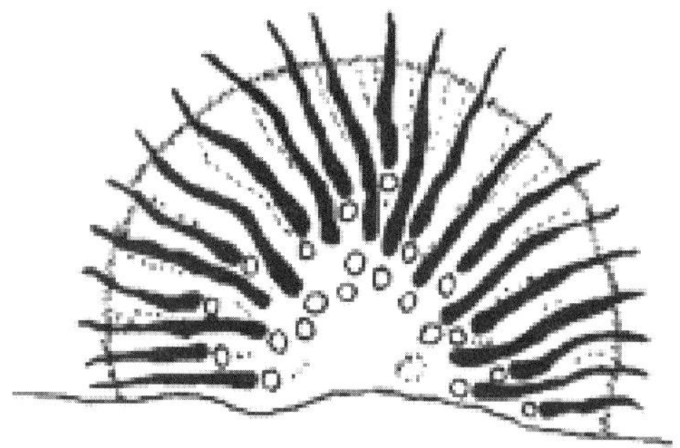

Fig. 7. Sketch of Figure 6 (not to scale).

Leiosphaeridia and the two genera *Pterospermopsimorpha* (Timofeev 1966) and *Trachysphaeridium* (Timofeev 1959) reported from the Sirban Limestone samples at Bidda (Fig. 5g–v).

In addition, the reworked Proterozoic assemblage of microflora recovered from the Muthal beds included 14 species belonging to nine genera, namely *Navifusa* (Combaz *et al.* 1967); *Chuaria* (Walcott 1899; Vidal & Ford 1985); *Tappania* (Yin 1997); *Kildinosphaera* (Vidal & Knoll 1983); *Myxococcoides* (Schopf 1968); *Megalytrum*; *Maculosphaera*; *Leptoteichos* and *Glenobotrydion* (Fig. 5f–v). Out of the 14 species *Leiosphaeridia anquiensis* (Yin 1991) and *L. tenuissima* (Eisenack 1958) belong to the genus *Leiosphaeridia*. Other species include *Leptoteichos golubici* sp.; *Navifusa majensis* (Pyatiletov 1980); *Chuaria circularis* (Walcott 1899); *Tappania tubata* (Yin 1997); *Kildinella* sp. (Timofeev 1966); *Maculosphaera* sp.; *Trachysphaeridium* sp.; *Megalytrum diacenum*; *Glenobotrydion aenigmatis* (Schopf 1968); *G. majorinum* (Schopf & Blacic 1971); *Pterospermopsimorpha* sp.; *Myxococcoides minor* (Schopf 1968) (Fig. 5g–v). The reworked species show comparatively lower colour index and better preservation than the *in situ* assemblage recovered from the Sirban Limestone. Earlier, Venkatachala & Kumar (1996, 1998) also recovered *Leiosphaeridia* and *Chuaria circularia* from shale beds of the Sirban Limestone in the Riasi–Katra area and from near Muthal. They also recovered *Myxococcoides* from the samples of chert beds of the Sirban Limestone near Bidda, and *Kildinosphaera* sp. from the shale beds of Sirban Limestone near Muthal.

Other microflora reported by Venkatachala & Kumar (1996, 1997, 1998) from the chert and shale beds of the Sirban Limestone at Bidda and Muthal localities include coccoid, chroococcacean cyanobacteria including *Eoentophysalis belcherensis* and *Palaeopleurocapsa wopfnerii*, *Gloeodiniopsis lamillosa*, *Eogloeocapsa avzyanica*, *Sphaerophycus cf.*, *S. medium*, *Eosynechococcus moorei*, *E. grandis*, *Siphonophycus kestron*, *S. robustum*, *Oscillatoriopsis* sp., *Circumvaginalis elongatus*, *Archaeollipsoides major* and *A. minor*. They also recovered an assemblage of microflora from Proterozoic shale beds near the Muthal section, including *S. kestron*, *S. robustum*, *Oscillatoriopsis* sp., *Calyptothrix* spp., *Eosynechococcus medius*, *Archaeotrichion contortum*, *Archaeollipsoides bactriformis*, *A. grandis*, *A. obesus*, *Arctacellularia ellipsoidea*, *Nostochomorpha prisca*, *Palenostocalia irregularia*, *Lophosphaeridum* spp., *Micrhystridium* spp., *Sphaerocongregus* sp., *Obruchevella* sp. and *Spiromorhpa* indet. However, they did not support their microfloral identification with photographic evidence. Many of the recovered taxa from the Sirban Limestone and the Muthal beds are common in the Bitter Springs Formation of Australia, which is approximately 830–800 Ma in age. This suggests that Neoproterozoic and younger units of the Sirban Limestone may have been the source rock for the Eocene Subathu Formation.

Source potential and porosity

The average amount of organic matter present in samples from seven stratigraphic levels in the Sirban Limestone is approximately 10% (thin section estimates). Most of the organic matter consists of cyanobacteria dominated by an alternation of densely calcified and loosely calcified zones. Thick micritic envelopes around the allochems may have been formed by growth on the particles by organic films. Algae range from 10 to 60% in the analysed samples. Sparite and micrite range from 10 to 40% and from 10 to 80%, respectively. Fe-calcite occurs in three samples and ranges from 10 to 20% of bulk volume, whereas intraclasts occur in one sample comprising 70% of the bulk volume. Pellets also occur in one sample, where they account for 30% of the volume. Quartz occurs in two samples with 10 and 20% of the bulk volume (Fig. 3). Vug porosity is preserved in certain sections. The pelletal limestone shows intergranular porosity, whereas the intraclastic facies possess both inter- and intragranular porosity. The cortoid facies exhibits fracture porosity as well as intergranular porosity and bored micritic rims. Stylolites are conical–columnar projections with intervening depressions. They cut across the grains and have produced new grain fabric in the rocks. Micrite contains both biogenic and abiogenic detritus. Some thin sections show inter- and intralayer porosity, particularly those sections characterized by alternate layers of intraclasts and algal mats. Some dolobiomicrites are dominated by organic matter. Microbial structures and oncoids with specks of dolosparite have been observed. This microfacies is highly porous. Dolomitization has increased matrix permeability in zones of leaching. The cortoid moulds have been connected to the surrounding matrix pore system through dolomitization.

Discussion

The Sirban Limestone beds of the measured section show a veritable explosion in the evolution of life forms, similar to the forms reported elsewhere from the Proterozoic sequences around the world (e.g. Fedorov 1994). These life forms are dominated by cyanobacteria and microbial stromatolites, which provide favourable conditions for accumulation of organic matter in Proterozoic rocks. The microfloral assemblage within the Sirban Limestone is similar to assemblages recovered from the DMH-A well near Damoh in the Proterozoic Vindhyan Supergroup (Prasad *et al.* 2005) and from the subsurface Vindhyan sequence encountered in Jabera Well-1 Madhya Pradesh (Shanmukhappa *et al.* 1996). Among other forms, Prasad *et al.* (2005) reported six genera, that is *Navifusa*, *Pterospermopsimorpha*, *Tappania*, *Vandalosphaeridium*, *Kildinosphaera* and *Leiosphaeridia*, which also occur in the Sirban Limestone assemblage. Shanmukhappa *et al.* (1996) reported 24 species of sphaeromorphic acritarchs from subsurface samples of the Proterozoic Charkaria Shale Member of the Kheinjua Formation and the Rohtas Formation of the Lower Vindhyan sequence (Jabera Well-1). Their genera, including *Leiosphaeridia* and *Trachysphaeridium*, are similar to the genera recovered in this study.

The occurrence of the reworked microfloral assemblage of Neoproterozoic and younger age within the limestone beds of the Eocene Subathu Formation suggests either upwards migration of hydrocarbons from deeper sediments or the existence of a positive terrain comprising the Proterozoic Sirban Limestone and possibly younger Neoproterozoic formations as the provenance of this reworked microflora during Eocene deposition. Evidence for upwards migration of hydrocarbons includes the presence of allochems enveloped by thin dark films in samples from both the stratigraphic sections (Bidda and Muthal). The possibility is further supported by the observation by Krayushkin *et al.* (2001), who recovered spore-pollen and other microflora of Devonian and Proterozoic ages from oil samples in shallow Permian and Carboniferous sandstone reservoirs in the Dnieper–Donets Basin in Russia. They recovered Proterozoic microflora from the oil samples including *Protoleiospheridium conglutinatum*, *Zonoleiospheridium larum*, *Leiominuscula rugosa*, *Margominuscula rugosa*, *Protoarchaeosacculina stava*, *Leiopsophosphaera giganteus*, *Asperatopsophosphaera magna*, *Strictosphaeridium implexum*, *Gloecapsomorpha hebeja*, *Turuchanica alara* and *Pulvinomorpha angulata*. However, keeping in view the diversity and varying Proterozoic ages of the reworked assemblage of the microflora recovered from the Muthal beds, it is more likely that Proterozoic Sirban Limestone and possibly the younger formations formed positive terrain and sourced the Proterozoic microflora to the Eocene Subathu Formation.

The close association of lamina-scale source rocks (microbial laminites, cyanobacteria, etc.) with stromatolitic dolomite in the Sirban Limestone is similar to the situation in the giant oil and gas fields in the Proterozoic carbonates in the Baikit Anticline of the Siberian Platform (e.g. Postnikov & Postnikova 2006). The Baikit Anticline reservoir rocks are composed of fractured and karstified stromatolitic dolomite similar to the Sirban Limestone. The Sirban Limestone could have similar source potential to the Siberian Platform sequence. The rich and altered organic matter (microbial sapropel) devoid of microfossils and mostly containing kerogenous matter encountered from the Sirban Limestone suggest maturation with dominantly gas-prone capability and generation of liquid hydrocarbon (e.g. Shanmukhappa *et al.* 1996). The Sirban Limestone possesses a system of fractures, vugs and caverns similar to the stromatolitic dolomite hydrocarbon reservoirs in the Baikit Anticline. The rock matrix itself is of moderate porosity and permeability. There are potential seal horizons at different stratigraphic levels in the area, and ample evidence of both potential source rocks and reservoirs with effective porosity for the entrapment of hydrocarbons in the Sirban Limestone succession. Further investigation is in progress.

Conclusions

- The Sirban Limestone of the Jammu area yields good evidence for its potential as a source rock for hydrocarbons.

- The occurrence of the microfloral assemblage in the Sirban Limestone and recovery of some microflora of this assemblage from the Eocene Subathu Formation suggests either upwards migration of hydrocarbons or a pre-existing positive terrain comprising the Proterozoic Sirban Limestone and possibly younger Neoproterozoic formations as the provenance of this microflora during the Eocene.
- The close association of lamina-scale source rocks with both good-quality potential reservoir and seal horizons indicate that the basic physical elements of a petroleum system are present in the Sirban Limestone.

We are grateful to Drs S. Sarkar and M. Sharma, Birbal Sahni Institute of Palaeobotany, Lucknow India and Dr B. Thusu of University College London for their help in identification of the microflora and discussions. Critical review and constructive suggestions by Dr K. Grey, Australia and Dr J. L. Etienne, UK are gratefully acknowledged.

References

BLOESER, B. 1979. Melanocyrillium – new acritarch genus from Kwagunt Formation (Late Precambrian) Chuar Group, Grand Canyon Supergroup, Arizona (Abstract). *AAPG Bulletin*, **63**, 420–421.

BLOESER, B. 1985. Melanocyrillium, a new genus of structurally complex late Proterozoic microfossils from the Kwagunt Formation (Chuar Group) Grand Canyon, Arizona. *Journal of Palaeontology*, **59**, 741–765.

COMBAZ, A., LANGE, F. W. & PANSART, J. 1967. Les Leiofusidae, Eisenack. 1938. *Review of Palaeobotany and Palynology*, **1**, 291–307.

EISENACK, A. 1958. Tasmanites Newton 1875 und Leiosphaeridla n.g als Gattungen der Hystrichosphaeridia. *Palaeontographica*, **110**, 1–19.

FEDOROV, D. L. 1994. The petroleum potential of ancient strata of the Russian platform. *Geologiya nefti i gaza*, **2**, 8–12.

FENSOME, R. A., WILLIAM, G. L., BARSS, M. S., FREEMAN, J. M. & HILL, J. M. 1990. *Acritarchs and Fossil Prasinophytes. An Index to Genera Species and Infraspecific Taxa.* American Association of Stratigraphic Palynologists Foundation, Contributions Series, **25**, 771.

JANKAUSKAS, T. V. 1989. *Mikrofossilii Dokembriya USSR [Precambrian Microfossils of the USSR]*. Trudy Institut Geologii ee Geokhronologii SSSR, Akademii Nauk, Leningrad.

KUMAR, S. 1984. Present status of Stromatolite biostratigraphy: a review. *Geophytology*, **14**, 96–110.

KRAYUSHKIN, V. A., TCHEBANENKO, T. I., KLOCHKO, V. P., DVORYANIN, YE. S. & KENNEY, J. F. 2001. The exploration and development of the twelve major and one giant oil and gas fields on the Northern Flank of the Dnieper–Donetsk Basin. *Energia*, **22**(3), 44–47.

LAMARCK, J. B. & DE CANDOLLE, A. P. 1806. *Synopsis Plantarum in Flora Gallica Descriptarum.* Apud H. Agasse, Paris.

NAUMOVA, S. N. 1949. *Obschchaya morfologicheskaya kharak teristika spor nizhnego kembriya.* Akademiya Nauk USSR Lzvestiya, Seriya Geologicheskaya, 49–56.

PYATILETOV, V. G. 1980. O nakhodkakh mikrofossiliy roda Navifusa v Lakhandinskoy svite. *Paleontologicheskii zhurnal*, **1980**, 143–145.

PYKHOVA, N. G. 1973. Dokembriskie akritarhi Moskovskogo graben I Yuzhnogo. *Obshchestva ispitateli prirody otdel geologicheskii novaya*, **48**, 91–107.

PRASAD, B., UNIYAL, S. N. & RAMSON, A. 2005. Organic-walled microfossils from the Proterozoic Vindhyan Supergroup of Son Valley, Madhya Pradesh, India. *Palaeobotanist*, **54**, 13–60.

POSTNIKOV, A. V. & POSTNIKOVA, O. V. 2006. Vendian–Riphean deposits as main objective hydrocarbon exploration on the Siberian Platform (Abstract). *In*: *Global Infracambrian Hydrocarbon Systems and the Emerging Potential in North Africa.* Geological Society, London, 39–40.

RAHA, P. K. 1984. *Stratigraphy of the Jammu Limestone (Great Limestone), Udhampur District, Jammu and Kashmir State, With Special Reference to Its Stromatolite Content and Age.* Geological Survey of India, Palaeontologia indica, **XLVII**, 1–103.

SCHOPF, J. W. 1968. Microflora of the Bitter Springs Formations, Late Precambrian Central Australia. *Journal of Palaeontology*, **42**, 651–688.

SCHOPF, J. W. & BLACIC, J. M. 1971. New microorganisms from the Bitter Springs Formation (Late Precambrian) of the north-central Amadeus Basin, Central Australia. *Journal of Palaeontology*, **45**, 925–961.

SHANMUKHAPPA, M., SINGH, R. N. & POOVENDAN, A. 1996. Precambrian (Proterozoic) acritarchs from subsurface Vindhyan sequence, Jabera Well-1, Madhya Pradesh. *In*: *Contributions to the XV Indian Colloquium Micropalaeontology and Stratratigraphy, Dehradun*, K. D. Malviya Institute of Petroleum Exploration, 541–549.

SHARMA, G. R. 2006. *Evaluation of Sedimentary and Palynofacies of the Palaeogene Sequence of the Jammu Region in the Context of India–Asia Collision Tectonics.* Unpublished PhD thesis, Jammu University, India.

STAPLIN, F. L., JANSONIUS, J. & POCOCK, S. A. 1965. Evaluation of some acritarchous hystrichosphere genera. *Neues Jahrbuch fur Geologie and Palaontologie*, **123**, 167–201.

TIMOFEEV, B. V. 1959. *Drevneishaya flora Prebaltiki l ee stratigraficheskoe znachenie [The Ancient Flora of Peribaltic and Its Stratigraphic Significance].* Vsesoyuznyi Neftyanoi Nauchno Issledovatelskii Geolgora Zvedochnyi Instit (VNIGRI), Leningrad, **129**, 1–136.

TIMOFEEV, B. V. 1966. *Mikropalaeofitologicheskoe issledovanie drevnikh svit.* Akademiya Nauk SSSR, Nauka, Moscow, 1–147.

TIMOFEEV, B. V. 1969. *Sferomorfidy proterozoya [Proterozoic Spheromorphs].* Nauka, Leningrad.

TIMOFEEV, B. V., GERMAN, T. N. & MIKHAYLOVA, N. S. 1976. *Mikrofitofossilii dokembriya, Kembriya i ordovika [Microfossils From the Precambrian, Cambrian and Ordovician].* Nauka, Leningrad.

VENKATACHALA, B. S. & KUMAR, A. 1996. Significant Microbiota from the Great Limestone of Jammu, Lesser Himalaya. *In*: *Contributions of the XV Indian Colloqulum on Micropalaeontology and Stratigraphy, Dehradun*, K. D. Malviya Institute of Petroleum Exploration, 551–557.

VENKATACHALA, B. S. & KUMAR, A. 1997. Cynobacterial mats from the Neoproterozoic Vaishnodevi Limestone, Jammu and Kashmir. *Current Science*, **73**, 83–87.

VENKATACHALA, B. S. & KUMAR, A. 1998. Fossil microbiota from the Vaishnodevi Limestone, Himalayan Foothills, Jammu: age and palaeoenvironmental implications. *Journal of the Geological Society of India*, **52**, 529–536.

VIDAL, G. 1976. Late Precambrian microfossils from the Visingso Beds in southern Sweden. *Fossils and Strata*, **9**, 57.

VIDAL, G. 1981. Micropalaeontology and biostratigraphy of the Upper Proterozoic and Lower Cambrian sequence in East Finnmark, northern Norway. *Norges Geologiske Undersokelse*, **362**, 1–53.

VIDAL, G. 1988. A palynological preparation method. *Palynology*, **12**, 215–220.

VIDAL, G. & KNOLL, A. H. 1983. Proterozoic plankton. *Memoirs of the Geological Society of America*, **161**, 265–277.

VIDAL, G. & FORD, T. D. 1985. Microbiotas from the Late Proterozoic Chuar Group (northern Arizona) and Uinta Mountain Group (Utah) and their Chronostratigraphic implications. *Precambrian Research*, **28**, 349–389.

WILSON, J. L. 1975. *Carbonate Facies in Geologic History*. Springer, Berlin.

WALCOTT, C. D. 1899. Precambrian fossiliferous formations. *Geological Society of America Bulletin*, **10**, 199–244.

YIN, L. 1991. Late Proterozoic microfossils from Tongjiazhung Formation, western Shandong, China. *Acta Micropalaeontol Sinica*, **8**, 253–269.

YIN, L. 1997. Acanthomorphic acritarchs from Meso-Neoproterozoic shales of the Ruyang Group in Shanxi, China. *Review of Palaeobotany and Palynology*, **98**, 15–25.

ZANG, W. 1995. Early Neoproterozoic sequence stratigraphy and acritarch biostratigraphy, eastern Officer Basin, South Australia. *Precambrian Research*, **74**, 119–175.

Late Cryogenian (Neoproterozoic) glacial and post-glacial successions at the southern margin of the Congo Craton, northern Namibia: facies, palaeogeography and hydrocarbon perspective

THILO BECHSTÄDT[1,2]*, HARTMUT JÄGER[1,2], GUY SPENCE[1] & GEORG WERNER[1]

[1]*Institute of Geosciences, University of Heidelberg, Heidelberg, Germany*

[2]*Steinbeis-Transfer Centre GeoResources, Heidelberg, Germany*

**Corresponding author (e-mail: bechstaedt@uni-hd.de)*

Abstract: The geological paradox of at least two Neoproterozoic glacial intervals at tropical latitudes intercalated within carbonates remains an unsolved puzzle. Several conceptual models have been proposed to explain these apparent rapid swings between climatic extremes and the associated isotopic changes in sea-water chemistry. In Oman, post-glacial transgressive sedimentary successions represent important hydrocarbon source rocks. Source rock characteristics of Neoproterozoic post-glacial successions in other parts of the world (even if not directly correlatable) are, therefore, of special economic interest.

This paper concentrates on the Ghaub Formation diamictite interval in northern Namibia and the major environmental change in the aftermath of the assumed glaciation. The relationship of the post-glacial sediments with the underlying different types of cap carbonate and diamictite successions is discussed, and a model of the succession of events is presented. The palaeotopography, caused mostly by ongoing tectonic activity including uplift on the scale of thousands of metres, strongly influenced the petroleum system created and played an important role for the hydrocarbon prospectivity of this post-glacial succession. Tectonic activity on the shelf of the southern margin of the Congo Craton was repeated, and different sub-basins were created before, during and after the Ghaub glaciation. The newly formed relief was flooded, and the different sub-basins were affected by restricted circulation for quite some time. This general scenario bears many similarities to the late Ordovician–early Silurian petroleum system, also formed during post-glacial sea-level rise.

The Neoproterozoic sedimentary basin of Oman is an important producer of hydrocarbons (e.g. Schröder *et al.* 2004) owing to the presence of prolific source rocks in the post-Fiq Formation succession (Allen & Leather 2006). Neoproterozoic glacial sediments are very prominent in Oman (e.g. Allen *et al.* 2004; Kilner *et al.* 2006; Rieu *et al.* 2006). It has been speculated (Craig *et al.* 2006) that here and elsewhere deposition of hydrocarbon source rocks was the result of strong, post-glacial sea-level rises. Thus, Neoproterozoic successions in many parts of the world have gained a distinct increase in interest and are ranked as hydrocarbon exploration targets with considerable reserve potential, although the Fiq Formation cannot be proven to equate with any other glaciation worldwide. In North Africa proven petroleum systems within Neoproterozoic and overlying Palaeozoic rocks occur in several different basins of Libya and Algeria that contain thick sedimentary fills. Most of the Palaeozoic-sourced hydrocarbons in North Africa are owing to the presence of Lower Silurian organic-rich ('hot') shales. According to Lüning *et al.* (1999, 2000*a*, *b*), these source rocks are related to the widespread Late Ordovician glaciation and were deposited in early Silurian time during marine transgression caused by the melting of the glaciers. The authors assume that, during early transgression, the flooded glacial palaeorelief caused emergent or shallow-water highs and intervening small basins with restricted circulation. In addition, high primary productivity conditions may have prevailed during flooding along the shelf edge because of strong nutrient supply. A small caveat might be that organic-rich graptolitic shales also occur in peri-Gondwanan areas apart from the formerly glaciated region of northern Africa. In the Palaeozoic post-glacial time interval, circulation apparently was sluggish and hampered in very large areas (Lüning *et al.* 1999, 2000*a*, *b*).

This Early Palaeozoic petroleum system is the result of a major environmental and sedimentological change, the passage from glaciation to post-glacial warming, sea-level rise and infill of the former glacial relief under anoxic conditions. In the Neoproterozoic, during the Cryogenian period, different but even more extended and severe glacial episodes occurred (Harland 1964), the glaciers

From: Craig, J., Thurow, J., Thusu, B., Whitham, A. & Abutarruma, Y. (eds) *Global Neoproterozoic Petroleum Systems: The Emerging Potential in North Africa.* Geological Society, London, Special Publications, **326**, 255–287.
DOI: 10.1144/SP326.15 0305-8719/09/$15.00

partly reaching low latitudes as indicated by palaeomagnetic studies (e.g. Embelton & Williams 1986; Sumner *et al.* 1987; Schmidt & Williams 1995; Sohl *et al.* 1999; Evans 2000; Macouin *et al.* 2004). The Neoproterozoic is also the period of likely transition from a methane-rich to a CO_2-dominated climate system (Pavlov *et al.* 2003; Catling & Claire 2005; Fairchild & Kennedy 2007). In the newly established CO_2-system, and in contrast to today, atmospheric CO_2 could not be buffered by variable dissolution of deep-sea carbonate ooze (Ridgwell & Kennedy 2004; Ridgwell *et al.* 2004) because of the lack of pelagic calcifiers. This change to an unbuffered CO_2-dominated climate system might, therefore, have enhanced climatic extremes. The different glacial intervals of disputed duration and synchroneity are intercalated between much longer lasting and, apparently, mostly warm-water settings.

According to the current view (Hurtgen *et al.* 2005; Halverson 2006) there are three groups of major glaciations, although age datings are often imprecise: 'Sturtian' (<740–647 Ma), 'Marinoan' (<660–635 Ma) and 'Gaskiers' (*c.* 580 Ma). The large time span of the 'Sturtian' and the partly overlapping age intervals to the 'Marinoan' clearly shows the uncertainties in age assignments as well as the possibility of different, local not global glaciations during the 'Sturtian'. The type areas for the 'Sturtian' and 'Marinoan' glaciations are located in South Australia. The type section for the 'Gaskiers' glaciation is in Avalonia in Newfoundland. The geographic distribution of alleged 'Marinoan' deposits is widest, including several low-palaeolatitude examples, and that of the 'Gaskiers' deposits is smallest. As the basal Ediacaran (top Cryogenian) GSSP (Global Stratigraphic Section and Point) is placed in Australia at the base of the Nuccaleena Formation–top of the Elatina Formation and the 'Marinoan' stratigraphic interval is much larger, the term 'Elatina glaciation' is preferred (MacGabhann *et al.* 2006).

In this paper we will deal with the extreme global change and concomitant petroleum potential brought by one of these glaciations, the late Cryogenian Ghaub glaciation of northern Namibia, a classical field location for research on this glaciation (e.g. Hoffman *et al.* 1998*a*, *b*; Hoffman & Schrag 2000, 2002), and compare this setting with other roughly coeval successions in the world. We will concentrate on the relief formed and on the proposed post-glacial sudden warming, possibly including strongly enforced tropical weathering in the immediate post-glacial intervals, stimulating both clay mineral formation and phosphorus delivery to the oceans, facilitating organic carbon burial (Fairchild & Kennedy 2007). For the different scenarios to explain low-latitude Neoproterozoic glaciations (e.g. 'Snowball Earth', 'Slushball Earth', 'Zipper Rift Earth', etc.), the reader is referred to the overviews and discussions by Hoffman & Schrag (2002), Allen (2006) and Fairchild & Kennedy (2007).

Geological overview

In northern Namibia continental rifting occurred in the Late Neoproterozoic at about 750 Ma. During rifting and thereafter, thick sedimentary successions of the Damara Group (Miller 1980, 1983) were deposited on the southern margin of the Congo Craton (Figs 1 & 2). These Late Proterozoic successions and their crystalline 'basement' are well exposed in a roughly east–west-trending belt, the Otavi Mountain Land (OML) being the easternmost outcrops of this belt (Figs 1 & 3). Further to the west, in the Kaokoveld area, the outcropping Proterozoic strikes in a coast-parallel direction (Fig. 1). Younger sediments cover the area further to the east and north of the Kaokoveld belt (Owambo Basin: Figs 1 & 4).

Damaran succession

The Neoproterozoic succession of the Otavi Mountain Land (OML) (Fig. 2) (see Kamona & Günzel 2007) is more than 5000 m thick. The mostly shallow-water platform carbonates of the Otavi Group follow unconformably, partly on the Palaeoproterozoic basement of the Grootfontein Inlier, but mostly on Cryogenian rift sediments of the Nosib Group (Miller 1983; Laukamp 2006). This continental rifting, dissecting a Palaeoproterozoic supercontinent (e.g. Condie 2002), was associated with distinct volcanism, and opened the Damara Ocean in the south and the Adamastor Ocean in the west (all directions given are those of today).

The Nosib Group comprises the siliciclastic Nabis Formation (main part) and the volcanic Askevold Formation (Fig. 2), which can be correlated with the Naauwpoort Formation in central Namibia, rendering an age of 780–740 Ma (Burger & Coertze 1973; Hoffman *et al.* 1996). Within the Nosib rift succession, diamictites, pyroclastics and local banded ironstones of the Chuos Formation (also called the Varianto Formation) occur. This formation is strongly variable in thickness, laterally discontinuous, and shows stratigraphic repetition of several diamictite intervals interbedded with carbonates and siliciclastics of the Nabis Formation (Laukamp 2006). The Nosib Group was deposited in a horst–graben system caused by late stages of the break-up of Rodinia (Hartnady *et al.* 1985; Frimmel *et al.* 1996*b*).

The overlying Otavi Group reaches up to a maximum thickness of about 5000 m (Beukes

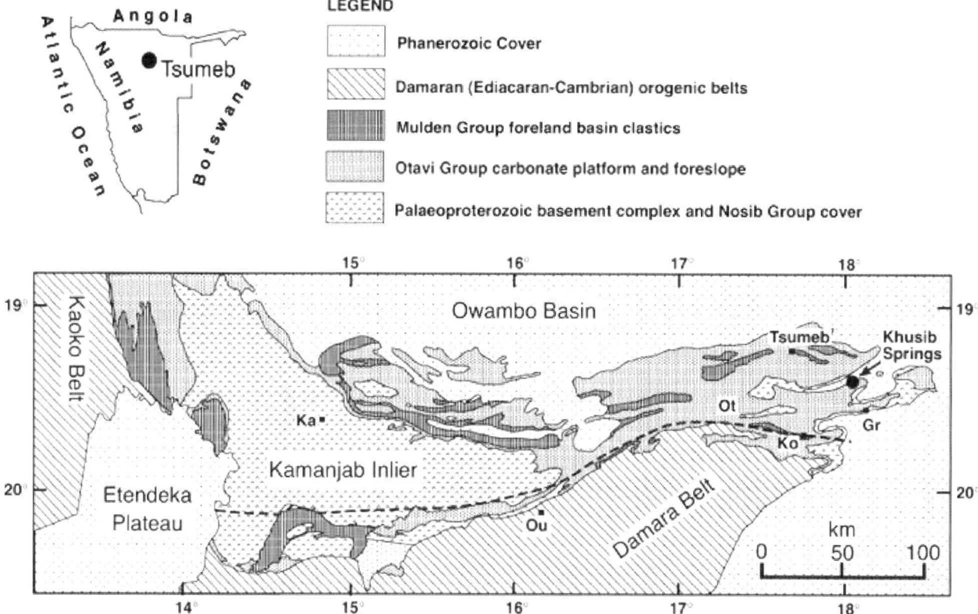

Fig. 1. Upper left: location and structural overview of the Otavi Mountainland in northern Namibia. Bottom: geological map of the area indicated by box in the overview map. The dashed line indicates platform foreslope–margin break. Place name abbreviations: Ts, Tsumeb; Gr, Grootfontein; Ko, Kombat; Ot, Otavi; Ou, Outjo, Ka, Kamanjab. Modified after Hofmann (http://www.snowballearth.org/slides/Ch1-16.gif).

1986), and represents a time span of approximately 200 Ma (Frimmel 2004; Hoffmann *et al.* 2004). The group is subdivided into the underlying Abenab Subgroup (Berg Aukas, Gauss and Auros formations) and the overlying Tsumeb Subgroup (Ghaub, Maieberg, Elandshoek and Hüttenberg formations) (Fig. 2) (see Kamona & Günzel 2007).

	Group	Subgroup	Formation
Damaran Supergroup	Karstification		
	Mulden		Kombat
			Tschudi
	Karstification		
	Otavi	Tsumeb	Hüttenberg
			Elandshoek
			Maieberg
		635 ma →	Ghaub
		Abenab	Auros
			Gauss
			Berg Aukas
	Nosib		Askevold
		742 ma →	Chuos
			Nabis
Basement Grootfontein Metamorphic Complex & Mafic Body			

Fig. 2. Stratigraphic succession of the Otavi Mountain Land. Stratigraphic position of diamictite intervals arrowed.

Cap carbonates of the basal Berg Aukas Formation unconformably overlie, with sharp contact, the diamictites of the Chuos Formation (Hoffmann *et al.* 2004). The cap carbonates consist of a finely laminated succession, followed by laminated microbial sediments. The overlying succession of the Gauss Formation is mainly of shallow-water platform origin (Frimmel *et al.* 1996*a*), and includes massive and bedded dolomites. The following Auros Formation consists of stromatolites and oolites, alternating with bedded limestones and shales.

The Tsumeb Subgroup starts with Ghaub diamictites, which are often missing, however. The Ghaub Formation has been frequently correlated with the late Cryogenian 'Marinoan' glaciation of Australia (Hoffmann & Prave 1996; Hoffman *et al.* 1998*a*, *b*; Hoffmann *et al.* 2004), although this correlation is far from robust. On top of the diamictites there is a cap carbonate succession, represented by limestones and dolomites at the base, and overlying often dark-coloured dolomitic shales. This succession belongs to the Maieberg Formation. In the central OML a shaly, partly dark-coloured and often pyrite-rich succession follows, containing intercalated debris flows. Also, the dolomites of the upper

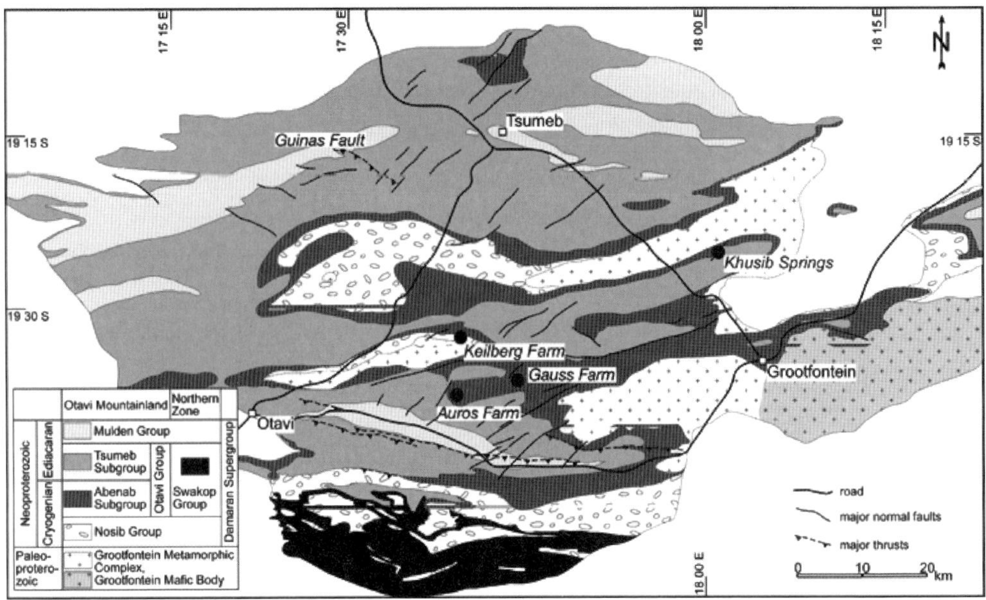

Fig. 3. Geological map of the Otavi Mountain Land (Laukamp 2006).

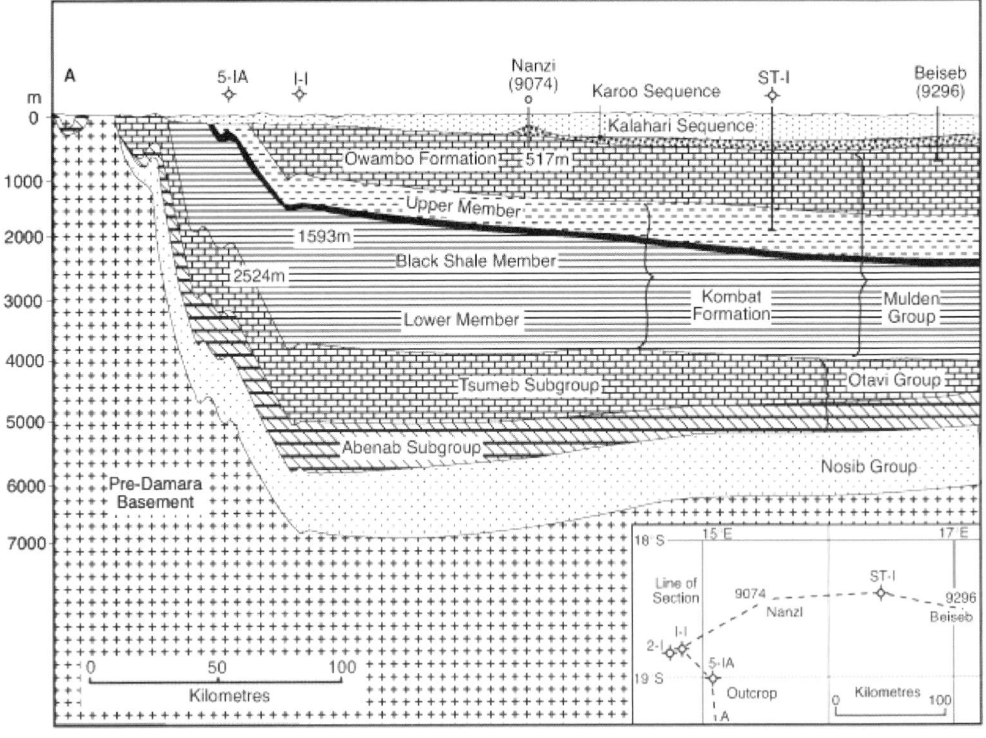

Fig. 4. Cross-section through the Owamboland Basin (after Miller 1997).

Maieberg Formation and the lower Elandshoek Formation are characterized by intense synsedimentary brecciation. Lagoonal peritidal deposits of the middle Hüttenberg Formation (Söhnge 1957) follow bedded dolomites of the upper Elandshoek and lower Hüttenberg formations, whereas the upper Hüttenberg Formation consists of stromatolitic and oolitic dolomites with intercalated cherts.

The late Neoproterozoic carbonates of the Otavi Group, rimming the southern margin of the Congo Craton (Prave 1996), were interpreted as a 'Bahama-type' relatively shallow-water platform (Hoffman *et al.* 1998*a*), although it is not clear whether this platform was attached to or detached from the Congo Craton. Deposition occurred at subtropical–tropical latitudes (Meert *et al.* 1995; Meert & Van Der Voo 1996). The Otavi Group succession is traditionally interpreted as indicative of a passive margin (Dürr & Dingeldey 1997; Hoffman *et al.* 1998*a*, *b*), although an active foreland basin setting has also been proposed (Prave 1996, 1997). Prave's assumption is corroborated by large-scale growth faults, synsedimentary breccias and local turbiditic sheddings in different areas, especially within the Tsumeb Subgroup (Frets 1969; Guj 1974; Schermerhorn 1974; Hedberg 1976, 1979; Downing 1983; Miller 1983; Porada & Wittig 1983*a*, *b*; Martin *et al.* 1985; Eyles & Januszczak 2007). These indicate ongoing tensional tectonics and part reactivation of pre-Damaran basement structures, as well as the evolution of local small-scaled basins, in the area of the Grootfontein basement high (Dürr & Dingeldey 1997) as well as in other areas.

The Otavi carbonate succession is overlain by the Mulden Group clastics, interpreted as foreland basin deposits (e.g. Dürr & Dingeldey 1997). The Mulden Group occurs predominantly in NE- to east-trending synclines in the northern OML, and consists of conglomerates, quartz sandstones, arkoses and shales. The angular unconformity between the Otavi Group and the Tschudi Formation often correlates with karst depressions, and is filled by the Tschudi conglomerates (Misiewicz 1988). At Tsumeb and Combat, sandstones were shed into large karstic pipes (e.g. Frimmel *et al.* 1996*a*). Pan-African thrusting of the carbonates over the siliciclastic rocks (Laukamp 2006) often occurs owing to the differences in competence. The thermal overprint (*c.* 550–459 Ma) of the Damaran Orogeny is said to reach from greenschist to lower greenschist facies, dependent on the location in the OML (see later in this chapter).

Post-Damaran successions

To the north, east and south of the OML, sediments of Palaeozoic–Cenozoic age follow, belonging to the Karoo and Kalahari successions. These are particularly thick in the Owambo Basin to the north (Ypma 1979*a*, *b*; Miller 1983). Locally, the mainly siliciclastic Kalahari succession occurs in deeply eroded, east-trending Pan-African anticlines (King 1951; GKW Consult & Bicon Namibia 2002). In the northern OML, single calc-alkaline lamprophyres (e.g. kersantites at the Tsumeb Pipe) and dyke swarms occur, being presumably of Cretaceous age (Söhnge 1957). Basalts of the Etendeka Group could have extended to the southern OML (Marsh *et al.* 2003), and are related to the breakup of Gondwana and the formation of the South Atlantic Ocean in the Early Cretaceous. Cenozoic and Quaternary deposits were accumulated in post-Damaran karst structures (Pickford 1993, 2000; Boni *et al.* 2007), and as aeolian Kalahari sands and calcrete cover.

Late Cryogenian successions in the central OML

Sedimentological investigations dealing with the Otavi Group of the OML comprise published and unpublished material (e.g. Stahl 1940; Söhnge 1957; Krüger 1969; Schermerhorn 1974; Hedberg 1979; Miller 1983; Martin *et al.* 1985; Beukes 1986; Hoffmann & Prave 1996; Hoffman *et al.* 1998*a*, *b*; Kamona & Günzel 2007). Many data are to be found in MSc and PhD theses carried out at South African universities (e.g. Smit 1959; Botha 1960; Grobler 1961; Misiewicz 1988; King 1990). Our ongoing research has concentrated on the area of the Keilberg Farm and the adjacent Auros Farm to the south (Fig. 3). The northern Keilberg area has been partly mapped by us (Fig. 5), and Werner (2005) investigated the Ghaub diamictite in detail. Farm names correspond to the Ministry of Mines and Energy, Namibian Geological Survey, Preliminary Open File, Geological Map of the Otavi Mountain Land, 1998.

Preglacial succession (Auros Formation)

The Auros Formation, 130–350 m thick on the Keilberg Farm, consists of dolomite, limestone and shaly dolomitic intervals, the latter being distinctly thicker in the west where the overall thickness also increases. Eastwards, the shales interfinger with carbonate intervals consisting of microcrystalline mudstone, partly laminated and often stromatolitic. Cross-bedded oolite intervals, with ooids distinctly smaller than in the underlying Gauss Formation, are frequent in the lower parts as well as at the top of the formation directly underneath the Ghaub diamictites. This interval, which weathers out distinctly, dies out eastwards. In eastern areas, the upper parts of the Auros Formation consist of

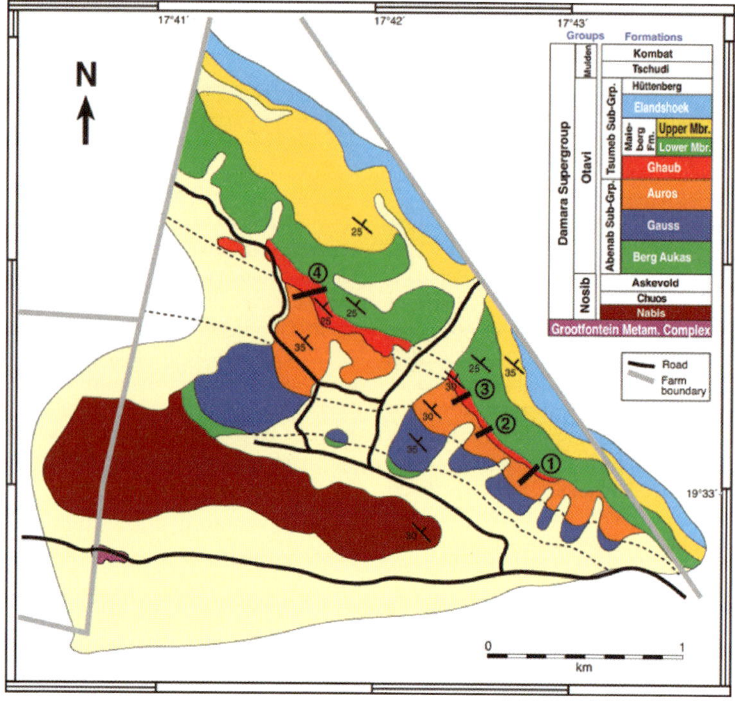

Fig. 5. Detailed geological map of the Keilberg Farm (after Werner 2005): the positions of the four sections through the Ghaub diamictites are marked.

several breccia horizons with rounded to well-rounded coarse clasts originating from the underlying Auros and also Gauss facies (e.g. containing oversized ooids). These beds are interpreted as debris flows. On maps of the mining companies this interval was assigned to the Auros Formation, probably because of micritic intercalations in-between debris flow sheddings, but it is difficult to draw a boundary to the Ghaub diamictites here.

At the southern Keilberg, Auros and Gauss farms, a few kilometres to the south of the diamictite outcrops, columnar stromatolites occur, several metres in size, in the upper part of the Auros succession above a shaly interval a few tens of metres thick (Fig. 6a, b). No oolites have been observed here.

Glacial succession (Ghaub Formation)

Ghaub diamictites often reach considerable thickness (>100 m), but are restricted to parts of the OML, especially western areas; being mostly absent in the east. According to Misiewicz (1988), the best outcrops of the Ghaub Formation are on the Keilberg and the neighbouring Maieberg farms. On the Keilberg Farm, the formation is well exposed in the northwestern and northern part, whereas in the east outcrops are moderately well exposed; no diamictites occur in southern parts of this farm, despite the presence of the same stratigraphic interval. The maximum thickness of the Ghaub Formation in the OML is 200 m according to Beukes (1986), and 1000 m in the Otavi area according to Söhnge (1957); in the Keilberg area we measured a maximum thickness of 110 m. Detailed descriptions of the formation are scarce. Le Roex (1941) was the first to detect diamictites between Otavi and Tsumeb. He described faceted and striated clasts, and interpreted the succession as glacial. This author also realized that these diamictites (now called the Ghaub Formation) are missing in the eastern OML. Söhnge (1957) gave a more detailed (unpublished) account, describing different types of carbonate clasts in the diamictites, and more rare sandstone, granite and gneiss components, which are all embedded in a blue-grey–brownish weathering matrix containing angular quartz clasts. Martin *et al.* (1985) described the Ghaub succession as unbedded, polymict mixtite with up to metre-sized clasts: the authors mentioned specifically, in contrast to Le Roex (1941) and Söhnge (1957), the lack of faceted and striated clasts, and the very well-rounded Nosib and basement clasts, in contrast to the carbonate components. Also, from Ghaub diamictites in NE

Fig. 6. Stromatolites and tubular structures exposed at the Auros Farm. (**a**) Densely packed columnar stromatolites in the upper part of the preglacial Auros Formation. Individual stromatolites exhibit internal synoptic surface relief with radial spatial symmetry in cross-sections horizontal to the vertical accretion direction. The synoptic depressions form 'tubes' infilled with quartz spar and chert. The pencil is 13 cm. (**b**) Detail of stromatolite surface relief shown in (A). The coin is 2.2 cm. (**c**) Quartz-spar-filled vertical tubes in the Maieberg cap carbonate dolostone exposed at the Auros Farm. The pencil is 13 cm. (**d**) Cross-sections of quartz-spar-filled vertical tubes in the Maieberg cap carbonate dolostone seen on a lamination-parallel surface. The subround cross-sections of the tubes protrude approximately 1 cm above the surface, reflecting that the quartz spar infill is more resistant to weathering than the cap carbonate rock. These quartz 'studs' are referred to in Namibia as the 'quartz clusters'. The pencil is 13 cm. (**e**) Quartz-spar-filled tubes and horizontal sheet cracks seen in vertical section in the microbial laminated Maieberg cap carbonate dolostone. Vertical and subvertical tubes and sheet cracks often link to one another. Tubes have no common horizon from which they originate, but individual tubes often begin and end at horizontal sheet cracks. Tubes can have irregular diameters along their length. Tubes sometimes coalesce or branch upwards. The coin is 2.2 cm. (**f**) Detail of coalescing tubes in (e). The coin is 2.2 cm.

Namibia, a lack of faceted and striated clasts is reported (Eyles & Januszczak 2007 and references therein). No glacially grooved and striated pavement has been found.

Interpretations of depositional environment differ stongly. In favour of a glacial origin were Le Roex (1941), Söhnge (1957) and Kröner & Rankama (1972). More recent publications, arguing for a glacial origin, are Kaufman *et al.* (1991, 1997), Hoffmann & Prave (1996) and Kennedy *et al.* (1998). Hoffmann (1990) suggested a continental to shallow-marine, glacial origin, while Hoffman *et al.* (1998*a*, *b*) and Hoffman & Schrag (1999) interpreted the Ghaub Formation as deposits of a grounded ice sheet above a minor topography. However, no evidence for a basal striated erosion surface was found, striated clasts are rare and water-laid sedimentation has been observed locally. Consequently, Eyles & Januszczak (2004, 2007) interpret most of the late Cryogenian ('Marinoan') successions from further to the west (Fransfontein Ridge) as debris flows in rift basins. This is in line with earlier interpretations: Schermerhorn (1974), Martin (1965), Martin *et al.* (1985), Miller (1983) and Porada & Wittig (1983*a*, *b*) interpreted the Ghaub Formation as mass flows deposited in deeper water.

Lateral variability of Ghaub diamictite

On the northern Keilberg Farm the Ghaub diamictite shows a wide variety of facies and is almost entirely dolomitic. In lower parts it appears to be iron free. In its upper part the Ghaub diamictite is relatively iron-rich (although no banded iron is present) with a characteristic red weathering colour. The matrix consists of dolomitic pseudosparite, probably original micrite with small millimetre-sized lithoclasts, embedding the centimetre- and larger-sized components. The amount and type of small components is important because they are a further characteristic of different subunits, for example quartz clasts or detritic mica. The Ghaub diamictite exhibits marked lateral facies changes in terms of composition and fabric, in some areas showing interfingering with clear debris flows.

Diamictite clasts

The spectrum of the clasts varies laterally and vertically in terms of size, type and rounding (see the description of the measured sections later). As mentioned earlier, the matrix contains a large amount of very small clasts, but clasts of up to metre size have also been found: the clasts larger than 20 cm belong entirely to the carbonate succession (Abenab Subgroup), and are mainly restricted to the central and western areas. With a few exceptions, the clasts within the Ghaub Formation are not sorted and are distributed irregularly. Carbonate clasts are mostly angular–subangular, in contrast to sandstone and basement clasts. This difference can be explained by the reworking of already well-rounded components from the Nosib Group that does not contain carbonate components, but, rather, sandstones, quartzites and clasts from metamorphic or plutonic basement. Basement clasts might result from direct erosion and reworking of the local basement, but the comparable degree of rounding of sandstone and of the (much less common) basement clasts argues that at least a large percentage of the latter represent reworked Nosib Group.

Overall, we estimate that about 85% of the components are carbonates from the Abenab Subgroup, 10% are clastic sediments, mainly sandstones from the Nosib Group, and 5% are basement clasts. No exotic components have been observed at all, so a larger contribution of clasts via dropstones originating from icebergs floating over large distances does not exist.

Some sandstone clasts show faceted surfaces, but only one striated clast, a quartzite, was found. The preservation of facets and striations is apparently highly lithology dependent; these were not observed in the much more common carbonate clasts, but this may reflect that they have been removed by subsequent alteration of carbonate clasts.

Lateral and vertical changes of diamictite succession

Four sections of the Ghaub diamictite were measured in the Keilberg area (Werner 2005) (Fig. 5). Clast types strongly differ laterally and vertically (Fig. 7).

- Section 1: in the 90 m-thick dolomitic succession, clasts up to 30 cm in size are derived from the Abenab Subgroup (Fig. 8b), and mainly consist of peloidal and oolitic wackestones. In contrast to the other sections, no finely laminated interbeds occur, with the exception of a mudstone at the base of the diamictites. Some parts of the section are slightly calcitic.
- Section 2: the 43 m-thick Ghaub Formation discordantly overlies oolitic grainstones of the Auros Formation. The diamictite shows a dark matrix at the base with a large amount of clasts. After about 13 m, the matrix colour changes to light grey and the amount of clasts decreases. At 17 m, a finely bedded interval with current ripples has been found (Fig. 8c). Towards the top, the matrix colour is yellowish, and the amount of clasts is again very high. Only clasts belonging to the Abenab Subgroup have been found in this section.

In between Sections 2 and 3 a graded, clast-supported interval (Fig. 8d) has been found.

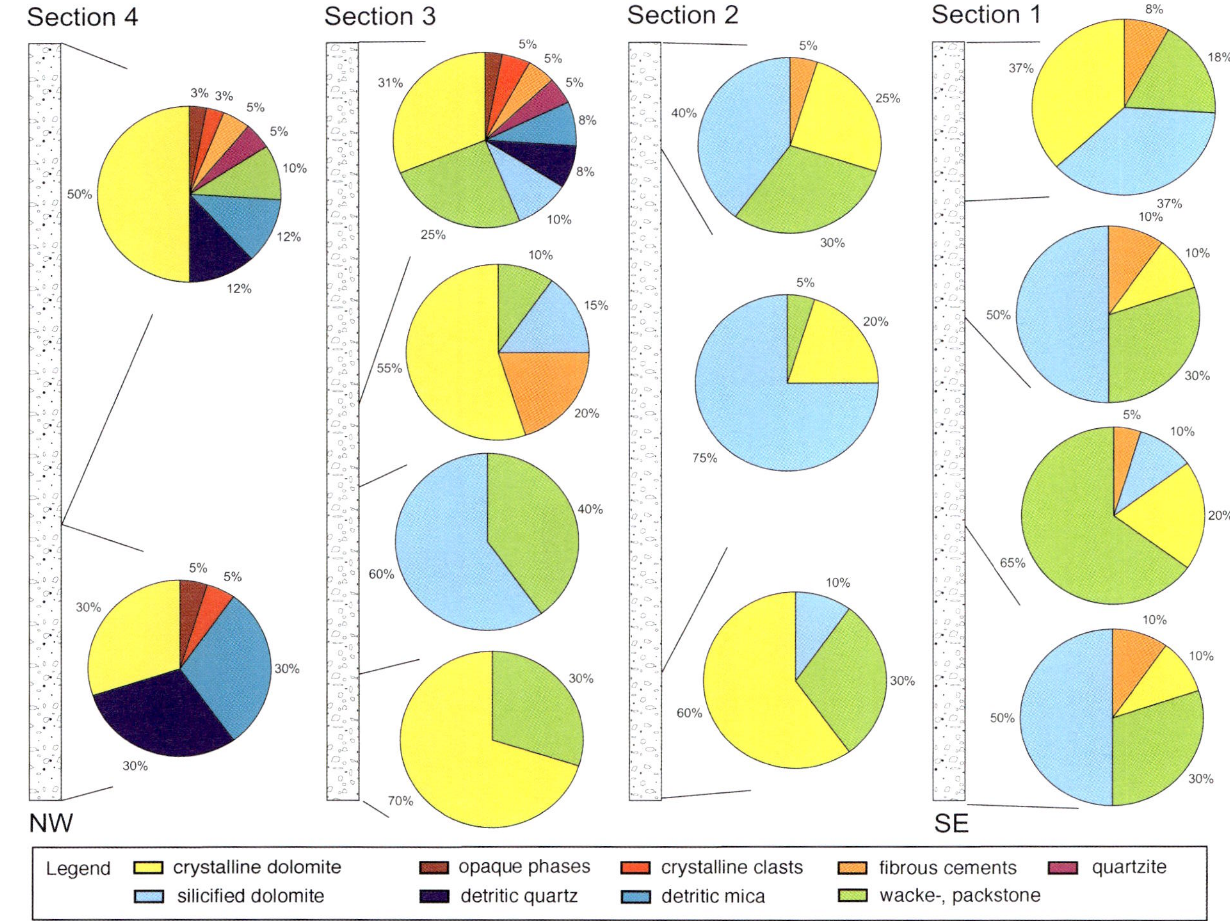

Fig. 7. Clast variations in the Ghaub diamictite exposed at the Keilberg Farm in Sections 1–4 (see Fig. 5), based on the analysis of thin sections.

Fig. 8. (**a**) Tubes with variable diameters along their length producing a pinch-and-swell appearance seen in vertical section in the Maieberg cap carbonate dolostone exposed at the Auros Farm. The coin is 2.5 cm. (**b**) Polished slab of Ghaub diamictite from the upper part of Section 1 (Fig. 5). Matrix-supported diamictite, containing clasts that are mainly locally derived limestone and carbonates. The matrix is mainly fine-grained dolomite. The pen scale-mark at the top of picture is 1 cm. (**c**) Fine laminated horizon, (30 cm thick) occurring within the Ghaub diamictite in Section 2 (Fig. 5). This bed shows current ripples in its upper part indicating a palaeocurrent direction from SW to NE. (**d**) Clast-supported interval in the Ghaub diamictite consisting of different sub-units. The scale rule is in millimetres. (**e**) Crystal fans pseudomorphic after aragonite in the Maieberg cap carbonate dolostone exposed at the Auros Farm. The coin is 2.2 cm. (**f**) Matrix-supported Ghaub diamictite containing well-rounded crystalline clasts derived from the basement.

- Section 3: the Ghaub Formation is 73 m thick. The facies of the underlying Auros Formation and the lower part of the diamictite with the mentioned colour change and differences in clast content correspond to Section 2. At 25 m, the diamictite deposits are interrupted by a 1 m-thick clast-rich interval. After 27 m, Nosib clasts occur in addition to the ones from the Abenab Subgroup, and the weathering colour of the matrix changes to rusty brown. Upsection of 40 m, the matrix contains a large amount of detritic quartz and mica, and some centimetre- to decimetre-sized clasts from the metamorphic basement occur (Fig. 8f) in addition to Nosib sandstones or quartzites. Three finely bedded intervals have been found in the upper Ghaub Formation of Section 3, one of these contains a lonestone, another interval (at 55 m) shows slump features. Close to the top of the formation a carbonate clast of about 1 m in size has been found; in topmost parts a waterlaid conglomeratic bed occurs; some elongated clasts within this interval show preferred orientation. The 2.5 m-thick finely bedded layer overlying this interval forms the transition to the cap dolomite, which is not well exposed.
- Section 4: the Ghaub succession is 115 m thick and overlies partly dedolomitized Auros oolites. This section differs from the others: (1) the lower contact of the diamictites is only slightly erosional; (2) several finely bedded and partly clast-free intervals (termed 'silt stringers' by Hoffman 2005) occur in the lower 20 m (in contrast to Section 3, where such interbeds were found in the upper part, see Fig. 9); (3) the type of clasts do not differ upsection – clasts from the Abenab Subgroup, Nosib Group and the basement occur throughout this section; and (4) the matrix contains a large amount of detritic quartz and mica, but also newly formed sericite.

Within the finely bedded intervals, one arenitic layer and one layer consisting entirely of dolomitic clasts, without matrix, occur. The topmost part of the Ghaub Formation is, again, represented by a

Fig. 9. Relative clast-free, finely laminated lens or 'silt stringer' (right of the main picture) within the Ghaub diamictite (left of the main picture), exposed at the northern Keilberg Farm slightly NW of Section 3. The insert shows detail of lithologies: diamictite (top of sample) and 'silt stringer' (bottom part of sample), and the sharp contact between them. The rucksack is for scale in the main picture. The coin for scale in the insert picture is 25 mm.

finely bedded interval containing slump features and lonestones.

Imbricated clast zones

Strong changes in the types of clasts (and of the matrix) have been observed. Three different 'clast zones' can be distinguished (Figs 7 & 10). The degree of rounding of the carbonates is similar in all three clast zones, and the better rounding of sandstones and basement clast has already been mentioned. We estimate that the overall ratio of matrix to clasts is about 2:3.

- Clast Zone 1, which is restricted to the east, contains clasts from the Abenab Subgroup only. This clast zone shows a tripartite (1a, 1b, 1c; see Figs 7 & 10) subdivision according to the type of carbonate clasts. In each of the three parts, wacke- and packstones form about one third of the components. The other percentage is formed by silicified dolomite (1a), fine–coarse crystalline dolomite (1b) and an equal mixture of both (1c). Dolomitic pseudosparite forms the matrix in each of these subunits. Clast sizes are similar in all three parts, but the percentage of clasts to matrix and the rock colours are very different: 1a and 1c have a light–medium grey colour, and are dominated by matrix (with a mean of about 70%), whereas 1b has a black colour and is clast dominated, with a matrix content of about one-third.
- Clast Zone 2, which is relatively thin (Fig. 10), has Abenab and Nosib sandstone–quartzite clasts incorporated. This clast zone is similar to Clast Zone 1, the only difference is the additional content of Nosib clasts. All other features, like matrix composition, maximum clast sizes and the higher matrix content, are similar to Clast Zone 1.
- Clast Zone 3 has rare basement clasts in addition to the predominant Abenab and Nosib clasts (Figs 7 & 10). Abenab clasts are up to 1 m in size, larger than in the other zones, but mostly up to a few tens of centimetres. Nosib clasts mostly consist of quartzitic sandstones of up to several centimetres in size. The basement clasts are

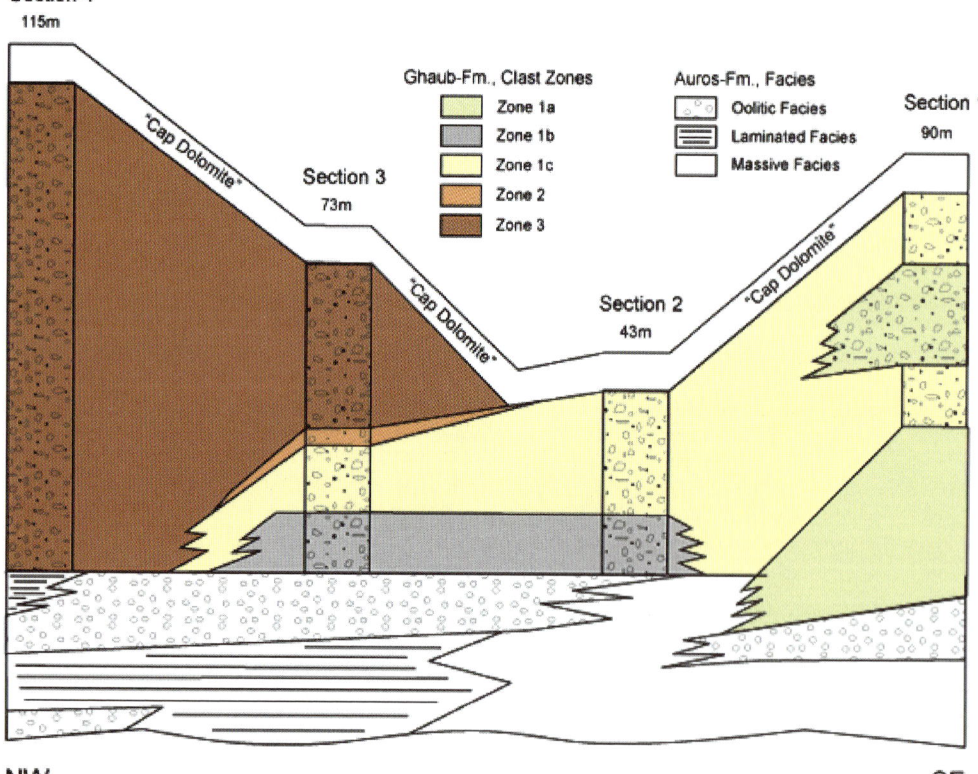

Fig. 10. Schematic diagram of lateral variations within the Ghaub Formation and at the top of the Auros Formation. For positions of the sections see Figure 5.

mainly granites or diorites from the local Grootfontein basement complex. Another difference is the matrix. In addition to the predominant dolomitic pseudosparite, detrital quartz, feldspar and mica occur (and some neomorphic sericite), which are missing in the other clast zones. In outcrop, the matrix shows a distinct red-brown, rusty colour, which might be mainly due to the weathering of the mica and of finely dispersed pyrite.

Clear trends of lateral and vertical distribution of the clast zones exist (Figs 7 & 10). Zone 1a is restricted to Section 1, Zone 1b occurs in basal parts of Sections 2 and 3, and passes laterally in Zone 1c. In general, Sections 1 and 2 are formed by Clast Zone 1, whereas Clast Zone 3 builds up all of Section 4. In Section 3, Clast Zone 1 is restricted to the lower third of the succession. Clast Zone 2 on top of it forms a relatively thin interval in Section 3 and is missing in the other sections. Section 3, where all three clast zones occur, represents a transitional area. Clast Zone 3 is restricted to this transitional area, and to the area further to the NW, where it builds up all of the Ghaub succession.

Locally (e.g. between Sections 2 and 3), graded, clast-supported intervals (Fig. 8d) of variable facies occur that cannot be correlated for longer distances. The two intercalations of Section 2 are fine laminated, arenitic interbeds with current ripples. Fine clastic intercalations (Fig. 9) are more frequent than coarser ones. Their number increases towards the west. Within Clast Zone 3, four finely (millimetre-scale) laminated and fine-grained intercalations occur in the upper part of Section 3, and six interbeds were found in the basal part of Section 4. Timewise, the upper part of Section 3 might be correlatable with the lower part of Section 4, as indicated by the common content in crystalline clasts and crystalline matrix. This highlights the assumed imbricated stacking pattern of different units. The previously mentioned clastic interbeds are much less frequent than in the Fransfontein area of northwestern Namibia: Eyles & Januszczak (2007) described numerous breccia intervals and conglomerate, as well as frequent sandstone intercalations.

Post-glacial succession (lower Maieberg Formation)

The post-glacial cap carbonate sequence (Maieberg Formation) occurs in wide areas of the OML. Söhnge (1957) subdivided the formation into different 'lithozones'. For the cap carbonates (also called cap dolomites) at the base of the formation, Hoffmann & Prave (1996) coined the term Keilberg Subformation. In this area the thickness is around 10 m in the NW and diminishes eastwards. Hedberg (1979), partly with reference to Söhnge (1957), describes this lowermost part as thinly bedded, light-grey–grey limestones with microlaminations resulting from a higher clay and silt content. Pyrite and hematite crystals are frequent according to this author, and the non-carbonate content is said to be around 10–12%. Together with pyritic dark shales above, this forms the lower lithozone of Söhnge (1957), whereas, in contrast, pyrite is largely missing in the upper lithozone. Hedberg (1979) describes these upper parts of the formation as fine crystalline, laminated dolomites. According to Söhnge (1957), the total Maieberg Formation is 106 m thick on the Gauss Farm, SE of Keilberg, and up to 1800 m on the Harasib Farm, about 5 km to the ENE. According to Beukes (1986), the maximum thickness is 900 m. In accordance with the SE–NW thickness trend of the Auros Formation, there is a distinct northwestwards thickening of the Maieberg Formation on the northern Keilberg Farm: being about 120 m thick in eastern parts (70 m for the lower lithozone), it measures about 400 m in western parts (230 for the lower lithozone). These strong thickness differences in the OML indicate strongly differential subsidence in post-Ghaub time.

The Maieberg cap carbonate dolostone shows both palaeogeographical and stratigraphic variability in the OML. The cap carbonate dolostone is described here from two depositional settings reflecting sea-floor relief: (1) above a palaeoslope, where the diamictite and overlying cap carbonate can be traced from the slope on to the platform (northern Keilberg Farm); and (2) a palaeogeographic high on the platform–ramp (southern Keilberg, Auros and Gauss farms).

Northern Keilberg Farm

In this area (Fig. 3) Maieberg outcrops are fair to very good because much of the formation occurs on steeper slopes above the more easily weathered diamictites. An unfortunate exception is the approximately 6 m-thick cap dolomite at the base and the overlying pyritic shaly dolomites, which show good outcrops only in the steeper west of the mapped area. The transition from Ghaub diamictites to the well-bedded cap dolomites is marked by an approximately 3 m-thick interval of laminated dolomite containing a variable number of clasts, slump features and fine clastic interbeds. The Maieberg cap carbonate is approximately 10 m thick in the NW, but thins southeastwards. It is wavy laminated and predominantly pale cream or buff coloured–light grey (see Fig. 11g). Wavy laminations typically have a relief of 2–5 cm and wavelengths of approximately 20 cm. Tubular structures (see

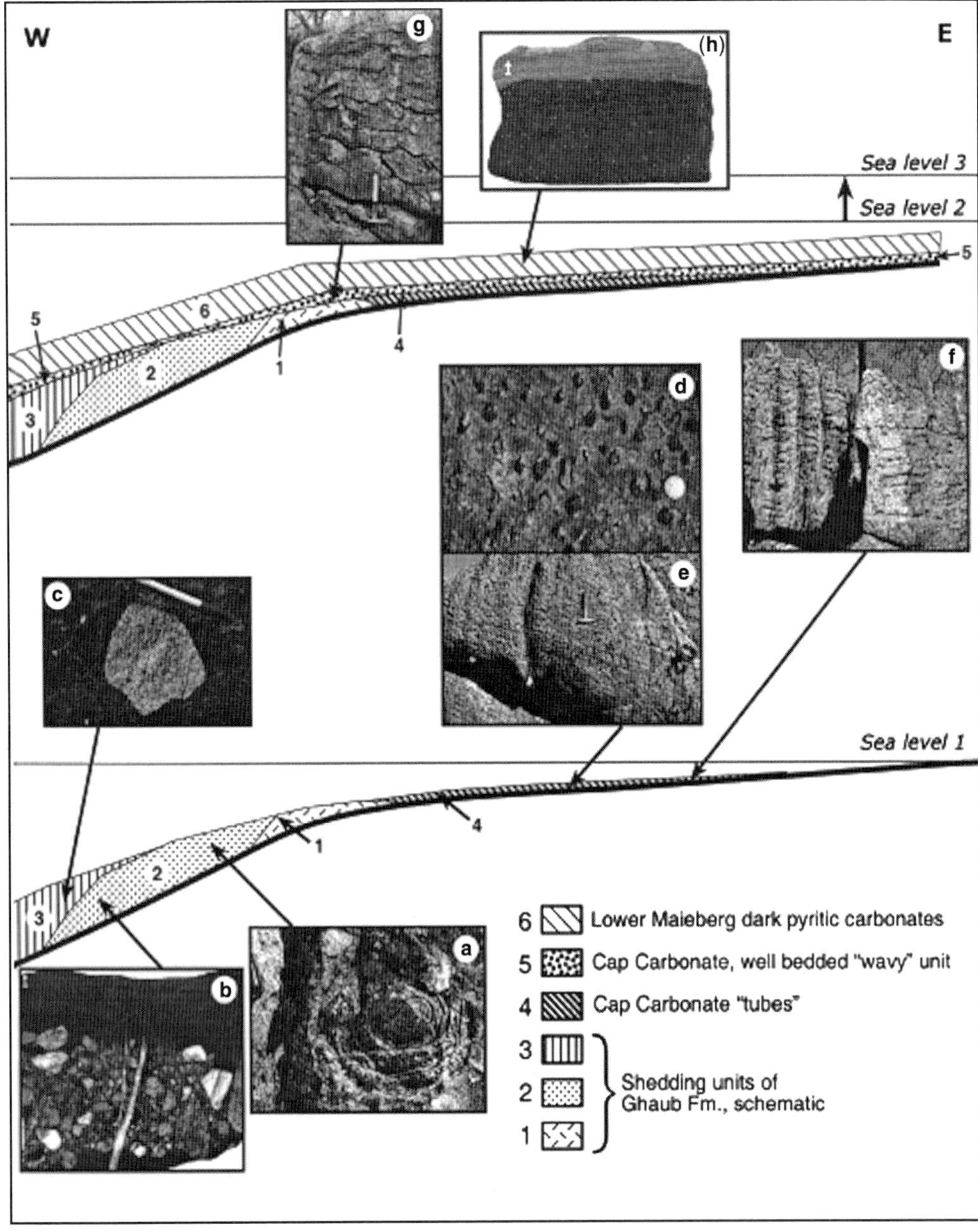

Fig. 11. Proposed depositional model for the 'Marinoan' diamictite and post-glacial cap carbonate sequence exposed in the Otavi Mountain Land. (**a**) Ghaub diamictite containing microbial stromatolites derived from the top of the Auros Formation in platform settings as at the southern Keilberg and Auros farms. The clast size is 20 cm. (**b**) Clast-rich horizon within the Ghaub diamictite. For scale, the arrow is 1 cm. (**c**) Matrix-supported crystalline clast in the Ghaub diamictite. The pencil is for scale. (**d**) Tubular facies in cap carbonate dolostone seen on the bedding plane. The coin is for scale. (**e**) Tubular facies with frequent horizontal sheet cracks in cap carbonate dolostone, vertical section. The hammer is for scale. (**f**) Tubular facies and horizontal sheet cracks in the cap carbonate dolostone viewed in vertical section (2 m high). (**g**) Wavy lamination in cap carbonate dolostone. The hammer is for scale. (**h**) Pyritic dark micritic limestone of the Maieberg Formation. See the text for a detailed discussion and explanation. For scale, the arrow is 1 cm.

later) are absent in the Maieberg cap carbonate of this area.

Dark-coloured pyritic micritic–microsparitic limestones–shales of the Maieberg cap carbonate sequence (nomenclature of Hoffman 2002) conformably overlie the cap carbonate dolostone. This sequence above the cap carbonate dolostone might represent the post-glacial highstand systems tract. In its lower part, this unit is frequently composed of pyritic black-brown–dark-grey finely laminated, occasionally cross-laminated, micritic–microcrystalline limestones. This unit often has a brown weathering colour. Towards the top of the unit there is a change to alternating grey and light-grey limestones on a sub-metre scale, sometimes separated by less than 1 cm-thick black organic-rich bands.

In the dark micritic limestone pyrite occurs as disseminated rhombs, and in veins and lenses. Pyrite is often concentrated along planar laminations, cross-lamination surfaces, within small black (<2 mm) organic lenses and along lamination-parallel para-unconformities. This distribution suggests synsedimentary deposition of pyrite in an anoxic reducing environment. Thin (<0.1 mm) cross-cutting pyrite veins also indicate that at least two separate generations of diagenetic pyrite are present.

These post-glacial organic-rich (Fig. 12) pyritic, dark-coloured micritic and microcrystalline limestones have total organic content (TOC) values comparable to potential hydrocarbon source rocks (see later). Similar transgressive late Cryogenian ('Marinoan') organic-rich carbonate deposits could represent a potentially globally distributed hydrocarbon source rock where they occur in the subsurface (e.g. Stewart 2007).

Southern Keilberg, Auros and Gauss farms

A few kilometres southwards of the diamictite outcrops, the Ghaub diamictite is absent and the Maieberg cap carbonate directly overlies shallow-water dolomites of the Auros Formation. The upper part of the Auros Formation is marked by a stromatolitic interval, some tens of metres thick. Densely packed columnar stromatolites are seen to contain internal synoptic depressions that exhibit radial spatial symmetry in cross-sections parallel to the vertical accretion direction (Fig. 6 a, b). Individual columnar stromatolites have diameters of up to 20 cm. Stromatolites of this type are present as reworked clasts in the Ghaub diamictite on the northern Keilberg Farm (Fig. 11a).

On top of these well-bedded stromatolitic successions the cap dolomites follow. These can be subdivided into two parts: (a) laminites with tube-like features at the base are overlain by (b) laminated and slightly wavy facies, which are comparable to the cap dolomites at the northern Keilberg Farm.

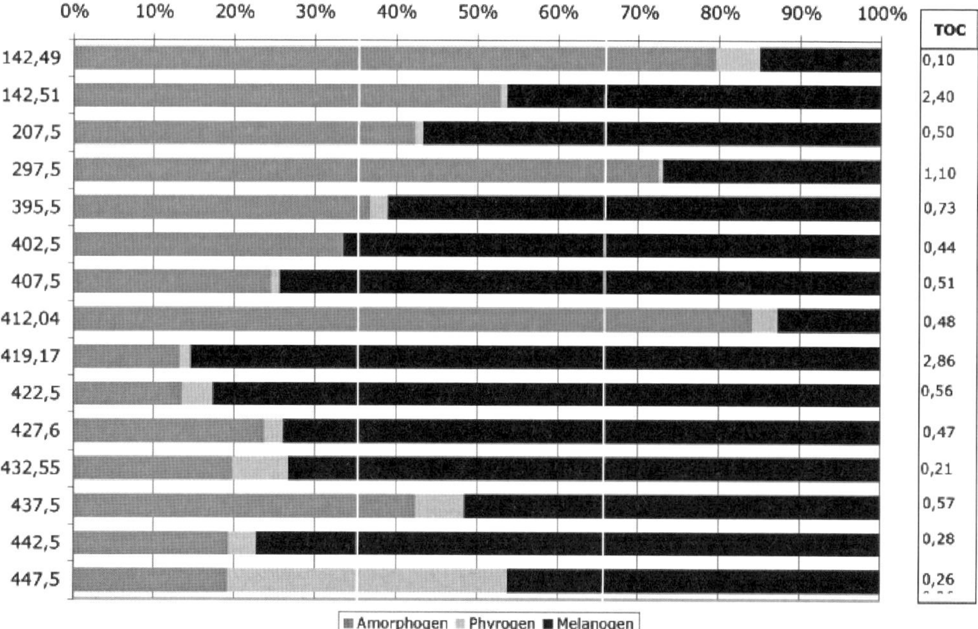

Fig. 12. Quantity (TOC) and composition of organic matter in the dark micritic limestones of the Maieberg cap carbonate sequence.

Tubular structures. Distinctive vertical–near-vertical tubular structures (compare Figs 6c–f & 8a) have been described globally, including Namibia (e.g. Hegenberger 1987; Kennedy *et al.* 2001*a*, *b*; Hoffman & Schrag 2002), Brazil (Rodrigues-Nogueira *et al.* 2003), the United States (Cloud *et al.* 1974; Marenco & Corsetti 2002; Corsetti & Grotzinger 2005) and Canada (James *et al.* 2001). In the OML, the tubular structures are exposed, for example, at the Auros Farm 595 and the Gumab Farm 538. In northern Namibia tubular structures have also been described on the Huab Ridge on the Tweelingskop Farm and the Hoanib shelf near Khowarib Schlucht (Hoffman 2002). Similar tubestones occur in central Namibia in the Bindha cap carbonate of the Witvlei Group (Hegenberger 1987, 1993), and in southern Namibia in the Bloeddrif cap carbonate of the Gariep Group (Fölling & Frimmel 2002).

This tubular facies was originally correlated stratigraphically to the uppermost part of the preglacial Auros Formation of the upper Abenab Subgroup (e.g. Beukes 1986; King 1994). In a major revison of the stratigraphic correlation of the Otavi Group, Hoffmann & Prave (1996) placed the tubular facies at the base of the (post-glacial) Maieberg Formation (Fig. 2).

The tubestones comprise an interval of pale-grey subparallel microbial laminites of about 10–20 m in thickness. It contains densely packed silicified tube-like features, arranged subvertical to the microbial laminations (Figs 6c–f & 8a), which partly deflects laterally, roughly parallel to the bedding. These are filled with dolomite cement and drusy quartz, but rare tubes unfilled at their tops are present as well (Fig. 6c–f). The bedding-parallel planar laminations between the tubes have concave upwards contacts with the outer walls of the tubes. Sheet cracks filled with dolomite cement and drusy quartz are also common. Sheet cracks and tubes often link to one another (Fig. 6e, f). The exposed tubes range in length from a few centimetres to approximately 50 cm, and could possibly be longer. Tubes can exhibit irregular diameters along their lengths ranging from approximately 1.5 up to 4 cm, giving them a pinch-and-swell appearance (Fig. 8a). Individual tubes may branch and/or coalesce with one another (Fig. 6e, f). There is no obvious common horizon from which tubes are rooted, but tubes frequently begin and terminate at sheet cracks (Fig. 6e, f). The subrounded cross-sections of siliceous-filled tubes sometimes protrude 1 cm or so above exposed bedding surfaces (Fig. 6d), because they are more resistant to weathering than the surrounding rock. These siliceous 'studs' on bedding planes are referred to in Namibia as 'quartz clusters' and have previously been used as a regional marker horizon (Söhnge 1957). When viewed on lamination surfaces in cross-section, the tubes are densely packed but do not exhibit obvious radial spatial symmetry (Fig. 6d). This contrasts with the radial arrangement of the stromatolite columns in upper parts of the underlying Auros Formation (Fig. 6a, b). No reworked clasts of this tubular facies were observed in the Ghaub diamicite, suggesting that the tubular facies was deposited at the same time as, or shortly after, the diamictites.

Laminated, wavy cap carbonate facies. The interval containing tube fabrics is overlain in turn by laminated and slightly wavy cap carbonates, which often contain cementstones (Fig. 8e) at their top (Hoffman 2002; compare also James *et al.* 2001). Bundles of former aragonitic rays were observed only in areas without underlying diamictites, in line with the tube facies. Both facies associations (columnar stromatolites, tube structures and laminated cap dolomites with 'rays', on one the hand, diamictites and laminated cap dolomites without 'rays', on the other) appear to be mutually exclusive in the OML.

Interpretation

The Maieberg cap carbonate dolostone has been interpreted as a transgressive systems tract deposited during the post-Ghaub relative sea-level rise (Hoffman *et al.* 1998*a*, *b*; Hoffman 2002, 2005; Hoffman & Schrag 2002). The Snowball Earth model proposes that cap carbonate formed as chemical precipitates (Hoffman *et al.* 1998*a*, *b*). Most documented cap carbonates are finely laminated microcrystalline dolostone, although dolomite probably did not precipitate directly from sea water, as shown by Shields (2005) in a review of competing models for deposition of cap carbonate dolostones. Stromatolitic microbial and peloidal rocks have also been described as cap carbonate dolostones (e.g. James *et al.* 2001; Hoffman & Schrag 2002), including the micobially laminated tubestone facies.

Several explanations have been proposed for the cap carbonate tubestones. The described features of the tubestones vary considerably (Figs 6c–f & 8a), and a combination of differing explanations might be possible therefore. A larger group of authors assume gas and/or fluid escape structures (Cloud *et al.* 1974; Hegenberger 1987; Kennedy *et al.* 2001*a*, *b*). Fluid or gas pressure opening horizontal fissures along surface of weakness between the microbial laminations would feed into the vertical tubes where pressure escaped vertically (Fig. 6e, f). In Hegenberger's model, gas was formed by the decay of algal mats mainly by desiccation during low tides. Gases, including CO_2, CH_4, NH_3, H_2S and H, would have built up along sheets cracks and escaped vertically when the pressure was sufficient to break through the overlying sediment. Compaction caused by sediment loading could also have generated pore-water overpressure that formed tubular dewatering structures, but this is unlikely in this case as sheet cracks would most probably not have been preserved. If this model is correct it is difficult to explain why such features do not occur in other parts of the succession, stratigraphically below or above the Ghaub interval. It has been suggested that tubes and sheet cracks could be methane escape structures resulting from the destabilization of clathrates (Kennedy *et al.* 2001*a*, *b*, 2002; Jiang *et al.* 2003).

Other interpretations, previously considered but now widely discounted, include metazoan burrows and root casts (Cloud *et al.* 1974). A more recent interpretation of the cap carbonate tubular structures is that the tubes have a microbial stromatolitic origin (James *et al.* 2001; Hoffman & Schrag 2002; Corsetti & Grotzinger 2005). These authors propose that depressions on the surfaces of stromatolites were preserved as the stromatolites aggraded. The tubes represent these sediment-filled isolated depressions. In Death Valley (USA) compartmentalization of tubes by bridging laminae traceable from the host rock into the tubes appears to support this interpretation (Corsetti & Grotzinger 2005). Cloud *et al.* (1974) had previously dismissed a stromatolitic origin for the tubes, as they did not resemble the normal pattern for columnar stromatolites. Corsetti & Grotzinger (2005) proposed that the unusual stromatolite morphology preserved by the tubestones was related to highly carbonate-supersaturated post-glacial sea water.

Typically, after the columnar stromatolites of the Auros Formation, the immediately overlying microbial laminites contain only a few tubular structures. The number of these strongly increases upwards after 1–2 m, but decreases again towards the top of the 'tube interval'. The laminites themselves, however, look identical throughout. This lower interval of the cap carbonate dolostone then passes into the finely laminated upper interval, which is similar in the areas with and without diamictites and tube facies. Locally, crystal fans pseudomorphic after aragonite occur near the top (Fig. 8e), which have been interpreted as sea-floor cements precipitated directly from sea water, indicating very high carbonate oversaturation (James *et al.* 2001; Hoffman & Schrag 2002). In the OML these have been found only in successions without diamictites; in other words, in platform settings containing the tubestone interval below. However, Hoffman & Schrag (2002) described these from the cap dolostone that overlies diamictites to the west of the OML. Other types of cementstones also occur in areas with diamictite facies.

The Maieberg cap carbonate conformably overlies the Ghaub diamictite, and the Ghaub Formation unconformably follows on top of the Auros Formation. A palaeoslope setting with an apparent dip to the present-day NW (Fig. 5) is indicated by facies and strong thickening of the Auros, Ghaub and Maieberg formations in this direction. When the diamictite is traced laterally to the SE, it strongly thins in this direction. The critical area of possible pinchout is covered by recent valley-floor sediments. In contrast to the diamictite, facies changes of the cap dolomite are less distinct in the SE–NW direction, suggesting a significant post-glacial sea-level rise and deposition on both the slope and former platform top.

Upper Maieberg Formation

The cap carbonates are overlain by the much thicker, and more variable, middle and upper part of the 'cap carbonate sequence'. As in many other early and late Cryogenian glacial intervals (e.g. SW Namibia, central Australia, Mackenzie Mountains in Canada, East Greenland and East Svalbard Caledonides), the sediments on top of the Maieberg cap carbonates are micritic limestones, which change in colour from brown to grey and/or black, but also contain dark-coloured shaly carbonates. In some beds 0.5–3 m-thick cross-bedding has been observed locally. In its lower part the succession is often rich in pyrite, distributed within the sediment in small aggregates, but also filling veins. Pyrite seems to be more frequent in the areas where the Ghaub diamictite is present in the underlying rocks. Upsection, light-grey, laminated dolomites become more numerous. The last intercalation of a pyrite-rich carbonate forms the boundary between the two earlier mentioned lithozones of Söhnge (1957). In the Keilberg area this interval is more than 200 m thick. In other areas of the OML, thicknesses of 1000 m and more are mentioned in the literature (Söhnge 1957; Beukes 1986).

The upper lithozone consists of grey, laminated dolomites, partly intercalated with reddish/brownish weathering beds. Stromatolitic laminations occur. From the Kaokoveld in NW Namibia, Prave (1996) described turbidites, slumps and intraformational breccias, and explained the depositional setting as gravitational mass transports into a deeper basin. A similar facies with different intercalations of breccia beds has been observed at the northern Keilberg Farm. Clast sizes reach up to 0.5 m. In contrast to the Kaokoveld (Prave 1996), no distinct coarser turbidites have been encountered, but many of the finely laminated beds represent microturbidites. The succession clearly indicates a transgressive trend, from higher oxygenated to anoxic conditions (cap carbonates to pyritic shales). With the upper lithozone this trend reverses.

Carbon isotope records

Up to now only two independent and reliable radiometric ages exist for late Cryogenian assumed glaciations: zircon dating of Ghaub-correlative tuffs in central Namibia (base of the diamictites) gave ages of 635.5 ± 1.2 Ma (Hoffmann *et al.* 2004). Exactly the same age (635.23 ± 0.57 Ma) was obtained for tuffs within Nantuo cap carbonates on top of diamictites in China (Condon *et al.* 2005). Owing to the scarcity of radiometric ages, isotope stratigraphy has been applied on the Cryogenian successions to correlate sections worldwide (Halverson *et al.* 2002).

Apparently globally traceable trends in the stratigraphic carbon isotope profiles of the early Cryogenian ('Sturtian') and late Cryogenian ('Marinoan') preglacial carbonates immediately below the diamictites, and in the cap carbonates and the cap carbonate sequence above the diamictites, have been identified (e.g. Halverson *et al.* 2002; Hoffman & Schrag 2002, p. 140, fig. 8). In the Ghaub interval of northern Namibia, a negative excursion of $\delta^{13}C$ begins approximately 10–20 m below the diamictites, in the Auros Formation (Kaufman *et al.* 1991; Hoffman *et al.* 1998*a*, *b*; Halverson *et al.* 2002; Hoffman & Schrag 2002). Comparable negative anomalies also occur in the Mackenzie Mountains of NW Canada (James *et al.* 2001) and in southern Australia (Trezona anomaly: Halverson *et al.* 2005). Correlation of these anomalies of up to −15‰ PDB led Halverson *et al.* (2005) to suggest a worldwide negative oceanic excursion caused by the glaciations. The shift to negative values underneath the diamictites is not present in all sections (Kennedy *et al.* 1998); this was supposed to be partly due to the removal of the interval by erosion, making chemostratigraphic trends in successions underneath glacial intervals more difficult to correlate. There is, however, the danger of circular reasoning: because the anomalies occur in post-glacial cap carbonate facies, which were compared with one another on a global basis, global and synchronous glacial events have been proposed.

Worldwide effective chemostratigraphic trends can be expected if a global ocean with ample connections between the different oceanic basins existed. A true worldwide contemporaneous (controlled by biostratigraphy) signal has been observed, for instance, for the Ordovician glaciations in Gondwana, also showing a positive $\delta^{13}C$ excursion in ice-free Baltica several thousands of kilometres away from the glaciations (Kaljo *et al.* 2003; Page *et al.* 2006; Smith 2009). In contrast, large differences in carbon isotope values developed in the last 8 Ma after the closure of oceanic gateways (Isthmus of Panama, Tethyan gateways) between open marine sediments of the Pacific and Atlantic oceans (Ravizza & Zachos 2003), which had uniform values in the preceding time. During the break-up of Rodinia (early Cryogenian; Sturtian interval) negative carbon isotope anomalies might, therefore, not necessarily represent worldwide signals, but might at least partly be related to local facies.

Fairchild & Hambrey (1995) pointed out that the Elbobreen glacial of eastern Greenland, possibly correlative to the late Cryogenian ('Marinoan'), shows a negative anomaly in basinal facies underlying glaciomarine facies. Laterally, platform limestones with positive $\delta^{13}C$ signatures overlie the basinal facies, and are followed themselves by glacial successions. Negative anomalies might, therefore, reflect at least partly a distinct facies, for example the signature of a basinal environment, as shown by Fairchild & Hambrey (1995) and Fairchild *et al.* (2000). Basinal sediments can differ isotopically from shallow-water sediments for different reasons. Cenozoic marine sediments show such isotopic differences owing to primary mineralogy, alteration effects (especially on the platform) and sediment being shed from shallow-water environments into deeper-water settings (Swart & Eberli 2005). In addition, fluctuations in $\delta^{13}C$ content might also be caused by differences in TOC.

In the Ghaub glacial successions, the negative $\delta^{13}C$ excursion persists for several tens of metres after cap carbonate deposition ceased, before returning slowly to Neoproterozoic background levels (e.g. Kaufman *et al.* 1991; Hoffman *et al.* 1998*a*, *b*; Hoffman & Schrag 2002). A direct connection of the negative $\delta^{13}C$ values with glaciation is not very likely, as these light values occur before, during and after glaciation. The assumption of local sedimentary facies not related to glaciation influencing the isotope anomalies is further corroborated by the carbon isotopic record given by Hoffman (2005): successions of the deeper foreslope (Sections A, B and C) in marly limestone rhythmites directly on top of the cap carbonates show an even more negative interval than in the underlying cap carbonates themselves, changing to heavier values (around 0) upsection. This trend of first further decreasing and then increasing isotope values is in contrast to sections of the outer and, especially, inner platform (Sections I, J and K of Hoffman 2005), where the equivalent part of the succession on top of the cap carbonates is more negative throughout (−5‰ for several tens of metres) than the cap carbonates below.

In the Ediacaran Nafun Group of Oman there is one very distinct negative carbon isotope anomaly (basal Shuram Formation), which is totally unrelated to glaciations and distinctly above horizons correlated with the late Cryogenian ('Marinoan') interval (Le Guerroué *et al.* 2006, 2007; Peltier *et al.* 2007). All these data indicate that different reasons for negative (and positive) shifts of carbon isotopes exist (see also Fairchild & Kennedy 2007), which might often not be related to glaciations or be a worldwide signal. General stratigraphic reproducibility of carbon isotope trends is often used as an argument for a primary marine origin resulting from secular variation of changes in sea water.

There is special concern because of the effects of dolomitization: the Neoproterozoic successions discussed are almost entirely dolomitic and originally there might have been an unstable dolomitic precursor phase with some Ca. In the probable absence of organisms mostly precipitating low-Mg calcite (as in the Phanerozoic), primary carbonate precipitates from Neoproterozoic ocean water are most likely to be metastable aragonite, vaterite, Mg-calcite or poorly crystalline dolomite. It is often argued that, during stabilization–recrystallization of these phases, the signatures do not change because carbonate acts as a relatively closed system and there is not sufficient dissolved carbon in stabilizing fluids to affect the isotope mass balance (Jacobsen & Kaufman 1999). Data on the stoichiometry of these dolomites are usually not given in the literature. To reach near stoichiometry, an open system is necessary and, in the case of dolomitization of calcium carbonate, needs to be pumping hundreds to thousands of times the pore volumes of fluid through the succession (Land 1973; Machel 2004). There are different possibilities for a dolomite reaction to take place. For a volume per volume reaction, it can be written (Morrow 1990*a*, *b*) as:

$$1.75CaCO_3 + Mg^{2+} + 0.25CO_3^{2-} \Longleftrightarrow CaMg(CO_3)_2 + 0.75Ca^{2+}.$$

According to this reaction, an external source for Mg^{2+}, as well as for CO_3^{2-}, is required and a different carbon signature might, therefore, mix with the original one (Fairchild *et al.* 2000). Carbon isotope excursions in dolomitized and altered successions might, therefore, not serve as an unquestionable correlation tool (e.g. Corsetti & Kaufman 2003; Allen 2006; Hoffman & Allen 2007) and might record a global event as well as a facies or diagenetic signal.

Recently, Peltier *et al.* (2007) argued that a Snowball stage was prevented by a drawdown of atmospheric oxygen into the ocean. Decreased surface temperatures increase the rate of remineralization of a large pool of dissolved organic carbon in the ocean. This prevents a Snowball state because of an increase in atmospheric carbon dioxide, causing an enhanced greenhouse warming of the Earth.

Organofacies of the post-Ghaub interval

The post-glacial succession above the cap carbonate is, to a large degree, made up of dark-grey, locally black, laminated micritic limestones, rich in pyrite and flakes of organic matter, which are strongly euxinic deposits. To test the hydrocarbon source rock potential, these dark micritic limestones were analysed for TOC, organofacies and organic maturation. Samples were taken from core KH1 (Khusib Springs, Fig. 3), where the post-Ghaub interval is well preserved, and from some surface outcrops from the northern Keilberg Farm. Analysis was focused on the basal part of the succession (448–390 m in the core): about 50 m of only dark-grey–black lithologies, without any light-grey interlayers, directly above the cap carbonate. Samples were taken in 5 m spacing across this interval. To compare the source rock potential of the grey and light-grey alternations of the upper part of the succession a few samples were analysed between 297.5 and 142.5 m, including one grey–light-grey couplet (142.49 and 142.51 m).

TOC values show wide variations from 0.1 to 2.8% (Fig. 12). The majority of the samples show TOC values of about 0.5%. Samples from just above the cap carbonate and in the light-grey layer of the alternations in the upper part (142.49 m) show lower TOC values of 0.1–0.3%. On the other side, few samples are rich in C_{org} with a maximum of 2.8% TOC, reaching TOC values of typical black shales (Tyson 1995). Hydrocarbon source rocks are classically defined by more than 0.5% TOC in clastic rocks and 0.3% TOC for carbonates (Littke & Welte 1992). Based on this definition, the dark micritic limestones of the whole studied interval show source rock quality. It has to be mentioned that the core analysed is situated on the platform, where diamictites are absent. The same stratigraphic interval overlying slope and basin settings has an even higher potential for source rocks with higher TOC values, but owing to the oxidation of surface outcrops the source rock quality of these basinal successions is difficult to prove with our limited dataset.

In the Proterozoic, organic matter is almost completely produced by marine planktonic organisms of possibly bacterial and algal affinity, widely distributed in the marine realm. These planktonic microfossils consist largely of acritarchs, a poorly defined group of uncertain affinities, presumably eukaryotic biota (Vidal & Moczydlowska 1992). Acritarchs are recorded continuously throughout the Ediacaran in Australia (Grey 2005). Other organic-walled microfossils (mainly bacteria) are recorded by biomarker analysis in age-equivalent oils and sedimentary rocks from the Huqf Group in Oman (Summons *et al.* 1999). Terrestrial-derived organic matter (OM: bacteria, lichens) is negligible. Therefore, for organofacies analysis, the classification scheme of Bujak *et al.* (1977) was modified: amorphogen (all unorganized, structureless, fluffy, translucent debris), phyrogen (non-opaque structured OM, organic tissues and filaments, and acritarchs), melanogen (all opaque OM). Hylogen (non-opaque, fibrous woody material) is not available at this time. Correlated to kerogen types (Tyson 1993), amorphogen corresponds to kerogen type I/II (oil prone–highly oil prone), phyrogen–kerogen type I/II (oil prone–highly oil prone) and melanogen–kerogen type IV (inert, barren). Melanogen is the dominant fraction in most samples (Fig. 12). Because of its strongly corroded shape, it is seen as a product of intense degradation (comparable to degrado-fusinite: Taylor *et al.* 1998), possibly partially reworked, and not as a product of high thermal overprint. The dominant fraction of the non-opaque kerogen is amorphogen. About two-thirds of the samples are dominated by very dark coloured amorphogen with a minor part of light- to medium-brown amorphogen, of between 5 and 25%. In the rest of the samples light–medium amorphogen dominates over dark-brown amorphogen by a factor of 2–3. The lowermost sample (447.5 m), directly above the cap carbonate, shows light- to medium-brown amorphogen only (mainly of bacterial and algal origin). Phyrogen is present in small numbers in all samples, except at 402.5 m where it is completely missing. Values range from less than 1 to 7%, with an average percentage of 1–2.5%. The majority of the phyrogen is made up of organic tissues and filaments (50–70%); acritarchs are less common. Phyrogen is mostly poorly preserved, showing intense degradation. The lowermost sample (447.5 m), directly above the cap carbonate, is exceptional again. Phyrogen is the second most abundant kerogen group with about 35%, dominated by organic tissues and filaments, typical for microbial mats. Based on classical hydrocarbon source rock diagrams, the organofacies of the majority of samples from the Khusib Springs core shows low hydrocarbon source rock potential (Fig. 13).

In general, the OM preserved in the dark micritic limestones (Fig. 14) of the Maieberg cap carbonate sequence is intensely degraded, as shown by the poor preservation of the kerogen and the large amount of amorphogen, which is typical for dys- to anoxic depositional conditions (Tyson 1993). The large amount of melanogen, the most stable kerogen group, also supports the assumption of intense degradation, where all less-resistant kerogen groups are widely destroyed by degradation. The high level of degradation observed in all samples provides evidence for a significant loss of OM, especially phyrogen and amorphogen, leading to the residual kerogen actually being preserved in the samples. Therefore it is supposed that the primary kerogen was much higher in non-opaque OM (phyrogen and amorphogen) and that the primary TOC values were also significantly higher. Considering that the OM in these rocks represents a strongly residual kerogen, this provides evidence that the hydrocarbon source rock potential of the primary kerogen of this succession might have been significantly better than that observed today.

Analysis of the thermal overprint of the Maieberg cap carbonate sequence was carried out using

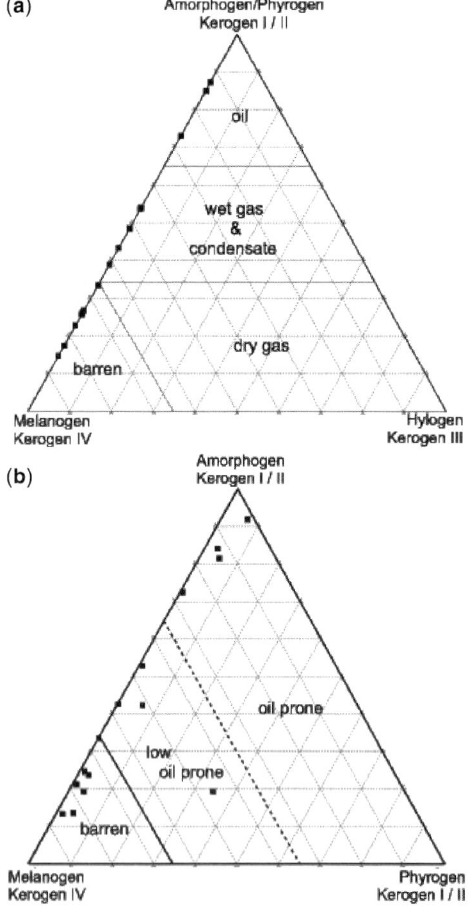

Fig. 13. Organofacies of dark micritic limestones of the Maieberg cap carbonate sequence and its relationship to hydrocarbon source-rock potential. (**a**). Standard scheme by Tyson (1995) and (**b**) the scheme modified for this study.

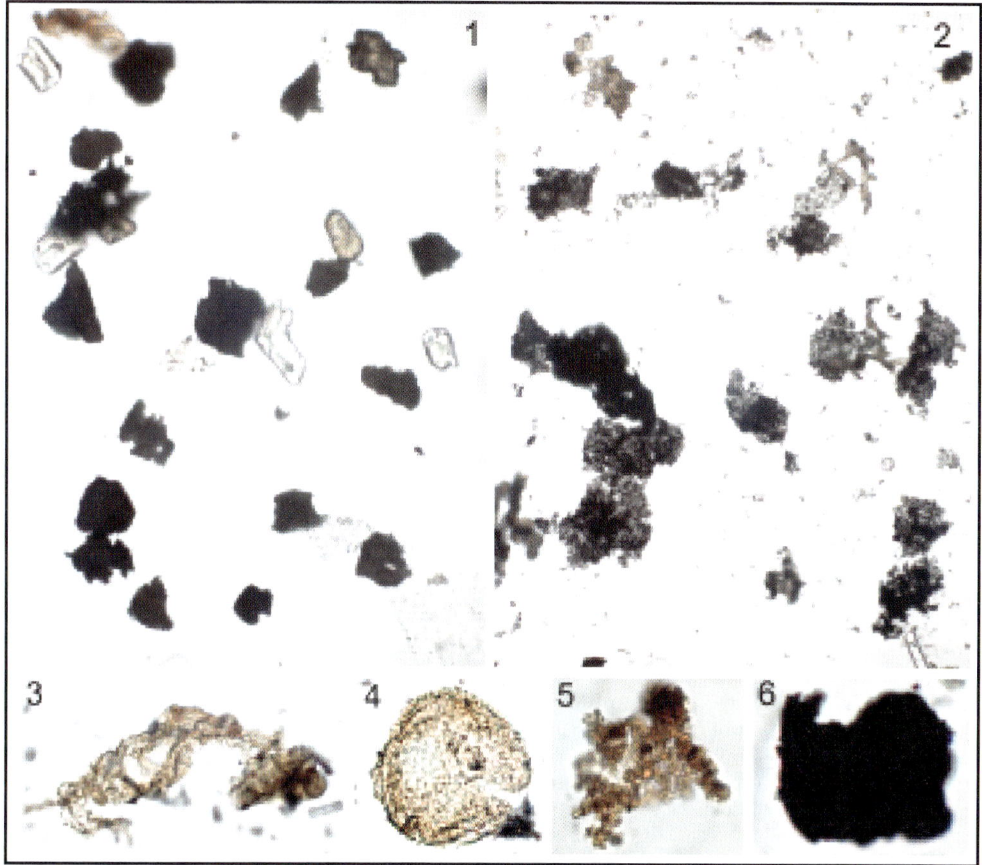

Fig. 14. Organofacies of the dark micritic limestones of the Maieberg cap carbonate sequence: 1, melanogen-dominated kerogen; 2, amorphogen-dominated kerogen, degraded phyrogen; 3, organic tissues; 4, acritarch; 5, amorphogen; 6, melanogen, corroded.

the Thermal Alteration Index (TAI). Therefore, the colours of the non-degraded phyrogen were analysed. Because most phyrogen is clearly degraded, only three samples could be used for maturation analysis (447.5, 432.5 and 395.5 m). TAI analysis was performed on organic tissues and filaments because all acritarchs were clearly degraded. The light- to medium-brown colour of these organic remains fit to the TAI 3− to 3. This can be correlated to a maximum long-term palaeotemperature of approximately 100–130 °C. It is assumed that this relates to the thermal increase caused by the burial of the formation falling into the oil window.

Based on the amount and composition of the organic matter and the maturation level observed, the dark micritic limestones of the Maieberg cap carbonate sequence provide fair–good hydrocarbon source rocks. The reason why only a few hydrocarbons produced by these source rocks are known from Namibia might be the large migration distance, probably to the north towards the southern margin of the Congo Craton. From the Owambo Basin, an oil seepage is described in Etosha well 5-1A (Fig. 4) (Ypma 1979*a*, *b*). Similar organic-rich carbonate successions representing post-glacial late Cryogenian ('Marinoan') highstand deposits are a potential hydrocarbon source rock worldwide, based on TOC and organofacies, but depending on local thermal history.

Tectonic control of the Otavi Group sediments

There are two strongly differing explanations for the setting of the Ghaub sedimentary succession in Namibia. According to Hoffman (2002, 2005), the diamictites are glacially induced (Snowball Earth theory) – for example, deposited by a grounded ice sheet above a minor topography, and a carbonate platform that mainly aggraded, but did not prograde much. In contrast, Eyles & Januszczak (2004, 2007) assumed distinct rift tectonics, promoting the uplift of rift shoulders, erosion and deposition of waterlaid conglomerates, and debris flows in rift basins, possibly unrelated to glaciation.

On the northern Keilberg Farm, the thickness of the Ghaub diamictites increases from a few metres to much more than 100 m towards the NW, in the approximate direction of strike of an intraplatform basin. Also, a distinct change of components is observed, including crystalline clasts and quartz, feldspar and mica as constituents of the matrix of the youngest sheddings in the NW (Figs 7 & 10). The stratigraphic position of the eroded clasts indicate that during the Ghaub (pre-Ghaub?) tectonic episode an uplift of more than 2000 m has to be assumed.

This tectonic episode in the middle part of the Otavi Group is not the only tectonic event in the area after the Nosib Group rifting episode. Just a few indications for tensional tectonics representing every formation of the Otavi Group (in stratigraphic order) are given below.

The *Berg Aukas* and *Gauss formations* contain many unsorted and chaotic breccia intercalations, as well as a network of fissures ('colloform structures') filled with radial sparry carbonate (Beukes 1986; own observations on the Olifantsfontein Farm west and east of the Tsumeb–Grootfontein road; Annshope west of the same road).

The *Auros Formation* contains thick breccia horizons underneath the Ghaub diamictites on the northern Keilberg Farm.

The *Maieberg Formation* shows clastic intercalations, which are mostly debris flows, that range in thickness from centimetres to a few metres. These have been found in the northern Keilberg area, above the occurrences of diamictites, but also further to the east. In the area of the Hambone Prospect, near the boundary between the Olifantsfontein and Jägerquelle farms, east of the Grootfontein–Tsumeb road, several metre thick, dark-coloured debris flows with different amounts of matrix occur, containing dark carbonate clasts overlain by slumped horizons and turbiditic intervals. These outcrops clearly represent deeper slope settings.

The *Elandshoek Formation* contains slump fabrics as well as massive, strongly brecciated dolomites, marked by an irregular network of fissures filled by quartz as well as carbonate cement. The fissures bound dolomite blocks often several metres in size. In the northern part of the Keilberg Farm, and the Tigerschlucht area further to the east, prominent slump features occur, and the brecciation is absolutely chaotic. The brecciation is marked by an irregular network of fissures filled by quartz as well as carbonate cement, bounding partly dolomite blocks of metre size. In the northern part of the Keilberg Farm, and further to the east, the brecciation is often chaotic, whereas on the Auros Farm, about 16 km further to the south, the breccias are mostly strata-bound (Beukes 1986). There is neither a detailed model for the brecciation nor a regional analysis of the lateral and vertical variability. Beukes (1986) and Misiewicz (1988) assumed creeping and sliding of semi-lithified sediment along a palaeoslope that resulted from a strong tectonic (seismic) event.

The *Hüttenberg Formation*, although to a large degree a shallow-water facies, contains, west of the Guinas Fault (Farm Bobos), a mixture of sediments deposited under non-turbulent and turbulent conditions: component-supported sheddings alternate with thicker intervals of matrix-rich debrites. The components originated from the slope and platform margin, and include oolites. In addition, dark, organic-rich shales are also intercalated, but often form the matrix of the debris flows.

We interpret all of the abovementioned intervals in the Otavi Supergroup as having been affected by distinct tectonics. Other than in the Ghaub interval, these debris flows are clearly non-glacial in origin and do not show any climatic significance, but indicate instead the repeated creation of small, fault-bounded basins, dissecting the shelf of the Congo Craton. Different types of carbonate–clastic and carbonate–shale deposits, originating from the directly underlying interval or neighbouring platform areas, fill these basins. The frequency of these synsedimentary tectonics might indicate that subduction started relatively soon after Nosib rifting and that a foreland basin was formed early in history, as assumed by Prave (1996, 1997) (in contrast to the assumption of a long-lasting 'passive' margin setting: Dürr & Dingeldey 1997; Hoffman *et al.* 1998*a*, *b*). At present, the geometries of these basins are largely unknown and the relationship of the tectonic events on the Congo shelf with subduction of the Khomas trough (Kukla 1992) is, therefore, strongly hypothetical.

Fabrics and clast composition of the Ghaub Formation are, however, different from the other basin infills. Only the Ghaub contains a wide spectrum of clasts, from the directly underlying Auros Formation down to the crystalline basement, indicating continuous unroofing of uplifted blocks. The only striated clast found by us so far comes from the Ghaub Formation; in addition, Hoffman & Schrag (2000, 2002) describe relatively well-constrained dropstones. The deposition of the Ghaub sediments was therefore tectonically induced, due to the uplift of the hinterland and creation of accommodation space in basinal areas, but it was also related to a glacial interval. The glaciation probably was mostly confined to the hinterland. Glacial deposits were reworked, transported in feeder channels via the platform, and were preserved in the tectonically active Ghaub slope and basins.

Structural evolution, fluid flow and temperature development

The OML was affected by different deformation events. Pre-Pan-African deformation is syndepositional, comprising slumping, growth faulting and brecciation (see earlier). Three major Pan-African (Damaran) deformation phases have been identified (see Laukamp 2006): a late Cryogenian–middle Ediacaran east–west shortening (D_1); a late Ediacaran–Early Cambrian north–south shortening (D_2); and an Early Palaeozoic uplift (D_3). These tectonic events are due to the collision of the Rio de la Plata Craton with the Kalahari and Sao Francisco–Congo cratons, followed by suturing along the Damaran–Lufilian–Zambezi Orogen.

In the OML the Cryogenian and Ediacaran succession was uplifted, whereas in the north, in the Owambo Basin (Figs 1 & 4), it continued to subside and was covered by younger sediments. This is the basinal area where hydrocarbon prospection was carried out in the past (Etosha Petroleum Company) and more recent activities are under way.

Many base metal ore deposits, belonging to two main types, have been found in the OML, with the majority structurally controlled (Pirajno & Joubert 1993; Cairncross 1997; Melcher *et al.* 2003, 2006; Laukamp 2006; Kamona & Günzel 2007 with further references). Melcher (2003) discussed hypothetical relationships of ore mineralizations with the Cryogenian glaciations. The hypogene Berg Aukas-type mineralizations, related to basinal brines, consist of strata-bound Zn–Pb sulphides, and are followed by the synorogenic Tsumeb-type Pb–Zn–Cu mineralizations, which are hosted to a large degree by breccia pipes, the Tsumeb pipe being the best known of these. The pipes formed during karstification of the pre-Mulden surface and often developed at cross-junctions of different tectonic trends, generated during D_2 and further extended during D_3. Pan-African syn-D_2 shear zones probably acted as important pathways for ascending metal-bearing hydrothermal fluids, which did not affect neighbouring areas.

The age of the mineralization is not well constrained. Based on preliminary Re–Os (rhenium–osmium) analyses of sulphide ores, Melcher *et al.* (2003) reported an age of 530 Ma for the main ore phase of the Tsumeb pipe. A minimum Rb–Sr (rubidium–strontium) age of 493–560 Ma was obtained by Schneider *et al.* (2007) from willemites replacing earlier sulphides. Model ages of ore samples from the Tsumeb pipe and the Kombat mine range from 600 to 530 Ma (Allsopp *et al.* 1981; Hughes *et al.* 1984; Kamona *et al.* 1999), but these ages have been questioned by other authors (Haack 1993; Frimmel 2004). Fluid inclusions of two–three phase inclusions measured at Tsumeb, Khusib Springs and Berg Aukas (see review by Kamona & Günzel 2007) indicate highly saline brines (around 20 wt% NaCl eq.) with homogenization temperatures of about 115–173 °C at Tsumeb, 190–266 °C at Khusib Springs and 92–210 °C at Berg Aukas (Misiewicz 1988; Chetty & Frimmel 2000; Melcher *et al.* 2006). Low-salinity fluids are regarded as secondary. Homogenization temperatures of two-phase (liquid + vapour) inclusions from different Pb–Zn quartz vein generations (King 1990) reflect a cooling history from high (180–215 °C), through to moderate (150–170 °C) to low (125–145 °C). All of these fluid inclusion data are related solely to the mineralized areas; no fluid inclusion data are available from areas further distant to ore mineralizations, except for the report by Ypma (1979*a*, *b*) that gave homogenization temperatures of 120–150 °C from Tsumeb subgroup dolomites.

To the north, east and south (Figs 1, 3 & 4) the OML is covered by younger sediments largely belonging to the Karoo and Kalahari successions; the latter also occurs locally inside the OML. Early Cretaceous basalts of the Etendeka Group, related to the opening of the South Atlantic, might have once extended to the OML (Marsh *et al.* 2003) and might have affected the temperature history. Calc-alkaline lamprophyres occur locally that are presumably Cretaceous in age (e.g. kersantites at the Tsumeb pipe infill: Söhnge 1957). Much younger are vanadate ores in the OML (Boni *et al.* 2007), which formed at temperatures of 40–50 °C and are related to erosional episodes of the 'African Erosional Cycle', controlled by the tectonomorphological evolution during the drifting stages of the Southern Atlantic.

Ypma (1979*a*, *b*) mentioned several indicators for the non-metamorphic nature of the OML succession, which are largely still valid today.

- Expandable clays (montmorillonite as well as smectite) are reported from carbonate successions of the OML, as well as from the Etosha 5-1A and 1-1 wells (Clauer & Kröner 1979).
- Delicate palynomorphs were reported from Proterozoic dolomites (internal Gulf report 1974 and own data; see above).
- Luminescence studies show 20 m-wide, higher temperature halos around veins, with a drop in temperature from 230 °C in the vein to 150 °C in the surrounding rock.

- Illite crystallinity data place the Owambo Formation of Stratigraphic Test No. 1 in the middle–upper anchizone, with Kubler indices of 4.0–5.5. This correlates to the bituminous coal, wet gas field.
- The reported oil seepage in Etosha well 5-1A (Fig. 4; see the earlier discussion) indicates the presence of hydrocarbons.

These data are supported by the results of the thermal alteration analysis of the organic matter from the dark micritic limestones of the Maieberg cap carbonate sequence in core KH1, as discussed above. The light- to medium-brown colours of non-degraded OM fit to TAI 3− to 3, which correlates to maximum temperatures of about 100–130 °C. Although well KH1 (Khusib Springs) is located in the area of strong hydrothermal activity, palaeotemperatures derived from OM alteration are much lower than temperatures obtained from fluid inclusion studies in this area. Thermal alteration of OM is related to long-term thermal overprint (10^6 years) controlled by the burial history of the whole area. Relatively short-lived and local hydrothermal events are not recorded in the OM, but in the fluid inclusions. Therefore, the lower palaeotemperatures from the organic maturation represent the thermal overprint resulting from the burial history of the OML, whereas the higher palaeotemperatures derived from fluid inclusions record locally restricted systems of hot fluid flow, associated with the ore mineralizations in the OML.

Ypma (1979*a*, *b*) pointed out that temperatures effective for maturation never exceeded 150 °C for extended periods of time. These temperatures are in contrast to those in the metamorphic Damara belt further to the south. Martin & Porada (1977*a*, *b*) describe the sedimentary facies of the Damara Belt and the OML as fault controlled, as is the temperature gradient between metamorphic rocks in the south and the largely non-metamorphic successions of the OML. Further to the north, in the Owambo Basin, the palaeotemperatures might have been even lower.

Discussion

Preglacial succession

The thick oolite successions underneath Ghaub diamictites in much of the northwestern Keilberg area are probably not *in situ* but were shed from the platform. This is indicated by the intercalation of oolite layers and microsparites, as well as breccia and slump successions. The oolites are not necessarily products of tropical warm waters, as indicated by the occurrence of glendonite within comparable oolites elsewhere (James *et al.* 2005).

Large-scale columnar stromatolites do not occur in the succession underneath the diamictites of the northern Keilberg Farm; they do occur, however, further to the south (southern Keilberg, Auros and Gauss farms, and many other areas) where diamictites are missing. These differences in facies show that in pre-Ghaub time a distinct relief was already present.

The Ghaub diamictite interval

The diamictites have only been found in western parts of the OML, and were deposited on a slope similar to outcrops in northwestern Namibia (Hoffman *et al.* 1998*a*, *b*). The clasts in the Ghaub Formation are locally derived. Exotic clasts are absent, while oversized clasts of up to metre size occur slightly underneath the cap carbonates. Some authors interpret the Ghaub Formation as debris flow sheddings, without indications of glacial influence (Schermerhorn 1974; Martin *et al.* 1985; Henry *et al.* 1986; Eyles & Januszczak 2007). This is questioned by possible dropstones that occur in beds transitional to and, rarely, also within the cap carbonates. Although faceted clasts are rare, they have been found in sandstone lithologies originating from the Nosib Group. One striated clast (quartzite) has been detected; however, striated pavements have never been observed.

Intercalated fine clastic layers clearly separate different diamictite units and probably represent time periods when matrix was winnowed away and the winnowed material was deposited locally. The intercalations thin out laterally and cannot be correlated, but increase in abundance westwards. Some of these show ripple marks (Fig. 8c), others show grain-support and graded bedding (Fig. 8d). These features indicate different stages of shedding and clearly show the activity of currents in the depositional area.

The sedimentary units within the Ghaub Formation indicate deeper and deeper erosion in the source area. The changing clast content (vertically as well as laterally) outlines different clast zones, which are imbricated on top of one another (Figs 7 & 10). In the east (= early sheddings) the spectrum of clast types is restricted to the carbonates of the directly underlying succession, whereas a wider spectrum including clasts from the Nosib Group and possibly the crystalline basement (or recycled lower Nosib Group, rich in crystalline material) is to be found in central and western areas, which were reached later by diamictite deposition. This indicates westwards progradation of different shedding events related to the unroofing of a source area. The crystalline clasts were supplied from parts of the succession, which are stratigraphically at least 2000 m underneath the Ghaub (1500 m of compacted Abenab, Gauss and Auros formations, several hundreds of metres of Nosib Group sediments, decompacted by about 20%). The sedimentary pile must have been strongly uplifted before erosion, distinctly above the area of deposition. An uplift of about 3000 m, probably at some distance to the depositional area, is a conservative value therefore. There is no indication that the depositional area on the slope was bordering a platform margin fault cutting into the underlying succession and exposing it.

A distinct uplift is also assumed by Eyles & Januszczak (2007). In the contrasting interpretation by Hoffman (e.g. 2005), horst blocks were created only during the Nosib rifting episode and, subsequently, draping of the relief filled an assumed small depression to the Congo Craton in the north. According to this author there is no progradation of the Otavi carbonate platform towards the basinal area to the south: the margin of the carbonate platform was stable throughout approximately 200 Ma. Despite the assumed lack of progradation, Bahama-type slope gradients are assumed (e.g. Hoffman 2005), although the Bahamas represent a strongly prograding platform. We also do not know whether the Otavi carbonate platform was attached to the Congo Craton in the north or if it was really a Bahama-like isolated platform, as assumed by Hoffman (e.g. 2005). The given estimates of water depths of the Ghaub diamictites, depending on the distance to the platform break and modelled according to today's slope gradients of the Bahamas, are speculative. It is even questionable whether a distinct platform-break occurred at all. A distally steepened ramp setting is also a possibility (Fig. 11). Ongoing tectonism affected the carbonate platform again and again, before and after the Ghaub interval, and the margin prograded and retreated.

Most of the sedimentological patterns argue for an origin of the Ghaub Formation as a product of different debris flows, interrupted by winnowing and hydrological activity. A 'rain' of material from melting marine ice would not have created imbricated sedimentary units with clear trends in clast content. The sheddings were clearly favoured by synsedimentary tectonism, creating the necessary relief. Tectonic uplift might have favoured glaciations in the source areas, providing material for the debris flows that might have been additionally promoted by a fall in base level. From the source area, the clastic material might have been transported in feeder channels (incised valleys?) via the platform–ramp to the sea. Ice might even have reached the sea locally. There is, however, no indication for an ice-covered carbonate platform.

The existence of potential dropstones inside these debris flow deposits is difficult to prove or disprove, but faceted and striated clasts occur. At the top, in the uppermost diamictites and the overlying cap carbonates, a few potential dropstones have been found, for example oversized (up to 1 m) carbonate clasts. This interval with larger clasts is overlain by waterlaid, sorted conglomerates, which themselves are covered by laminated cap carbonates, locally containing some clasts at their base. Clasts of a few centimetres in size are embedded in these lowermost cap carbonates in northwestern parts of the Keilberg area, above diamictites, and might represent dropstones. Some floating ice might have been present, therefore, at the end of the Ghaub Formation, when the glaciation receded and ice masses potentially broke apart.

Areas without Ghaub diamictites

Large parts of the Auros succession, which contain large stromatolites and other shallow-water fabrics, are not covered by diamictites. Diamictites are restricted, as already mentioned, to slope settings. Cap carbonates cover slope settings with diamictites, as well as the carbonate platform bearing no diamictites.

Two points are of interest in this respect.

(a) Terrestrially deposited glacial sediments should be expected to occur on the platform, if a global 'Snowball Earth' with a concomitant sea-level drawdown existed. This is not the case. In such a scenario a pavement of diamictite clasts and fluvioglacial sediments, deposited to a large part during the retreat of the ice, is to be expected. Such clasts should have been reworked and incorporated into the basal cap carbonate as palimpsest sediment during the post-glacial flooding, even if larger parts of the original pavement had been eroded away. The absence of diamictite clasts underneath and within lower parts of the cap deposits covering the platform strongly questions an ice cover for the platform: it is very unlikely that a large-scale grounding ice sheet covered the platform. No extensive grounded ice existed on the platform, and the diamictite clasts were transported via feeder channels to the depositional areas on the slope. After melting and flooding, no sediment-laden icebergs travelled over the platform–ramp (in contrast to assumptions of the Snowball Earth model). One might assume that the top of the platform–ramp was flooded and cap carbonates were deposited at a time when all remaining sea ice nearby had already melted away. The lack of dropstones on the platform might, therefore, be taken as an argument that flooding needed some time.

(b) The lack of erosional and karstic features is puzzling, even assuming scenario (a). If shallow platform areas were emergent during an assumed glacial sea-level drawdown, and diamictites were transported and partly deposited in feeder channels as well as on the slope and in the basin, karstic and erosional features on the platform are to be expected: these have not been found nor described in the literature until now. The absence of either diamictite veneers on the platform or karstification and erosion of the platform are difficult to reconcile with a worldwide long-lasting 'Snowball Earth' glaciation associated with a distinct fall of sea level. These facts are better in line with local glaciations and a Slushball Earth model, causing a less pronounced drop in sea level. In addition, the platform–ramp was probably compartmentalized, allowing the diamictites to bypass the shallower areas.

Cap carbonates and overlying succession

On the platform–upper ramp, diamictites are missing and no break in sedimentation was observed that might correspond to the diamictite interval. Only in these settings do enigmatic tube-like fabrics, more or less perpendicular to bedding, occur within microbial laminites. These fabrics overlie the Auros Formation, as at the southern Keilberg, Auros and Gauss farms (see below). When the Ghaub diamictite is present, tubular structures are absent almost everywhere. Only a few outcrops are known where tubular structures occur on thin remnants of diamictites (Hoffmann & Schrag 2002; K. H. Hoffmann oral comm. 2007). These can be interpreted as interfingering between the normally separated depositional areas of tubestones and diamictites. These observations further constrain palaeogeographical setting and regional variations in palaeowater depth.

We observed no evidence for an unconformity underneath or on top of this tubular facies, in contrast to the basal contacts of the diamictites in slope settings. Although the tubular facies is incorporated stratigraphically in the cap carbonate interval (Fig. 11) (Hoffmann & Prave 1996), the laminites might be a syndiamictite facies, whereas the 'tubes' within these laminites probably formed subsequently and represent an early diagenetic syn- or post-diamictite facies. During synglacial laminite deposition clathrates might have been emplaced within the laminae, which melted during glacial retreat, causing the tubular facies. This assumption is largely in line with the model of Kennedy *et al.* (2001*a*, *b*, 2002). Alternatively, the tubes within the laminites have a synglacial stromatolitic origin (James *et al.* 2001; Hoffman & Schrag 2002; Corsetti & Grotzinger 2005) and were formed during growth of the microbial laminae. Because of the increase and decrease in the number of tubes within an interval containing apparently the same type of stromatolitic laminites we favour the former explanation.

The probability that diamictites and finely laminated tubestone rocks are lateral equivalents, and the lack of karstic features underneath or on top of the tubestone facies, allow assumptions to be made on the number of the assumed sea-level fluctuations.

- The outer part of the platform or ramp was not exposed for any long period of time, sea-level drawdown directly preceding the diamictites might therefore have been weak or short-lived as indicated by the lack of karstic fabrics. Inner parts of the platform–ramp might, however, have been exposed. Detailed fieldwork is necessary to prove the existence of incised valleys or feeder channels in inner parts of a compartmentalized platform–ramp.
- Diamictites and microbial facies with tube-like fabrics are both covered by wavy laminated 'classical' cap carbonates, indicating relatively uniform conditions during their deposition. The deposition of cap carbonates might have taken a substantial period of time, as was assumed by Trindade *et al.* (2003) and Raub *et al.* (2007), because of the presence of magnetic reversals in South American and Australian post-glacial cap carbonate successions. Our interpretation of a weak glacial relative sea-level fall (see above) is in contrast to the 'Snowball Earth' model, assuming a distinct sea-level drop. Because of the extent of this assumed fall, during the subsequent slow but long-lasting rise, the deposition of the same type of facies is said to have occurred in the basin much earlier than on the platform (Hoffman 2005). If, however, the sea-level rise was sudden, as assumed by earlier versions of the 'Snowball Earth' model (Hoffman *et al.* 1998*b*), then deposition of deep-water facies and more shallow-water facies of cap carbonates would have occurred at almost the same time and the uniformity of the cap carbonate facies in basinal and platform settings would be even more difficult to explain.
- In different parts of the world there is an association of early as well as late Cryogenian cap carbonates with overlying fine-grained and often dark, partly shaly, pyritic carbonates, containing a greater amount of organic material. In the OML this succession is more than 100 m thick. These long-lasting worldwide post-glacial transgressions are arguments for a large volume of ice, which melted over a considerable amount of time. Partly cyclic facies repetitions within the late Cryogenian post-glacial succession

might indicate sea-level fluctuations and concomitant fluctuations of the ice volume. A part of the deepening of the succession might also be the result of increased subsidence and/or tilting, as indicated by thick debris flows.

- The assumption of a long-lasting late Cryogenian post-glacial transgression is in contrast to the mentioned lack of indications of a preceding synglacial regression. Ice build-up might, therefore, have taken quite some time (again, in contrast to 'Snowball Earth'), allowing subsidence to counterbalance the slow eustatic sea-level fall caused by glaciation. Alternatively, a faster sea-level drop was balanced by increased subsidence resulting from tensional tectonics. This asymmetry (slow ice build-up, fast melting) is in agreement with the situation during the Quaternary glaciations. In any one of these scenarios, the tubestone facies at the outer platform probably represents a shelf margin systems tract and not a lowstand systems tract.
- During the Late Proterozoic different types of glaciations (ice forming at sea level, ice formed at higher altitudes, with glaciers reaching the sea) affected the Earth (Leather *et al.* 2002), even in low latitudes. For the initiation of glaciers transporting larger masses of debris to the sea, factors like rifting and uplift were favourable (see Eyles & Januszczak 2004, 2007). The frequent association of diamictites with rift basins could be due, however, to the combination of accommodation and preservation in the newly formed rift troughs, and also to a more favourable setting for glaciations caused by the uplift of rift shoulders, whereas other types of glaciation are less well documented in the sediments. There is no indication of one, uniform, longer-lasting Snowball event.

The organic-rich post-Ghaub succession

The early Cryogenian ('Sturtian') deposits are especially influenced by rifting and might be strongly diachronous in different parts of the world, whereas the late Cryogenian ('Marinoan') successions seem to be better correlatable. From an applied point of view, the 'academic' question of tectonic-related or glacial-related deposits (or, more probably, a combination of both) is less important than the creation of a petroleum system and its clear linkage to the aftermath of diamictite deposition. A relief was created before and during the glacial periods, which was then flooded, and the different sub-basins were affected by restricted circulation for quite some time. This general scenario bears many resemblances to the extremely economically important early Silurian source rocks formed during post-glacial sea-level rise (Lüning *et al.* 2000*a*).

Dark, fine-grained carbonate successions, which in the OML are very rich in pyrite and bear a considerable amount of OM, were deposited during this transgression. These are found on top of cap carbonates covering diamictites, interpreted as slope setting, but also in areas deprived of the diamictites. The content in OM might be higher in slope settings, but this is far from being proven at this moment because no core material was available from these areas and TOC from outcrops is diminished during weathering. Most of the published research has concentrated on the diamictites and the cap carbonates, whereas the overlying succession was largely neglected. This change to a deeper-water setting on top of assumed glacial sediments in the late Cryogenian is present in many parts of the world. The trinity of diamictites, cap carbonates and suboxic–anoxic often-pyritic dark carbonates–shales (Shields 2005; Stewart 2007) is characteristic of the late Cryogenian ('post-Marinoan') transgressive, shelf margin and highstand systems tracts in several areas, even when their correlatibility is questionable. Some examples are the Mackenzie Mountains (James *et al.* 2001), Greenland and Svalbard (Halverson *et al.* 2004), and the SW Amazon Craton in Brazil (Noguiera *et al.* 2003). This co-occurrence is an argument against a 'rift only' hypothesis (Eyles & Januszczak 2007). In Namibia, the probably post-glacial transgression was enhanced by increased subsidence resulting from long-lasting tensional tectonics.

Conclusion

The assumed succession of events is as follows (see also Fig. 11):

- Preglacial successions are characterized by different facies, indicating a strong relief. Sea-level fall might have been started distinctly before the signs of glaciation in the area. The sea-level drop was largely balanced by subsidence, keeping some accommodation potential on the platform (shelf margin systems tract).
- The diamictite clasts were the result of an uplift of several thousands of metres in the hinterland and subsequent erosion down to the crystalline basement. There are no indications of grounded ice on the platform; the diamictite sediment bypassed the compartmentalized platform probably via feeder channels or incised valleys (not observed so far) and was deposited on the slope. Different imbricated debris flows are distinguished, only the later ones contain crystalline material. These debris shedding events were separated by hydrodynamic activity, partly winnowing fine-grained material from the top of the sedimentary intervals.

- Laminites were deposited on the platforms as a sedimentary deposit coeval to the diamictites. The tubes within the laminites might be a later diagenetic feature, occurring during post-glacial warming and sea-level rise, or formed (less likely) during stromatolitic growth.
- Subsequently, wavy bedded cap carbonates of relatively uniform facies covered the diamictites on the slope, as well as the tubestone facies of laminites on the platform.
- The mainly tectonically enhanced relief was flooded possibly due to a combination of post-glacial sea-level rise and strong subsidence. The different sub-basins were affected by restricted circulation for quite some time. This general scenario for the late Cryogenian potential hydrocarbon system bears many resemblances to the economically extremely important North African Early Silurian system, also formed during post-glacial sea-level rise.

We thank K.-H. Hoffmann, V. Petzel, G. Schneider, U. Schreiber, E. Shiweda, J. van Tonder and R. Wackerle of the Namibian Geogical Survey, and A. Günzel of the former Ongopolo Mining Ltd, for valuable discussions, assistance and logistical help over several field seasons in Namibia. We are especially indebted to the Volkmann family, Mr and Mrs Freyer, and Mr Schneider for allowing us generous access to their farms for fieldwork and for their support of our work. Our special thanks go to the late W. Volkmann. We wish to thank P. Allen and G. Shields for constructive reviews that helped improve the manuscript.

We gratefully acknowledge support by Deutsche Forschungsgemeinschaft (DFG Graduate College 273 on Fluid–rock Interaction, as well as grant Be 641/41). This is a contribution to IGCP 512. The IGCP network helped us to shape some of our ideas.

References

ALLEN, P. A. 2006. Snowball Earth on trial. *Eos, Transactions of the American Geophysics Union*, **87**, 495.

ALLEN, P. A. & LEATHER, J. 2006. Post-Marinoan siliciclastic sedimentation: The Masirah Bay Formation, Neoproterozoic Huqf Supergroup of Oman. *Precambrian Research*, **144**, 167–198.

ALLEN, P. A., LEATHER, J. & BRASIER, M. D. 2004. The Neoproterozoic Fiq glaciation and its aftermath, Huqf Supergroup of Oman. *Basin Research*, **16**, 507–534.

ALLSOPP, H., WELKE, H. J. & HIGHES, M. J. 1981. Shortening the odds in exploration. *Nuclear Active*, **24**, 8–12.

BEUKES, N. J. 1986. *A Field Introduction to the Geology of the Otavi Mountainland, Northern Namibia*. Workshop on Precambrian Carbonate Sedimentology. Tsumeb Corp. Ltd, Tsumeb.

BONI, M., TERRACCIANO, R., EVANS, N. J., LAUKAMP, C., SCHNEIDER, J. & BECHSTÄDT, T. 2007. Genesis of vanadium ores in the Otavi Mountainland (Namibia). *Economic Geology*, **102**, 441–469.

BOTHA, B. J. B. 1960. *The Arenaceous Rocks and the Pseudo-aplite of the Otavi Mountainland, South West Africa*. PhD thesis, University of Pretoria.

BUJAK, J. P., BARSS, M. S. & WILLIAMS, G. L. 1977. Offshore east Canada's organic type and colour and hydrocarbon potential. *Oil and Gas Journal*, **75**, 198–201.

BURGER, A. J. & COERTZE, F. J. 1973. Radiometric age measurements of rocks from Southern Africa to the end of 1971. *Geological Survey of South Africa, Bulletin*, **58**.

CAIRNCROSS, B. 1997. The Otavi Mountain Land Cu–Pb–Zn–V deposits. *Mineralogical Record*, **28**, 109–157.

CATLING, D. C. & CLAIRE, M. W. 2005. How Earth's atmosphere evolved to an oxic state: A status report. *Earth and Planetary Science Letters*, **237**, 1–20.

CHETTY, D. & FRIMMEL, H. E. 2000. The role of evaporites in the genesis of base metal sulphide mineralization in the northern platform of the Pan-African Damara Belt, Namibia; geochemical and fluid inclusion evidence for carbonate wall rock interaction. *Mineralium Deposita*, **35**, 364–376.

CLAUER, N. & KRÖNER, A. 1979. Strontium and argon isotopic homogenization of pelitic sediments during low grade regional metamorphism: The Pan-African Upper Damara sequence of Northerrn Namibia (South West Africa). *Earth and Planetary Science Letters*, **43**, 117–131.

CLOUD, P. E., WRIGHT, L. A., WILLIAMS, E. G., DIEHL, P. E. & WALTER, M. R. 1974. Giant stromatolites and associated vertical tubes from the upper Proterozoic Noonday Dolomite, Death Valley region, east California. *Geological Society of America Bulletin*, **85**, 1869–1882.

CONDIE, K. C. 2002. Breakup of a Paleoproterozoic supercontinent. *Gondwana Research*, **5**, 41–43.

CONDON, D., ZHU, M., BOWRING, S., WANG, W., YANG, A. & JIN, Y. 2005. U–Pb ages from the Neoproterozoic Doushantuo Formation, China. *Science*, **308**, 95–98.

CORSETTI, F. A. & GROTZINGER, J. P. 2005. Origin and significance of the tube structures in Neoproterozoic post-glacial cap carbonates: Examples from the Noonday Dolomite, Death Valley, United States. *Palaios*, **20**, 348–362.

CORSETTI, F. A. & KAUFMAN, A. J. 2003. Stratigraphic investigations of carbon isotope anomalies and Neoproterozoic ice ages in Death Valley, California. *Geological Society of America Bulletin*, **115**, 916–932.

CRAIG, J., THUSU, B., LÜNING, S., MECIANI, L., TROMBETTI, A. & ERCHI, G. 2006. Snowball Earth and global Neoproterozoic petroleum Systems. *In*: Snowball Earth 2006, 16–21 July, Monte Verita, Ticino, Switzerland, Abstract Volume. ETH Zurich, 25.

DOWNING, K. N. 1983. The stratigraphy and palaeoenvironment of the Damara Sequence in the Okahandja Lineament Area. *In*: MILLER, R. MCG. (ed.) *Evolution of the Damara Orogen*. Geological Society of South Africa, Special Publications, **11**, 37–42.

DÜRR, S. B. & DINGELDEY, D. P. 1997. Tale of three cratons: Tectonostratigraphic anatomy of the Damara orogen in northwestern Namibia and the assembly of Gondwana: Comment. *Geology*, **25**, 1149–1151.

EMBELTON, B. J. J. & WILLIAMS, G. E. 1986. Low palaeolatitude of deposition for late Precambrian varvities in South Australia: Implications for palaeoclimatology. *Earth and Planetary Science Letters*, **79**, 419–430.

EVANS, D. A. D. 2000. Stratigraphic, geochronological and palaeomagnetic constraints upon the Neoproterozoic climatic paradox. *American Journal of Science*, **300**, 347–433.

EYLES, N. & JANUSZCZAK, N. 2004. 'Zipper Rift': a tectonic model for Neoproterozoic glaciations during the break-up of Rodinia after 750 Ma. *Earth-Science Reviews*, **65**, 1–73.

EYLES, N. & JANUSZCZAK, N. 2007. Syntectonic subaqueous mass flows of the Neoproterozoic Otavi Group, Namibia: where is the evidence of global glaciation? *Basin Research*, **19**, 179–198, doi:10.1111/j.1365-2117.2007.00319.x.

FAIRCHILD, I. J. & HAMBREY, M. J. 1995. Vendian basin evolution in East Greenland and NE Svalbard. *Precambrian Research*, **73**, 217–233.

FAIRCHILD, I. F. & KENNEDY, M. J. 2007. Neoproterozoic glaciation in the Earth System. *Journal of the Geological Society, London*, **164**, 895–921.

FAIRCHILD, I. J., SPIRO, B., HERRINGTON, P. M. & SONG, T. 2000. Controls on Sr and C isotope compositions of Neoproterozoic Sr-rich limestones of E Greenland and N China. *In*: GROTZINGER, J. P. & JAMES, N. P. (eds) *Carbonate Sedimentation and Diagenesis in the Evolving Precambrian World*. SEPM, Special Publications, **67**, 297–313.

FÖLLING, P. G. & FRIMMEL, H. E. 2002. Chemostratigraphic correlation of carbonate successions in the Gariep and Saldania belts, Namibia and south Africa. *Basin Research*, **14**, 69–88.

FRETS, D. C. 1969. Geology and structure of the Welwitschia area, South West Africa. *Precambrian Research Unit, University of Cape Town Bulletin*, **5**, 1–235.

FRIMMEL, H. E. 2004. Neoproterozoic sedimentation rates and timing of glaciations – a southern African perspective. *In*: ERIKSSON, P. G., ALTERMANN, W., NELSON, W. U., MUELLER, O. & CATUNEANU, O. (eds) *The Precambrian Earth: Tempos and Events*, Developments in Precambrian Geology, 12. Elsevier, Amsterdam, 460–472.

FRIMMEL, H. E., DEANE, J. G. & CHADWICK, P. J. 1996*a*. Pan-African tectonism and the genesis of base metal sulphide deposits in the northern foreland of the Damara Orogen, Namibia. *In*: SANGSTER, D. F. (ed.) *Carbonate Hosted Lead–Zinc Deposits*. Society of Exploration Geophysicists, Special Publications, **4**, 204–217.

FRIMMEL, H. E., KLÖTZLI, U. S. & SIEGFRIED, P. R. 1996*b*. New Pb–Pb single zircon constraints on the timing of Neoproterozoic glaciation and continental break-up in Namibia. *Journal of Geology*, **104**, 459–469.

GKW CONSULT & BICON NAMIBIA. 2002. *Tsumeb Groundwater Study, Final Report Volume 2 – Geological/Geophysical Surveys and Borehole Siting*. Unpublished report, file No.12/1/2/16/2. Department of Water Affairs (DWA), Windhoek.

GREY, K. 2005. *Ediacaran Palynology of Australia*. Memoirs of the Association of Australasian Palaeontologists, **31**.

GREY, K., WALTER, M. R. & CALVER, C. R. 2003. Neoproterozoic biotic diversification: Snowball Earth or aftermath of the Acraman impact? *Geology*, **31**, 459–462.

GROBLER, N. J. 1961. *The Geology of the Western Otavi Mountainland, South West Africa*. MSc thesis, University of O.F.S, Bloemfontein.

GUJ, P. 1974. A revision of the Damara stratigraphy along the southern margin of the Kamanjab Inlier, South West Africa. *Precambrian Research Unit, University of Cape Town Bulletin*, **5**, 167–176.

HAACK, U. 1993. Critical note on lead-lead model ages. *In*: MÖLLER, P. & LÜDERS, V. (eds) *Formation of Hydrothermal Vein Deposits – A Case Study on the Pb–Zn, Barite and Fluorite Deposits of the Harz Mountains*. Monograph Series on Mineral Deposits, **30**, 115–116.

HALVERSON, G. P. 2006. A Neoproterozoic chronology. *In*: XIAO, S. & KAUFMAN, A. J. (eds) *Neoproterozoic Geobiology and Palaeobiology*. Springer, Berlin, 231–271.

HALVERSON, G. P., HOFFMAN, P. F., SCHRAG, D. P. & KAUFMAN, A. J. 2002. A major perturbation of carbon cycle before the Ghaub glaciation (Neoproterozoic) in Namibia: prelude to snowball Earth? *Geochemistry, Geophysics, Geosystems*, **3**, 1035.

HALVERSON, G. P., HOFFMAN, P. F., SCHRAG, D. P., MALOOF, A. C. & RICE, A. H. N. 2005. Toward a Neoproterozoic composite carbon-isotope record. *Geological Society of America Bulletin*, **117**, 1181–1207.

HALVERSON, G. P., MALOOF, A. C. & HOFFMAN, P. F. 2004. The Marinoan glaciation (Neoproterozoic) in northeast Svalbard. *Basin Research*, **16**, 297–324.

HARLAND, W. B. 1964. Critical evidence for a great infra-Cambrian glaciation. *Geologische Rundschau*, **54**, 45–61.

HARTNADY, C., JOUBERT, P. & STOWE, C. 1985. Proterozoic crustal evolution in southwestern Africa. *Episodes*, **8**, 236–244.

HEDBERG, R. M. 1976. *Stratigraphy of the Owamboland Basin, South West Africa*. PhD thesis, Harvard University.

HEDBERG, R. M. 1979. Stratigraphy of the Owamboland Basin, South West Africa. *Bulletin of the Precambrian Research Unit, University of Cape Town, South Africa*, **24**.

HEGENBERGER, W. 1987. Gas escape structures in Precambrian peritidal carbonate rocks. *Communications of the Geological Survey SW Africa/Namibia*, **3**, 49–55.

HEGENBERGER, W. 1993. *Stratigraphy and Sedimentology of the Late Precambrian Witvlei and Nama Groups, East of Windhoek*. Geological Survey of Namibia, Memoirs, **17**.

HENRY, G., STANISTREET, I. G. & MAIDEN, K. J. 1986. Preliminary results of a sedimentological study of the Chuos formation in the central zone of the Damara orogeny: Evidence for mass flow processes and glacial activity. *Communications of the Geological Survey of SW Africa/Namibia*, **2**, 75–92.

HOFFMAN, P. F. 2002. Carbonate bounding glacial deposits: Evidence for Snowball Earth episodes and greenhouse aftermaths in the Neoproterozoic Otavi Group of Northern Namibia (Field Guide). *In*: *16th International Sedimentological Conference, 1–7 July, South Africa.* Rand Africaans University, Johannesburg.

HOFFMAN, P. F. 2005. On Cryogenian (Neoproterozoic) ice sheet dynamics and the limitations of the glacial sedimentary record. *In*: *28th DeBeers Alex. Du Toit Memorial Lecture, 2004. South African Journal of Geology*, **108**, 557–577.

HOFFMAN, P. F. & ALLEN, P. A. 2007. Snowball Earth on trial; discusson and reply. *EOS, Transactions of the American Geophysics Union*, **88**, 110.

HOFFMAN, P. F. & SCHRAG, D. P. 1999. Considering a Neoproterozoic Snowball Earth. Response. *Science*, **284**, 1087–1088.

HOFFMAN, P. F. & SCHRAG, D. P. 2000. Snowball Earth. *Scientific American*, **285**, 50–57.

HOFFMAN, P. F. & SCHRAG, D. P. 2002. The snowball Earth hypothesis: testing the limits of global change. *Terra Nova*, **14**, 129–155.

HOFFMAN, P. F., HAWKINS, D. P., ISACHSEN, C. E. & BOWRING, S. A. 1996. Precise U–Pb zircon ages for Early Damaran magmatism in the Summas Mountains and Welwitschia Inlier, northern Damara Belt, Namibia. *Communications of the Geological Survey of Namibia*, **11**, 47–52.

HOFFMAN, P. F., KAUFMAN, A. J. & HALVERSON, G. P. 1998*a*. Comings and goings of global glaciations on a Neoproterozoic tropical platform in Namibia. *GSA Today*, **8**, 1–9.

HOFFMAN, P. F., KAUFMAN, A. J., HALVERSON, G. P. & SCHRAG, D. P. 1998*b*. A Neoproterozoic Snowball Earth. *Science*, **281**, 1342–1346.

HOFFMANN, K. H. 1990. Sedimentary depositional history of the Damara Belt related to continental breakup, passive margin to active margin transition and foreland basin development. *Extended Abstracts*. Geocongress 90. Geological Society of South Africa, Cape Town, 250–253.

HOFFMANN, K. H. & PRAVE, A. R. 1996. A preliminary note on a revised subdivision and regional correlation of the Otavi Group based on glaciogenic diamictites and associated cap dolostones. *Communications of the Geological Survey of Namibia*, **11**, 77–82.

HOFFMANN, K. H., CONDON, D. J., BOWRING, S. A. & CROWLEY, J. L. 2004. U–Pb zircon date from the Neoproterozoic Ghaub Formation, Namibia: Constraints on Marinoan glaciation. *Geology*, **32**, 817–820.

HUGHES, M. J., WELKE, H. J. & ALLSOPP, H. L. 1984. Lead isotopic studies of some Late Proterozoic stratabound ores of central Africa. *Precambrian Research*, **25**, 137–139.

HURTGEN, M. T., ARTHUR, M. A. & HALVERSON, G. P. 2005. Neoproterozoic sulphur isotopes, the evolution of microbial sulphur species, and the burial efficiency of sulphide as sedimentary pyrite. *Geology*, **33**, 41–44.

JACOBSEN, S. B. & KAUFMAN, A. J. 1999. The Sr, C and O isotopic evolution of Neoproterozoic seawater. *Chemical Geology*, **161**, 37–57.

JAMES, N. P., NARBONNE, G. M. & KYSER, T. K. 2001. Late Neoproterozoic cap carbonates, Mackenzie Mountains, northwestern Canada: precipitation and global ice meltdown. *Canadian Journal of Earth Sciences*, **38**, 1229–1262.

JAMES, N. P., NARBONNE, G. M., DALRYMPLE, R. W. & KYSER, T. K. 2005. Glendonites in Neoproterozoic low-latitude, interglacial, sedimentary rocks, northwest Canada: Insights into the Cryogenian ocean and Precambrian cold-water carbonates. *Geology*, **33**, 9–12.

JIANG, G., KENNEDY, M. J. & CHRISTIE-BLICK, N. 2003. Stable isotopic evidence for methane seeps in Neoproterozoic post-glacial cap carbonates. *Nature*, **426**, 822–826.

KALJO, D., MARTMA, T., MÄNNIK, P. & VIIRA, V. 2003. Implications of Gondwana glaciations in the Baltic late Ordovician and Silurian and a carbon isotopic test of environmental cyclicity. *Bulletin de la Société Geologique de France*, **174**, 59–66.

KAMONA, A. F. & GÜNZEL, A. 2007. Stratigraphy and base metal mineralization in the Otavi Mountain Land, Northern Namibia – a review and regional interpretation. *Gondwana Research*, **11**, 396–413.

KAMONA, A. F., LÉVEQUE, J., FRIEDRICH, G. & HAACK, U. 1999. Lead isotopes of the carbonate-hosted Kabwe, Tsumeb and Kipushi Pb–Zn–Cu sulphide deposits in relation to Pan-African orogenesis in the Damaran–Lufilian fold belt of central Africa. *Mineralium Deposita*, **34**, 273–283.

KAUFMAN, A. J., HAYES, J. M., KNOLL, A. H. & GERMS, G. J. B. 1991. Isotopic composition of carbonates and organic carbon from upper Proterozoic successions in Namibia. Stratigraphic variations and effects of diagenesis and metamorphism. *Precambrian Research*, **49**, 301–327.

KAUFMAN, A. J., KNOLL, A. H. & NARBONNE, G. M. 1997. Isotopes, ice ages and terminal Proterozoic Earth history. *National Academy of Sciences Proceedings*, **94**, 6600–6605.

KENNEDY, M. J., CHRISTIE-BLICK, N. & PRAVE, A. R. 2001*a*. Carbon isotopic composition of the Neoproterozoic glacial carbonates as a test of palaeoceanographic models for snowball Earth phenomena. *Geology*, **29**, 1135–1138.

KENNEDY, M. J., CHRISTIE-BLICK, N. & SOHL, L. R. 2001*b*. Are Proterozoic cap carbonates and isotopic excursions a record of gas hydrate destabilization following Earth's coldest intervals? *Geology*, **29**, 443–446.

KENNEDY, M. J., CHRISTIE-BLICK, N. & SOHL, L. R. 2002. Are Proterozoic cap carbonates and isotopic excursions a record of gas hydrate destabilization following Earth's coldest intervals? Comment and reply. *Geology*, **30**, 287–288.

KENNEDY, M. J., RUNNEGAR, B., PRAVE, A. R., HOFFMANN, K. H. & ARTHUR, M. A. 1998. Two or four Neoproterozoic glaciations? *Geology*, **26**, 1059–1063.

KILNER, B., CONALL, MCN. & BRASIER, M. 2006. Low-latitude glaciation in the Neoproterozoic of Oman. *Geology*, **33**, 413–416.

KING, C. H. M. 1990. *The geology of the Tsumeb carbonate sequence and associated lead–zinc occurrences on the farm Olifantsfontein, Otavi Mountainland, Namibia*. MSc thesis, Rand Afrikaans University.

KING, C. H. M. 1994. *Carbonates and Mineral Deposits of the Otavi Mountainland. Excursion 4. Proterozoic Crustal and Metallic Evolution*. Geological Society and Geological Survey of Namibia, Windhoek.

KING, L. C. 1951. *South African Scenery: A Textbook of Geomorphology*. Oliver & Boyd, Edinburgh.

KRÖNER, A. & RANKAMA, K. 1972. Late Precambrian glaciogenic sedimentary rocks in southern Africa: a compilation with definitions and correlations. *Bulletin of the Precambrian Research Unit, University of Cape Town Bulletin*, **11**, 37.

KRÜGER, L. 1969. Stromatolites and oncolites in the Otavi Series, South West Africa. *Journal of Sedimentary Petrology*, **39**, 1046–1056.

KUKLA, P. 1992. *Tectonics and Sedimentation of a Late Proterozoic Damaran Convergent Continental Margin, Khomas Hochland, Central Namibia*. Geological Survey of Namibia, Memoir, **12.**

LAND, L. S. 1973. Holocene meteoric dolomitization of Pleistocene limestones, North Jamaica. *Sedimentology*, **20**, 411–424.

LAUKAMP, C. 2006. *Structural and fluid system evolution in the Otavi Mountainland (Namibia) and its significance for the genesis of sulphide and nonsulphide mineralization*. PhD thesis, University of Heidelberg.

LEATHER, J., ALLEN, P. A., BRASIER, M. D. & COZZI, A. 2002. Neoproterozoic snowball Earth under scrutiny: Evidence from the Fiq glaciation of Oman. *Geology*, **30**, 891–894.

LE GUERROUÉ, E., ALLEN, P. A. & COZZI, A. 2006. Parasequence development in the Ediacaran Shuram Formation (Nafun Group, Oman): primary origin stratigraphic test of negative carbon isotope ratios. *Basin Research*, **18**, 205–219.

LE GUERROUÉ, E., ALLEN, P. A. & COZZI, A. 2007. Chemostraigraphic and sedimentological framework of the largest negative carbon isotopic excursion in Earth history: The Neoproterozoic Shuram Formation (Nafun Group, Oman). *Precambrian Research*, **153**, 74–81.

LE ROEX, H. D. 1941. A tillite in the Otavi Mountains, S.W.A. *Transactions of the Geological Survey of South Africa*, **44**, 207–218.

LITTKE, R. & WELTE, D. H. 1992. Hydrocarbon source rocks. *In*: BROWN, G., HAWKESWORTH, CH. & WILSON, C. (eds) *Understanding the Earth*. Cambridge University Press, Cambridge, 364–374.

LÜNING, S., CRAIG, J. *ET AL*. 1999. Re-evaluation of the petroleum potential of the Kufra Basin (SE Libya, NE Chad): Does the source rock barrier fall? *Marine and Petroleum Geology*, **16**, 693–718.

LÜNING, S., CRAIG, J., LOYDELL, D. K., STORCH, P. & FITCHES, W. R. 2000*a*. Lowermost Silurian 'hot shales' in North Africa and Arabia: Regional distribution and depositional model. *Earth-Science Reviews*, **49**, 121–200.

LÜNING, S., LOYDELL, D. K., SUTCLIFFE, O., AIT SALEM, A., ZANELLA, E., CRAIG, J. & HARPER, D. A. T. 2000*b*. Silurian–Lower Devonian black shales in Morocco, which are the organically richest horizons? *Journal of Petroleum Geology*, **23**, 293–311.

MACGABHANN, B. A., ETIENNE, J. L., SHIELDS, G. & HALVERSON, G. P. 2006. Integrated global correlation of the Late Neoproterozoic: exploring the glacial roots of the metazoa. *In*: Snowball Earth 2006, 16–21 July, Monte Verita, Ticino, Switzerland, Abstract Volume, 69–70.

MACOUIN, M., BESSE, J., ADER, M., GILDER, S., YANG, Z., SUN, Z. & AGRINIER, P. 2004, Combined palaeomagnetic and isotopic data from the Doushantuo carbonates, south China; implications for the 'Snowball Earth' hypothesis. *Earth and Planetary Science Letters*, **224**, 387–389.

MACHEL, H. G. 2004. Concepts and models of dolomitization: a critical reappraisal. *In*: BRAITHWAITE, C. J. R., RIZZI, G. & DARKE, G. (eds) *The Geometry and Petrogenesis of Dolomite A. Hydrocarbon Reservoirs*. Geological Society, London, Special Publications, **235**, 7–63.

MARENCO, P. J. & CORSETTI, F. A. 2002. Noonday tubes: observation and reinterpretation based on better preservation from a new locality. *In*: CORSETTI, F. A. (ed.) *Proterozoic–Cambrian of the Great Basin and Beyond*. SEPM, Pacific Section Book, **93**, Fullerton, CA, 31–42.

MARSH, J. S., SWART, R. S. & PHILIPS, D. 2003. Implications of a new $^{40}Ar/^{39}Ar$ age for a basalt flow interbedded with thc Etjo Formation, northeast Namibia. *South African Journal of Geology*, **106**, 281–286.

MARTIN, H. 1965. *The Precambrian Geology of South West Africa and Namaqualand*. Precambrian Research Unit, University of Cape Town.

MARTIN, H. & PORADA, H. 1977*a*. The intercratonic branch of the Damara Orogen in South West Africa. I. Discussion of geodynamic models. *Precambrian Research*, **5**, 311–338.

MARTIN, H. & PORADA, H. 1977*b*. The intercratonic branch of the Damara Orogen in South West Africa. II. Discussion of relationships with the Pan-African Mobile Belt system. *Precambrian Research*, **5**, 339–357.

MARTIN, H., PORADA, H. & WALLISER, O. H. 1985. Mixtite deposits of the Damara sequence, Namibia, problems of interpretation. *Palaeogeography, Palaeoclimatology, Palaeoecology*, **51**, 159–196.

MEERT, J. G. & VAN DER VOO, R. 1996. Palaeomagnetic and $^{40}Ar/^{39}Ar$ study of the Sinyai dolerite, Kenya: Implications for Gondwana assembly. *Journal of Geology*, **104**, 131–142.

MEERT, J. G., VAN DER VOO, R. & AYUB, S. 1995. Palaeomagnetic investigation of the Neoproterozoic Gagwelavas and Mbozi complex, Tanzania and the assembly of Gondwana. *Precambrian Research*, **74**, 225–244.

MELCHER, F. 2003. The Otavi Mountain Land in Namibia: Tsumeb, Germanium and Snowball Earth. *Mitteilungen Österreichischer Geologischer Gesellschaft*, **148**, 413–435.

MELCHER, F., OBERTHÜR, T. & RAMMLMAIR, D. 2006. Geochemical and mineralogical distribution of germanium in the Khusib Springs Cu–Zn–Pb–Ag sulfide deposit, Otavi Mountain Land, Namibia. *Ore Geology Reviews*, **28**, 32–56.

MELCHER, F., OBERTHÜR, T., VETTER, U., GROSS, C., VOLLBRECHT, A., BRAUNS, M. & HAACK, U. 2003. Germanium in carbonate-hosted Cu–Pb–Zn mineralization in the Otavi Mountain Land, Namibia. *In*: ELIOPOULOS, D. (ed.) *Mineral Exploration and Sustainable Development. Proceedings of the 7th Biennial SGA Meeting, Athens*. Millpress, Rotterdam, 701–704.

MILLER, R. McG. 1980. *Geology of a portion of central Damarland, South West Africa/Namibia*. Memoirs of the Geological Survey of Southern Africa, SW Africa Series, **6**.

MILLER, R. McG. 1983. The Pan-African Damara orogen of South West Africa/Namibia. *In*: MILLER, R. MCG. (ed.) *Evolution of the Damara Orogen*. Geological Society of South Africa, Special Publications, **11**, 431–515.

MILLER, R. McG. 1997. The Owambo Basin of northern Namibia. *In*: SELLEY, R. C. (ed.) *African Basins (Sedimentary Basins of the World, Volume 3)*. Elsevier, Amsterdam, 237–268.

MISIEWICZ, J. E. 1988. *The Geology and Metallogeny of the Otavi Mountain Land, Damara Orogen, SWA/Namibia, With Particular Reference to the Berg Aukas Zn–Pb–V Deposit–A Model of Ore Genesis*. MSc thesis, Rhodes University, Grahamstown.

MORROW, D. W. 1990*a*. Dolomite–Part 1: the chemistry of dolomitization and dolomite precipitation. *In*: MCIIREATH, I. A. & MORROW, D. W. (eds), *Diagenesis*. Geoscience Canada, Reprint series, **4**, 113–124.

MORROW, D. W. 1990*b*. Dolomite – Part 2: dolomitization models and ancient dolostones. *In*: MCIIREATH, I. A. & MORROW, D. W. (eds) *Diagenesis*. Geoscience, Canada, Reprint Series, **4**, 125–139.

NOGUIERA, A. C. R., RICCOMINI, C., MOURA, C. A. V. & FAIRCHILD, T. R. 2003. Soft sediment deformation at the base of the Neoproterozoic Puga cap carbonate (southwestern Amazon craton, Brazil): confirmation of a rapid icehouse greenhouse transition in snownball Earth. *Geology*, **31**, 613–616.

PAGE, A., WILLIAMS, M. & ZALASIEWICZ, J. 2006. Were transgressive black shales a negative feedback modulating glacioeustasy in the Early Palaeozoic icehouse? *Geophysical Research Abstracts*, **8**, 09210. SRef-ID: 1607-7962/gra/EGU06-A-09210.

PAVLOV, A. A., HURTGEN, M. T., KASTING, J. F. & ARTHUR, M. A. 2003. Methane-rich Proterozoic atmosphere? *Geology*, **31**, 87–90.

PELTIER, W. R., LIU, Y. & CROWLEY, J. W. 2007. Snowball Earth prevention by dissolved organic carbon remineralization. *Nature*, **450**, 813–818.

PICKFORD, M. 1993. Age of supergene ore bodies at Berg Aukas and Harasib 3a, *Namibia. Communications of the Geological Survey of Namibia*, **8**, 147–150.

PICKFORD, M. 2000. Neogene and Quaternary vertebrate biochronology of the Sperrgebiet and Otavi Mountainland, Namibia. *Communications of the Geological Survey of Namibia*, **12**, 359–365.

PIRAJNO, F. & JOUBERT, B. D. 1993. An overview of carbonate-hosted mineral deposits in the Otavi Mountain Land, Namibia: implications for ore genesis. *Journal of African Earth Science*. **16**, 265–272.

PORADA, H. & WITTIG, R. 1983*a*. Turbidites in the Damara Orogen. *In*: MARTIN, H. & EDER, F. W. (eds) *Intracratonic Fold Belts, Case Studies in the Variscan Belt of Europe and the Damara Belt in Namibia*. Springer, Berlin, 543–576.

PORADA, H. & WITTIG, R. 1983*b*. Turbidites and their significance in the geosynclinal evolution of the Damara Orogen, South West Africa/Namibia. *In*: MILLER, R. MCG. (ed.) *Evolution of the Damara Orogen*. Geological Society of South Africa, Special Publications, **11**, 21–36.

PRAVE, A. R. 1996. Tale of three cratons: Tectonostratigraphic anatomy of the Damara orogen in northwestern Namibia and the assembly of Gondwana. *Geology*, **24**, 1115–1118.

PRAVE, A. R. 1997. Tale of three cratons: Tectonostratigraphic anatomy of the Damara orogen in northwestern Namibia and the assembly of Gondwana, Reply. *Geology*, **25**, 1149–1151.

RAUB, T. D., EVANS, D. A. D. & SMIRNOV, A. V. 2007. Siliciclastic prelude to Elatina-Nuccaleena deglaciation: lithostratigraphy and rock magnetism through the base of the Ediacaran system. *In*: VICKERS-RICH, P. & KOMAROWER, P. (eds) *The Rise and Fall of Ediacaran Biota*. Geological Society, London, Special Publications **286**, 53–76.

RAVIZZA, G. E. & ZACHOS, J. C. 2003. Records of Cenozoic ocean chemistry. *In*: HOLLAND, H. D. & TUREKIAN, K. K. (eds) *The Oceans and Marine Geochemistry*. Treatise on Geochemistry, **6**, Elsevier, Amsterdam, 551–582.

RIDGWELL, A. & KENNEDY, M. J. 2004. Secular changes in the importance of neritic carbonate deposition as a control on the magnitude and stability of Neoproterozoic ice ages. *In*: JENKINS, G. S., MCMENAMIN, M., SOHL, L. E. & MCKAY, C. P. (eds) *The Extreme Proterozoic: Geology, Geochemistry and Climate*. American Geophysical Union, Geophysical Monograph, **146**, 55–72.

RIDGWELL, A., KENNEDY, M. J. & CALDEIRA, K. 2004. Carbonate deposition, climate stability, and Neoproterozoic ice ages. *Science*, **302**, 859–862.

RIEU, R., ALLEN, P. A., ETIENNE, J.L, COZZI, A. & WIECHERT, U. 2006. A Neoproterozoic glacially influenced basin margin succession and 'atypical' cap carbonate associateds with bedrock palaeovalleys, Mirbat area, southern Oman. *Basin Research*, **18**, 471–496.

RODRIGUES-NOGUEIRA, A. C., RICCONIMI, C., SIAL, A. N., VELOSO-MOURA, C. A. & FAIRCHILD, T. R. 2003. Soft sediment deformation at the base of the Neoproterozoic Puga cap carbonate (southwestern Amazon craton, Brazil): confirmation of rapid icehouse to greenhouse transition in a snowball Earth. *Geology*, **31**, 613–616.

SCHERMERHORN, L. J. G. 1974. Late Precambrian mixtites: glacial and/or non glacial? *American Journal of Science*, **274**, 673–824.

SCHMIDT, P. W. & WILLIAMS, G. E. 1995. The Neoproterozoic climatic paradox: equatorial palaeolatitude for Marinoan glaciations near sea level in South Australia. *Earth and Planetary Science Letters*, **134**, 107–124.

SCHNEIDER, J., BONI, M., LAUKAMP, C., BECHSTÄDT, T. & PETZEL, V. 2007. Willemite (Zn_2SiO_4) as a possible Rb–Sr geochronometer for dating nonsulfide

Zn–Pb mineralization: examples from the Otavi Mountainland (Namibia). *Ore Geology Reviews*, **33**, 152–167, doi:10.1016/j.oregeorev.2006.05.012.

SCHRÖDER, S., SCHREIBER, B. C., AMTHOR, J. E. & MATTER, A. 2004. Stratigraphy and environmental conditions of the terminal Neoproterozoic–Cambrian Period in Oman: evidence from sulphur isotopes. *Journal of the Geological Society, London*, **161**, 489–499.

SHIELDS, G. A. 2005. Neoproterozoic cap carbonates: a critical appraisal of existing models and the plume-world hypothesis. *Terra Nova*, **17**, 299–310.

SMIT, J. M. 1959. *The Geology of the Southern Part of the Otavi Mountainland, South West Africa.* MSc thesis, University of Pretoria.

SMITH, A. B. 2009. Neoproterozoic timescales and stratigraphy. *In*: CRAIG, J., THUROW, J., THUSU, B., WHITHAM, A. & ABUTARRUMA, Y. (eds) *Global Neoproterozoic Petroleum Systems: The Emerging Potential in North Africa.* Geological Society, London, Special Publications, **326**, 25–52.

SOHL, L. E., CHRISTIE-BLICK, N. & KENT, D. V. 1999. Palaeomagnetic polarity reversals in Marinoan (ca. 600 Ma) glacial deposits of Australia: implications for the duration of low-amplitude glaciation in Neoproteozoic time. *Geological Society of America Bulletin*, **111**, 1120–1139.

SÖHNGE, P. G. 1957. *Revision of Geology of the Otavi Mountain Land, South West Africa.* Unpublished report, Tsumeb Corporation Ltd, Tsumeb, Namibia.

STAHL, A. 1940. Die Otaviformation des Etoshabogens (Südwest-Afrika). *Beiträge geologischen Erforschung deutscher Schutzgebiete*, **22**, 66.

STEWART, J. H. 2007. *World Map Showing Surface and Subsurface Distribution, and Lithologic Character of Middle and Late Neoproterozoic Rocks.* US Geological Survey Open-File Report, 2007-1087, Version 1.0. http://pubs.usgs.gov/of/2007/1087/.

SUMMONS, R. E., JAHNKE, L. L., HOPE, J. M. & LOGAN, G. A. 1999. 2-Methylhopanoids as biomarkers for cyanobacterial oxygenic photosynthesis. *Nature*, **400**, 554–557.

SUMNER, D. Y., KIRSCHVINK, J. K. & RUNNEGAR, B. N. 1987. Soft sediemnt palaeomagnetic field test of late Precambrian glaciogenic sediments. (Abstract). *Eos, Transactions of the American Geophysical Union*, **68**, 1251.

SWART, P. K. & EBERLI, G. 2005. The nature of the δ^{13}C of periplatform sediments. Implications for stratigraphy and the global carbon cycle. *Sedimentary Geology*, **175**, 115–129.

TAYLOR, G. H., TEICHMÜLLER, M., DAVIS, A., DIESSEL, C. F. K., LITTKE, R. & ROBERT, P. 1998. *Organic Petrology.* Gebrudes Borntraeger, Berlin.

TRINDADE, R. I. F., FONT, E., D'AGRELLA-FILHO, A. S., NOGUEIRA, A. C. R. & RICCOMINI, C. 2003. Low-latitude and multiple geomagnetic reversals in the Neoproterozoic Puga cap carbonate, Amazon craton. *Terra Nova*, **15**, 441–446.

TYSON, R. V. 1993. Palynofacies analysis. *In*: JENKINS, D. J. (ed.) *Applied Micropalaeontology.* Kluwer, Dordrecht, 91–153.

TYSON, R. V. 1995. *Sedimentary Organic Matter Organic Facies and Palynofacies.* Chapman & Hall, London.

VIDAL, G. & MOCZYDLOWSKA, M. 1992. Patterns of phytoplankton radiation across the Precambrian–Cambrian boundary. *Journal of the Geological Society, London*, **149**, 647–654.

WERNER, G. 2005. *Geologische Kartierung und laterale Faziesanalyse von Mixtit-Ablagerungen im Otavi Bergland, Namibia.* Diploma thesis, University of Heidelberg.

YPMA, P. J. M. 1979*a*. Mineralogical and geological indications for the petroleum potential of the Etosha Basin, Namibia (S.W. Africa). I. *Proceedings of the Koninklijke Nederlandse Akademie van Wetenschappen, Series B: Palaeontology, Geology, Physics and Chemistry*, **82**, 91–103.

YPMA, P. J. M. 1979*b*. Mineralogical and geological indications for the petroleum potential of the Etosha Basin, Namibia (S.W. Africa). II. *Proceedings of the Koninklijke Nederlandse Akademie van Wetenschappen, Series B: Palaeontology, Geology, Physics and Chemistry*, **82**, 104–112.

The 'Infracambrian System' in the southwestern margin of Gondwana, southern South America

J. C. HLEBSZEVITSCH*, I. GEBHARD, C. E. CRUZ & V. CONSOLI

Pluspetrol SA, Lima 339 C1073AAG, Buenos Aires, Argentina

**Corresponding author (e-mail: jhlebszevitch@pluspetrol.net)*

Abstract: The classical definition of 'Infracambrian' that refers strictly to sequences of Proterozoic age is not applicable in southwestern Gondwana. In this paper the term 'Infracambrian' is used to define the sequences deposited during the Pampean orogenic cycle, which extends until the Cambrian period.

A classification of Infracambrian basins is proposed based on location, level of preservation and perceived petroleum potential. Only the eastern basins of the South American plate have potentially significant exploration potential.

Two Neoproterozoic petroleum systems have been identified in the eastern basins: a Riphean system, developed on the western margin of the San Francisco Craton, in the San Francisco Basin of Brazil; and a Vendian system, developed on the eastern margin of the Amazonia–Río de la Plata cratons, in the Corumbá Basin of Brazil and Paraguay. The Riphean system is reportedly proven by a well test drilled by Petrobras. An active Vendian petroleum system is proven by the presence of oil seeps within fractured limestones. A Vendian petroleum system is proposed for the Claromecó Basin of Argentina based on the correlation of the Vendian and Riphean sequences.

During the Neoproterozoic–Cambrian period, the southwestern region of the Gondwana continent was structured by a complex history of riftings, oceanic apertures and collisions. Infracambrian sediments are either encountered as platform sequences upon the underlying Archaean–Mesoproterozoic cratons or are involved in the mobile belts configured during the Pampean Orogeny (Aceñolaza & Toselli 1973, 1988), also called Pan-American (Harrington 1972, 1975), Pan-African (Kennedy 1964; Caby *et al.* 1981; Plumb & James 1986; Caby 1987; Unrug 1996) or Brazilian (de Almeida 1945, 1968, 1971, 1978; de Almeida *et al.* 1976; de Brito Neves & Cordani 1991). The presence of limestones and organic-rich shales within these sedimentary sequences could suggest the existence of petroleum systems, but which, owing to the limited existing data, have been rarely studied. The purpose of this paper is to provide an overview of the Infracambrian basins and of their possible hydrocarbon potential.

Definition of the term 'Infracambrian' in southwestern Gondwana

The classical 'Infracambrian' term refers to the sedimentary sequences encountered prior to the first 'Shelly Faunas' of Cambrian age. However, this definition is not applicable in southwestern Gondwana where the term 'Infracambrian' defines the sequences deposited during the Pampean cycle of Neoproterozoic–Cambrian age. Two reasons led to the change in the definition: first, the tectono-stratigraphical evolution of the Neoproterozoic basins ends in the Cambrian period and, second, in South America the 'Shelly Faunas' are not older than Middle–Late Cambrian.

The Pampean cycle is composed of three distinct orogenic events (da Silva *et al.* 2005).

Brasiliano I: with two principal stages, at 790 Ma (Embu Domain: Campos Neto 2000; Cordani *et al.* 2002) and at 730–700 Ma (Sao Gabriel Orogen: Hartmann *et al.* 2000).

Brasiliano II: major activity at 640–620 Ma (Don Feliciano Orogen: Basei *et al.* 2000) and at 600 Ma. (Paranapiacaba and Río Pien Orogens: Basei *et al.* 2000).

Brasiliano III: with a climax between 590–560 Ma (Araçuai Orogen: de Almeida *et al.* 1981) and 520–500 Ma (Buzios Orogen: Schmitt 2000).

From a basin and tectonic evolution point of view, the Brasiliano III event is considered to be part of the Neoproterozoic evolution and not part of the Gondwanan cycle.

Geotectonic framework and basin types

According to the location, degree of preservation and perceived petroleum potential, two groups of basins have been identified in southwestern Gondwana. The eastern basins (Corumbá, San Francisco and Claromecó) are located either nearby or on the passive margin of the South American plate (Figs 1 & 2). These basins developed around the

From: Craig, J., Thurow, J., Thusu, B., Whitham, A. & Abutarruma, Y. (eds) *Global Neoproterozoic Petroleum Systems: The Emerging Potential in North Africa.* Geological Society, London, Special Publications, **326**, 289–302.
DOI: 10.1144/SP326.16 0305-8719/09/$15.00

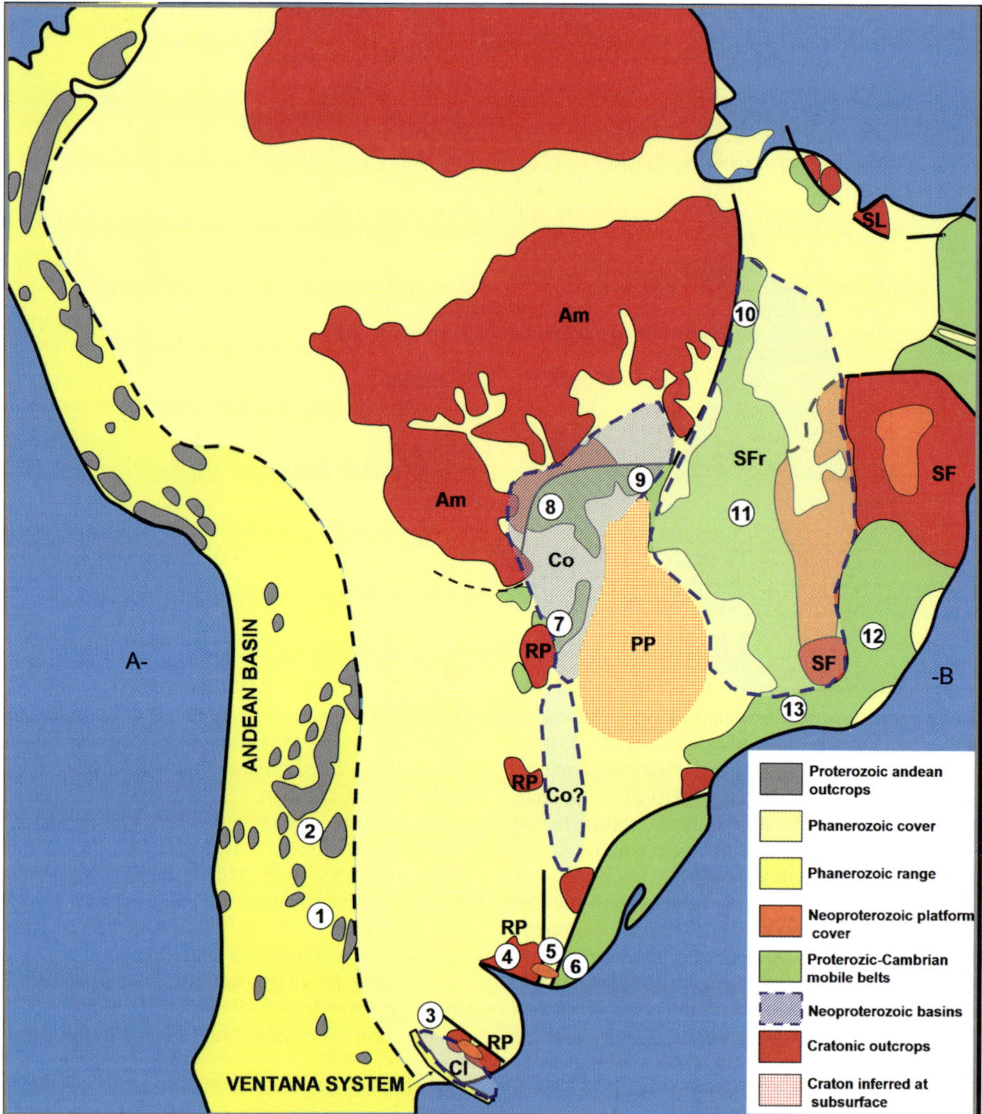

Fig. 1. Distribution of Archaean–Mesoproterozoic cratons and Neoproterozoic terranes. Cratons: Am, Amazonia; SF, San Francisco; SL, San Luis; LA, Luis Alves; PP, Paranapanema; RP, Río de la Plata. Neoproterozoic basins: Co, Corumba Basin; Cl, Claromecó Basin; SF, San Francisco Basin. Mobile belts, outcrops and terranes: 1, Pampean Range, Difunta Correa Sequence; 2, Andean and Subandean, Puncoviscana Formation; 3, Claromecó Basin, Sierras Bayas Group; 4, Piedra Alta Terrane; 5, Nico Perez Terrane; 6, Don Feliciano Belt; 7, Corumbá Group; 8, Paraguay Belt; 9, Alto Paraguay Belt; 10, Araguaia Belt; 11, Brasilia Belt; 12, Araçuaí Belt; 13, Ribeira Belt. A– –B, location of the schematic regional cross-section in Figure 2 (modified from Cordani & Sato 1999).

Amazonia, San Francisco, Paranapampeana and Río de la Plata cratons, and were subjected in general to a low degree of deformation and diagenesis after the Neoproterozoic. Owing to the fairly high degree of preservation of the possible source rocks and reservoirs these basins could hold significant petroleum potential.

An exception to the aforementioned statement is the case of the sedimentary sequences involved in the Neoproterozoic orogenic belts (e.g. Paraguay Belt in the Corumbá Basin or the Brazilia Belt in the San Francisco Basin), which suffered a more intense deformation and were poorly preserved. Two tectonic styles can be recognized within these ancient

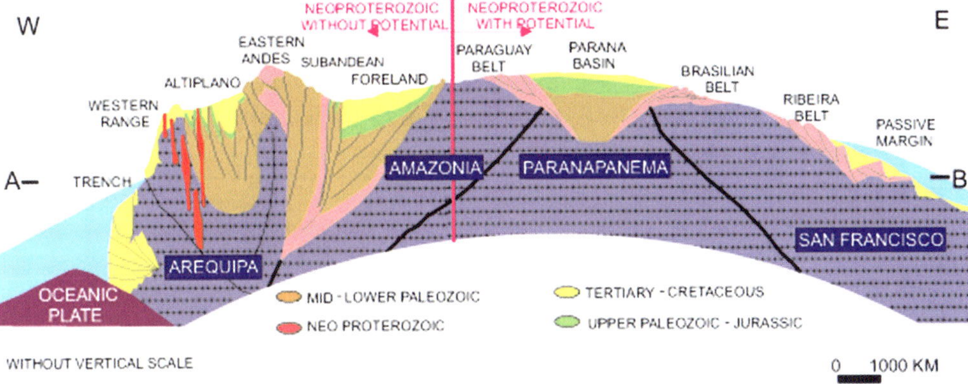

Fig. 2. West–east schematic regional cross-section of southern South America and the distribution of Neo- and Phanerozoic terranes (modified from Franca *et al.* 1995).

mobile belts: frontal collision-type (Brazilia, Paraguay Araçuai belts and, partially, the Araguaia Belt); and oblique collision-type (Ribeira and Don Feliciano belts) (Guinaldo *et al.* 2004). The frontal collision-type generated thrust and fold belts with a clear vergence towards the cratons and the associated foreland basins (Guinaldo *et al.* 2004). The oblique collision-type generated wrench zones and pull-apart basins (Guinaldo *et al.* 2004).

The Claromecó Basin (Argentina) was affected by a later deformation pulse during the Triassic period, which resulted in the generation of the Ventana System, with its African counterpart, the Cape Fold Belt (South Africa).

The western basins such as Puncoviscana and western Pampean Range (*sensu lato*) are located along the active margin of the South American Plate and have been intensively deformed during the Phanerozoic Andean compression (Figs 1 & 2). These basins have undergone middle- to high-grade metamorphism and, even though the presence of organic-rich shales and limestones can be locally recognized, the perceived petroleum potential of these basins is fairly low.

Neoproterozoic glacial events and the 'Snow Ball' hypothesis

As in many other Neoproterozoic basins, the sedimentary record of the southwestern Gondwana continent includes glacial events followed by cap carbonates. The diamictite deposits were traditionally correlated with the Sturtian and the Marinoan glaciations. However, additional glacial events are described in South America: Caigas (771–741 Ma) within the Jequitai Group, San Francisco Basin (Frimmel & Fölling 2004); Sturtian (717–684 Ma) within the Jacarecica/Ribeirópolis Formation, northern margin of the San Francisco Craton (Sial *et al.* 2006); Marinoan (635 Ma) within the Jacadigo Group, Paraguay Belt (Alvarenga *et al.* 2004); Gaskiers (584–582 Ma) within the Serra Azul Formation, Paraguay Belt (Figueiredo *et al.* 2006); Moelv (560 Ma) within the Playa Hermosa and Middle Polanco Formation, Nico Perez Terrane (Gaucher *et al.* 2004); and Vingerbreek (548 Ma) within the Barriga Negra Formation, Nico Perez Terrane (Gaucher *et al.* 2004) (Fig. 3).

The 'Snow Ball' hypothesis postulates the occurrence of a major worldwide glaciation at the end of the Proterozoic, which caused the Earth to be frozen like a snowball. However, Poiré (2004) proposed that certain areas of the Río de la Plata Craton, certainly known for having been affected by glaciation, do not show any evidence of glacially related deposits in their sedimentary records. These observations, plus the existence of tropical 'paradises', suggest that the Snowball Earth hypothesis is questionable.

These glacial sedimentary sequences have potentially significant economic interest because they are associated with carbonates and transgressive shales with high TOC (total organic carbon) content and source rock potential.

Stratigraphy: eastern domain

Two main stratigraphic sequences can be recognized in this domain: the Riphean System and the Vendian System (Fig. 3).

Riphean sequences

The most complete record of the Riphean sequences has been identified in the central and eastern areas of the San Francisco Basin (Fig. 4). These sequences

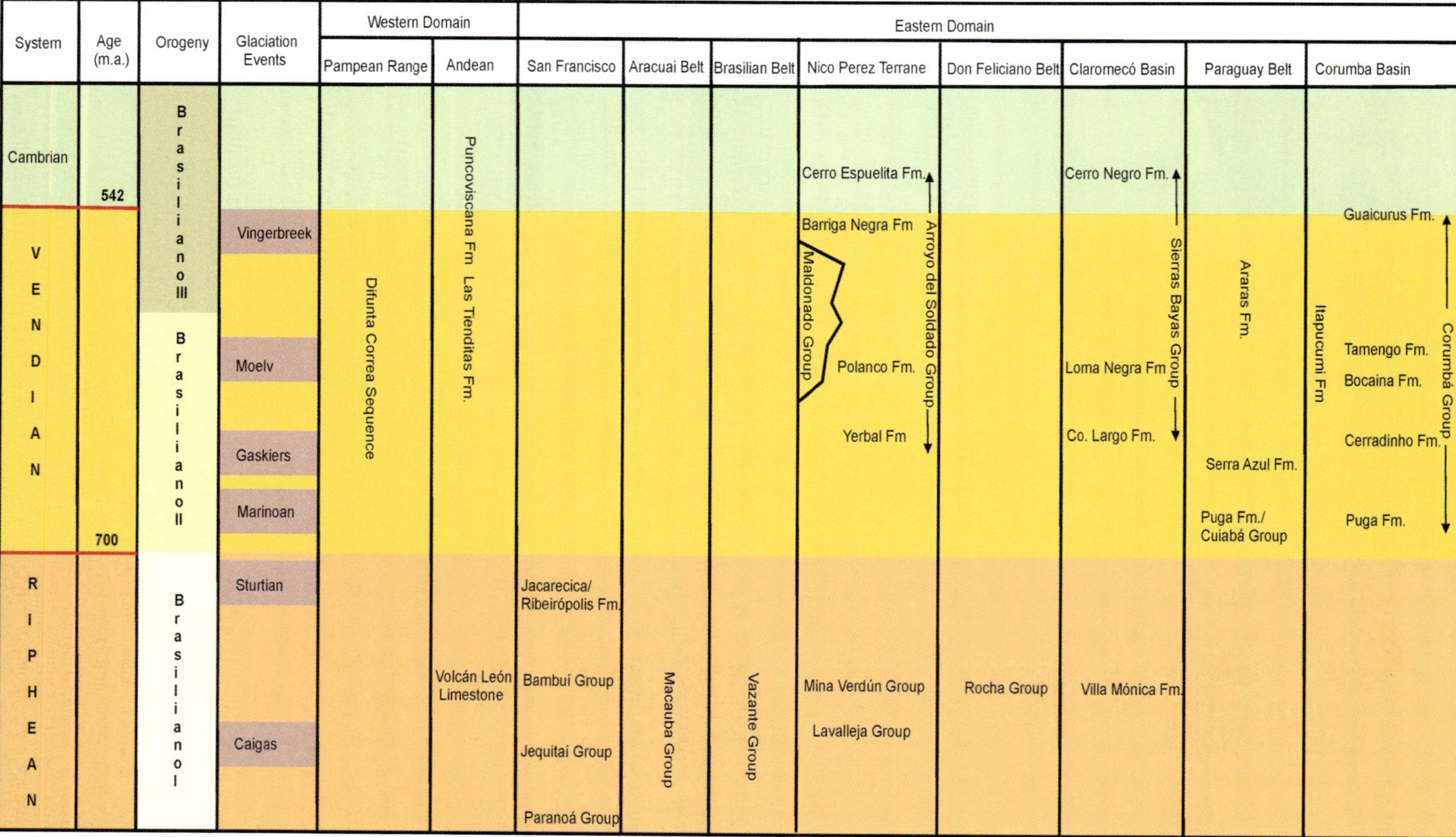

Fig. 3. Correlation chart for the Infracambrian basins and belts of southern South America.

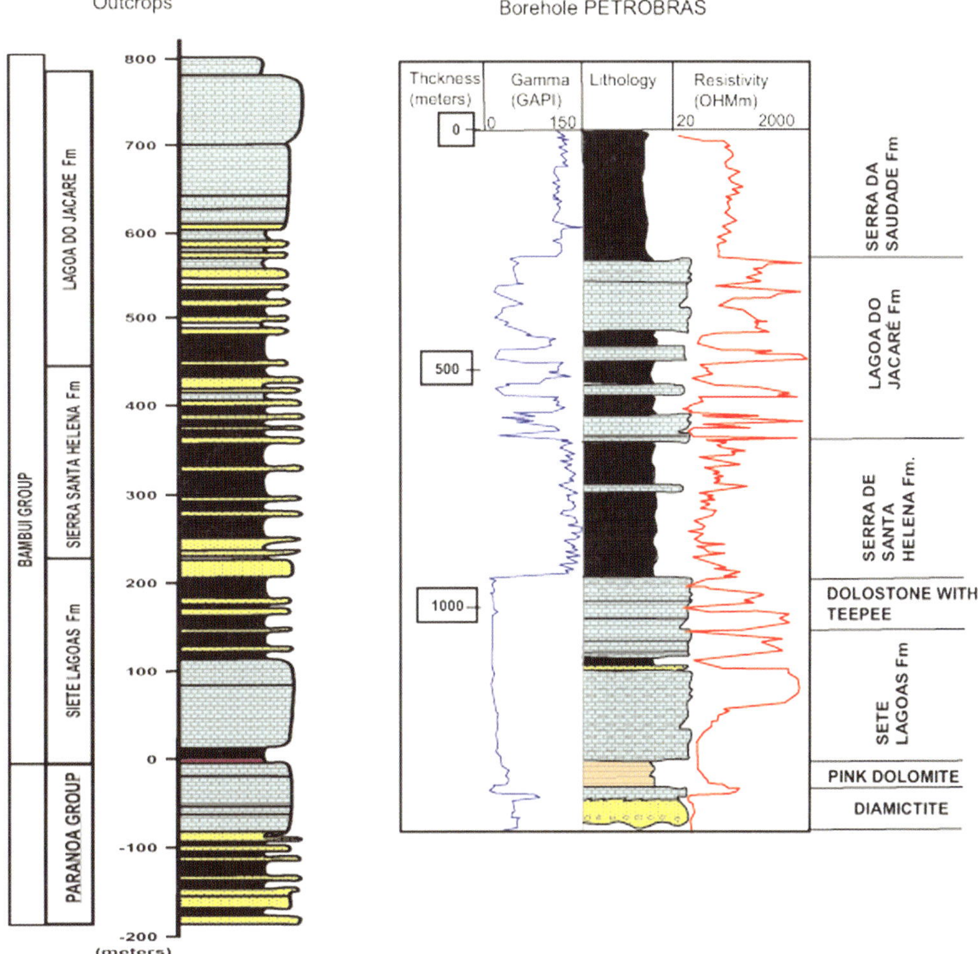

Fig. 4. Riphean sequences. Typical section of the Bambuí Group in outcrops (left) and wells drilled by Petrobras (right) (after Missi *et al.* 2007).

comprise two groups, Paranoá and Bambuí. The Paranoá Group of Meso-Neoproterozoic age (1170–950 Ma according to Fairchild *et al.* 1996) was deposited in a marine shelf environment and consists of a series of silicoclastic transgressive–regressive cycles, which culminate with the deposition of shales and stromatolitic dolomites. Afterwards, the diamictites of the Jequitaí Group were deposited during the Caigas glaciation. The Bambuí Group comprises a series of dolomitic limestones, evaporates, shales and marls, with a significant content of organic matter.

In the area of the Araçuai Belt (Brazil) the aforementioned sedimentary sequences (Paranoá, Jequitaí and Bambuí) are commonly referred to as the Macauba Group. In the area of the Brazilian Belt, the Vazante Group is an equivalent of the Bambuí Group (for the location see Fig. 1).

Within the Claromecó Basin (Río de la Plata Craton, Argentina) the La Villa Monica Formation has a wide distribution, but it crops out over a limited area. It comprises laminated limestones, biostromes and bioherms dated with Rb/Sr as 793 ± 32 Ma (Cingolani & Bonhomme 1988). The Mina Verdún Group (Poiré *et al.* 2003) crops out over the Nico Perez Terrane (Uruguay), and is composed of organic-rich black shales and limestones (see the location in Fig. 1).

The Lavalleja and the Rocha groups (Uruguay) are identified in the Nico Perez and the Don Feliciano terranes, respectively. These sequences mainly consist of metamorphic rocks originating from siliciclastic and volcanic units (for the location see Fig. 1).

Vendian sequences

The Vendian sequences were deposited over the passive margin of the Amazonian and the Río de la Plata cratons (for the location see Fig. 1). They are represented by the Sierras Bayas Group in Argentina (Poiré 1993), the Arroyo del Soldado Group in Nico Perez Terrane, Uruguay (Gaucher *et al.* 2003), the Itapucumi Group in Brazil and Paraguay (Boggiani & Gaucher 2004), the Corumbá Group in Brazil (Boggiani *et al.* 1993; Gaucher *et al.* 2003) and the Ararás Formation in Brazil (Luz & Abreu Filho 1978) (Figs 5 & 6). The basal part is constituted by a Marinoan glaciomarine diamictite locally known as the Puga Formation (Corumba Group, Brazil), which is followed by laminated limestones ('cap carbonates'). A transgressive sequence composed predominantly of fine-grained sediments was developed on top of the Puga carbonates. The sequence is known as the Cerradinho Formation in Brazil, the Yerbal Formation in Uruguay and the Cerro Largo Formation in Argentina, and it crops out intermittently over a distance of approximately 2500 km along the eastern margin of the Amazonian and the Río de La Plata cratons.

In some areas the Cerradinho Formation is absent, and is replaced by alluvial fans and glacial outwash deposits that pass transitionally to carbonates known as the Bocaina or Tamengo Formation in Brazil, the Polanco Formation in Uruguay and the Loma Negra Formation in Argentina (see Fig. 3). The Polanco Formation is characterized by dolomitic stromatolites with strong lateral facies variations, and the Loma Negra Formation comprises dark limestones and marls with high organic matter content. The Tamengo Formation is covered by the shales of the Guaicurus Formation.

Equivalent units of the Corumbá Group (Cuiabá Group, Serra Azul and Ararás formations) strongly deformed by the Brazilian Orogeny can be recognized in the Paraguay Belt (Fig. 1).

Other Vendian sequences are encountered in the Nico Perez Terrane and are known as the Maldonado Group (Uruguay) (Fig. 1). The basal unit is the Playa Hermosa Formation, characterized by conglomerates, diamictites, ritmites and shales deposited during the Moelv glaciation (570–560 Ma). The overlying unit is the volcaniclastic La Ventana Formation, deposited as sheetflood fluxes, alluvial fans and fan deltas (Pecoits *et al.* 2005).

Stratigraphy: western domain

The Puncoviscana Basin extends between 23°S (Tarija, Bolivia) and 27°S (Tucumán, Argentina) along the Eastern Cordillera (Figs 1 & 3). This basin was originally considered as the southwestern branch of an aulacogenic structure developed between the Arequipa–Antofalla cratons on the west and the Guaporé–Río de la Plata cratons on the east (Aceñolaza & Alonso 2001). An alternative interpretation postulates a foreland basin developed between the Río de La Plata and the Arequipa–Antofalla–Pampia cratons (Kraemer *et al.* 1995; Zimmerman 2003). The Proterozoic sequence consists of red and green shales, wackes, diamictites, turbidites and conglomerates with intercalated basalts (Fig. 7). The overlaying Las Tienditas Formation, of Marinoan–Cambrian age, is composed of fine-grained black limestones. The psammopelitic Puncoviscana Formation of Upper Vendian age was deposited unconformably on top of the Las Tienditas Formation (Aceñaloza & Alonso 2001). Recent studies (Toselli *et al.* 2005) support the correlation of the Las Tienditas Formation with limestones of the Corumbá Basin (Brazil).

Another calcareous unit with intercalated shaly sandstone levels is also recognized in the Puncoviscana Basin (the Volcán León limestones). These limestones represent a sedimentary cycle older than the Las Tienditas Formation, probably of Riphean age.

The carbonates of the Difunta Correa Sequence in the Pampean Range are correlatable with the Corumbá and Arroyo Soldado groups (Missi *et al.* 2007). These sequences mainly consist of metamorphic rocks (Figs 1 & 3).

The first deformation pulse affecting the western domain occurred during the Middle Cambrian as a result of the collision between the Arequipa and the Río de la Plata cratons, and culminated in the closure of the Puncoviscana Basin.

Geodynamic evolution

The break-up of the Rodinia Supercontinent (1–0.9 Ga; Ga is 10^9 years) and the tectonic formation of southwestern Gondwana involved the drift and collision of cratons and magmatic arcs, and the accretion of sedimentary prisms. The palaeogeographic model for the beginning of the Neoproterozoic Erathem is presented in Figure 8 to explain the geological evolution of southwestern Gondwana.

Numerous terranes recognized in southwestern Gondwana can be seen in Figure 8. The main units are the Amazonas, Río de La Plata, San Francisco and Kalahari cratons. The Brazilia–Araguaia Belt was developed on the western and southwestern margin of the San Francisco Craton in response to the closure of the Goniánides Ocean. The development of the Don Feliciano Belt on the eastern margin of the Río de La Plata Craton is associated with the subduction of the Adamastor Ocean.

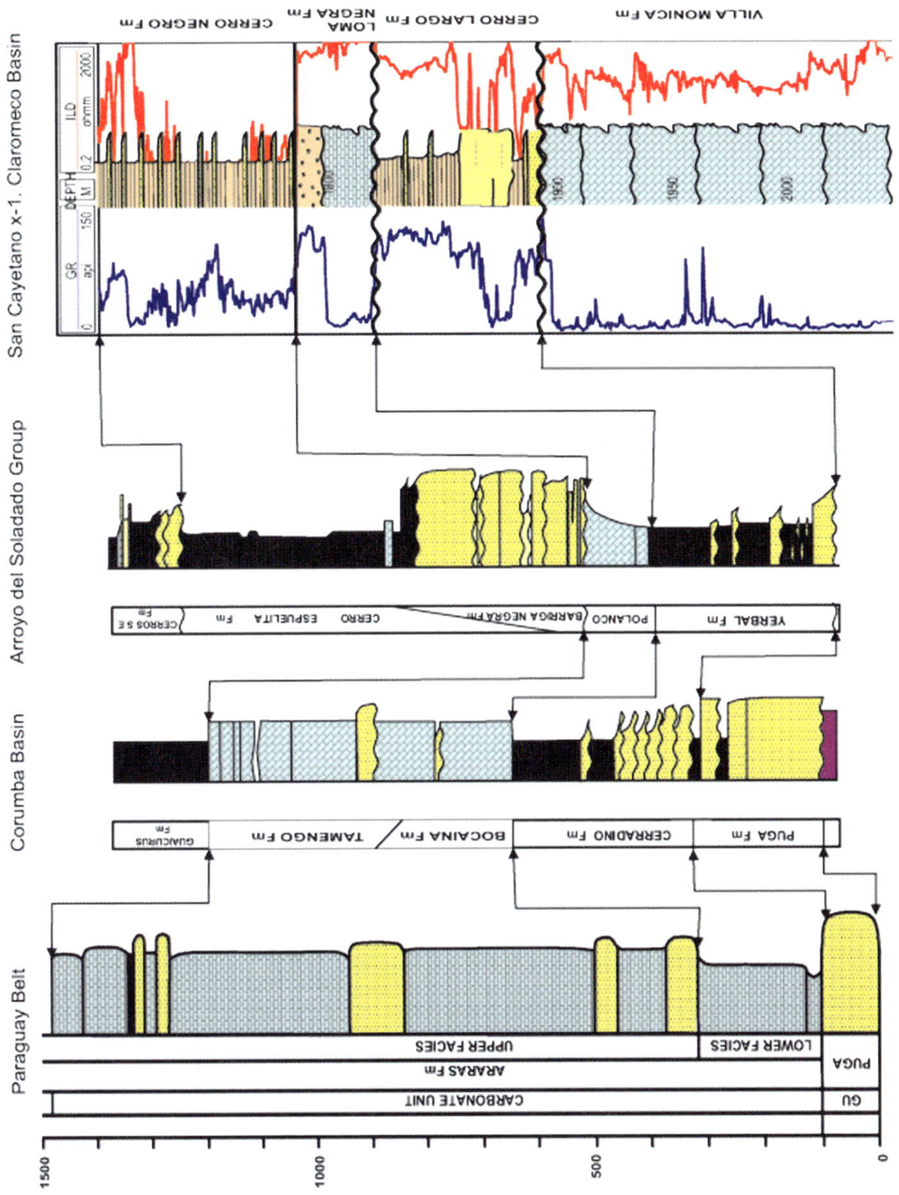

Fig. 5. Composite cross-section of the Corumbá Group and equivalent units (Paraguay Belt, Brazil; Corumbá Basin, Brazil; Arroyo del Soldado Basin, Nico Perez Terrane, Uruguay; well San Cayetano 1, Claromecó Basin, Argentina) (after Gaucher *et al.* 2003).

Fig. 6. Sierras Bayas Group, Claromecó Basin, Olavarría, Argentina. Quartzitic sandstone deposited in a shallow-marine environment.

Fig. 7. The Puncoviscana Formation, strongly deformed and superimposed by the Cretaceous, Salta, Argentina.

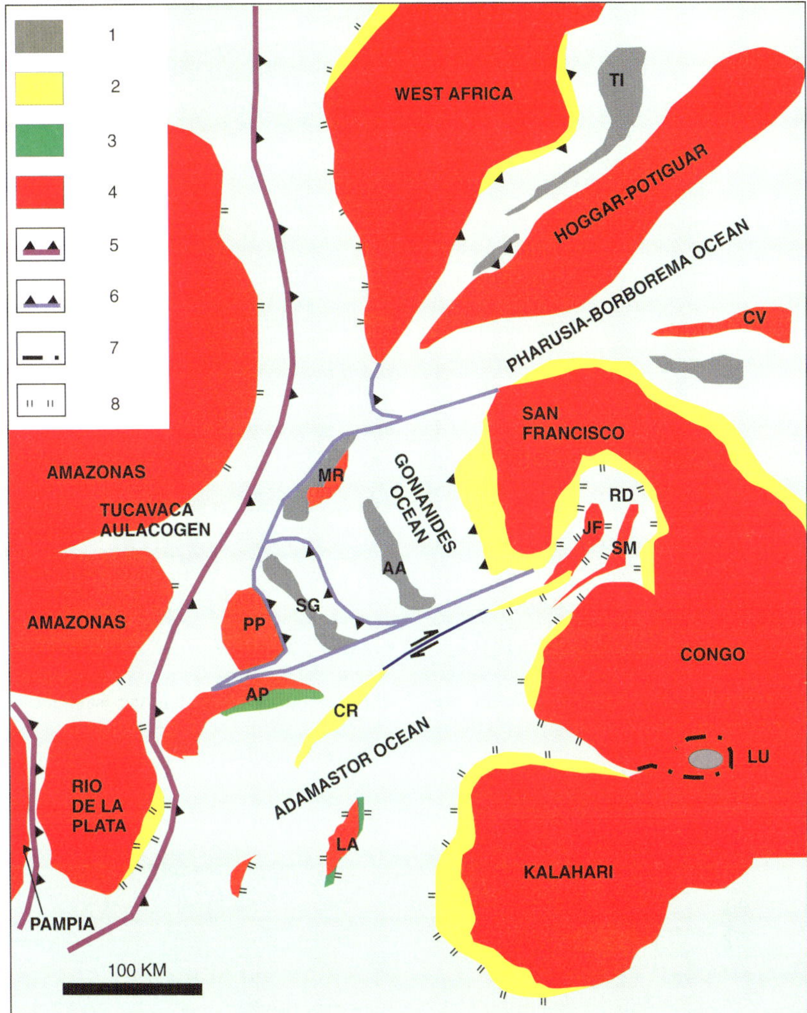

Fig. 8. Geodynamical framework of the southwestern Gondwana continent in Proterozoic times. General references: 1, magmatic arcs; 2, passive continental margin deposits and thinner continental crust domains; 3, mesoproterozoic belts; 4, continental plate segments; 5, western subduction zones with sense of dip; 6, eastern subduction zones with sense of dip; 7, Mesoproterozoic sutures; 8, ancient rift lines. Cratons and terranes: AA, Anápolis–Andrelândia; AP, Río Apas; CR, Curitiba; CV, Cariris Velhos; JF, Juiz de Fora; LA, Luis Alves; LU, Lufilian; MR, Goías Block; PP, Paranapanema; RD, Río Doce Gulf; SG, Socorro Guaxupé; SM, Serra do Mar; TI, Tilemsi. Modified from Brito Neves *et al.* (1999).

The collision of the San Francisco–Río Apas–Paranapanema cratons with the Amazonas–Río de La Plata cratons initiated the formation of the Paraguay Belt.

The configuration of the Brazilia–Araguaia Belt started at 630 Ma with the subduction of the Socorro Guaxupé and the Anápolis–Andrelândia magmatic arcs under the San Francisco Craton and ended at 605 Ma (Valeriano *et al.* 2004) with the collision of the Paranapanema Craton and the Goiás Block.

Simultaneously with the development of the Brazilia Belt, subduction and the collision of the San Francisco Craton with the Juiz de Fora, Serra do Mar, Luis Alves and Congo cratons resulted in the formation of the Ribeira–Araçuaí Belt between 595 and 560 Ma.

The interaction between the Kalahari and the Río de La Plata cratons originated the Don Feliciano Belt. The subduction began with the development of a back-arc basin (Gariep Basin) on the western side of the Kalahari Craton, which was filled with shelf and turbidite sediments of the Rocha and Orajemund groups (600–550 Ma). The subduction led to the closure of the Adamastor Ocean and of the Gariep retroarc Basin (Bossi & Gaucher 2004; Basei *et al.* 2005). The suture is represented by the Sierra Ballena–Major Gercino lineament, located between Nico Perez Terrane and the Don Feliciano Belt.

A calcareous and siliciclastic marine platform deposited between 600 and 543 Ma was developed on the eastern margin of the Amazonas–Río de la Plata cratons. These sequences were deformed as a result of the collision of the San Francisco–Río Apa–Paranapanema and the Amazonas–Río de La Plata cratons, during the configuration of the Paraguay–Alto Paraguay Belt. According to Gaucher *et al.* (2003), the collision culminated in the Cambrian–Ordovician.

Vendian sequences equivalent to the Corumbá group were deposited over the southern edge of the Río de La Plata Craton and are partially included within the Sierras Bayas Group. These sequences were not deformed until the Triassic period (see the section on 'Geotectonic framework and basin types' earlier in this paper).

Fig. 9. Limestones of the Itapucumi Formation in Cerro Paiva, Corumbá Basin (northern Paraguay). Bituminous carbonates that have not undergone strong deformation.

Source rocks

Source rocks and seeps have been identified within the Vendian and Riphean shales and carbonates deposited within restricted platforms and sabkhas.

Oil seeps found in the Corumbá Basin (Chiquitanas area, Bolivia) prove the existence of an active Neoproterozoic petroleum system. The geochemical analysis reveals the presence of a marine carbonate source rock deposited in a sabhka environment. Gaucher *et al.* (2003) report organic-rich

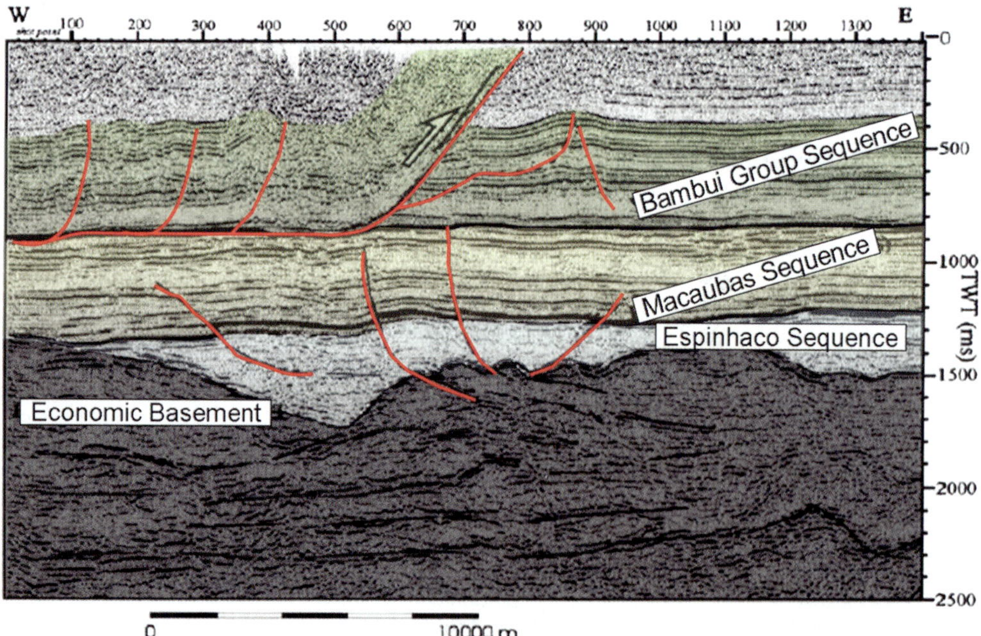

Fig. 10. West-east seismic line from the San Francisco Basin (in Brazil), showing potential trap styles (modified from Brazil Round 4, Technical Workshop 2001). TWT, two-way time.

levels with up to 11% TOC. Bitumen shows and stylolites were found within fractures of the Araras Formation (Paraguay Belt) in the area of Matto Grosso, Brazil (Nogueira *et al.* 2001). Hydrocarbon gas was tested from a well drilled by Petrobras in the San Francisco Basin in Brazil.

Geochemical analyses of black shales of the Poço Verde Formation (Vazante Group) in the San Francisco Basin of Brazil revealed TOC values of 3% (Olcott *et al.* 2005). In Argentina, limestones of the Villa Mónica Formation have TOC values of 0.8% (Gomez Peral *et al.* 2004).

Reservoirs

Neoproterozoic carbonates affected by fracturing and/or dissolution could constitute potential reservoirs (Fig. 9). Poiré (2002) reported karstic structures in Argentina and Uruguay. Vendian sandstones such as the Puga Formation in Brazil

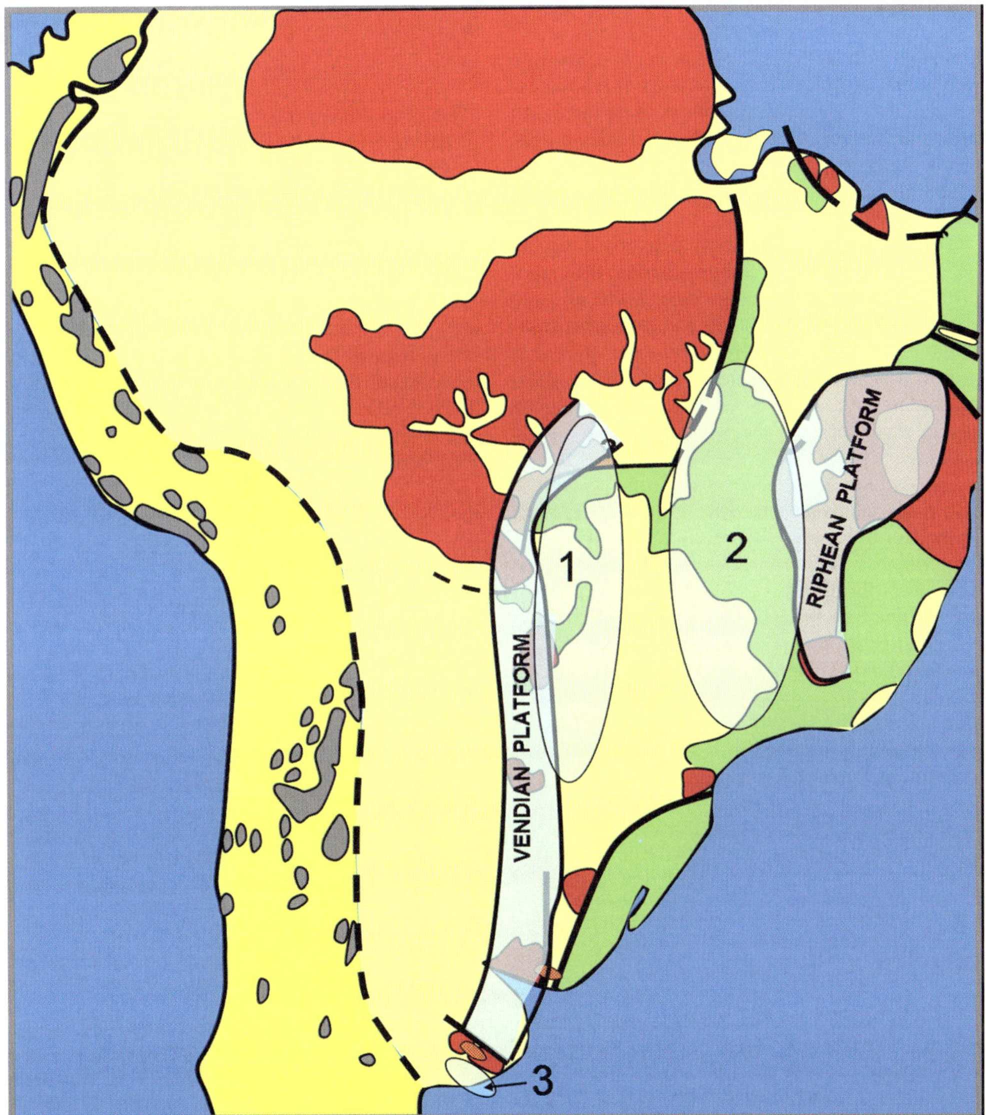

Fig. 11. Infracambrian basins with exploration potential. 1, Corumbá Basin; 2, San Francisco Basin; 3, Claromecó Basin (for the key see Fig. 1).

or the Sierras Bayas Group in Argentina could also constitute potential reservoirs (Fig. 6).

Traps

The collision and amalgamation of cratons that took place during the Pampean cycle generated numerous anticlinal structures that could constitute potential traps. Thrust- and wrench-related anticlines (e.g. inversion structures) were identified in the San Francisco Basin (Fig. 10) (Brazil Round 4, Technical Workshop 2001).

Conclusions

The Infracambrian basins have been classified into two groups according to their location, level of preservation and perceived petroleum potential. The eastern basins are located nearby or on the current passive margin of South America; they are better preserved and, consequently, could hold significant exploration interest.

The western basins are located nearby or on the current active margin of South America; they have suffered intense deformation during Phanerozoic times and have most probably lost their petroleum potential.

Two Proterozoic petroleum systems are proposed for southwestern Gondwana: a Riphean System developed on the western margin of the San Francisco Craton (in the San Francisco Basin) and a Vendian System developed on the eastern margin of the Amazonia–Río de la Plata cratons (in the Corumbá and Claromecó basins) (Fig. 11). The potential source rocks could have reached maturation during the subsidence of the Phanerozoic foreland basins. Oil seeps, shales and marls with good generation potential have been identified. The reservoirs could be fractured limestones or carbonates with secondary porosity due to dissolution effects. Thrust- and wrench-related anticlines were created during the Pampean cycle and could constitute potential hydrocarbon traps.

The authors would like to thank Pluspetrol for supporting this research. We are also grateful to G. Vergani, P. Lesta and E. Kozlowski (Pan American Energy) for the information they provided and helpful discussions.

References

Aceñolaza, F. G. & Alonso, R. 2001. Icnoasociaciones de la transición Precámbrico–Cámbrico en el noroeste de Argentina. *Journal of Iberian Geology*, **27**, 11–22.

Aceñolaza, F. G. & Toselli, A. J. 1973. Consideraciones estratigráficas y tectónicas sobre el Paleozoico Inferior del noroeste argentino. *II Congreso Latinoamericano de Geología*, **2**, 755–763.

Aceñolaza, F. G. & Toselli, A. J. 1988. El sistema de Famatina, Argentina: su interpretación como orógeno de margen continental activo. *V Congreso Geológico Chileno, Actas*, **1**, A55–A67.

Alvarenga, C. J. S. de, Santos, R. V. & Dantas, E. L. 2004. C–O–Sr stratigraphy of cap carbonates overlying Marinoan-age glacial diamictites in the Paraguay belt, Brazil. *Precambrian Research*, **131**, 1–21.

Basei, M. A. S., Frimmel, H. E., Nutran, A. P., Preciozi, F. & Jacob, J. 2005. A connection between the Neoproterozoic Don Feliciano (Brasil/Uruguay) and Gariep (Namibia/South Africa) orogenic belts – evidence from reconnaissance provenance study. *Precambrian Research*, **139**, 195–221.

Basei, M. A. S., Siga, O., Jr., Masquelin, H., Harara, O. M., Reis Neto, J. M. & Preciozzi, P. F. 2000. The Don Feliciano Belt of Brazil and Uruguay and its foreland domain, the Rio de la Plata Craton. *In*: Cordani, U. G., Milani, E. J., Thomaz Filho, A. & Campos, D. A. (eds) *Tectonic Evolution of South America, 31st International Congress, Rio de Janeiro, Brazil*, 313–334.

Boggiani, P., Fairchild, T. & Coimbra, A. 1993. O grupo Corumbá (Neoproterozóico–Cambriano) na região central da Serra da Bodoquena (Faixa Paraguai), Mato Grosso do Sul. *Revista Brasilera de Geociências*, **23**, 301–305.

Boggiani, P. & Gaucher, C. 2004. Cloudina from the Itapucumi Group (Ediacaran) SW Brazil, South America. *In*: *1st Symposium on Neoproterozoic–Early Paleozoic in SW-Gondwana. Extended Abstract. IGCP Project 478, Sao Paulo, Brazil*, 13–15.

Bossi, J. & Gaucher, C. 2004. The Cuchilla Dionisio Terrane, Uruguay: an allochthonus block accreted in the Cambrian to SW-Gondwana. *Gondwana Research*, **7**, 661–674.

Brazil Round 4, Technical Workshop. 2001. Sâo Francisco Basin. World Wide Web Address: http://www.anp.gov.br/brasil-rounds/round4/round4/workshop/restrito/SaoFrancisco_ing.pdf.

Caby, R. 1987. The Pan-African belt of West Africa from the Sahara desert to the Gulf of Benin. *In*: Schear, J. P. & Rodger, J. (eds) *Anatomy of Mountain Ranges*, Volume 1. Princeton University Press, Princeton, NJ, 129–170.

Caby, R., Bertrand, J. M. L. & Black, R. 1981. Pan-African ocean closure and continental collision in the Hoggar–Desiforas segment, Central Sahara. *In*: Kroner, A. (ed.) *Precambrian Plate Tectonics*, Volume 1. Elsevier, Amsterdam, 407–434.

Campos Neto, M. C. 2000. Orogenic systems from southwestern Gondwana. *In*: Cordani, U. G., Milani, E. J., Thomaz Filho, A. & Campos, D. A. (eds) *Tectonic Evolution of South America, 31st Internacional Geological Congreso, Rio de Janeiro, Brazil*, 355–365.

Cingolani, C. & Bonhomme, M. 1988. Resultados geocronológicos en niveles pelíticos intercalados en las dolomías de Sierras Bayas (Grupo La Tinta), provincia de Buenos Aires. *Segundas Jornadas Geológicas Bonaerenses, Actas*, 283–289.

Cordani, U. G. & Sato, K. 1999. Crystal evolution of the South American Platform, based on Nd isotopic systematics on granitoid rocks. *Episodes*, **22**, 167–173.

CORDANI, U. G., COUTINHO, J. M. V. & NUTMAN, A. 2002. Geochronological constraints for the age of the Embu Complex, São Paulo, Brazil. *Journal of South American Earth Science*, **14**, 903–910.

DA SILVA, L. C., MCNAUGHTON, N. J., ARMSTRONG, R., HARTMANN, L. A. & FLETCHER, I. R. 2005. The neoproterozoic Mantiqueira Province and its African connections: a zircon-based U–Pb geochronologic subdivision for the Brasiliano/Pan-African system of orogens. *Precambrian Research*, **136**, 203–240.

DE ALMEIDA, F. F. M. 1945. Geologia do sudoeste Matogrossense, Brasil. *Ministerio de Agricultura, Boletim da Divisão de Geologia e Mineralogia*, **116**, 9–115.

DE ALMEIDA, F. F. M. 1968. Evoluçao tectonica do centro oeste Brasileiro no Proterozoico Superior. *Annales Academia Brasileira do Ciencias*, **40**, 285–294.

DE ALMEIDA, F. F. M. 1971. Geochronological division of the Precambrian of South America. *Revista Brasilera de Geociências*, **1**, 13–21.

DE ALMEIDA, F. F. M. 1978. Chronotectonic boundaries for Precambrian time divisions in South America. *Annales Academia Brasileira do Ciencias*, **50**(4), 527–535.

DE ALMEIDA, F. F. M., HASUI, Y. & BRITO NEVES, B. B. 1976. The Upper Precambrian of South America. *Boletim Instituto de Geociencias, Universidade de Sao Paulo, Brasil*, **7**, 45–80.

DE ALMEIDA, F. F. M., HASUI, Y., DE BRITO NEVES, B. B. & FUCK, R. A. 1981. Brazilian structural provinces: an introduction. *Earth-Science Reviews*, **17**, 1–29.

DE BRITO NEVES, B. B. & CORDANI, U. G. 1991. Tectonic evolution of South America during the Late Proterozoic. *Precambrian Research*, **53**, 23–40.

FAIRCHILD, T. R., SCHOPF, J. W. ET AL. 1996. Recent Discoveries of Proterozoic microfossils in the southcentral Brazil. *Precambrian Research*, **80**, 125–152.

FIGUEIREDO, M. M., BABINSKI, M., ALVARENGA, C. J. S. de, PINHO, F. E. C. & SIMON, C. M. 2006. Chemostratigraphy (C, O and Sr) of Ediacaran post-glacial carbonates of Paraguay Belt, Matto grosso State, Brazil. *In*: *V South American Symposium on Isotope Geology, Short Papers, IGCP Project 478, Punta del Este, Uruguay*, 240–244.

FRANCA, A. B., MILANI, E. J. ET AL. 1995. Phanerozoic correlation in southern South America. *In*: TANKARD, A. J., SUAREZ, S. R. & WELSINK, H. J. (eds) *Petroleum Basins of South America*. AAPG Memoirs, 62, 129–161.

FRIMMEL, H. E. & FÖLLING, P. G. 2004. Late Vendian closure of the Adamastor Ocean: timing of tectonic inversion and syn-orogenic sedimentation in the Gariep basin. *Gondwana Research*, **7**, 685–699.

GAUCHER, C., BOGGIANI, P., SPRECHMANN, P., NOBREGA SIAL, N. & FAIRCHILD, T. 2003. Integrated correlation of the Vendian to Cambrian Arroyo del Soldado and Corumba Groups (Uruguay and Brazil): paleogeographic, paleoclimatic and paleobiologic implications. *Precambrian Research*, **120**, 241–278.

GAUCHER, C., SIAL, A. N., PIMENTEL, M. M. & FERREIRA, V. P. 2004. Impact of a late Vendian, non-global glacial vent on a carbonate platform, Polanco Formation, Uruguay. *In*: *First Symposium on Neoproterozoic–Early Paleozoic Events in SW-Gondwana. Extended Abstracts, IGCP Project 478, Sao Paulo, Brazil*, 21–23.

GOMEZ PERAL, L., POIRÉ, D., ZIMMERMANN, U. & STRAUSS, H. 2004. Chemostratigraphy and diagenetic constrains on Neoproterozoic carbonate successions from the Sierra Bayas Group, Tandilia System, Argentina. *In*: *First Symposium on Neoproterozoic–Early Paleozoic Events in SW-Gondwana. Extended Abstracts, IGCP Project 478, Second Meeting, Sao Paulo, Brazil*, 30–32.

GUINALDO, A., DA CAMPANHA, C. & DE BRITO NEVES, B. B. 2004. Frontal and oblique tectonics in the Brazilian Shield. *Episodes*, **27**, 255–259.

HARRINGTON, H. J. 1972. La paleogeografía de América del Sur. Conferencia invitada V Congreso Geologico Argentino, CM-unas, Rep. Int., 4.

HARRINGTON, H. J. 1975. South America. *In*: FAIRBRIDGE, R. W. (ed.) *The Encyclopedia of World Regional Geology. Western Hemisphere*, Volume 1. Reinhold Book Co., New York, 456–465.

HARTMANN, L. A., LEITE, J. A. D. ET AL. 2000. Advances in SHRIMP geochronology and their impact on understanding the tectonic and metallogenic evolution of southern Brazil. *Australian Journal of Earth Sciences*, **47**, 829–844.

KENNEDY, W. Q. 1964. The structural differentiation of Africa in the Pan-African (±500 m.y.) tectonic episode. *Annual Report of the Research Institute of African Geology, University of Leeds*, **8**, 48–49.

KRAEMER, P., ESCAYOLA, M. P. & MARTINO, R. D. 1995. Hipótesis sobre la evolución tectónica neoproterozoica de las sierras Pampeanas de Córdoba (30°40′–32°40′), Argentina. *Revista de la Asociación Geológica Argentina*, **50**, 47–59.

LUZ, J. DA S. & ABREU FILHO, W. 1978. Aspectos Geológicos-enonómicos da Formação Araras do Grupo Alto Paraguai-MT. *In*: *XXX Congreso Brasileiro de Geologia. Recife-PE, SBG*, **4**, 1816–1826.

MISSI, A., KAUFMAN, A. ET AL. 2007. Chemostratigraphic correlation of Neoproterozoic successions in South America. *Chemical Geology*, **237**, 161–185.

NOGUEIRA, A., RICCOMINI, C., KERKIS, A., FAIRCHILD, T. & HIDALGO, R. 2001. Hydrocarbons in carbonate rocks of the Neoproterozoic Alto Paraguay basin, Mato Grosso, Brazil. Summary of communications. *Annales Academia Brasileira do Ciencias*, **73**, 464.

OLCOTT, A., SESSIONS, A., CORSETTI, F., KAUFFMAN, A. & DE OLIVIERA, T. 2005. Biomarker evidence for photosynthesis during Neoproterozoic glaciation. *Science*, **310**, 471–474.

PECOITS, E., AUBET, N., OYHANTÇABAL, P. & SANCHEZ BETTUCI, L. 2005. Estratigrafía de Sucesiones Sedimentarias y volcanosedimentarias Neoproterozoicas del Uruguay. *Revista de la Sociedad Uruguaya de Geología*, **11**, 18–27.

PLUMB, K. A. & JAMES, H. L. 1986. Subdivision of Precambrian time: recommendations and suggestions by the Subcommission on Precambrian Stratigraphy. *Precambrian Research*, **32**, 65–92.

POIRÉ, D. 1993. Estratigrafía del Precámbrico sedimentario de Olavaria, Sierras Bayas, Provincia de Buenos Aires, Argentina. *In*: *XII Congreso Geológico Argentino y II Congreso de Exploración de Hidrocarburos, Mendoza, Argentina*, **2**, 1–11.

POIRÉ, D. 2002. The Precambrian sedimentary sequences from Olavarría, Tandilia System, Argentina: their biological evidence. *In*: *II International Colloquium Vendian–Cambrian of W-Gondwana. Expanded Abstracts, Montevideo, Uruguay*, 27–30.

POIRÉ, D. 2004. Sedimentary history of the Neoproterozoic of Olavarria, Tandilia System, Argentina: new evidence from their sedimentary sequences and unconformities – a 'snowball earth' or a 'phantom' glacial? *In*: *First Symposium on Neoproterozoic-Early Paleozoic Events in SW-Gondwana. Extended Abstracts, IGCP Project 478, Second Meeting, Sao Paulo, Brazil*, 46–48.

POIRÉ, D., GONZÁLEZ, P., CANALICCHIO, J. & GARCÍA REPETTO, F. 2003. Litoestratigrafía y estromatolitos de la sucesión sedimentaria precámbrica de la cantera Mina Verdún, Minas, Uruguay. *In*: PEÇOIS, E. (ed.) *Estratigrafía del Precámbrico del Uruguay*. Revista de la Sociedad Uruguaya de Geología, Publicación Especial, **1**, 108–123.

SCHMITT, R. S. 2000. *Um evento tectono-metamórfico Cambro-Ordoviciano caracterisado no dominio tectônico Cabo Fria, faixa Ribeira-sudeste do Brasil*. Dr thesis, IG/UFJR, Rio de Janeiro.

SIAL, A. N., FERREIRA, V. P. *ET AL*. 2006. Chemostratigraphy of two Neoproterozoic cap carbonates from the Sergipano Belt (Northeastern Brazil). *In*: *V South American Symposium on Isotope Geology, Short Papers, Punta del Este, Uruguay*, 314–317.

UNRUG, R. 1996. The assembly of Gondwanaland. *Episodes*, **19**, 11–20.

VALERIANO, C. M., DARDENNE, M. A., FONSECA, M. A., SIMOES, L. S. A. & SEER, H. J. 2004. *A evolução tectônica da Faixa Brasilia*. *In*: Geologia do Continente Sul-Americano: Evolução da Obra de Fernando Flavio Marques de Almeida. Mantesso-Neto, V., Bartorelli, A., Carneiro, C. D. R., Brito-Neves, B. B. (Eds), Editora Beca, São Paulo, 575–593.

ZIMMERMAN, U. 2003. Provenance study on Neoproterozoic rocks of NW Argentina: Puncoviscana Formation – first results. *In*: *III International Colloquium Vendian–Cambrian of W-Gondwana. Programme and Extended Abstracts, Cape Town, South Africa*, 41–44.

Index

Page numbers in *italic* denote figures. Page numbers in **bold** denote tables.